Practical Guide to ICP-MS and Other Atomic Spectroscopy Techniques

Written by one of the very first practitioners of ICP-MS, *Practical Guide to ICP-MS and Other Atomic Spectroscopy Techniques: A Tutorial for Beginners* presents ICP-MS in a completely novel and refreshing way. By comparing it with other complementary atomic spectroscopy (AS) techniques, it gives the trace element analysis user community a glimpse into why the technique was first developed and how the application landscape has defined its use today, 40 years after it was first commercialized in 1983.

What's new in the 4th edition:

- Updated chapters on the fundamental principles and applications of ICP-MS
- New chapters on complementary AS techniques including AA, AF, ICP-OES, MIP-AES, XRF, XRD, LIBS, LALI-TOFMS
- Strategies for reducing errors and contamination with plasma spectrochemical techniques
- Comparison of collision and reaction cells including triple/multi quad systems
- Novel approaches to sample digestion
- Alternative sample introduction accessories
- Comprehensive glossary of terms used in AS
- New vendor contact information

The book is not only suited to novices and beginners, but also to more experienced analytical scientists who want to know more about recent ICP-MS developments, and where the technique might be heading in the future. Furthermore, it offers much needed guidance on how best to evaluate commercial AS instrumentation and what might be the best technique, based on your lab's specific application demands.

PRACTICAL SPECTROSCOPY A SERIES

Ultraviolet Spectroscopy and UV Lasers
Edited by Prabhakar Misra and Mark A. Dubinskii

Pharmaceutical and Medical Applications of Near-Infrared Spectroscopy, Second Edition
Emil W. Ciurczak and James K. Drennen III

Applied Electrospray Mass Spectrometry
Edited by Birendra N. Pramanik, A. K. Ganguly, and Michael L. Gross

Practical Guide to ICP-MS
Written by Robert Thomas

NMR Spectroscopy of Biological Solids
Edited by A. Ramamoorthy

Handbook of Near Infrared Analysis, Third Edition
Edited by Donald A. Burns and Emil W. Ciurczak

Coherent Vibrational Dynamics
Edited by Guglielmo Lanzani, Giulio Cerullo, and Sandro De Silvestri

Practical Guide to ICP-MS: A Tutorial for Beginners, Second Edition
Robert Thomas

Practical Guide to ICP-MS: A Tutorial for Beginners, Third Edition
Robert Thomas

Pharmaceutical and Medical Applications of Near-Infrared Spectroscopy
Emil W. Ciurczak and Benoit Igne

Measuring Elemental Impurities in Pharmaceuticals: A Practical Guide
Robert Thomas

Measuring Elemental Contaminants in Cannabis and Hemp: A Practical Guide
Robert Thomas

Handbook of Near Infrared Analysis, Fourth Edition
Edited by Emil W. Ciurczak, Benoit Igne, Jerry Workman Jr., Donald A. Burns

Practical Guide to ICP-MS and Other Atomic Spectroscopy Techniques: A Tutorial for Beginners, Fourth Edition
Robert Thomas

For more information about this series, please visit: www.routledge.com/Practical-Spectroscopy/
book-series/CRCPRASPECTR

Practical Guide to ICP-MS and Other Atomic Spectroscopy Techniques

A Tutorial for Beginners

4th Edition

Robert Thomas

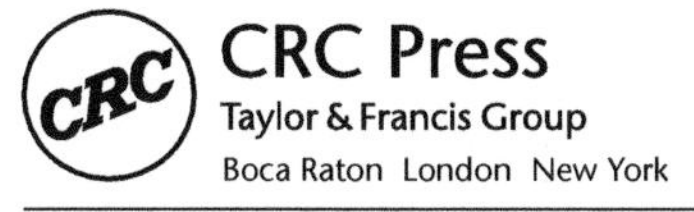

CRC Press
Taylor & Francis Group
Boca Raton London New York

CRC Press is an imprint of the
Taylor & Francis Group, an **informa** business

Cover design credit: © Glenna Thomas

Fourth edition published 2023
by CRC Press
6000 Broken Sound Parkway NW, Suite 300, Boca Raton, FL 33487–2742

and by CRC Press
4 Park Square, Milton Park, Abingdon, Oxon, OX14 4RN

CRC Press is an imprint of Taylor & Francis Group, LLC

© 2024 Robert Thomas

First edition published by CRC Press 2004
Third edition published by CRC Press 2014

ISBN: 978-1-032-03502-4 (hbk)
ISBN: 978-1-032-03504-8 (pbk)
ISBN: 978-1-003-18763-9 (ebk)

DOI: 10.1201/9781003187639

Typeset in Times New Roman
by Apex CoVantage, LLC

Contents

Foreword

ICP-MS and its complementarity with other atomic spectroscopy techniques has grown rapidly since 2013, when the 3rd edition of this book was published. This has underscored the need for a further revision. Major driving forces for publication of the current volume have been the fundamental developments in collision/reaction cell technology with single and multiple quadrupole analysers and the development of software and hardware for time resolved analysis monitoring of single events. These are highlighted in the updated chapters on the fundamental principles and applications of ICP-MS, the development of new sample introduction systems and the efforts introduced by instrument manufacturers in producing IP-MS instrumentation with either shorter dwell times or with multi-elemental character in transient analysis mode (e.g. improved ICP-ToFMS instrumentation).

The author's vision with regards to the market needs and applications of ICP-MS driving instrumental developments is enlightening (Chapter 20). In particular, I have highly valued the gap analysis made with regards to the features and disadvantages of ICP-MS in comparison with other atomic spectroscopy techniques and the complementary role that it plays (Chapter 24). This provides a comprehensive guide on how to select and apply these techniques in the most effective manner.

As a metrologist, I would emphasise the importance of the guidance provided in this volume towards strategies that help practitioners reduce errors using plasma spectrochemical techniques. This is achieved through examples of applications where there is a need to meet accreditation and/or regulatory requirements to demonstrate fitness for purpose (Chapter 17). In this context, the author rightly pays special attention to the provision of practical guidance for use of standards and reference materials for instrument calibration and method validation.

Finally, I have found invaluable the guidance on how best to evaluate commercial instrumentation and what might be the best technique, based on the laboratory specific application(s). This makes the book suited not only to beginners but also to more experienced analytical scientists who want to know the advances in plasma spectrochemistry instrumentation and related future opportunities.

I have very much enjoyed reading this book and look forward to future editions of *Practical Guide to ICP-MS and Other Atomic Spectroscopy Techniques* to keep me up to date in this exciting field.

Heidi Goenaga-Infante, BSc PhD, FRSC has more than twenty-five years of experience on elemental and speciation analysis, beginning with her PhD from Oviedo University, Spain. Currently, she is a Science Fellow of LGC. She is also the Principal Scientist and Team Leader of the Inorganic Analysis team of seventeen PhDs and postgraduate scientists at the National Measurement Laboratory hosted at LGC. Her group expertise currently lies in trace element speciation and metallomic analysis, the characterisation of nanomaterials, high accuracy isotope ratio analysis, quantitative elemental bio-imaging and the characterisation of "speciated" reference materials and standards. Heidi Goenaga-Infante is the chair of the Editorial Board of the Journal of Analytical Atomic Spectrometry. She is the UK representative at the Inorganic Analysis Working Group of the CCQM, the international Consultative Committee for Metrology in Chemistry. She is the Government Chemist representative on the Nanomaterials Environment and Health Government Group chaired by DEFRA and the LGC representative at ISO

TC 24 (Particle characterisation). She is the EURAMET representative for inorganic analysis at the CCQM Key Comparison Working Group and a member of IUPAC. She has acted as the coordinator of the EU EUROPEAN Metrology Research Proposal (EMRP) NanoChop "Chemical, Optical and Biological Characterisation of Nanomaterials in Biological Samples". Heidi was awarded the 2020 Lester W. Strock Award from SAS and more recently the 2023 European Award for Plasma Spectrochemistry. She is currently an Honorary Senior Research Fellow of the University of Liverpool and a Visiting Professor of the University of Strathclyde, UK.

Heidi Goenaga Infante
LGC, UK

Preface

I cannot believe it has been almost 20 years since I published the first edition of this textbook in 2004, and 8 years since the 3rd edition was launched in 2014. But I have not been idle, as I have written two additional books since, one on elemental impurities in pharmaceuticals published in 2018, and the other on heavy metals in cannabis in 2021. However, what was originally intended as a series of ICP-MS tutorials on the basic principles for *Spectroscopy Magazine* in 2001 (1) quickly grew into a textbook focusing on the practical side of the technique. With over 6000 copies sold, including a Chinese (Mandarin) edition, I'm very honored that the book has gained the reputation of being the reference book of choice for novices and beginners to the technique all over the world. Sales have exceeded my wildest expectations. Of course, it helps when it is "recommended reading" for a Pittsburgh Conference ICP-MS Short Course I teach every year on "How to Select an ICP-MS", up until COVID had major impact on our lives in 2019. It also helps when you get the visibility of your book being displayed at 15 different vendors' booths at PittCon every year. But there is no question in my mind that the major reason for its success is that it presents ICP-MS in a way that is very easy to understand for beginners, and also shows the practical benefits of the technique for carrying out routine trace element analysis.

During the time the book has been in print, the application landscape has slowly changed. When the technique was first commercialized in 1983, it was mainly the environmental and geological communities that were the first to realize its benefits. Then, as its capabilities became better known, and its performance improved, other application areas like clinical, toxicological and semiconductor markets embraced the technique. Today, ICP-MS, with all its performance and productivity enhancement tools, is the most dominant trace element technique, and besides the traditional application fields mentioned, it is now being utilized in other exciting areas including nanoparticle research, oil exploration, pharmaceutical manufacturing and, most recently, for cannabis and hemp testing labs. As a result, with a street price tag for a single quadrupole-based ICP-MS on the order of $120, 000, it is being purchased for applications that were previously being carried out by Inductively Coupled Plasm Optical Emission Spectrometry (ICP-OES) and Atomic Absorption/ Fluorescence (AA/AV). It is truly an exciting time for users of the ICP-MS technique, but it also means that the other suitable techniques, which are often overlooked, now have lower price tags that make them a lot more attractive for labs with a limited budget, less demanding detection limit requirements or with lower sample throughput demands.

However, eight years is a long time for a book to remain current, even though sales of the book actually increased during COVID. But these were unusual times. So, even though the COVID-19 pandemic impacted the timing of a 4th edition of the book, the trace element using community was probably using the time to broaden their knowledge, by investing in educational resources. For that reason, it made sense to not only write an updated version to represent the current state of the technology and applications being carried out, but also to incorporate all the great feedback I received from users and vendors over the past few years. And one of the most common requests was to expand the book beyond ICP-MS and include other AS techniques. So, with that in mind, this edition is a going to be a little different. It will of course have comprehensive chapters on the fundamental principles of ICP-MS that have been covered in the previous editions of the book. However, it will also include other AS techniques that are complementary to ICP-MS, such as flame and graphite furnace Atomic Absorption (FAAS/GFAAS), Atomic Fluorescence (AF), Inductively Coupled Plasma Optical Emission Spectrometry (ICP-OES), Microwave Induced Plasma Atomic Emission Spectrometry (MIP-AES), X-Ray Fluorescence (XRF), Laser Induced Breakdown Spectrometry (LIBS) and Laser Ablation, Laser Ionization Time-of-Flight Mass Spectrometry (LALI-TOF-MS). However, the target audience will still be beginners and novice users, who want to better understand the basics and how to apply them to solve real-world analytical problems. So, with that in mind, I present to you the 4th edition of *Practical Guide to ICP-MS and other AS Techniques: A Tutorial for Beginners.*

The following is a summary of the major new chapters in this edition.

Atomic Absorption Spectrometry (AAS): This chapter will present the fundamental principles of both flame and furnace atomic absorption together with the complementary technique, atomic fluorescence, and summarize the benefits of both approaches and, in particular, their strengths and weaknesses.

Flame AAS is predominantly a single-element technique that uses a flame to generate ground-state atoms. When light is passed through this atom cloud, specific wavelengths of light are absorbed which are characteristic of the metal of interest. The amount of light absorbed is indicative of the amount of analyte present. Flame AAS is capable of parts per million (ppm) detection limits. Graphite furnace AAS is also a single-element technique, although some multielement instrumentation is now available. Graphite furnace AAS works on exactly the same principle as flame AAS, except that the flame is replaced by a heated graphite tube to generate the source of atoms. Because the ground-state atoms are concentrated in a smaller area than a flame, more light absorption takes place, resulting in detection limits approximately 100 times lower than for flame AAS.

Atomic Fluorescence Spectrometry (AFS): A similar technique to atomic absorption is atomic fluorescence, where ground-state atoms created in a flame are excited by focusing a beam of light into the atomic vapor. However, instead of looking at the amount of light absorbed in the process, the emission resulting from the decay of the atoms excited by the source light is measured. The intensity of this "fluorescence" increases with increasing atom concentration, providing the basis for quantitative determination. The source lamp for atomic fluorescence is mounted at an angle to the rest of the optical system, so that the light detector sees only the fluorescence in the flame and not the light from the lamp itself.

Inductively coupled Plasma Optical Emission Spectrometry (ICP-OES): Since its introduction over half a century ago, ICP-OES has significantly changed the capabilities of elemental analysis. This technique combined the energy of an argon-based plasma with an optical spectrometer and detection system capable of measuring low-level emission signals, which allowed laboratories to perform rapid, automatic, multielement analyses of trace concentrations. This was approximately 20 years before the introduction of the first commercial, inductively coupled plasma-mass spectrometer, so ICP-OES became the workhorse instrument in many laboratories required to perform elemental analysis at trace-level concentrations. The sensitivity, speed, ease of use and tolerance to high levels of dissolved solids make ICP-OES a technique that laboratories continue to rely heavily on, decades after its inception.

This chapter will focus on the methodologies that are developed for the application work being conducted. While ICP-OES instruments provide automatic, intuitive operation with ppb-level detection limits and working ranges that extended over several orders of magnitude, developing a useful, well-optimized analytical method can be a manual, labor-intensive and time-consuming process. The quality of any developed and optimized analytical method can be determined by a number of different figures of merit, including accuracy, precision, sensitivity, speed, working range, ease of use and reproducibility. Both the instrument settings and method parameters affect these performance characteristics, and it is important to be familiar with key instrument and method parameters, and to understand their effects, if any, on the analytical data being collected.

Microwave Induced Plasma Optical Emission Spectroscopy: This chapter will focus on a more recent development of microwave induced plasma atomic emission spectrometer (MIP-AES). Microwave-generated plasmas have been used as gas chromatography detectors for decades and were explored as potential excitation sources for emission studies back in the 1970s, when the ICP was being investigated. However, it was found they had limited applicability, because they were not considered robust enough for introducing liquids.

However, one vendor has now developed an instrument, which appears to have resolved many of those earlier limitations.

Inside an MP-AES instrument, microwave energy from an industrial magnetron is used to form a nitrogen plasma that has been extracted from compressed air. Using a magnetic field rather than an electric one for excitation generates a very robust plasma—capable of handling a wide range of sample types. An optimized microwave waveguide creates concentrated electromagnetic fields at the torch. Then an axial magnetic field and a radial electrical field focus and contain the microwave energy to create a plasma, which is used for excitation. Because the excitation temperature is significantly lower than an ICP-OES, detection capability is somewhere between flame AA and ICP-OES.

X-Ray Analytical Techniques: This chapter will focus on two different X-ray analytical techniques. X-ray fluorescence (XRF), in both the wavelength and energy dispersive configurations, has traditionally been used for the determination of major and minor elemental concentrations in solid samples such as minerals, metals, polymers, cement, pharmaceuticals and foodstuffs. Additionally, over the past couple of decades, breakthroughs in miniaturization of optical components, electronics and solid-state detectors have resulted in the commercialization of handheld devices to carry out measurements in the field where it is very difficult to bring the sample to the laboratory and, more recently, in the development of compact, online spectrometers for industrial process control applications.

On the other hand, X-ray diffraction (XRD), which is typically utilized to get a better understanding of the crystalline structure of both man-made and natural materials, is better applied in a laboratory environment, because the sample has to be in a suitable form to ensure the interaction of the X-rays with a uniformly flat surface. This is particularly true for mining or geological matrices, where the samples have to be ground to a fine homogeneous powder to be analyzed.

Laser Induced Breakdown Spectroscopy (LIBS): This chapter will focus on Laser Induced Breakdown Spectroscopy, which has been commercially available since the early 2000s, but it has only been the past five years where its application potential has been fully explored. Initially developed by a group of government scientists in the early 1980s, it was first applied to the analysis of soils and hazardous-waste sites, because of its remote sampling capabilities. However, since it has been in the hands of the routine analytical community, it is now being used to solve real-world application problems. Although the analytical capability, performance and elemental range of LIBS compares quite favorably with other AS techniques, it should not be considered a direct competitor to either ICP-OES, or XRF.

The unique benefits of LIBS are in its ability to sample a very diverse range of matrices, both in a laboratory environment and at remote locations out in the field. For example, it was taken to Mars on NASA's Curiosity robotic rover, where it was used to sample and ablate rocks and geological formations from a distance of a few meters and process the spectral onboard the rover to get an assessment of the chemical nature of the mineral being sampled.

Laser Ionization Time-of-Flight Mass Spectrometry (LALI-TOF-MS): Traditionally, the most widely used commercial techniques for the analysis of solid samples include Laser Induced Breakdown Spectroscopy (LIBS), arc/spark Optical Emission Spectrometry (A/S OES), X-ray Fluorescence (XRF) and Laser Ablation coupled with ICP Mass Spectrometry (LA-ICP-MS). While useful for many diverse applications, each of these techniques has known limitations, including matrix suppression, diffusion/transport effects, spectral overlaps and varying degrees of calibration challenges. In addition, many labs that determine trace impurities in solid materials have no way of reaching the required limits of quantitation by direct analysis, so they are faced with the dilemma of digesting the samples and using a solution technique like ICP-MS.

However, recent advances in Laser Ablation Laser Ionization (LALI) significantly reduce many of the drawbacks that plague other techniques, thereby simplifying the quantification of elemental impurities in solid samples. This chapter examines the fundamental principles of the LALI technique coupled with TOF mass spectrometry, and compares/contrasts application figures of merit with other, more traditional approaches of analyzing solid materials.

This chapter will therefore focus on the principles of laser ablation, laser ionization coupled with TOF mass spectrometry (LALI-TOFMS), which is a brand-new development that offers virtually the entire periodic table of the elements at mass spec detection limits. It has applicability to any type of trace element application area that is either looking to directly analyze solid samples, wants to avoid sample digestion procedures or has been faced with inherent shortcomings of a solution technique such as Inductively Coupled Plasma Mass Spectrometry (ICP-M), because of its well-recognized matrix suppression effects, polyatomic spectral interferences and signal instability.

I have also included another brand-new chapter on:

Reducing Errors and Contamination using Plasma Spectrochemistry: There are many factors that influence the ability to get the correct result with any trace element technique, particularly in today's high throughput cannabis testing laboratories. Unfortunately, with plasma spectrochemistry techniques, the problem is magnified even more because of their extremely high sensitivity. So, in order to ensure that the data reported is an accurate reflection of the sample in its natural state, the analyst must be aware of all the potential sources of contamination that could negatively impact the quality of the measurements. This chapter addresses this topic and suggests ways to minimize sources of errors in the generated data.

Furthermore, there are additions and supplemental information to the following chapters:

- **Collision/Reaction Cell Technology including Triple/Multi Quad Systems**
- **Sampling and Sample Preparation Techniques**
- **Alternative Sample Introduction Approaches**
- **ICP-MS Application Landscape**
- **How to Select an ICP-MS System**
- **Glossary of ICP-MS Terms**

ICP-MS: RESEARCH OR ROUTINE?

There are approximately five times more ICP-OES instruments installed globally than ICP-MS instrumentation, depending on the configuration. Today, the ICP-MS marketplace represents approximately 2000 units per year with a total revenue of around $400M (instruments and accessories). The price, of course, is a big consideration, with the cost of an ICP-MS roughly twice as much as an ICP-OES. A single quadrupole system is typically $150–200K, while an ICP-OES is typically $50–100K, again depending on the configurations. But as mentioned, in a competitive world, the "street price" of an ICP-MS system is much closer to a top-of-the-line ICP-OES with sampling accessories So, if ICP-MS is not significantly more expensive than ICP-OES in the "real world", why has it not been more widely accepted by the analytical community? It is still my firm opinion that the major reason ICP-MS has not gained the popularity of the other trace element techniques lies in the fact that 40 years after its commercialization, it is still considered a complicated research-type technique, requiring a very skilled person to operate it. Manufacturers of ICP-MS equipment are constantly striving to make the systems easier to operate, the software easier to use and the hardware easier to maintain, but it is still not perceived as a mature, routine technique like FAA or ICP-OES. The picture is even fuzzier now that most instruments are sold with collision/reaction cells/interfaces and, more recently, triple and multi quadrupole ICP-MS instruments have been introduced into the marketplace to "muddy the waters". This means that even though this exciting new technology is making ICP-MS more powerful and flexible, the method development process for unknown samples is generally still more complex. In addition, vendors of this type of equipment are very skilled at emphasizing the capabilities of their technology while at the same time pointing out the limitations of other approaches, making it even more confusing for the inexperienced user.

The bottom line is that ICP-MS has still not gained the reputation as a technique that you can allow a complete novice to use with no supervision, for the fear of generating erroneous data. This makes for all the more reason why there is still a need for a practical textbook explaining the basic principles and application benefits of ICP-MS in a way that is interesting, unbiased and easy to understand for a novice who has limited knowledge of the technique. There is no question that there are some excellent books out there, but they are mainly compiled by academics or research scientists who are not approaching the subject from a beginner's perspective. So, they tend to be technically "heavy" and more biased towards fundamental principles, and less on how ICP-MS is being applied to solve real-world application problems.

SO, WHAT'S INSIDE?

As with all the other editions, the fundamental principles of ICP-MS occupy a significant amount of the content of the book since the 1st edition was published in 2004. When you eventually get through these chapters, you'll be taken through practical issues like sample preparation, contamination control and routine maintenance. Next, I address the many different kinds of alternative sampling accessories available today, such as productivity-enhancing tools to maximize sample throughput and performance-improvement approaches like laser ablation and chromatographic separation techniques, which can optimize ICP-MS performance and extend its capabilities when dealing with complex, real-world sample matrices. And to emphasize how real-world samples are being addressed by the technique, I also felt it was important to give an overview of its application landscape.

The next few chapters of the book cover all the complementary AS techniques that I mentioned earlier in my introduction, including AA, ICP-OES, MIP-AES, XRF, LBS and LALI-TOF-MS. To complete this section of the book, I compare ICP-MS with some of these other trace element techniques, focusing on performance criteria like elemental range, detection capability, sample throughput, analytical working range, interferences, sample preparation, maintenance issues, operator skill level and running costs. This kind of head-to-head comparison will enable the reader to relate both the advantages and disadvantages of ICP-MS to other atomic spectroscopy instrumentation they are more familiar with. I included this because there is still a role for the other complementary techniques. And for those of you who might be interested in purchasing an instrument, I have included a chapter on the most important selection criteria. In my opinion, this is a critical ingredient in presenting ICP-MS to a novice, because there is very little information in the public domain to help someone carry out an evaluation of commercial instrumentation. Very often, people go into this evaluation process completely unprepared and, as a result, may end up with an instrument that is not ideally suited for their needs . . . something I am very well aware of, based on teaching my "How to Select an ICP-MS Short Course" at the Pittsburgh Conference for the past 20 years.

Hopefully, after having completed the book, there is still a serious interest in investing in the technique—unless, of course, you have purchased one before reading the final chapter! Even though this might sound a little ambitious, the main objective is to make ICP-MS a little more compelling to purchase, compare its capabilities with other AS techniques and ultimately open up its potential to the vast majority of the trace element analysis community who have not yet realized its full potential. So, with this in mind, please feel free to come and share in my thoughts on a *Practical Guide to ICP-MS and Other Atomic Spectroscopy Techniques*—edition number four.

Acknowledgments

Having worked in the field of ICP-MS for almost 40 years, my incentive for writing this book was based on a realization that there were no textbooks being written specifically for beginners with a very limited knowledge of the technique. I quickly came to the conclusion that the only way this was going to happen was to write it myself. So, in 2003, I set myself the objective of putting together a reference book that could be used by both analytical chemists and senior management who were experienced in the field of trace-metal analysis, but only had a limited knowledge of ICP-MS. The fruit of this endeavor was the publication of the first edition of the book in 2004. Then in 2007, my publisher convinced me it was time to update the book, which eventually resulted in the 2nd edition published in 2008 and a 3rd edition in 2014.

So now in 2023, as the fourth edition hits the "book stands", I would like to take this opportunity once again to thank some of the people and organizations that have helped me put the book together.

First, I would like to thank the three people who contributed to the content of this book.

Patricia (Patti) Atkins has written the chapter on the importance in understanding contamination issues in the modern trace element laboratory and, in particular, how to minimize those potential sources of contamination to ensure the generated data is representative of what's in the original sample.

Patti is a scientist and Global Product Manager at SPEX Certiprep (an Antylia company) in New Jersey. She is a graduate of Rutgers University in NJ and was a laboratory supervisor for Ciba Specialty Chemicals in the Water Treatment Division. Patti later accepted a position conducting research and managing an air-pollution research group within Rutgers University's Civil & Environmental Engineering Department. In 2008, Patricia joined SPEX CertiPrep as a senior application scientist in their certified reference material's division, and spends her time researching industry trends and developing new reference materials. She is a frequent presenter and speaker at numerous conferences, including NACRW, NEMC, PittCon and AOAC, and a published author, with her work appearing in various journals and trade publications including *Spectroscopy*, *LCGC* and *Cannabis Science and Technology*, where she is a columnist for analytical issues in cannabis testing.

Deborah (Debbie) Bradshaw is an analytical chemist who has been working in the field of atomic spectroscopy for over 35 years. Working as a chemist in the environmental field, she started using flame atomic absorption in the 1970s, and then migrated into graphite furnace. Using a Zeeman background-corrected instrument, in the early 1980s she developed methods for a number of elements for the analysis of seawater samples. Taking a position with Perkin Elmer Corporation in 1989, she migrated into analyzing a variety of sample types using not only AAS, but also the plasma techniques, both optical and mass spectrometry. For the past 20 years, she has been working as a consultant in the field of atomic spectroscopy, conducting training classes and giving technical support for AA, ICP-OES and ICP-MS.

Debbie is currently on the Editorial Advisory Board for *Spectroscopy* magazine. She has been a member of the Society for Applied Spectroscopy (SAS) since 1980 and is now a fellow of the Society. For 14 years, she was the news-column editor for the *Journal of Applied Spectroscopy*, SAS's monthly publication. From 2002–2004, she was Treasurer of SAS and has sat on several of the society committees. She was the recipient of the SAS Distinguished Service Award in 2008.

In efforts to promote education and to support analysts using atomic spectroscopy techniques, Debbie was FACSS (now SciX) Atomic Spectroscopy Symposia Chair in 2007 and 2008, has organized technical symposia at both FACSS and PittCon and has been a short-course instructor for SAS. Debbie continues to be a short-course instructor at the Winter Plasma Conference on Spectrochemical Analysis since 2000.

Dr. Maura Rury (PhD) has compiled the comprehensive chapter on ICP-OES, where she has extensive knowledge in the fundamental principles, as well as developing robust methods for various applications of the technique. She has published widely on atomic spectroscopic applications and has contributed a number of articles on plasma spectrochemistry to my Atomic Perspectives column in *Spectroscopy* magazine.

Maura is a Global Product Manager for Mettler Toledo where she manages the Total Organic Carbon (TOC) product line, along with related support products, tools, accessories, services and training. Her prior roles include portfolio marketing management at PerkinElmer Inc., and product marketing and technical applications at Thermo Fisher Scientific, supporting atomic absorption and inductively coupled plasma-based instruments for a wide range of applications. This provided her with a significant amount of experience with atomic spectroscopy applications in specific markets, including food and beverage, pharmaceutical, environmental, industrial and clinical. Maura holds a PhD in analytical chemistry from the University of Massachusetts, Amherst, and studied under the tutelage of professors Julian Tyson and Ramon Barnes.

Next, I would like to thank the editorial staff of *Spectroscopy* magazine, who gave me the opportunity to write a monthly tutorial on ICP-MS back in the spring of 2001. This was most definitely the spark I needed to start the original project. They also allowed me to use many of the figures from the series, together with material from other ICP-MS articles I wrote for the magazine. In particular, I would like to acknowledge the current editorial director, Laura Bush, who has given me the opportunity to publish the quarterly Atomic Perspectives (AP) column for the magazine, and who has given me the freedom to write on just about anything I think is compelling and interesting in the field of atomic spectroscopy

Second, I would like to thank all the folks at CRC Press/Taylor and Francis for having faith in my writing skills and subject matter expertise by giving me the opportunity to write and publish five textbooks in 20 years. In particular, I'd like to give a special mention to the chemistry editor, Barbara Knott (who has since retired), who has guided me through four of those books, and George Kenney, from the sales group, who allowed me to give out numerous complementary copies to increase the visibility of the book in certain strategic markets, and who also sent out bulk order quotes to any organization who wanted to order customized copies with their own logo and marketing message.

And of course, an acknowledgement section would not be complete without thanking the manufacturers of ICP-MS instrumentation, ancillary equipment, sampling accessories, consumables, calibration standards, chemical reagents and high-purity gases, who supplied me with the information, data, drawings, figures and schematics, etc., and particularly their willingness over the past 20 years to display the book at their Pittsburgh Conference exhibition booths. There are way too many people to name, but at the Pittsburgh Conference this year (2023) in Philadelphia 10 vendors displayed the book. This alone has made a huge difference to the visibility of the book, and its success would not have been possible without their help.

I would finally like to thank Dr. Heidi Goenaga-Infante, Chief Scientist, National Measurement Laboratory at LGC Standards in the UK for writing the Foreword to the book. Her endorsement is invaluable and means the world to me and the credibility of my book.

So let me wrap up this preface by wishing you the best of luck with your laboratory endeavors. I'm hoping my book will be placed next to your AS equipment to use as a reference source for all your colleagues, or better still, convince them that they should have their own copy! As always, thanks for your continued support.

Robert Thomas, CSci, CChem, FRSC

March, 2023

About the Author

 **Robert (Rob) Thomas** is the Principal Scientist at Scientific Solutions, a consulting company that serves the educational needs of the trace element user community. He has worked in the field of atomic and mass spectroscopy for almost 50 years, including 24 years for a manufacturer of atomic spectroscopic instrumentation. He has served on the American Chemical Society (ACS) Committee on Analytical Reagents (CAR) for the past 20 years as leader of the plasma spectrochemistry, heavy metals task force, where he has worked very closely with the United States Pharmacopeia (USP) to align ACS heavy metal testing procedures with pharmaceutical guidelines. Rob has written over 100 technical publications, including a 15-part tutorial series on ICP-MS. He is also the editor of and frequent contributor to the Atomic Perspectives column in *Spectroscopy* magazine, as well as serving on the editorial advisory board of *Analytical Cannabis*. In addition, Rob has authored five textbooks on the fundamental principles and applications of ICP-MS. Prior to this new edition of *Practical Guide to ICP-MS and Other Atomic Spectroscopy Techniques*, his most recent book was a paperback version of *Measuring Heavy Metal Contaminants in Cannabis and Hemp*, published in December 2021. Rob has an advanced degree in analytical chemistry from the University of Wales, UK, and is also a fellow of the Royal Society of Chemistry (FRSC) and a Chartered Chemist (CChem).

1 An Overview of ICP Mass Spectrometry

ICP-MS not only offers extremely low detection limits in the sub-parts per trillion (ppt) range, but also enables quantitation at the high parts per million (ppm) level. This unique capability makes the technique very attractive compared to other trace metal techniques, such as graphite furnace AA, which is limited to determinations at the trace level, or Flame AAS and ICP-OES, which are traditionally used for the detection of higher concentrations. In Chapter 1, we will present an overview of ICP-MS and explain how its characteristic low detection limits are achieved.

Inductively coupled plasma mass spectrometry (ICP-MS) is undoubtedly the fastest growing trace element technique available today. Since its commercialization in 1983, when only a handful of instruments were sold, today the annual worldwide market size is approximately 3000 systems used for many varied and diverse applications. The most common ones, which represent approximately 90% of the ICP-MS analysis being carried out today, include environmental, geological, semiconductor, biomedical, food and nuclear market segments. However, over the past few years, with new heavy metal regulations being implemented, the analysis of pharmaceuticals and cannabis consumer products is rapidly catching them up. There is no question that the major reason for its unparalleled growth is its ability to carry out rapid multielement determinations at the ultratrace level. Even though it can broadly determine the same suite of elements as other atomic spectroscopic techniques, such as flame atomic absorption (FAA), electrothermal atomization (ETA), atomic fluorescence (AF) and inductively coupled plasma optical emission (ICP-OES), ICP-MS has clear advantages in its multielement characteristics, speed of analysis, detection limits and isotopic capability. Figure 1.1 shows approximate detection limits of all the elements that can be detected by ICP-MS, together with their isotopic abundance.

1.1 PRINCIPLES OF OPERATION

There are a number of different ICP-MS designs available today that share many similar components, such as nebulizer, spray chamber, plasma torch, interface cones, vacuum chamber, ion optics, mass analyzer and detector. However, the engineering design and implementation of these components can vary significantly from one instrument to another. Instrument hardware is described in greater detail in the subsequent chapters on the basic principles of the technique. So let us begin here by giving an overview of the principles of operation of ICP-MS. Figure 1.2 shows the basic components that make up an ICP-MS system. The sample, which usually must be in a liquid form, is pumped at 1 mL/min, usually with a peristaltic pump into a nebulizer, where it is converted into a fine aerosol with argon gas at about 1 L/min. The fine droplets of the aerosol, which represent only 1–2% of the sample, are separated from larger droplets by means of a spray chamber. The fine aerosol then emerges from the exit tube of the spray chamber and is transported into the plasma torch via a sample injector.

It is important to differentiate between the roles of the plasma torch in ICP-MS compared to ICP-OES. The plasma is formed in exactly the same way, by the interaction of an intense magnetic field (produced by radio frequency (RF) passing through a copper coil) on a tangential flow of gas (normally argon), at about 12–15 L/min flowing through a concentric quartz tube (torch). This has the effect of ionizing the gas, which, when seeded with a source of electrons from a high-voltage spark, forms a very high-temperature plasma discharge (~10,000 K) at the open end of the tube.

DOI: 10.1201/9781003187639-1

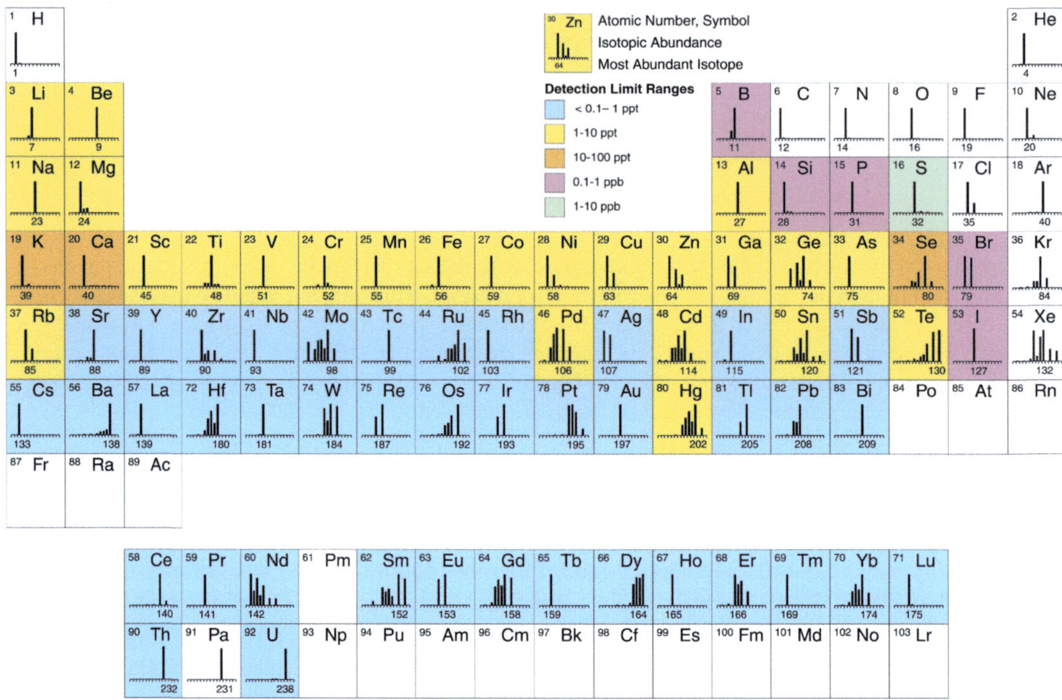

FIGURE 1.1 Approximate detection capability of ICP-MS, together with elemental isotropic abundances. (Reproduced with permission from PerkinElmer Inc., All Rights Reserved)

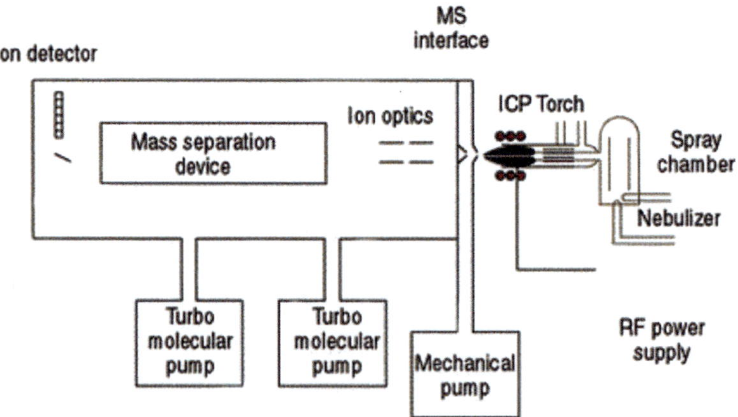

FIGURE 1.2 Basic instrumental components of an ICP mass spectrometer.

However, this is where the similarity ends. In ICP-OES, the plasma, which is normally vertical (but can be horizontal with dual-view designs), is used to generate photons of light by the excitation of electrons of a ground-state atom to a higher energy level. When the electrons "fall" back to ground state, wavelength-specific photons are emitted that are characteristic of the element of interest. In ICP-MS, the plasma torch, which is positioned horizontally, is used to generate positively charged ions and not photons. In fact, every attempt is made to stop the photons from reaching the detector because they have the potential to increase signal noise. It is the production and detection of large quantities of these ions that gives ICP-MS its characteristic low-ppt detection capability—about three to four orders of magnitude lower than ICP-OES.

Once the ions are produced in the plasma, they are directed into the mass spectrometer via the interface region, which is maintained at a vacuum of 1–2 torr with a mechanical roughing pump. This interface region consists of two or three metallic cones (depending on the design), called the sampler and a skimmer cone, each with a small orifice (0.6–1.2 mm) to allow the ions to pass through to the ion optics, where they are guided into the mass separation device.

The interface region is one of the most critical areas of an ICP mass spectrometer, because the ions must be transported efficiently and with electrical integrity from the plasma, which is at atmospheric pressure (760 torr), to the mass spectrometer analyzer region, which is at approximately 10^{-6} torr. Unfortunately, there is the potential of capacitive coupling between the RF coil and the plasma, producing a potential difference of a few hundred volts. If this is not eliminated, an electrical discharge (called a secondary discharge or pinch effect) between the plasma and the sampler cone would occur. This discharge would increase the formation of interfering species and also dramatically affect the kinetic energy of the ions entering the mass spectrometer, making optimization of the ion optics very erratic and unpredictable. For this reason, it is absolutely critical that the secondary charge be eliminated by grounding the RF coil. There have been a number of different approaches used over the years to achieve this, including a grounding strap between the coil and the interface, balancing the oscillator inside the RF generator circuitry, a grounded shield or plate between the coil and the plasma torch or the use of a double interlaced coil where RF fields go in opposing directions. They all work differently, but basically achieve a similar result, which is to reduce or eliminate the secondary discharge.

Once the ions have been successfully extracted from the interface region, they are directed into the main vacuum chamber by a series of electrostatic lenses, called ion optics. The operating vacuum in this region is maintained at about 10^{-3} torr with a turbomolecular pump. There are many different designs of the ion optic region, but they serve the same function, which is to electrostatically focus the ion beam towards the mass separation device, while stopping photons, particulates, and neutral species from reaching the detector.

The ion beam, containing all the analyte and matrix ions, exits the ion optics and now passes into the heart of the mass spectrometer—the mass separation device, which is kept at an operating vacuum of approximately 10^{-6} torr with a second turbomolecular pump. There are many different mass separation devices, all with their strengths and weaknesses. Three of the most common types are discussed in this book—quadrupole, magnetic sector and time-of-flight technology—but they basically serve the same purpose, which is to allow analyte ions of a particular mass-to-charge ratio through to the detector and to filter out (reject) all the non-analyte, interfering and matrix ions. Depending on the design of the mass spectrometer, this is either a scanning process where the ions arrive at the detector in a sequential manner, or a simultaneous process where the ions are either sampled or detected at the same time. Most quadrupole instruments nowadays are also sold with collision/reaction cells or interfaces. This technology offers a novel way of minimizing polyatomic spectral interferences by bleeding a gas into the cell or interface and using ion–molecule collision and reaction mechanisms to reduce the impact of the ionic interference.

The final process is to convert the ions into an electrical signal with an ion detector. The most common design used today is called a discrete dynode detector, which contains a series of metal dynodes along the length of the detector. In this design, when the ions emerge from the mass filter, they impinge on the first dynode and are converted into electrons. As the electrons are attracted to the next dynode, electron multiplication takes place, which results in a very high stream of electrons emerging from the final dynode. This electronic signal is then processed by the data-handling system in the conventional way and converted into analyte concentration using ICP-MS calibration standards. Most detection systems can handle up to ten orders of dynamic range, which means they can be used to analyze samples from low/sub ppt levels up to a few hundred ppm, depending on the analyte mass.

It is important to emphasize that because of the enormous interest in the technique, most ICP-MS instrument companies have very active R&D programs in place, in order to get an edge in a very competitive marketplace. This is obviously very good for the consumer, because not only does it drive down instrument prices, but the performance, applicability, usability and flexibility of the technique is being improved at a dramatic rate. Although this is extremely beneficial for the ICP-MS user community, it can pose a problem for a writer of textbooks who is attempting to present a snapshot of instrument hardware and software components at a particular moment in time. Hopefully, I have struck the right balance in not only presenting the fundamental principles of ICP-MS, but also making the reader aware of what the technique is capable of achieving and where new developments might be taking it, particularly with regard to its application landscape.

2 Principles of Ion Formation

Chapter 2 gives a brief overview of the fundamental principles of ion formation in ICP-MS—the use of a high-temperature argon plasma to generate positively charged ions. The highly energized argon ions that make up the plasma discharge are used to first produce analyte ground-state atoms from the dried sample aerosol, and then to interact with the atoms to remove one or more electrons and generate positively charged ions, which are then steered into the mass spectrometer for detection and measurement.

In ICP-MS, the sample, which is usually in liquid form, is delivered into the sample introduction system, comprising of a spray chamber and nebulizer. It emerges as an aerosol, where it eventually finds its way, via a sample injector, into the base of the plasma. As it travels through the different heating zones of the plasma torch, it is dried, vaporized, atomized and ionized. During this time, the sample is transformed from a liquid aerosol to solid particles, and then into a gas. When it finally arrives at the analytical zone of the plasma, at approximately 6000–7000 K, it exists as ground-state atoms and ions, representing the elemental composition of the sample. The excitation of the outer electrons of a ground-state atom to produce wavelength-specific photons of light is the fundamental basis of atomic emission. However, there is also enough energy in the plasma to remove one or more electrons from its orbitals to generate a free ion. The energy available in an argon plasma is ~13 eV, which is high enough to ionize most of the elements in the periodic table (the majority have first ionization potentials on the order of 4–12 eV). It is the generation, transportation and detection of significant numbers of positively charged ions that gives ICP-MS its characteristic ultratrace detection capabilities. It is also important to mention that although ICP-MS is predominantly used for the detection of positive ions, negative ions are also produced in the plasma. However, because the extraction and transportation of negative ions is different from that of positive ions, most commercial instruments are not designed to measure them. The process of the generation of positively charged ions in the plasma is conceptually shown in greater detail in Figure 2.1.

2.1 ION FORMATION

The actual process of conversion of a neutral ground-state atom to a positively charged ion is shown in Figures 2.2 and 2.3. Figure 2.2 shows a very simplistic view of the chromium atom Cr^0, at mass 52, consisting of a nucleus with 24 protons (p^+) and 28 neutrons (n), surrounded by 24 orbiting electrons (e^-). It must be emphasized that this is not meant to be an accurate representation of the electron's shells and subshells, but just a conceptual explanation for the purpose of clarity. From this we can conclude that the atomic number of chromium is 24 (number of protons) and its atomic mass is 52 (number of protons + neutrons).

If energy is then applied to the chromium ground-state atom in the form of heat from a plasma discharge, one or more orbiting electrons will be stripped off the outer shell. This will result in only 23 (or less) electrons left orbiting the nucleus. Because the atom has lost a negative charge (e^-) but still has 24 protons (p^+) in the nucleus, it is converted into an ion with a net positive charge. It still has an atomic mass of 52 and an atomic number of 24, but is now a positively charged ion and not a neutral ground-state atom. This process is shown in Figure 2.3.

DOI: 10.1201/9781003187639-2

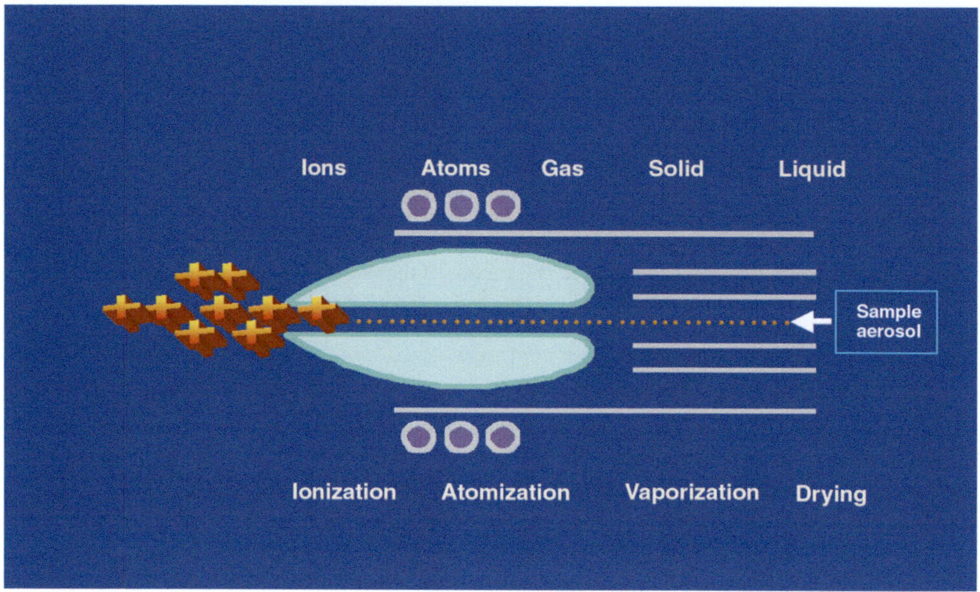

FIGURE 2.1   Generation of positively charged ions in the plasma.

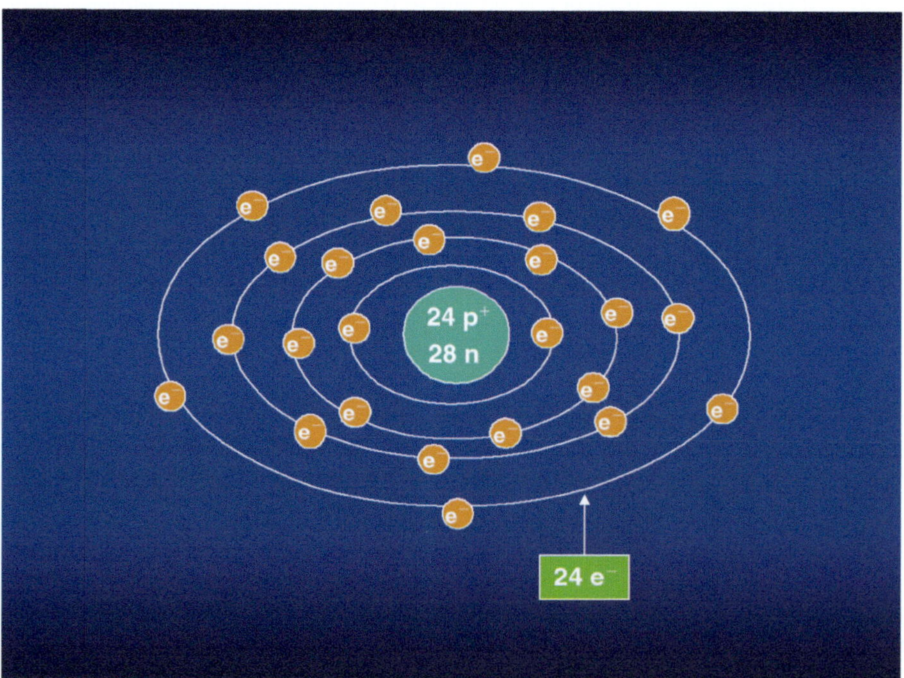

FIGURE 2.2   Simplified schematic of a chromium ground-state atom (Cr^0).

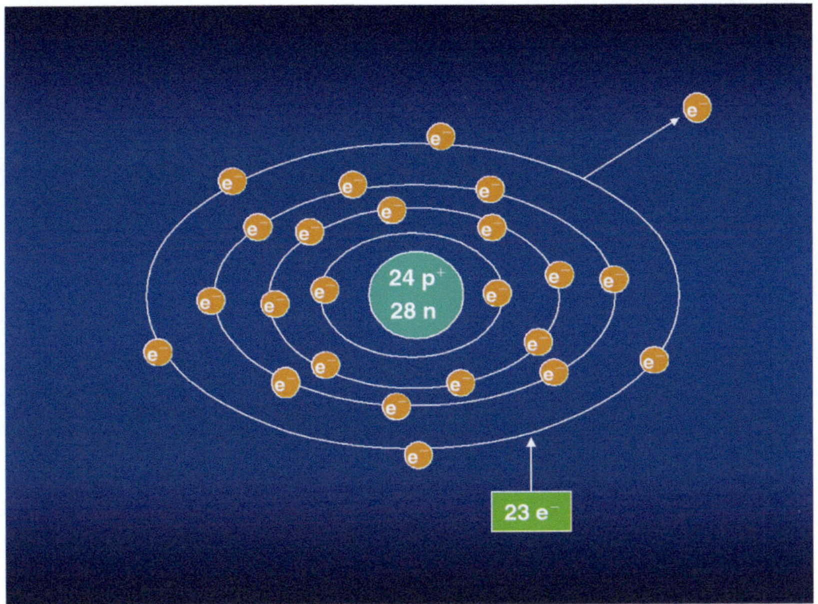

FIGURE 2.3 Conversion of a chromium ground-state atom (Cr^0) to an ion (Cr^+).

TABLE 2.1

Breakdown of the Atomic Structure of Copper Isotopes

	^{63}Cu	^{65}Cu
Number of protons (p^+)	29	29
Number electrons (e^-)	29	29
Number of neutrons (n)	34	36
Atomic mass ($p^+ + n$)	63	65
Atomic number (p^+)	29	29
Natural abundance	69.17%	30.83%
Nominal atomic weight	63.55[a]	

[a] The nominal atomic weight of copper is calculated using the formula 0.6917n (^{63}Cu) + 0.3083n (^{65}Cu) + p^+ and referenced to the atomic weight of carbon.

2.2 NATURAL ISOTOPES

This is a very basic look at the process, because most elements have more than one form (isotope). In fact, chromium has four naturally occurring isotopes, which means that the chromium atom exists in four different forms, all with the same atomic number of 24 (number of protons), but with different atomic masses (numbers of neutrons). So besides the $^{52}Cr^+$ ion, chromium can also generate the $^{50}Cr^+$, $^{53}Cr^+$ and $^{54}Cr^+$ positively charged ions.

To make this a little easier to understand, let us take a closer look at an element such as copper, which only has two different isotopes—one with an atomic mass of 63 (^{63}Cu) and another with an atomic mass of 65 (^{65}Cu). They both have the same number of protons and electrons, but differ in the number of neutrons in the nucleus. The natural abundances of ^{63}Cu and ^{65}Cu are 69.1% and 30.9%, respectively, which gives copper a nominal atomic mass of 63.55—the value you see for copper in atomic weight reference tables. Details of the atomic structure of the two copper isotopes are shown in Table 2.1.

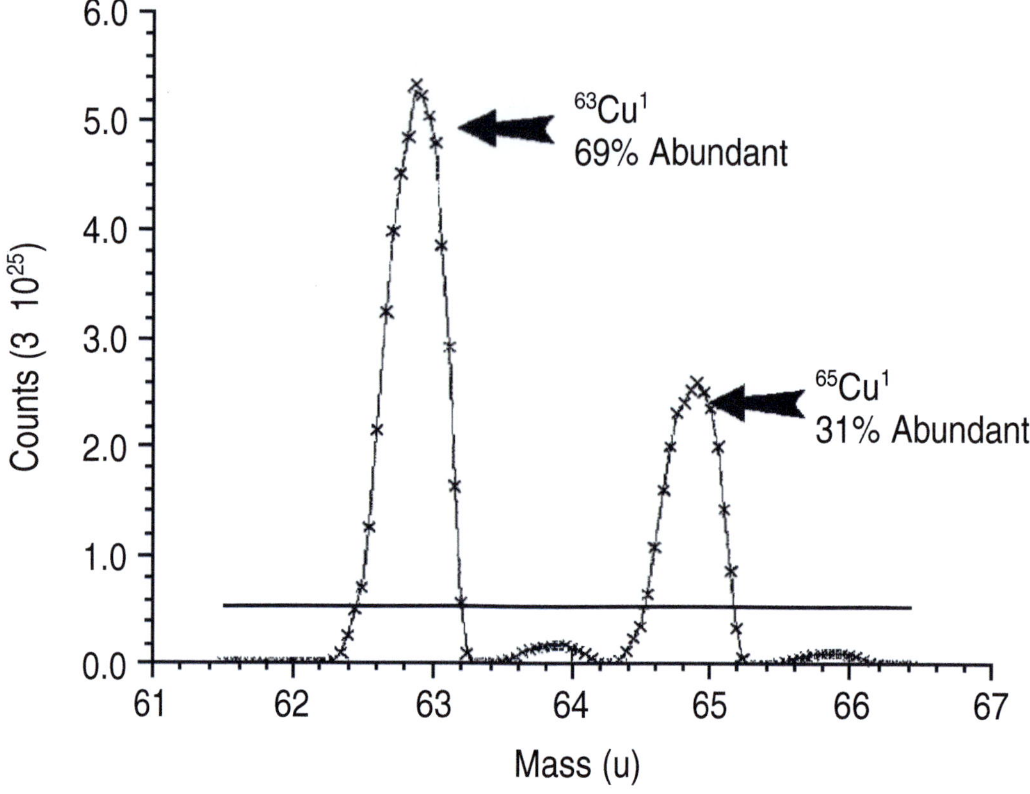

FIGURE 2.4	Mass spectra of the two copper isotopes—$^{63}Cu^+$ and $^{65}Cu^+$.

When a sample containing naturally occurring copper is introduced into the plasma, two different ions of copper, $^{63}Cu^+$ and $^{65}Cu^+$, are produced that generate two different masses—one at mass 63 and another at mass 65. This can be seen in Figure 2.4, which is an actual ICP-MS spectral scan of a sample containing copper, showing a peak for the $^{63}Cu^+$ ion on the left, which is 69.17% abundant, and a peak for $^{65}Cu^+$ at 30.83% abundance on the right.

You can also see small peaks for two Zn isotopes at mass 64 ($^{64}Zn^+$) and mass 66 ($^{66}Zn^+$) (Zn has a total of five isotopes at masses 64, 66, 67, 68 and 70). In fact, most elements have at least two or three isotopes, and many elements, including zinc and lead, have four or more isotopes. Figure 2.5 is a chart showing the relative abundance of the naturally occurring isotopes of all the elements.

Relative Abundance of the Natural Isotopes

Isotope	El	%	El	%	El	%
1	H	99.985				
2	H	0.015				
3			He	0.000137		
4			He	99.999863		
5						
6					Li	7.5
7					Li	92.5
8	Be	100				
9						
10			B	19.9		
11			B	80.1		
12					C	98.90
13					C	1.10
14	N	99.643				
15	N	0.366				
16			O	99.762		
17			O	0.038		
18			O	0.200		
19					F	100
20	Ne	90.48				
21	Ne	0.27				
22	Ne	9.25				
23			Na	100		
24					Mg	78.99
25					Mg	10.00
26					Mg	11.01
27	Al	100				
28			Si	92.23		
29			Si	4.67		
30			Si	3.10		
31					P	100
32	S	95.02				
33	S	0.75				
34	S	4.21				
35			Cl	75.77		
36	S	0.02			Ar	0.337
37			Cl	24.23		
38					Ar	0.063
39	K	93.2581				
40	K	0.0117	Ca	96.941	Ar	99.600
41	K	6.7302				
42			Ca	0.647		
43			Ca	0.135		
44			Ca	2.086		
45					Sc	100
46	Ti	8.0	Ca	0.004		
47	Ti	7.3				
48	Ti	73.8	Ca	0.187		
49	Ti	5.5				
50	Ti	5.4	V	0.250	Cr	4.345
51			V	99.750		
52					Cr	83.789
53					Cr	9.501
54	Fe	5.8			Cr	2.365
55			Mn	100		
56	Fe	91.72				
57	Fe	2.2				
58	Fe	0.28			Ni	68.077
59			Co	100		
60					Ni	26.223

Isotope	El	%	El	%	El	%
61					Ni	1.140
62					Ni	3.634
63	Cu	69.17				
64			Zn	48.6	Ni	0.926
65	Cu	30.83				
66			Zn	27.9		
67			Zn	4.1		
68			Zn	18.8		
69					Ga	60.108
70	Ge	21.23	Zn	0.6		
71					Ga	39.892
72	Ge	27.66				
73	Ge	7.73				
74	Ge	35.94	Se	0.89		
75					As	100
76	Ge	7.44	Se	9.36		
77			Se	7.63		
78	Kr	0.35	Se	23.78		
79					Br	50.69
80	Kr	2.25	Se	49.61		
81					Br	49.31
82	Kr	11.6	Se	8.73		
83	Kr	11.5				
84	Kr	57.0	Sr	0.56		
85					Rb	72.165
86	Kr	17.3	Sr	9.86		
87			Sr	7.00	Rb	27.835
88			Sr	82.58		
89					Y	100
90	Zr	51.45				
91	Zr	11.22				
92	Zr	17.15	Mo	14.84		
93					Nb	100
94	Zr	17.38	Mo	9.25		
95			Mo	15.92		
96	Zr	2.80	Mo	16.68	Ru	5.52
97			Mo	9.55		
98			Mo	24.13	Ru	1.88
99					Ru	12.7
100			Mo	9.63	Ru	12.6
101					Ru	17.0
102	Pd	1.02			Ru	31.6
103			Rh	100		
104	Pd	11.14			Ru	18.7
105	Pd	22.33				
106	Pd	27.33	Cd	1.25		
107					Ag	51.839
108	Pd	26.46	Cd	0.89		
109					Ag	48.161
110	Pd	11.72	Cd	12.49		
111			Cd	12.80		
112	Sn	0.97	Cd	24.13		
113			Cd	12.22	In	4.3
114	Sn	0.65	Cd	28.73		
115	Sn	0.34			In	95.7
116	Sn	14.53	Cd	7.49		
117	Sn	7.68				
118	Sn	24.23				
119	Sn	8.59				
120	Sn	32.59	Te	0.096		

Isotope	El	%	El	%	El	%
121					Sb	57.36
122	Sn	4.63	Te	2.603		
123			Te	0.908	Sb	42.64
124	Sn	5.79	Te	4.816	Xe	0.10
125			Te	7.139		
126			Te	18.95	Xe	0.09
127	I	100				
128			Te	31.69	Xe	1.91
129					Xe	26.4
130	Ba	0.106	Te	33.80	Xe	4.1
131					Xe	21.2
132	Ba	0.101			Xe	26.9
133			Cs	100		
134	Ba	2.417			Xe	10.4
135	Ba	6.592				
136	Ba	7.854	Ce	0.19	Xe	8.9
137	Ba	11.23				
138	Ba	71.70	Ce	0.25	La	0.0902
139					La	99.9098
140			Ce	88.48		
141					Pr	100
142	Nd	27.13	Ce	11.08		
143	Nd	12.18				
144	Nd	23.80	Sm	3.1		
145	Nd	8.30				
146	Nd	17.19				
147			Sm	15.0		
148	Nd	5.76	Sm	11.3		
149			Sm	13.8		
150	Nd	5.64	Sm	7.4		
151					Eu	47.8
152	Gd	0.20	Sm	26.7		
153					Eu	52.2
154	Gd	2.18	Sm	22.7		
155	Gd	14.80				
156	Gd	20.47	Dy	0.06		
157	Gd	15.65				
158	Gd	24.84	Dy	0.10		
159					Tb	100
160	Gd	21.86	Dy	2.34		
161			Dy	18.9		
162	Er	0.14	Dy	25.5		
163			Dy	24.9		
164	Er	1.61	Dy	28.2		
165					Ho	100
166	Er	33.6				
167	Er	22.95				
168	Er	26.8	Yb	0.13		
169					Tm	100
170	Er	14.9	Yb	3.05		
171			Yb	14.3		
172			Yb	21.9		
173			Yb	16.12		
174			Yb	31.8	Hf	0.162
175	Lu	97.41				
176	Lu	2.59	Yb	12.7	Hf	5.206
177					Hf	18.606
178					Hf	27.297
179					Hf	13.629
180	Ta	0.012	W	0.13	Hf	35.100

Isotope	El	%	El	%	El	%
181	Ta	99.988				
182					W	26.3
183					W	14.3
184	Os	0.02			W	30.67
185					Re	37.40
186	Os	1.58			W	28.6
187	Os	1.6			Re	62.60
188	Os	13.3				
189	Os	16.1				
190	Os	26.4			Pt	0.01
191			Ir	37.3		
192	Os	41.0			Pt	0.79
193			Ir	62.7		
194					Pt	32.9
195					Pt	33.8
196	Hg	0.15			Pt	25.3
197			Au	100		
198	Hg	9.97			Pt	7.2
199	Hg	16.87				
200	Hg	23.10				
201	Hg	13.18				
202	Hg	29.86				
203					Tl	29.524
204	Hg	6.87	Pb	1.4		
205					Tl	70.476
206			Pb	24.1		
207			Pb	22.1		
208			Pb	52.4		
209	Bi	100				
210						
211						
212						
213						
214						
215						
216						
217						
218						
219						
220						
221						
222						
223						
224						
225						
226						
227						
228						
229						
230						
231	Pa	100				
232	Th	100				
233						
234	U	0.0055				
235	U	0.7200				
236						
237						
238	U	99.2745				

"Isotopic Compositions of the Elements 1989" Pure Appl. Chem., Vol. 63, No. 7, pp. 991-1002, 1991. © 1991 IUPAC

FIGURE 2.5 Relative abundance of the naturally occurring isotopes of the elements.

(From UIPAC Isotopic Composition of the Elements, *Pure and Applied Chemistry* **75**[6], 683–799, 2003)

3 Sample Introduction

Chapter 3 examines one of the most critical areas of the ICP-MS instrument—the sample introduction system. It discusses the basic principles of converting a liquid into a fine-droplet aerosol suitable for ionization in the plasma, and presents an overview of the different types of commercially available nebulizers and spray chambers. Although this chapter briefly touches upon some of the newer sampling components introduced in the past few years, such as microflow nebulizers and aerosol dilution systems, the new advancements in desolvating nebulizers, chilled spray chambers, online chemistry approaches, auto dilution/auto calibration, intelligent autosamplers and productivity enhancement systems are specifically described in Chapter 18.

The majority of current ICP-MS applications involve the analysis of liquid samples. Even though the technique has been adapted over the years to handle solids and slurries, it was developed in the early 1980s primarily to analyze solutions. There are many different ways of introducing a liquid into an ICP mass spectrometer, but they all basically achieve the same result, which is to generate a fine aerosol of the sample so that it can be efficiently ionized in the plasma discharge. The sample introduction area has been called the Achilles' heel of ICP-MS, because it is considered the weakest component of the instrument. Only about 2% of the sample finds its way into the plasma, depending on the matrix and method of introducing the sample.[1] Although there has recently been significant innovation in this area, particularly in instrument-specific components custom-built by third-party vendors, the fundamental design of a traditional ICP-MS sample introduction system has not dramatically changed since the technique was first introduced in 1983.

Before I discuss the mechanics of aerosol generation in greater detail, let us look at the basic components of a sample introduction system. Figure 3.1 shows the location of the sample introduction area relative to the rest of the ICP mass spectrometer, whereas Figure 3.2 represents a more detailed view showing the individual components.

The traditional way of introducing a liquid sample into an analytical plasma can be considered as two separate events: aerosol generation using a nebulizer and droplet selection using a spray chamber.[2]

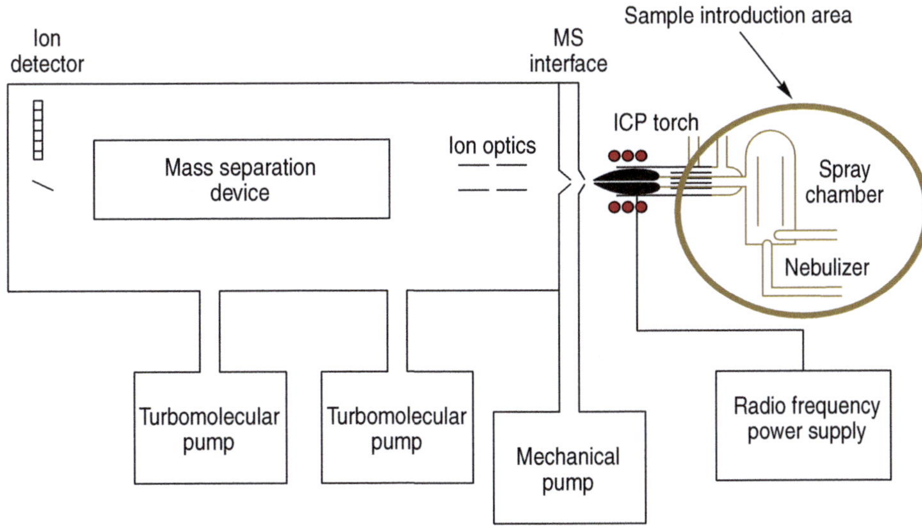

FIGURE 3.1 Location of the ICP-MS sample introduction area.

DOI: 10.1201/9781003187639-3

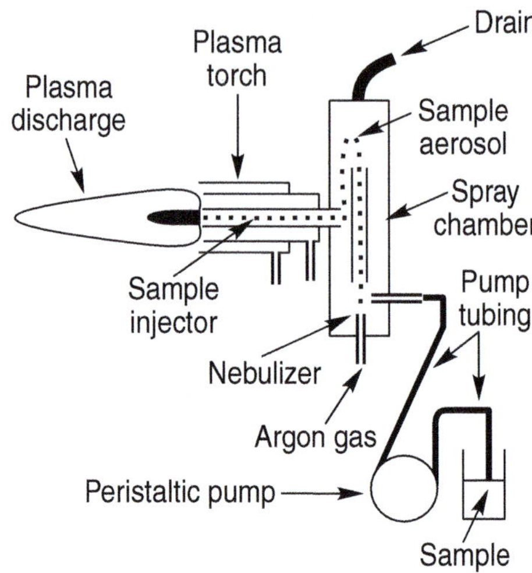

FIGURE 3.2 More detailed view of the ICP-MS sample introduction area.

3.1 AEROSOL GENERATION

As mentioned previously, the main function of the sample introduction system is to generate a fine aerosol of the sample. It achieves this with a nebulizer and a spray chamber. The sample is normally pumped at a maximum of 1 mL/min (for low-pressure nebulizers) or between 100 and 600 μL/min (for more efficient, higher-pressure nebulizers) via a peristaltic or syringe pump into the nebulizer. A peristaltic pump is a small pump with mini rollers mounted between circular disks that are turned by a motor with adjustable speed. The pump speed is adjusted depending on the pump-tubing diameter to deliver the desired liquid flow rate. The constant motion and pressure of the rollers on the pump tubing feeds the sample through to the nebulizer.

A syringe pump delivers the sample via a pneumatic piston, which imparts smoother flow in the absence of rollers. Both syringe and peristaltic pumps can also be equipped with a switching valve to switch the liquid flow path so that sample uptake path rinse may take place during previous sample analysis, followed by rapid sequential sample loading in the sample loop. This type of system is typically two to three times faster than standard autosampler rinse sequences, which means that not only is sample throughput increased, but also faster rinse-out is achieved and therefore less carryover to the next sample. The benefits of using a syringe-type pump for auto dilution and the online addition of internal standards are well recognized in ICP-MS and will be discussed in greater detail in Chapter 20 on "Performance and Productivity Enhancement Tools". However, in its basic configuration, a syringe pump, with switching valve, significantly improves precision by eliminating the pulsations of a peristaltic pump, allowing shorter measurement times to achieve the same performance levels. The stability of a continuous ICP-MS signal with a syringe pump compared to a peristaltic pump is shown in Figure 3.3.

Once the sample enters the nebulizer, the liquid is then broken up into a fine aerosol by the pneumatic action of a flow of gas (~1 L/min) "smashing" the liquid into tiny droplets, very similar to the spray mechanism in a can of deodorant. It should be noted that although pumping the sample is the most common approach to introducing the sample, some pneumatic designs such as concentric nebulizers do not require a pump, because they rely on the natural "Venturi effect" of the positive pressure of the nebulizer gas to suck the sample through the tubing. This sample uptake without the use of a pump is called "self-aspiration". Solution nebulization is conceptually represented in Figure 3.4, which shows aerosol generation using a cross-flow-designed nebulizer.

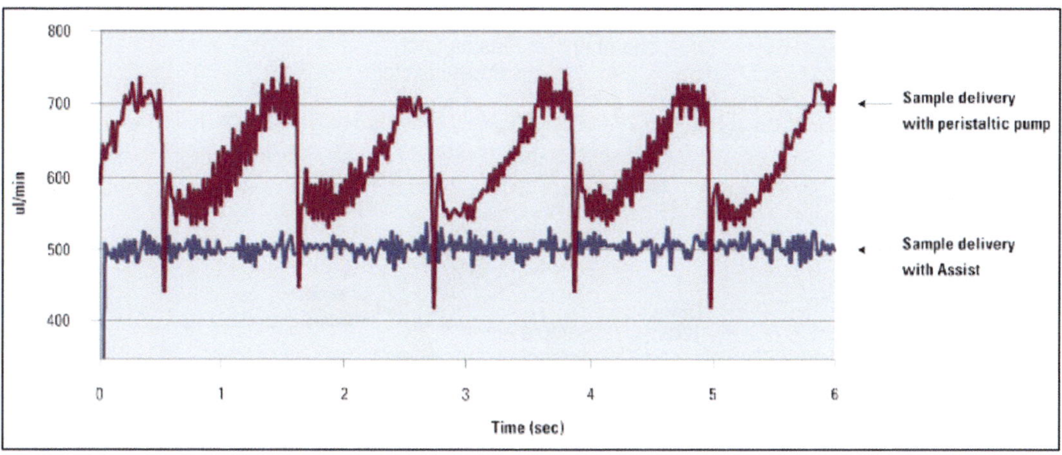

FIGURE 3.3   The stability of a continuous ICP-MS signal with a syringe pump compared to a peristaltic pump.

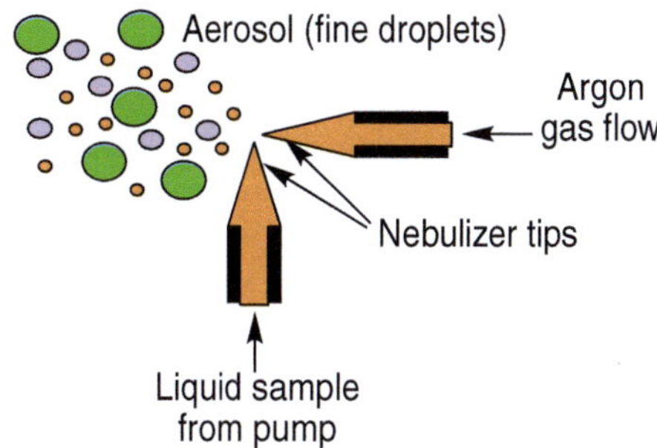

FIGURE 3.4   Conceptual representation of aerosol generation using a cross-flow nebulizer.

3.2   DROPLET SELECTION

Because the plasma discharge is not very efficient at dissociating large droplets, the function of the spray chamber is primarily to allow only the small droplets to enter the plasma. Its secondary purpose is to smooth out pulses that occur during the nebulization process, mainly from the peristaltic pump. Spray chambers are discussed in greater detail later in this chapter, but the most common type is the double-pass design, where the aerosol from the nebulizer is directed into a central tube running the entire length of the chamber. The droplets then travel the length of this tube, where the large droplets (> 10 μm dia.), which have greater momentum than small droplets, will hit the wall of the spray chamber and exit through the drain tube at the end of the spray chamber. The fine droplets (< 10 μm dia.) then pass between the outer wall and the central tube, where they eventually emerge from the spray chamber and are transported into the sample injector of the plasma torch.[3] Although there are many different designs available, the spray chamber's main function is to allow only the smallest droplets into the plasma for dissociation, atomization and, finally, ionization of the sample's elemental components. A simplified schematic of this process using a double-pass-designed spray chamber is shown in Figure 3.5.

Let us now look at the most common nebulizers and spray chamber designs used in ICP-MS. We cannot cover every conceivable design, because over the past few years there has been a huge demand for application-specific solutions, which has generated a number of third-party manufacturers that sell sample introduction components directly to ICP-MS users.

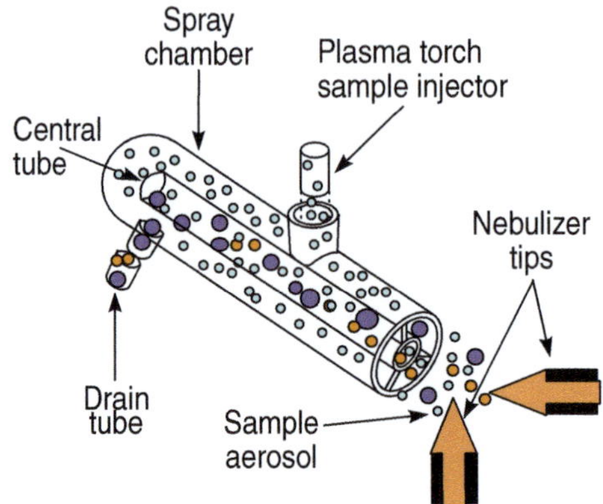

FIGURE 3.5   Simplified representation of the separation of large and fine droplets in a double-pass spray chamber.

3.3 NEBULIZERS

By far the most common design used for ICP-MS is the pneumatic nebulizer, which uses mechanical forces of a gas flow (normally argon at a pressure of 20–30 psi) to generate the sample aerosol. This type of nebulizer is generally used for higher dissolved solids, but unfortunately is very inefficient. Some of the most popular designs of pneumatic nebulizers include the concentric, micro-concentric, micro-flow, and cross-flow. They are usually made from glass or quartz, but other nebulizer materials, such as various kinds of polymers and plastics, are becoming more popular, particularly for highly corrosive samples and specialized applications.

It should be emphasized at this point that nebulizers designed for use with ICP-OES are far from ideal for use with ICP-MS. This is the result of a limitation in the quantity of total dissolved solids (TDS) that can be put into the ICP-MS interface area. Because the orifice sizes of the sampler and skimmer cones used in ICP-MS are so small (~0.6–1.2 mm), the matrix components must generally be kept below 0.2%, although higher concentrations of some matrices can be tolerated (refer to Chapter 5).[4] This means that general-purpose ICP-OES nebulizers that are designed to aspirate 1–2% dissolved solids, or high-solids nebulizers such as the Babbington, V-groove or cone-spray, which are designed to handle up to 20% dissolved solids, are not ideally suited to analyzing solutions by ICP-MS.

Some researchers have attempted to analyze slurries by ICP-MS using this approach. However, this is not recommended for high-throughput, routine work because of the potential of blocking the interface cones, but as long as the particle size of the slurry is kept below 10 μm in diameter, some success has been achieved using these types of nebulizers.[5] Additionally, there are researchers who are attempting to characterize engineered nanoparticles by using a technique called single-particle (SP) ICP-MS. This is in its early stages, but it is a very exciting development, which uses a novel way to separate out individual nanoparticles in suspension, typically in environmental samples, and then detecting them by optimizing the measurement electronics of the ICP-MS detection system.[6] This technique will be described in greater detail in Chapter 20.

The most common of the pneumatic nebulizers used in commercial ICP mass spectrometers are the concentric and cross-flow design types. The concentric design is the most widely used nebulizer, whereas the cross-flow is generally more tolerant to samples containing higher solids and particulate matter. However, the cross-flow design is not widely used or provided with instruments anymore unless specifically requested, because recent advances in the concentric design have allowed for the aspiration of many different types of samples, including those high solids.

3.4 CONCENTRIC DESIGN

In traditional concentric nebulization, a solution is introduced through a fine-bore capillary tube, where it comes into contact with a rapidly moving flow of argon gas at a pressure of approximately 30–50 psi. The high-speed gas and the lower-pressure sample combine to create a Venturi effect, which results in the sample being sucked through to the end of the capillary, where it is broken up into a fine-droplet aerosol. Most concentric nebulizers being used today are manufactured from borosilicate glass or quartz. However, polymer-based materials are now being used for applications that require corrosion resistance. Typical sample flow rates for a standard concentric nebulizer are on the order of 0.1–1 mL/min, although lower flows can be used to accommodate more volatile sample matrices, such as organic solvents. A schematic of a glass concentric nebulizer with the different parts labeled is shown in Figure 3.6, and the aerosol generated by the nebulization process is shown in Figure 3.7

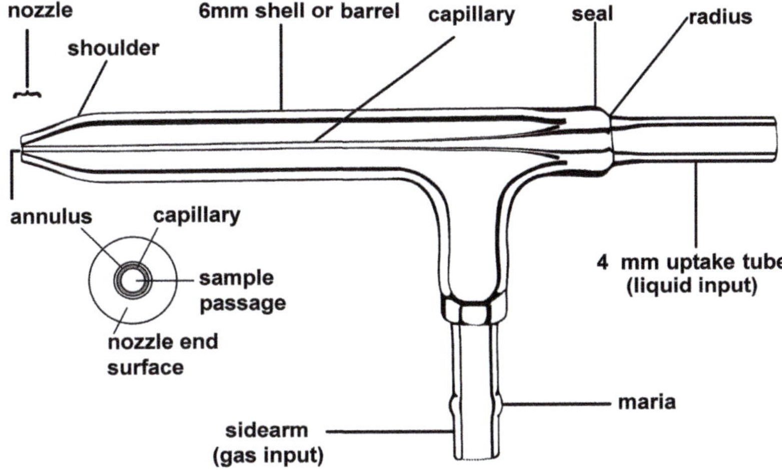

FIGURE 3.6 Schematic of a glass concentric nebulizer.

(Courtesy of Meinhard Glass Products, a part of Elemental Scientific Inc)

FIGURE 3.7 Aerosol generated by a concentric nebulizer.

(Courtesy of Meinhard Glass Products, a part of Elemental Scientific Inc)

The standard concentric pneumatic nebulizer will give excellent sensitivity and stability, particularly with clean solutions. However, the narrow capillary can be plagued by blockage problems, especially if heavier-matrix samples are being aspirated. For that reason, manufacturers of concentric nebulizers offer modifications to the basic design utilizing different-size capillary tubing and recessed tips to allow aspiration of samples with higher dissolved solids and particulate matter. There are even specially designed concentric nebulizers with a smaller-bore input capillary to significantly reduce the dead volume for better coupling of a high-performance liquid chromatography (HPLC) system to the ICP-MS when carrying out trace element speciation studies.

3.5 CROSS-FLOW DESIGN

For the routine analysis of samples that contain a heavier matrix, or maybe small amounts of undissolved matter, the cross-flow design is the more rugged design. With this nebulizer, the argon gas is directed at right angles to the tip of a capillary tube, in contrast to the concentric design, where the gas flow is parallel to the capillary. The solution is either drawn up through the capillary tube via the pressure created by the high-speed gas flow, or, as is most common with cross-flow nebulizers, fed through the tube with a peristaltic pump. In either case, contact between the high-speed gas and the liquid stream causes the liquid to break up into an aerosol. Cross-flow nebulizers are generally not as efficient as concentric nebulizers at creating the very small droplets needed for ionization in the plasma. However, the larger-diameter liquid capillary and longer distance between liquid and gas injectors reduces the potential for clogging problems. Some analysts feel that the penalty to be paid in analytical sensitivity and precision with cross-flow nebulizers compared to the concentric design is compensated by the fact that they are better suited for high-throughput, routine applications. In addition, they are typically manufactured from plastic materials, which make them far more rugged than a glass concentric nebulizer. A cross section of a cross-flow nebulizer is shown in Figure 3.8

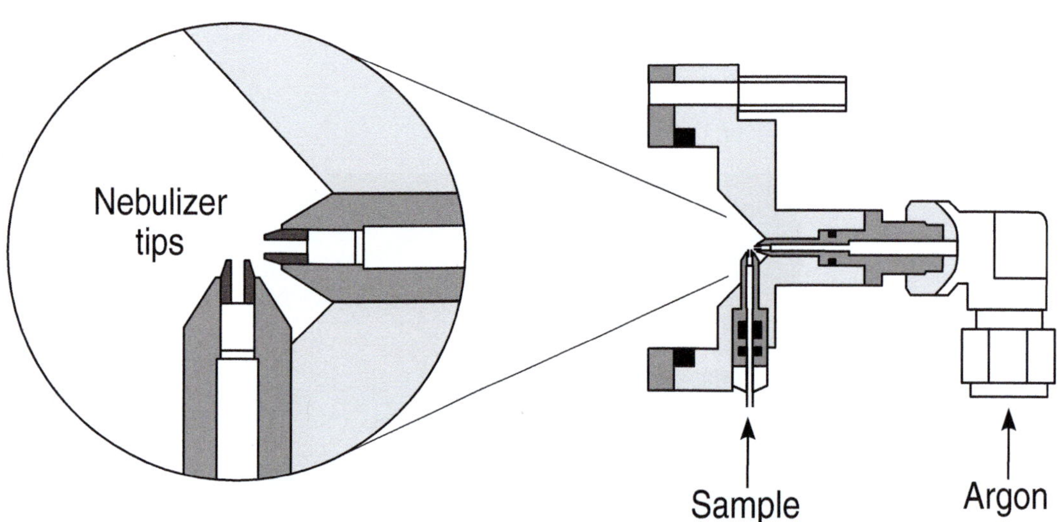

FIGURE 3.8   Schematic of a cross-flow nebulizer.

(Courtesy of PerkinElmer Inc.)

3.6 MICROFLOW DESIGN

More recently, microflow or high-efficiency nebulizers have been designed for ICP-MS to operate at much lower sample flows. Whereas conventional nebulizers have a sample uptake rate of about 1 mL/min, microflow/high-efficiency nebulizers typically run at less than 0.1 mL/min. They are based on the concentric principle, but usually operate at higher gas pressure to accommodate the lower sample flow rates. The extremely low uptake rate makes them ideal for applications where sample volume is limited or where the sample or analyte is prone to sample-introduction memory effects. The additional benefit of this design is that it produces an aerosol with smaller droplets, and as a result it is generally more efficient than a conventional concentric nebulizer.

These nebulizers and their components are typically constructed from polymer materials, such as polytetrafluoroethylene (PTFE), perfluoroalkoxy (PFA), or polyvinylfluoride (PVF), although some designs are available in borosilicate glass or quartz. The excellent corrosion resistance of the polymer nebulizers means they have naturally low blank levels. This characteristic, together with their ability to handle small sample volumes found in applications such as vapor phase decomposition (VPD), makes them an ideal choice for semiconductor laboratories that are carrying out ultratrace element analysis.[6,7] A microflow concentric nebulizer made from PFA is shown in Figure 3.9, and a typical spray pattern of the nebulization process is shown in Figure 3.10.

The disadvantage of microconcentric nebulizers is that they use an extremely fine-capillary, which makes them not very tolerant to high concentrations of dissolved solids or suspended particles. Their high efficiency also means that most of the sample makes it into the plasma, but the low liquid flow decreases plasma loading, and tends to provide greater freedom from matrix suppression problems. For these reasons, they have been found to be most applicable for the analysis of aqueous-type samples or samples containing low levels of dissolved solids. However, the development of higher-pressure (60–70 psi) concentric nebulizers with wider capillary internal diameters has resulted in higher tolerance to dissolved solids with higher nebulization efficiency as well.

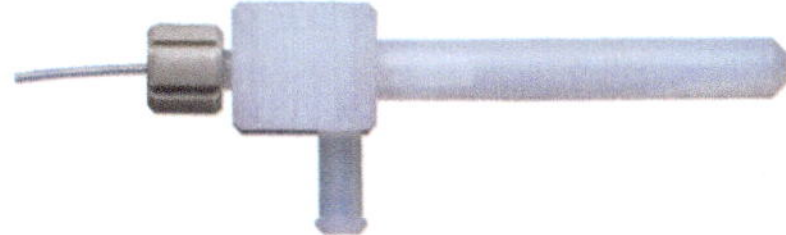

FIGURE 3.9 The OpalMist™ microflow concentric nebulizer made from PFA.

(Courtesy of Glass Expansion Inc.)

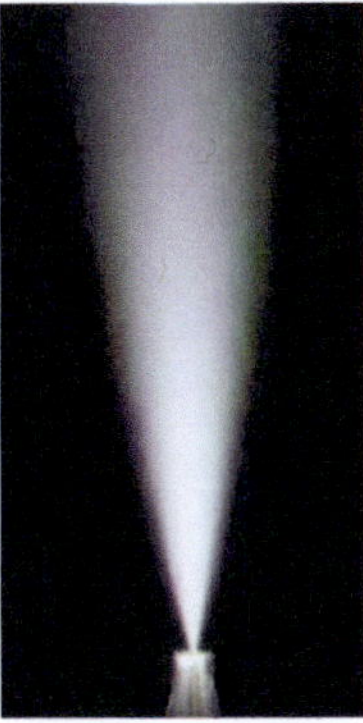

FIGURE 3.10 Spray pattern of a PFA microflow concentric nebulizer.

(Courtesy of Elemental Scientific Inc)

One of the application areas in which high-efficiency nebulizers are well suited is in the handling of extremely small volumes being eluted from an HPLC or flow injection analyzer (FIA) system into an ICP-MS for doing speciation or microsampling work. The analysis of discrete sample volumes encountered in these types of applications allows for detection limits equivalent to a standard concentric nebulizer, while consuming 10–20 times less sample.

3.7   SPRAY CHAMBERS

Let us now turn our attention to spray chambers. There are basically two designs that are used in today's commercial ICP-MS instrumentation: double-pass and cyclonic spray chambers. The double-pass is by far the most common, with the cyclonic type rapidly gaining in popularity. As mentioned earlier, the function of the spray chamber is to reject the larger aerosol droplets and also to smooth out nebulization pulses produced by the peristaltic pump, if it is used. In addition, some ICP-MS spray chambers are externally cooled for thermal stability of the sample and to reduce the amount of solvent going into the plasma. This can have a number of beneficial effects, depending on the application, but the main advantages are to reduce oxide species, minimize signal drift and reduce the solvent loading on the plasma, particularly when aspirating volatile organic solvents.

3.8   DOUBLE-PASS SPRAY CHAMBER

By far the most common design of the double-pass spray chamber is the Scott design, which selects the small droplets by directing the aerosol into a central tube. The larger droplets impact the spray-chamber wall, and exit the spray chamber via a drain tube. The liquid in the central tube is kept at positive pressure (usually by way of a loop), which forces the small droplets back between the outer wall and the central tube, and emerges from the spray chamber into the sample injector of the plasma torch. Double-pass spray chambers come in a variety of shapes, sizes and materials, and are generally considered the most rugged design for routine use. Figure 3.11 shows a Scott double-pass spray chamber made of a polysulfide-type material, coupled to a cross-flow nebulizer.

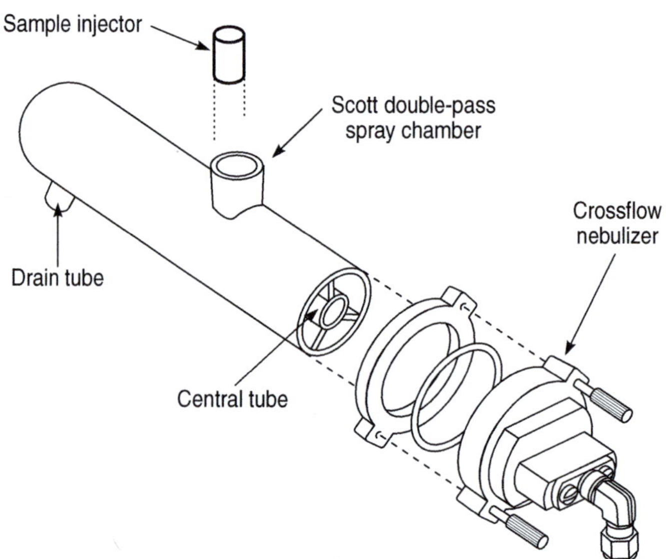

FIGURE 3.11   A Scott double-pass spray chamber with cross-flow nebulizer.

(Courtesy of PerkinElmer Inc.)

3.9   CYCLONIC SPRAY CHAMBER

The cyclonic spray chamber operates by centrifugal force. Droplets are discriminated according to their size by means of a vortex produced by the tangential flow of the sample aerosol and argon gas inside the chamber. Smaller droplets are carried with the gas stream into the ICP-MS, whereas the larger droplets impinge on the walls and fall out through the drain. It is generally accepted that a cyclonic spray chamber has a higher sampling efficiency, which, for clean samples, translates into higher sensitivity and lower detection limits. However, the droplet size distribution appears to be different from a double-pass design, and for certain types of samples can give slightly inferior precision. Beres and coworkers published a very useful study describing the capabilities of a cyclonic spray chamber.[8] Figure 3.12 shows a cyclonic spray chamber connected to a concentric nebulizer.

The cyclonic spray chamber is growing in popularity, particularly as its potential is getting realized in more and more application areas. Just as there is a wide selection of nebulizers available for different applications, there is also a wide choice of customized cyclonic spray chambers, manufactured from glass, quartz and different polymer materials. Depending on the application being carried out, modifications to the cyclonic design are available for low sample flows, high dissolved solids, fast sample washout, corrosion resistance and organic solvents. Figure 3.13 shows one of the many variations of cyclonic spray chamber, called the jacketed Cinnabar™, which is a water-cooled borosilicate glass spray chamber optimized for aspirating small sample volumes with a microflow concentric nebulizer.

It is worth emphasizing that cooling the spray chamber is generally beneficial in ICP-MS, because it reduces the solvent loading on the plasma. This has three major benefits. First, because very little plasma energy is wasted vaporizing the solvent, more is available to excite and ionize the analytes. Second, if there is less water being delivered to the plasma, there is less chance of forming oxide and hydroxide species, which can potentially interfere with other analytes. Finally, if the spray chamber is kept at a constant temperature, it leads to better long-term signal stability,

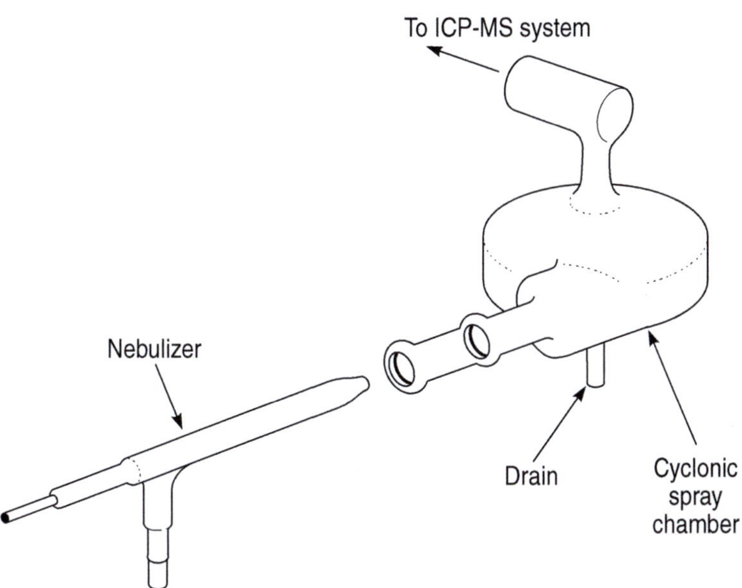

FIGURE 3.12   A cyclonic spray chamber (shown with a concentric nebulizer).

(From S. A. Beres, P. H. Bruckner, and E. R. Denoyer, *Atomic Spectroscopy*, **15**[2], 96–99, 1994)

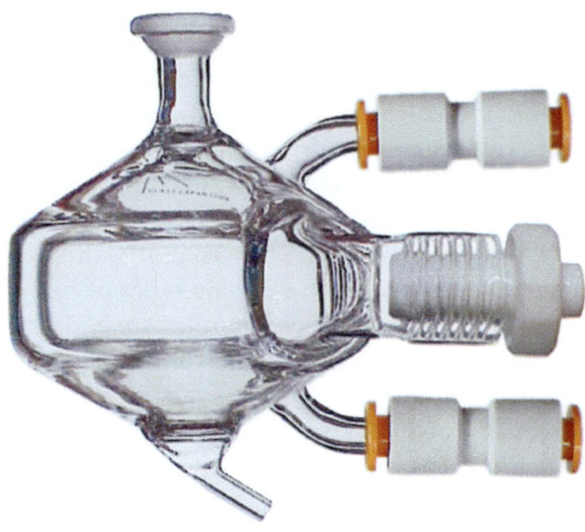

FIGURE 3.13 The low-flow Cinnabar™, a water-cooled cyclonic spray chamber for use with a microflow concentric nebulizer.

(Courtesy of Glass Expansion Inc.)

especially if there are environmental temperature changes over the time period of the analysis. For these reasons, some manufacturers supply cooled spray chambers as standard, whereas others offer the capability as an option. There is also a wide variety of cooled and chilled spray chambers available from third-party vendors.

3.10 AEROSOL DILUTION

To address the limitation of the total dissolved solids capability of ICP-MS, some vendors offer an aerosol dilution system, which introduces a flow of argon gas between the nebulizer and the torch to carry our aerosol dilution of the sample. This has the effect of reducing the sample's solvent loading on the plasma, so it can tolerate much higher total dissolved solids (TDS) levels than the < 0.2%, which is typical for most ICP-MS instrumentation. However, it's important to emphasize that the dilution is done after the nebulizer, so care must be taken in selecting the optimum nebulizer if the sample contains high levels of dissolved solids. The principles of aerosol dilution are shown in Figure 3.14. Some of the benefits of this novel type of dilution include:

- Enabling the direct analysis of samples containing medium to high percentage levels of dissolve solids, assuming the nebulizer can handle them.
- Significantly improving plasma robustness compared to conventional sample-introduction methods.
- Because less solvent/matrix is entering the plasma, it reduces oxide interferences to very low levels, providing better accuracy and more stable sampling conditions.
- Eliminating the need for conventional liquid dilution of high matrix samples prior to analysis, which has several disadvantages, including increased risk of sample contamination, dilution errors and sample prep time.

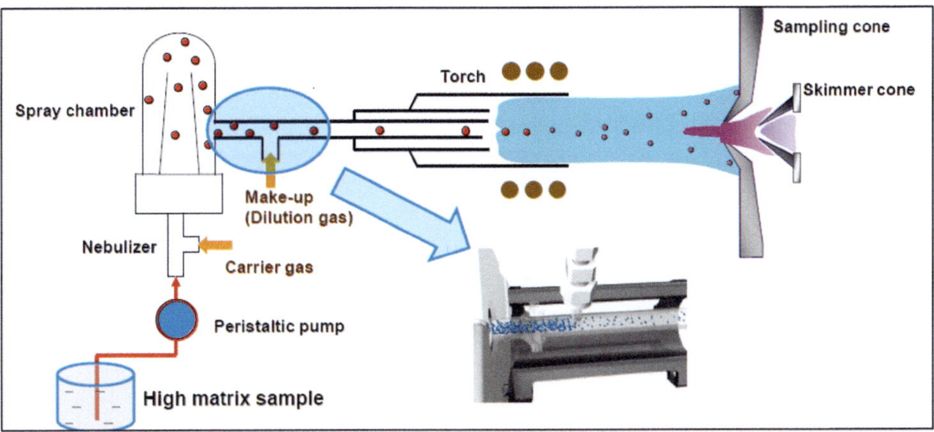

FIGURE 3.14 Principle of aerosol dilution.

(From "Today's Agilent—New Atomic Spectroscopy Solutions for Environmental Laboratories" online webinar)

3.11 FINAL THOUGHTS

There are many other nonstandard sample-introduction devices such as laser ablation, ultrasonic nebulizers, desolvation devices, direct injection nebulizers, flow injection systems, enhanced productivity systems, autodilution and online chemistry techniques, which are not described in this chapter. However, because they are becoming more and more important, particularly as ICP-MS users are demanding higher performance, more productivity and greater flexibility, they are covered in greater detail in a later chapter (Chapter 20).

FUTHER READING

1. R. A. Browner and A. W. Boorn, *Analytical Chemistry*, **56**, 786–798A, 1984.
2. B. L. Sharp, *Analytical Atomic Spectrometry*, **3**, 613, 1980.
3. L. C. Bates and J. W. Olesik, *Journal of Analytical Atomic Spectrometry*, **5**(3), 239, 1990.
4. R. S. Houk, *Analytical Chemistry*, **56**, 97A, 1983.
5. J. G. Williams, A. L. Gray, P. Norman, and L. Ebdon, *Journal of Analytical Atomic Spectrometry*, **2**, 469–472, 1987.
6. J. Ranville, K. Neubauer, and R Thomas, *Spectroscopy*, **27**(8), 20–27, August, 2012.
7. E. Debrah, S. A. Beres, T. J. Gluodennis, R. J. Thomas, and E. R. Denoyer, *Atomic Spectroscopy*, **16**(7), 197–202, 1995.
8. R. A. Aleksejczyk and D. Gibilisco, *Micro*, September 1997.
9. S. A. Beres, P. H. Bruckner, and E. R. Denoyer, *Atomic Spectroscopy*, **15**(2), 96–99, 1994.

4 Plasma Source

Chapter 4 takes a look at the region of the ICP-MS where the ions are generated—the plasma discharge. It gives a brief historical perspective of some of the common analytical plasmas used over the years and discusses the components used to create the inductively coupled plasma (ICP). It also explains the fundamental principles of formation of a plasma discharge and how it is used to convert the sample aerosol into a stream of positively charged ions of low kinetic energy required by the ion-focusing system and the mass spectrometer.

ICPs are by far the most common type of plasma sources used in today's commercial ICP optical emission (ICP-OES) and ICP mass spectrometric (ICP-MS) instrumentation. However, it was not always that way. In the early days, when researchers were attempting to find the ideal plasma source to use for spectrometric studies, it was not clear which approach would prove to be the most successful. In addition to ICPs, some of the other novel plasma sources developed were direct current plasmas (DCPs) and microwave-induced plasmas (MIPs). Before I go on to describe the ICP, let us first take a closer look at these other two excitation sources.

A DCP is formed when a gas (usually argon) is introduced into a high current flowing between two or three electrodes. Ionization of the gas produces a Y-shaped plasma. Unfortunately, early DCP instrumentation was prone to interference effects and also had some usability and reliability problems. For these reasons, the technique never became widely accepted by the analytical community.[1] However, its one major benefit was that it could aspirate high dissolved or suspended solids because there was no restrictive sample injector for the solid material to block. This feature alone made it very attractive for some laboratories, and once the initial limitations of DCPs were better understood, the technique became more accepted. Limitations in the DCP approach led to the development of electrodeless plasma, of which the microwave-induced plasma (MIP) was the simplest form. MIP technology has mainly been used as an ion source for mass spectrometry (MS)[2] and also as emission-based detectors for gas chromatography. It is only recently that that the technology has advanced and been viewed as a possible alternative to the ICP for elemental analysis.

An MIP basically consists of a quartz tube surrounded by a microwave waveguide or cavity. Microwaves produced from a magnetron fill the cavity and cause the electrons in the plasma support gas to oscillate. The oscillating electrons collide with other atoms in the flowing gas to create and maintain a high-temperature plasma. As in the inductively coupled plasmas, a high-voltage spark is needed to create the initial electrons to create the plasma, which achieves temperatures of approximately 5,000 K.

The limiting factor to their use was that with the low power and high frequency of the MIP, it was very difficult to maintain the stability of the plasma when aspirating liquid samples containing high levels of dissolved solids. Various attempts had been made over the years to couple desolvation techniques to the MIP, but only managed to achieve limited success. However, an MIP-AES system using nitrogen gas has been developed, which appears to have overcome many of the limitations of the earlier designs and is achieving detection limit better than flame atomic absorption and is only slightly inferior to ICP-OES for many elements.[3]

Because of the limitations of the DCP and MIP approaches, ICPs became the dominant area of research for both optical emission and mass spectrometric studies. As early as 1964, Greenfield and coworkers reported that an atmospheric pressure ICP coupled with OES could be used for elemental analysis.[4] Although crude by today's standards, it showed the enormous possibilities of the ICP as an excitation source and most definitely opened the door in the early 1980s to the even more exciting potential of using the ICP to generate ions.[5]

DOI: 10.1201/9781003187639-4

4.1   THE PLASMA TORCH

Before we take a look at the fundamental principles behind the creation of an ICP used in ICP-MS, let us take a look at the basic components used to generate the source—a plasma torch, radio-frequency (RF) coil and power supply. Figure 4.1 shows their proximity compared to the rest of the instrument, and Figure 4.2 is a more detailed view of the plasma torch and RF coil relative to the MS interface.

The plasma torch consists of three concentric tubes, which are normally made from quartz. In Figure 4.2, these are shown as the outer tube, middle tube and sample injector. The torch can either be one piece, in which all three tubes are connected, or it can employ a demountable design in which the tubes and the sample injector are separate. The gas (usually argon) that is used to form the plasma (plasma gas) is passed between the outer and middle tubes at a flow rate of ~12–17 L/min. A

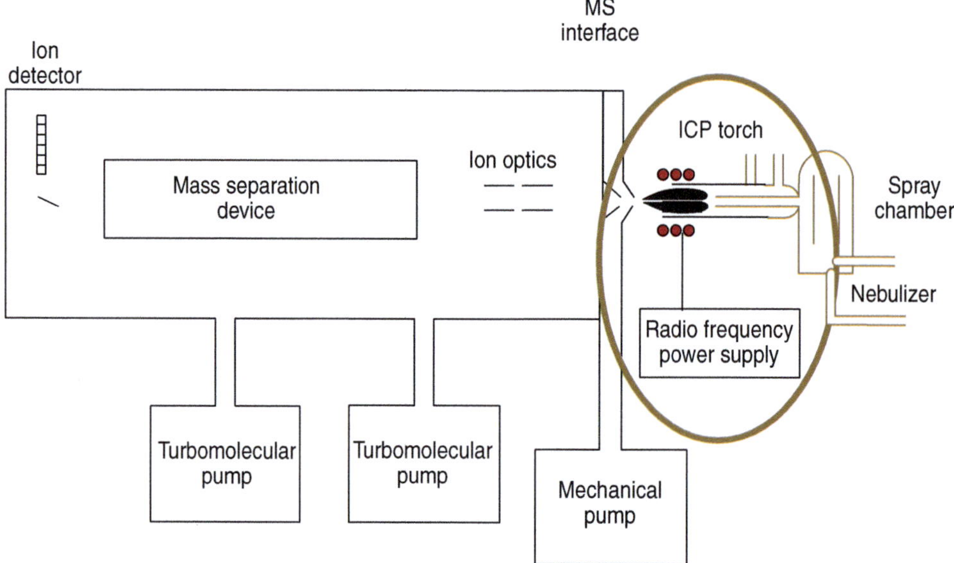

FIGURE 4.1   Inductively coupled plasma mass spectrometry (ICP-MS) system showing location of the plasma torch and RF power supply.

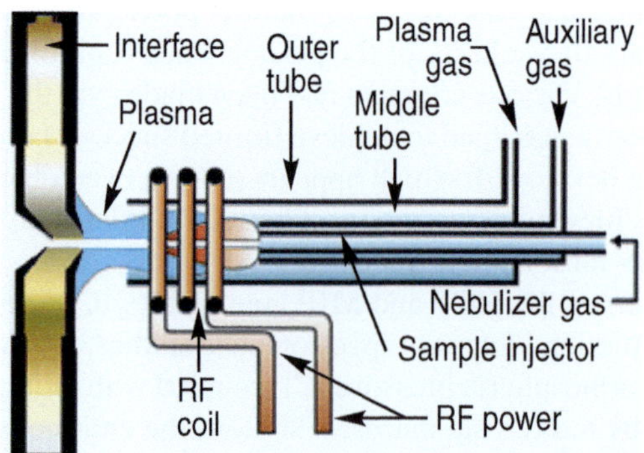

FIGURE 4.2   Detailed view of plasma torch and radio-frequency (RF) coil relative to the inductively coupled plasma mass spectrometry (ICP-MS) interface.

second gas flow (auxiliary gas) passes between the middle tube and the sample injector at ~1 L/min and is used to change the position of the base of the plasma relative to the tube and the injector. A third gas flow (nebulizer gas), also at ~1 L/min, brings the sample, in the form of a fine-droplet aerosol, from the sample-introduction system and physically punches a channel through the center of the plasma. The sample injector is often made from other materials besides quartz, such as alumina, platinum and sapphire—if highly corrosive materials need to be analyzed. It is worth mentioning that although argon is the most suitable gas to use for all three flows, there are analytical benefits in using other gas mixtures, especially in the nebulizer flow.[6] The plasma torch is mounted horizontally and positioned centrally in the RF coil, approximately 10–20 mm from the interface. This can be seen in Figure 4.3, which shows a photograph of a plasma torch mounted in an instrument.

Ceramic components are also available for most ICP-MS torches. The outer, inner and sample injector tubes are normally made of quartz, but for some applications, it is beneficial to consider using an alternative material like ceramic. And if a demountable torch is being used, any or all of the tubes can be replaced. Some of the applications that might benefit from a ceramic torch include:

- When silicon is one of the analytes, because quartz outer tubes often produce high Si background signals
- The analysis of fusion mixtures or samples with high levels of dissolved solids, which might cause devitrification of quartz tubes
- The analysis of organic-based samples where quartz outer tubes often suffer from short lifetime

FIGURE 4.3 Photograph of a plasma torch mounted in an instrument.

(Courtesy of PerkinElmer Inc)

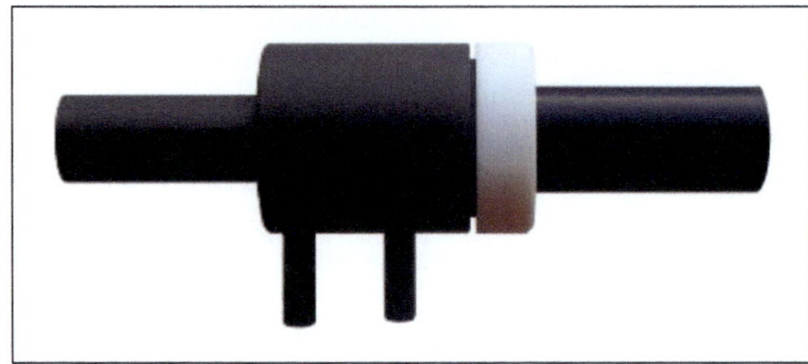

FIGURE 4.4   A fully demountable ceramic torch.

(Courtesy of Glass Expansion Inc)

A fully demountable ceramic torch is shown in Figure 4.4.

It's also worth emphasizing that the coil used in an ICP-MS plasma is slightly different from the one used in ICP-OES, the reason being that in a plasma discharge, there is a potential difference of a few hundred volts produced by capacitive coupling between the RF coil and the plasma. In an ICP mass spectrometer, this would result in a secondary discharge between the plasma and the interface cone, which can negatively affect the performance of the instrument. To compensate for this, the coil must be grounded to keep the interface region as close to zero potential as possible. The full implications of this are discussed in greater detail in the next chapter on the mass spectrometer interface.

4.2   FORMATION OF AN ICP DISCHARGE

Let us now discuss the mechanism of formation of the plasma discharge in greater detail. First, a tangential (spiral) flow of argon gas is directed between the outer and middle tube of a quartz torch. A load coil (usually copper) surrounds the top end of the torch and is connected to an RF generator. When RF power (typically 750–1500 W, depending on the sample) is applied to the load coil, an alternating current oscillates within the coil at a rate corresponding to the frequency of the generator. In most ICP generators, this frequency is either 27 MHz, 34 MHz or 40 MHz (commonly known as megahertz or million cycles/second). This RF oscillation of the current in the coil causes an intense electromagnetic field to be created in the area at the top of the torch. With argon gas flowing through the torch, a high-voltage spark is applied to the gas, causing some electrons to be stripped from their argon atoms. These electrons, which are caught up and accelerated in the magnetic field, then collide with other argon atoms, stripping off still more electrons. This collision-induced ionization of the argon continues in a chain reaction, breaking down the gas into argon atoms, argon ions, and electrons forming what is known as inductively coupled plasma (ICP) discharge. The ICP discharge is then sustained within the torch and load coil as RF energy is continually transferred to it through the inductive coupling process. The amount of energy required to generate argon ions in this process is on the order of 15.8 eV (first ionization potential), which is enough energy to ionize the majority of the elements in the periodic table. The sample aerosol is then introduced into the plasma through a third tube called the sample injector. The entire process is conceptually shown in Figure 4.5.[7]

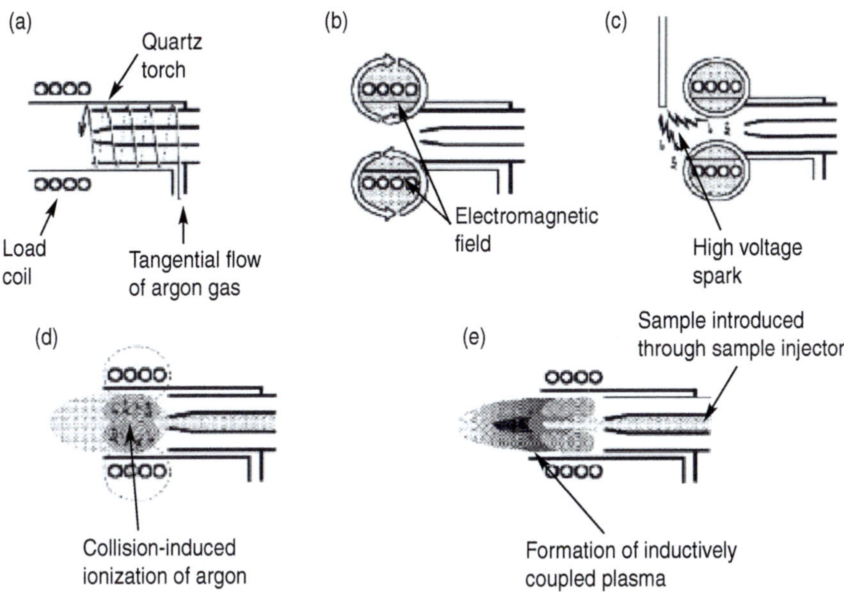

FIGURE 4.5 Schematic of an ICP torch and load coil showing how the inductively coupled plasma (ICP) is formed. (a) A tangential flow of argon gas is passed between the outer and middle tube of the quartz torch. (b) RF power is applied to the load coil, producing an intense electromagnetic field. (c) A high-voltage spark produces free electrons. (d) Free electrons are accelerated by the RF field, causing collisions and ionization of the argon gas. (e) The ICP is formed at the open end of the quartz torch. The sample is introduced into the plasma via the sample injector.

(From C. B. Boss and K. J. Fredeen, *Concepts, Instrumentation and Techniques in Inductively Coupled Plasma Optical Emission Spectrometry*, 3rd edition, Perkin Elmer Corporation, 2004)

4.3 THE FUNCTION OF THE RF GENERATOR

Although the principles of an RF power supply have not changed since the work of Greenfield, the components have become significantly smaller. Some of the early generators that used nitrogen or air required 5–10 kW of power to sustain the plasma discharge—and literally took up half the room. Most of today's generators use solid-state electronic components, which means that vacuum power amplifier tubes are no longer required. This makes modern instruments significantly smaller and, because vacuum tubes were notoriously unreliable and unstable, far more suitable for routine operation.

As mentioned previously, three frequencies have typically been used for ICP RF generators —27, 34 or 40 MHz. These frequencies have been set aside specifically for RF applications of this kind, so that they will not interfere with other communication-based frequencies. There has been much debate over the years as to which frequency gives the best performance.[8,9] I think it is fair to say that although there have been a number of studies, no frequency appears to give a significant analytical advantage over the other. In fact, of all the commercially available ICP-MS systems, there seems to be roughly an equal number of 27 and 40 MHz generators. More recently, a 34 MHz generator has been developed, which appears to offer benefits for certain applications.

The more important consideration is the coupling efficiency of the RF generator to the coil. The majority of modern solid-state RF generators are on the order of 70–75% efficient, which means that 70–75% of the delivered power actually makes it into the plasma. This was not always the case, and some of the older vacuum-tube-designed generators were notoriously inefficient, with some of them experiencing over a 50% power loss. Another important criterion to consider is the way the matching network compensates for changes in impedance (a material's resistance to the flow of an electric current) produced by the sample's matrix components or differences in solvent volatility, or both. In earlier-designed crystal-controlled generators, this was usually done with servo-driven capacitors. They worked very well with most sample types but, because they were mechanical devices, they struggled to compensate for very rapid impedance changes produced by some samples. As a result, it was fairly easy to extinguish the plasma, particularly when aspirating volatile organic solvents.

These problems were partially overcome by the use of free-running RF generators, in which the matching network was based on electronic tuning of small changes in frequency brought about by the sample solvent or matrix components or both. The major benefit of this approach was that compensation for impedance changes was virtually instantaneous, because there were no moving parts. This allowed for the successful analysis of many sample types, which would most probably have extinguished the plasma of a crystal-controlled generator. However, because of improvements in electronic components over the years, the more recent crystal-controlled generators appear to be as responsive as free-running designs.

However, it should be mentioned that the recent development of a novel 34 MHz free-running designed RF generator, using solid-state electronics, has enhanced the capability of ICP-MS to analyze some real-world samples, particularly when using cool plasma conditions (see Chapter 14 on Review of Interferences). This new design, which is based on an air-cooled plasma load coil, allows the matching network electronics to rapidly respond to changes in the plasma impedance produced by different sampling conditions and sample matrices, while still maintaining low plasma potential at the interface region.[10]

4.4   IONIZATION OF THE SAMPLE

To better understand what happens to the sample on its journey through the plasma source, it is important to understand the different heating zones within the discharge. Figure 4.6 shows a cross-sectional representation of the discharge along with the approximate temperatures for different regions of the plasma.

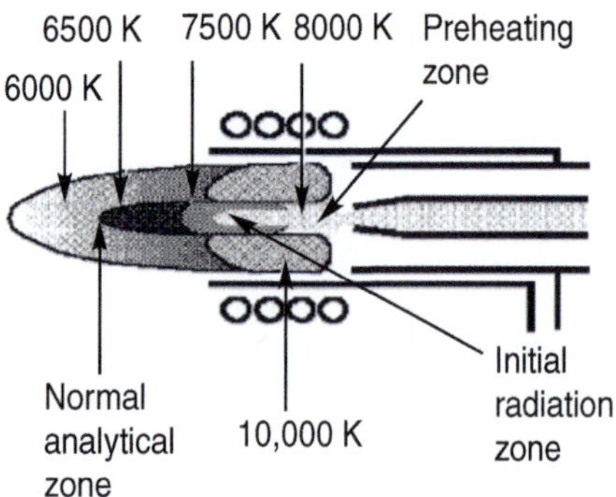

FIGURE 4.6   Different temperature zones in the plasma.

(From C. B. Boss and K. J. Fredeen, *Concepts, Instrumentation and Techniques in Inductively Coupled Plasma Optical Emission Spectrometry*, 3rd edition, Perkin Elmer Corporation, 2004)

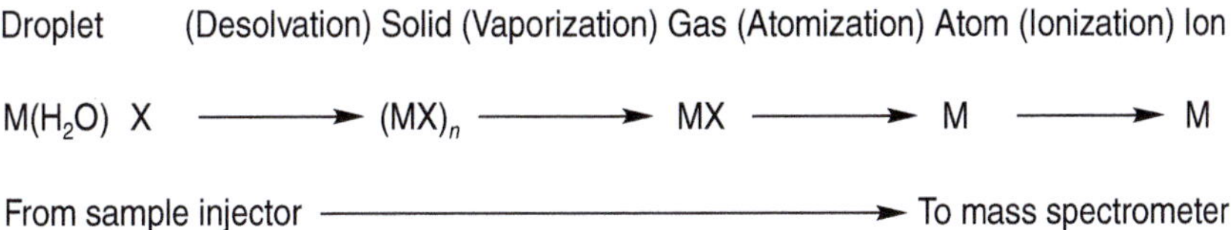

FIGURE 4.7 Mechanism of conversion of a droplet to a positive ion in the inductively coupled plasma (ICP).

As mentioned previously, the sample aerosol enters the injector via the spray chamber. When it exits the sample injector, it is moving at such a velocity that it physically punches a hole through the center of the plasma discharge. It then goes through a number of physical changes, starting at the preheating zone and continuing through the radiation zone, before it eventually becomes a positively charged ion in the analytical zone. To explain this in a very simplified way, let us assume that the element exists as a trace metal salt in solution. The first step that takes place is desolvation of the droplet. With the water molecules stripped away, it then becomes a tiny solid particle. As the sample moves further into the plasma, the solid particle changes first into gaseous form and then into a ground-state atom. The final process of conversion of an atom to an ion is achieved mainly by collisions of energetic argon electrons (and to a lesser extent by argon ions) with the ground-state atom.[11] The ions then emerge from the plasma and is directed into the interface of the mass spectrometer (for details on the mechanisms of ion generation, refer to Chapter 2 on Principles of Ion Formation). This process of conversion of droplets into ions is represented in Figure 4.4.

FURTHER READING

1. A. L. Gray, *Analyst*, **100**, 289–299, 1975.
2. D. J. Douglas and J. B. French, *Analytical Chemistry*, **53**, 37–41, 1981.
3. R. J. Thomas, Emerging Technology Trends in Atomic Spectroscopy Are Solving Real-World Application Problems, *Spectroscopy*, **29**(3), 42–51, 2014.
4. S. Greenfield, I. L. Jones, and C. T. Berry, *Analyst*, **89**, 713–720, 1964.
5. R. S. Houk, V. A. Fassel, and H. J. Svec, *Dynamic Mass Spectrometry*, **6**, 234–238, 1981.
6. J. W. Lam and J. W. McLaren, *Journal of Analytical Atomic Spectrometry*, **5**, 419–424, 1990.
7. C. B. Boss and K. J. Fredeen, *Concepts, Instrumentation and Techniques in Inductively Coupled Plasma Optical Emission Spectrometry*, 2nd edition, Perkin Elmer Corporation, Shelton, CT, 1994.
8. K. E. Jarvis, P. Mason, T. Platzner, and J. G. Williams, *Journal of Analytical Atomic Spectrometry*, **13**, 689–696, 1998.
9. G. H. Vickers, D. A. Wilson, and G. M. Hieftje, *Journal of Analytical Atomic Spectrometry*, **4**, 749–754, 1989.
10. K. Neubaeur, The Use of Ion-Molecule Chemistry Combined with Optimized Plasma Conditions to Meet SEMI Tier C Guidelines for Elemental Impurities in Semiconductor Grade Hydrochloric Acid by ICP-MS, *Spectroscopy*, **32**(9), 17–26, October 2017.
11. T. Hasegawa and H. Haraguchi, *ICPs in Analytical Atomic Spectrometry*, A. Montasser, D. W. Golightly (eds), 2nd edition, VCH, New York, 1992.

5 Interface Region

Chapter 5 takes a look at the ICP-MS interface region, which is probably the most critical area of the entire ICP-MS system. It gave the early pioneers of the technique the most problems to overcome. Although we take all the benefits of ICP-MS for granted, the process of taking a liquid sample, generating an aerosol that is suitable for ionization in the plasma and then sampling a representative number of analyte ions, transporting them through the interface, focusing them via the ion optics into the mass spectrometer and finally ending up with detection and conversion to an electronic signal is not a trivial task. Each part of the journey has its own unique problems to overcome, but probably the most challenging is the extraction of the ions from the plasma into the mass spectrometer.

The role of the interface region, which is shown in Figure 5.1, is to transport the ions efficiently, consistently and with electrical integrity from the plasma, which is at atmospheric pressure (760 torr), to the mass spectrometer analyzer region at approximately 10^{-6} torr.

This is first achieved by directing the ions into the interface region. The interface consists of two or three metallic cones (depending on the design) with very small orifices, which are maintained at a vacuum of 1–2 torr with a mechanical roughing pump. After the ions are generated in the plasma, they pass into the first cone, known as the sampler cone, which has an orifice of 0.8–1.2 mm i.d. From there, they travel a short distance to the skimmer cone, which is generally smaller and more pointed than the sampler cone. The skimmer also has a much smaller orifice (typically 0.4–0.8 mm i.d.) than the sampler cone. In some designs, there is a third cone called the hyper skimmer cone, which is used to reduce the vacuum in smaller steps and provide less dispersion of the ion beam. Whether the system uses two cones or incorporates a triple-cone interface, they are usually made of nickel, but can be made of other materials such as platinum, which is far more tolerant to corrosive liquids. To reduce the effects of high-temperature plasma on the cones, the interface housing is water-cooled and made from a material that dissipates heat easily, such as copper or aluminum. The ions then emerge from the skimmer cone, where they are directed through the ion optics and, finally, guided into the mass separation device. Figure 5.2 shows the interface region in greater detail, and Figure 5.3 shows a close-up of a platinum sampler cone on the left and a platinum skimmer cone on the right.

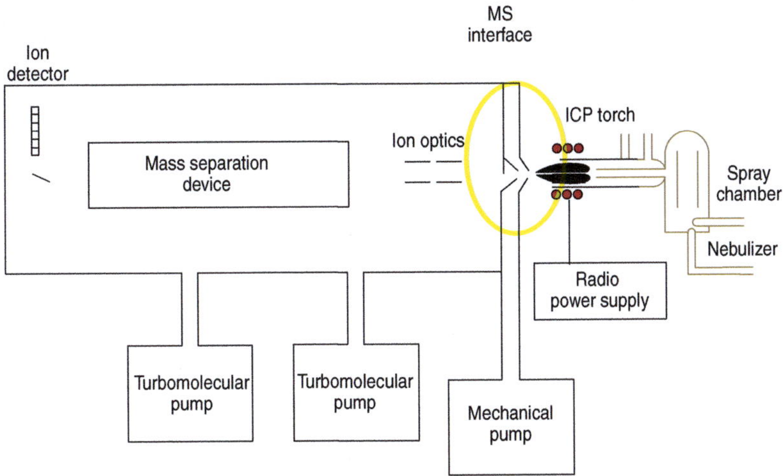

FIGURE 5.1 Schematic of an inductively coupled plasma mass spectrometer (ICP-MS), showing proximity of the interface region.1

DOI: 10.1201/9781003187639-5

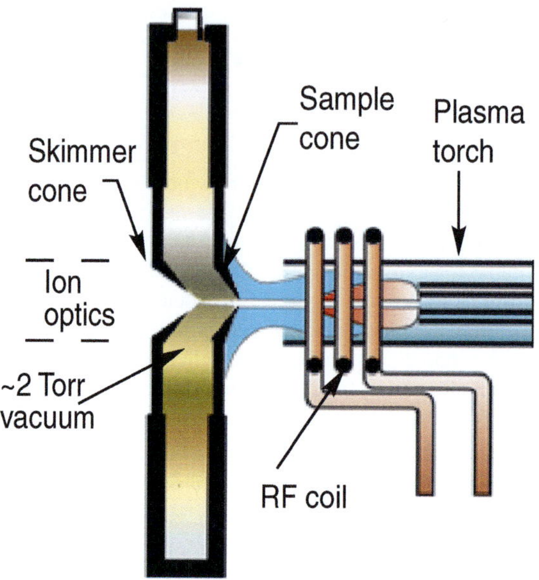

FIGURE 5.2   Detailed view of the interface region.

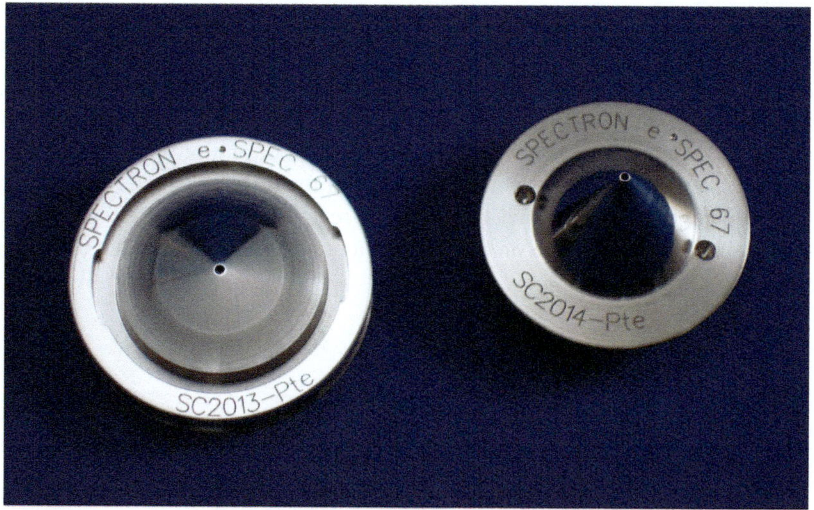

FIGURE 5.3   Close-up of a platinum sampler cone (left) and a platinum skimmer cone (right). (Courtesy of Spectron Inc.)

It should be noted that for most sample matrices, it is desirable to keep the TDS below 0.2%, because of the possibility of deposition of the matrix components around the sampler cone orifice. This is not such a serious problem with short-term use, but it can lead to long-term signal instability if the instrument is being run for extended periods of time. The TDS levels can be higher (0.5–1%) when analyzing a matrix that forms a volatile oxide such as sodium chloride because, once deposited on the cones, the volatile sodium oxide tends to re-vaporize without forming a significant layer that could potentially affect the flow through the cone orifice. By careful optimization of the plasma RF power, sampling depth and extraction lens voltage, it is therefore possible to run a 1:1 dilution of seawater (1.5% NaCl) for extended periods with no significant cone blockage.

More recently, a novel technique for the handling of high-matrix samples has been developed that introduces a make-up gas between the spray chamber and the torch. This technique, which has been termed "aerosol dilution", does not require the introduction of more water to the system and protects the mass spectrometer components from the high levels of matrix components in the sample. This technique enables the ICP-MS system to directly aspirate 2–3% total dissolved solids, with the added benefit of reducing oxide levels, because the sample is not being diluted with water in the traditional way. This technique was described in greater detail in the chapter on "Sample Introduction".

5.1   CAPACITIVE COUPLING

The coupling of the plasma to the mass spectrometer proved to be very problematic during the early development of ICP-MS because of an undesirable electrostatic (capacitive) coupling between the voltage on the load coil and the plasma discharge, producing a potential difference of 100–200 V. Although this potential is a physical characteristic of all ICP discharges, it was more serious in an ICP mass spectrometer, because the capacitive coupling created an electrical discharge between the plasma and the sampler cone. This discharge, commonly called the "pinch effect" or secondary discharge, shows itself as arcing in the region where the plasma is in contact with the sampler cone.[1] This is shown in a simplified manner in Figure 5.4.

If not addressed, this arcing can cause all kinds of problems, including an increase in doubly charged interfering species, a wide kinetic energy spread of sampled ions, formation of ions generated from the sampler cone and decreased orifice lifetime. These were all problems reported by many of the early researchers into the technique.[2,3] In fact, because the arcing increased with sampler cone orifice size, the source of the secondary discharge was originally thought to be the result of an electro-gas-dynamic effect, which produced an increase in electron density at the orifice.[4] After many experiments, it was eventually realized that the secondary discharge was a result of electrostatic coupling of the load coil to the plasma. The problem was first eliminated by grounding the induction coil in the center, which had the effect of reducing the RF potential to a few volts. This can be seen in Figure 5.5, which is taken from one of the early papers and shows the reduction in plasma potential as the coil is grounded at different positions (turns) along its length.[5]

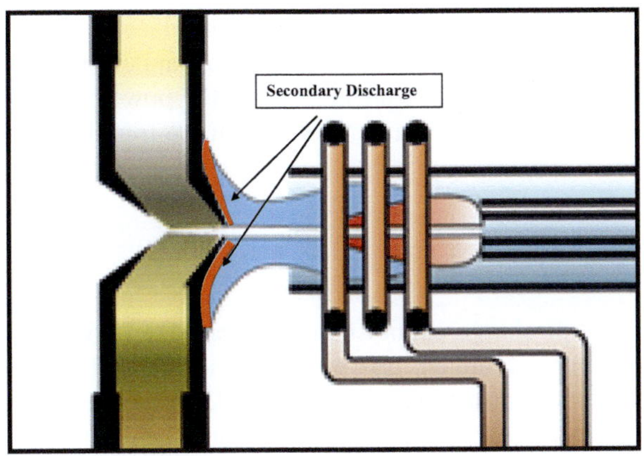

FIGURE 5.4   Interface showing area affected by a secondary discharge.

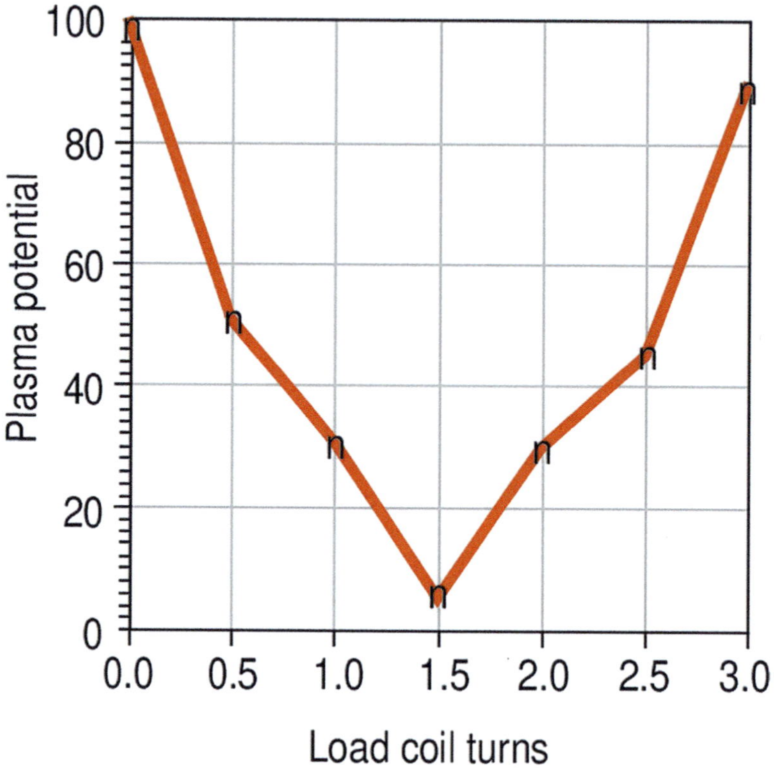

FIGURE 5.5   Reduction in plasma potential as the load coil is grounded at different positions (turns) along its length.[5]

This work has since been supported by other researchers who carried out Langmuir probe measurements, the results indicating that plasma potential was lowest with a center-tapped coil as opposed to the grounding elsewhere on the coil.[6,7] In today's instrumentation, the "grounding" is implemented in a number of different ways, depending on the design of the interface. Some of the most popular designs include balancing the oscillator inside the circuitry of the RF generator,[8] positioning a grounded shield or plate between the coil and the plasma torch,[9] and using two interlaced coils where the RF fields go in opposite directions.[10] They all work differently, but many experts believe that the center-tapped coil and the interlaced coil achieve the lowest plasma potential compared to the other designs. However, they all appear to work equally well when it comes to using cool plasma conditions requiring higher RF power and lower nebulizer gas flow. Further details about cool and cold plasma technology can be found in Chapter 14 on "Review of Interferences".

5.2   ION KINETIC ENERGY

The impact of a secondary discharge cannot be overemphasized with respect to its effect on the kinetic energy of the ions being sampled. It is well documented that the energy spread of the ions entering the mass spectrometer must be as low as possible to ensure they can all be focused efficiently and with full electrical integrity by the ion optics and the mass separation device. When the ions emerge from the argon plasma, they will all have different kinetic energies, depending on their mass-to-charge ratio. Their velocities should all be similar, because they are controlled by rapid expansion of the bulk plasma, which will be neutral as long as it is maintained at zero potential. As the ion beam passes through the sampler cone into the skimmer cone or cones, expansion will take place, but its composition and integrity will be maintained, assuming the plasma is neutral. This can be seen in Figure 5.6.

Electrodynamic forces do not play a role as the ions enter the sampler or the skimmer, because the distance over which the ions exert an influence on one another (known as the Debye length) is small (typically 10^{-3}–10^{-4} mm) compared to the diameter of the orifice (0.5–1.0 mm),[5] as shown in Figure 5.7.

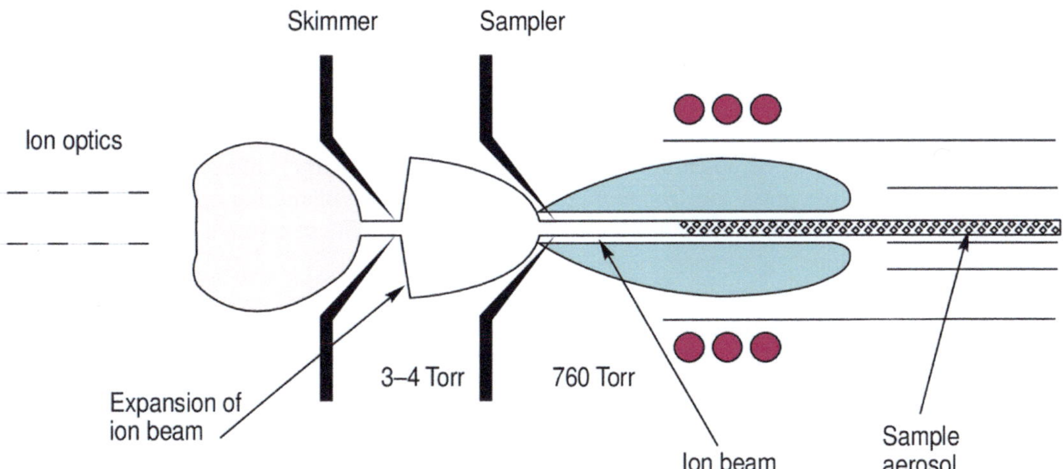

FIGURE 5.6 The composition of the ion beam is maintained as it passes through the interface, assuming a neutral plasma is acheived.

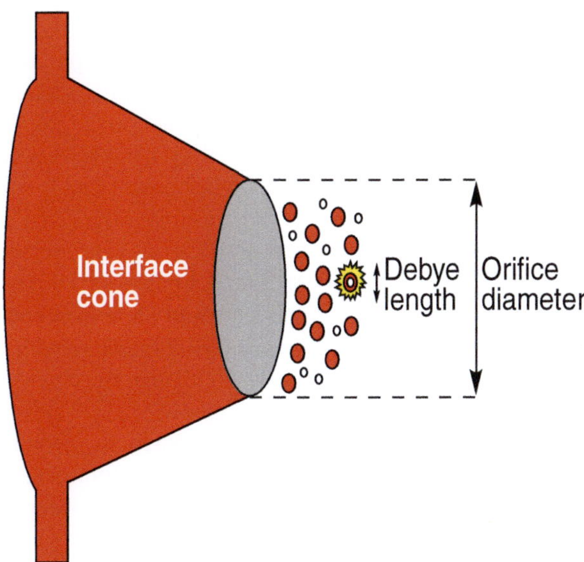

FIGURE 5.7 Electrodynamic forces do not affect the composition of the ion beam entering the sampler or skimmer cone.

It is therefore clear that maintaining a neutral plasma is of paramount importance to guarantee the electrical integrity of the ion beam as it passes through the interface region. If a secondary discharge is present, the electrical characteristics of the plasma change, which will affect the kinetic energy of the ions differently, depending on their mass-to-charge ratio. If the plasma is at zero potential, the ion energy spread is on the order of < 5 eV. However, if a secondary discharge is present, it results in a much wider spread of ion energies entering the mass spectrometer (typically 20–40 eV), which makes ion focusing far more complicated.[5]

5.3 BENEFITS OF A WELL-DESIGNED INTERFACE

The benefits of a well-designed interface are not readily obvious if simple aqueous samples are being analyzed using only one set of operating conditions. However, it becomes more apparent when many different sample types are being handled, requiring different operating parameters. The design of the interface is really put to the test when plasma conditions need to be changed, when the sample matrix changes or when ICP-MS is being used to analyze solid materials. Analytical scenarios such as these have the potential to induce a secondary discharge, change the kinetic energy of the ions entering the mass spectrometer, and affect the tuning of the ion optics. It is therefore critical that the interface grounding mechanism be able to handle these types of real-world analytical situations, including:

- **Using cool plasma conditions.** Although not utilized so much (but still useful for some applications) since the development of collision/reaction cells and interfaces, all instruments today have the ability to use cool plasma conditions. By using a combination of one or more of the following: reducing RF power to 500–700 W, reducing plasma coolant flow to ~ 10 L/min and increasing nebulizer gas flow to 1.0–1.3 L/min, the plasma temperature is lowered, which reduces argon-based polyatomic interferences such as $^{40}Ar^{16}O^+$, $^{40}Ar^+$ and $^{38}ArH^+$ in the determination of difficult elements such as $^{56}Fe^+$, $^{40}Ca^+$ and $^{39}K^+$. Such dramatic deviations from normal operating conditions (RF Power-1000 W, Plasma Coolant Flow-15 L/min, Neb Flow-0.8 L/min) will affect the electrical characteristics of the plasma (note: for some applications, it can be beneficial to combine cool/cold plasma conditions with collision/reaction cell technology to reduce some polyatomic spectral interferences).
- **Running organic solvents.** Analyzing oil or organic-based samples requires a chilled spray chamber or a membrane desolvation system to reduce the solvent loading on the plasma. In addition, higher RF power (~1300–1500 W) and lower nebulizer gas flow (~0.4–0.8 L/min) are required to dissociate the organic components in the sample. A reduction in the amount of solvent entering the plasma combined with higher power and lower nebulizer gas flow translates into a hotter plasma, and a change in its ionization mechanism.
- **Optimizing conditions for low oxides.** The formation of oxide species can be problematic in some sample types. For example, in geochemical applications it is quite common to sacrifice sensitivity by lowering the nebulizer gas flow and increasing the RF power to reduce the formation of rare earth oxides, which can spectrally interfere with the determination of other analytes. Unfortunately, these conditions will change the electrical characteristics of the plasma, which can induce a secondary discharge.
- **Using sampling accessories.** Sampling accessories such as membrane desolvators and laser-ablation systems are being used more routinely to improve performance/productivity and enhance the flexibility of ICP-MS. Some of these sampling devices, such as laser ablation, or membrane desolvation, generate a "dry" sample aerosol, which requires completely different operating conditions compared to conventional "wet" plasma. An aerosol that contains no solvent can have a dramatic effect on the ionization conditions in the plasma.

5.4 FINAL THOUGHTS

Even though most modern ICP-MS interfaces have been designed to minimize the effects of the secondary discharge, it should not be taken for granted that they can all handle changes in operating conditions and matrix components with the same ease. The most noticeable problems that have been reported include spectral peaks of the cone material appearing in the blank, erosion/discoloration of the sampling cones, widely different optimum plasma conditions (neb flow/RF power) for different masses and frequent retuning of the ion optics.[11,12] Chapter 26 on how best to evaluate ICP-MS instrumentation goes into this subject in greater detail, but there is no question that the plasma discharge, interface region and ion optics have to be designed in concert to ensure the instrument can handle a wide range of operating conditions and sample types. In addition, the interface cones have to be cleaned on a regular basis to ensure the secondary discharge is kept to a minimum. Neufeld wrote an excellent article on how to clean interface cones that may have been impacted by long-term aspiration of matrix components or discolored by acid erosion.[13]

FURTHER READING

1. A. L. Gray and A. R. Date, *Analyst*, **108**, 1033–1041, 1983.
2. R. S. Houk, V. A. Fassel, and H. J. Svec, *Dynamic Mass Spectrometry*, **6**, 234–238, 1981.
3. A. R. Date and A. L. Gray, *Analyst*, **106**, 1255–1259, 1981.
4. A. L. Gray and A. R. Date, *Dynamic Mass Spectrometry*, **6**, 252–256, 1981.
5. D. J. Douglas and J. B. French, *Spectrochimica Acta*, **41B**(3), 197–201, 1986.
6. A. L. Gray, R. S. Houk, and J. G. Williams, *Journal of Analytical Atomic Spectrometry*, **2**, 13–20, 1987.
7. R. S. Houk, J. K. Schoer, and J. S. Crain, *Journal of Analytical Atomic Spectrometry*, **2**, 283–286, 1987.
8. S. D. Tanner, *Journal of Analytical Atomic Spectrometry*, **10**, 905–911, 1995.
9. K. Sakata and K. Kawabata, *Spectrochimica Acta*, **49B**, 1027, 1994.
10. S. Georgitus and M. Plantz, *Winter Conference on Plasma Spectrochemistry*, **FP4**, Fort Lauderdale, 1996.
11. D. J. Douglas, *Canadian Journal of Spectroscopy*, **34**, 2–8, 1989.
12. J. E. Fulford and D. J. Douglas, *Applied Spectroscopy*, **40**, 7–12, 1986.
13. L. Neufeld, ICP-MS Interface Cones: Maintaining the Critical Interface between the Mass Spectrometer and the Plasma Discharge to Optimize Performance and Maximize Instrument Productivity, *Spectroscopy*, **34**(7), 12–17, 2019.

6 Ion-Focusing System

Chapter 6 takes a detailed look at the ICP-MS ion-focusing system—a crucial area of the ICP mass spectrometer—where the ion beam is focused before it enters the mass analyzer. Sometimes known as the ion optics, it comprises one or more ion lens components, which electrostatically steer the analyte ions in an axial (straight) or orthogonal (right-angled) direction from the interface region into the mass separation device. The strength of a well-designed ion-focusing system is its ability to produce a flat signal response over the entire mass range, low background levels, good detection limits and stable signals in real-world sample matrices.

Although the detection capability of ICP-MS is generally recognized as being superior to any of the other atomic spectroscopic techniques, it is probably most susceptible to the sample's matrix components. The inherent problem lies in the fact that ICP-MS is relatively inefficient—out of a million ions generated in the plasma, only a few ions actually reach the detector. One of the main contributing factors to the low efficiency is the higher concentration of matrix elements compared to the analyte, which has the effect of defocusing the ions and altering the transmission characteristics of the ion beam. This is sometimes referred to as a space charge effect, and can be particularly severe when the matrix ions are of a heavier mass than the analyte ions.[1] The role of the ion-focusing system is therefore to transport the maximum number of analyte ions from the interface region to the mass separation device, while rejecting as many of the matrix components and non-analyte-based species as possible. Let us now discuss this process in greater detail.

6.1 ROLE OF THE ION OPTICS

The ion optics, shown in Figure 6.1, are positioned between the skimmer cone (or cones) and the mass separation device. They typically consist of one or more electrostatically controlled lens components, maintained at a vacuum of approximately 10^{-3} torr with a turbomolecular pump.

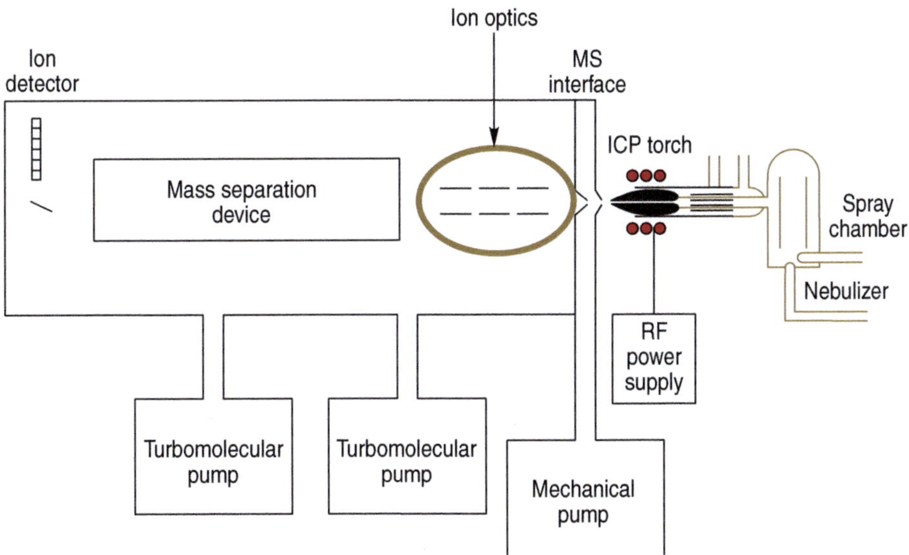

FIGURE 6.1 Position of ion optics relative to the plasma torch and interface region.

DOI: 10.1201/9781003187639-6

They are not traditional optics that we associate with ICP emission or atomic absorption, but are made up of a series of metallic plates, barrels or ion mirrors, which have a voltage placed on them. A recent commercial design implements a quadrupole deflector, which turns the ion beam at right angles into the mass spectrometer. Whatever the design, the function of the ion optic system is to take ions from the hostile environment of the plasma at atmospheric pressure via the interface cones and steer them into the mass analyzer, which is under high vacuum. The nonionic species such as particulates, neutral species and photons are prevented from reaching the detector by using some kind of physical barrier, positioning the mass analyzer off axis relative to the ion beam, or electro-statically bending the ions by 90° into the mass analyzer.

As mentioned in the previous chapter, the plasma discharge and interface region have to be designed in concert with the ion optics. It is absolutely critical that the composition and electrical integrity of the ion beam be maintained as it enters the ion optics. For this reason, it is essential that the plasma be at zero potential to ensure that the magnitude and spread of ion energies are as low as possible.[2]

A secondary, but also very important, role of the ion optic system is to stop particulates, neutral species and photons from getting through to the mass analyzer and the detector. These species cause signal instability and contribute to background levels, which ultimately affect the performance of the system. For example, if photons or neutral species reach the detector, they will elevate the noise of the background and therefore degrade detection capability. In addition, if particulates from the matrix penetrate further into the mass spectrometer region, they have the potential to deposit on lens components, and in extreme cases, get into the mass analyzer. In the short term, this will cause signal instability, and in the long term, increase the frequency of cleaning and routine maintenance.

There are basically four different approaches of reducing the chances of these undesirable species entering the mass spectrometer. The first method is to place a grounded metal stop (disk) behind the skimmer cone. This stop allows the ion beam to move around it and physically block the particulates, photons and neutral species from traveling "downstream".[3] Although implemented very successfully in earlier instruments, this design has not been utilized for a number of years, mainly because it can have a negative impact on instrument sensitivity, by reducing the number of ions reaching the detec-tor. The second approach is to set the mass analyzer off axis to the ion lens system (in some systems this is called a chicane design). The positively charged ions are then steered with the lens components into the mass analyzer, while the photons, neutral and nonionic species are ejected out of the ion beam.[4] The third development is to deflect the ion beam 90° with an ion mirror.[5] This allows the pho-tons, neutrals and solid particles to pass through, whereas the ions are deflected at right angles into an off-axis mass analyzer that incorporates curved fringe rod technology.[6] The fourth and most recent approach is to deflect the ion beam emerging from the plasma by 90°. This has the effect of changing direction and focusing the ion beam into the mass spectrometer, while allowing the neutral species, photons and particulate matter to go straight through and be ejected.

It is also worth mentioning that some lens systems incorporate an extraction lens after the skimmer cone to electrostatically "pull" the ions from the interface region. This has the benefit of improving the transmission and detection limits of the low-mass elements (which tend to be pushed out of the ion beam by the heavier elements), resulting in a more uniform response across the full mass range. In an attempt to reduce these space charge effects, some older designs have utilized lens components to accelerate the ions downstream. Unfortunately, this can have the effect of degrad-ing the resolving power and abundance sensitivity (ability to differentiate an analyte peak from the wing of an interference) of the instrument, because of the much higher kinetic energy of the acceler-ated ions as they enter the mass analyzer.[7]

6.2 DYNAMICS OF ION FLOW

To fully understand the role of the ion optics in ICP-MS, it is important to get an appreciation of the dynamics of ion flow from the plasma through the interface region into the mass spectrom-eter. When the ions generated in the plasma emerge from the skimmer cone (or cones), there is a rapid expansion of the ion beam as the pressure is reduced from 760 torr (atmospheric pressure) to approximately 10^{-3} to 10^{-4} torr in the lens chamber with a turbomolecular pump. The composition of the ion beam immediately behind the cone is the same as in front of the cone because the expansion at this stage is controlled by normal gas dynamics and not by electrodynamics. One of the main reasons for this is that in the ion-sampling process, the Debye length (the distance over which ions exert influence on one another) is small compared to the orifice diameter of the sampler or skimmer cone. Consequently, there is little electrical interaction between the ion beam and the cone, and relatively little interaction between the individual ions in the beam. In this way, the compositional integrity of the ion beam is maintained throughout the interface region.[8] With the rapid drop in pressure in the lens chamber, electrons diffuse out of the ion beam. Because of the small size of the electrons relative to the positively charged ions, the electrons diffuse further from the beam than the ions, resulting in an ion beam with a net positive charge. This is represented schematically in Figure 6.2.

The generation of a positively charged ion beam is the first stage in the charge-separation process. Unfortunately, the net positive charge of the ion beam means that there is now a natural tendency for the ions to repel each other. If nothing is done to compensate for this, ions of higher mass-to-charge ratio will dominate the center of the ion beam and force the lighter ions to the outside. The degree of loss will depend on the kinetic energy of the ions—those with high kinetic energy (high-mass elements) will be transmitted in preference to ions with medium (mid-mass elements) or low kinetic energy (low-mass elements). This is shown in Figure 6.3.

The second stage of charge separation, therefore, consists of electrostatically steering the ions of interest back into the center of the ion beam using the ion lens system. It should be emphasized that this is only possible if the interface is kept at zero potential, which ensures a neutral-gas dynamic flow through the interface, maintaining the compositional integrity of the ion beam. It also guarantees that the average ion energy and energy spread of each ion entering the lens systems are at levels optimum for mass separation. If the interface region is not grounded correctly, stray capacitance will generate a discharge between the plasma and sampler cone and increase the kinetic energy of the ion beam, making it very difficult to optimize the ion lens voltages (refer to Chapter 5 on the Interface Region for details).

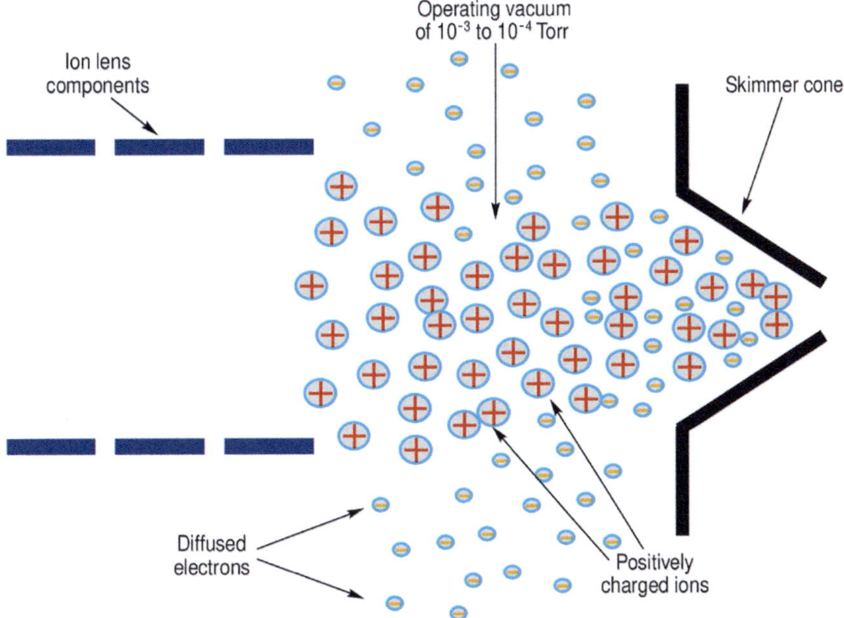

FIGURE 6.2 Extreme pressure drop in the ion optic chamber produces diffusion of electrons, resulting in a positively charged ion beam.

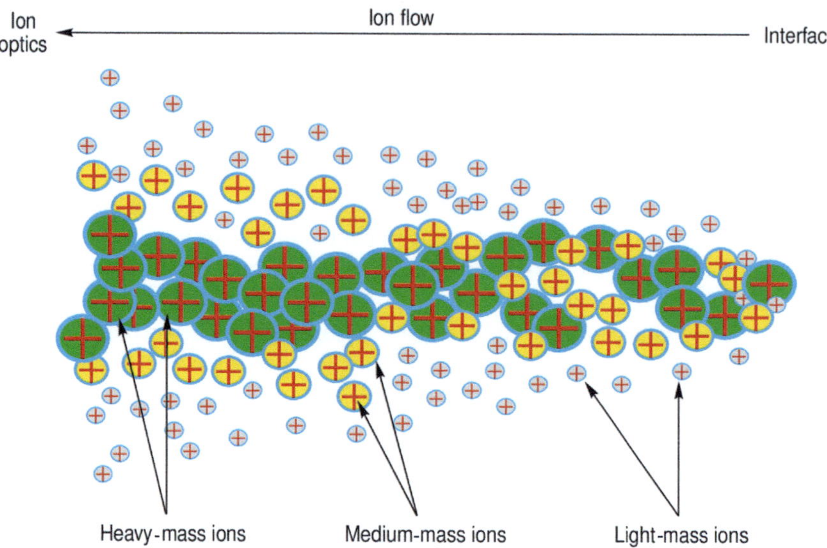

FIGURE 6.3 The degree of ion repulsion will depend on the kinetic energy of the ions—those with high kinetic energy (heavy masses) will be transmitted in preference to those with medium (medium masses) or low kinetic energy (light masses).

6.3 COMMERCIAL ION OPTIC DESIGNS

Over the years, there have been many different ion optic designs. Although they have their own individual characteristics, they perform the same basic function of allowing the maximum number of analyte ions through to the mass analyzer, while at the same time rejecting the undesirable matrix- and solvent-based ions. The oldest and most mature design of ion optics in use today consists of several lens components, all of which have a specific role to play in the transmission of the analyte ions with a minimum of mass discrimination. With these multicomponent lens systems, the voltage can be optimized on every lens of the ion optics to achieve the desired ion specificity. This type of lens configuration has been used in commercial instrumentation for almost 30 years and has proved to be very durable. One of its main benefits is that it produces a uniform response across the mass range with very low background levels, particularly when combined with an off-axis mass analyzer.[9] A schematic of a commercially available multicomponent lens systems is shown in Figure 6.4.

It should be emphasized that because of the interactive nature of parameters that affect the signal response, the more complex the lens system, the more the variables that have to be optimized. For this reason, if many different sample types are being analyzed, extensive lens optimization procedures have to be carried out for each matrix or group of elements. This is not such a major problem, because most of the lens voltages are computer controlled and methods can be stored for every new sample scenario. However, it could be a factor if the instrument is being used for the routine analysis of many diverse sample types, all requiring different lens settings.

Another well-established approach was the use of a cylinder lens, combined with a grounded stop—positioned just inside the skimmer cone. With this design, the voltage is dynamically ramped "on the fly", in concert with the mass scan of the analyzer. The benefit is that the optimum lens voltage is placed on every mass in a multielement run to allow the maximum number of analyte ions through, while keeping the matrix ions down to an absolute minimum.[10] This

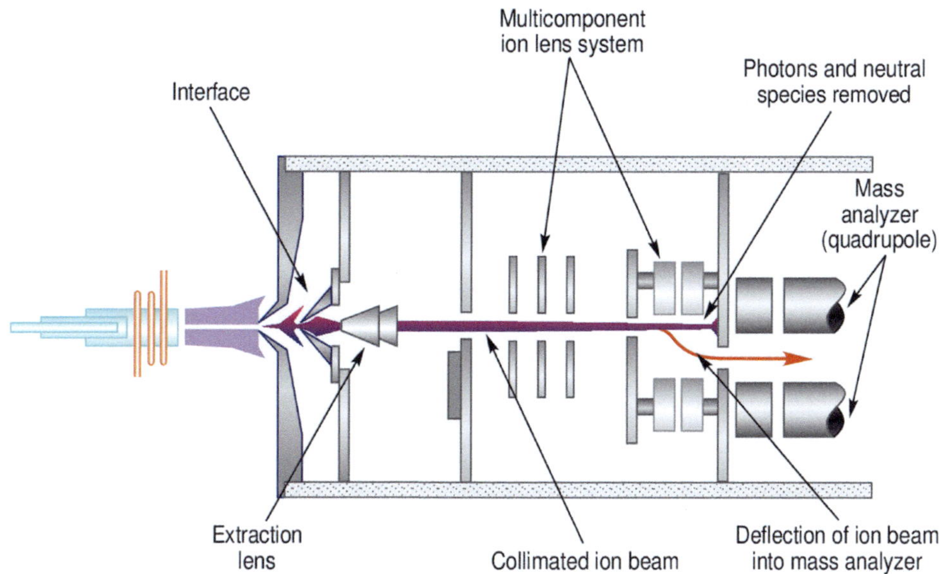

FIGURE 6.4 Schematic of a multicomponent lens system.

(From Y. Kishi, *Agilent Technologies Application Journal*, August 1997)

design is typically used in conjunction with a grounded stop to act as a physical barrier to reduce the chances that particulates, neutral species and photons will reach the mass analyzer and detector. Although this design does not generate such a uniform mass response across the full range as an off-axis multi-lens system with an extraction lens, it appears to offer better long-term stability with real-world samples. It works well for many sample types but is most effective when low-mass elements are being determined in the presence of high-mass matrix elements.

A more recent variation of the single-lens approach uses a right-angled cylinder lens. This is a very simple design that utilizes a single, fixed-voltage ion lens to eliminate particulates and photons from reaching the detector. This is done by deflecting the positive ion beam 90°, thus allowing the neutral species to go straight through and be pumped out of the mass spectrometer. The benefit of this design is simplicity, reduced maintenance, very low background levels and it's easy to clean. This lens system is unique to one instrument design and is used in conjunction with a low-mass cut-off collision/reaction cell. A photograph of this ion optic lens system is shown in Figure 6.5.

Another design in ion-focusing optics utilizes a parabolic electrostatic field created with an ion mirror to reflect and refocus the ion beam at 90° to the ion source.[5] This ion mirror incorporates a hollow structure, which allows photons, neutrals and solid particles to pass through it, while allowing ions to be reflected at right angles into the mass analyzer. The major benefit of this design is the very efficient way the ions are refocused, offering the capability of extremely high sensitivity across the mass range, with very little sacrifice in oxide performance. In addition, there is very little contamination of the ion optics, because a vacuum pump sits behind the ion mirror to immediately remove these particles before they have a chance to penetrate further into the mass spectrometer. Removing these undesirable species and photons before they reach the detector, in addition to incorporating curved fringe rods prior to an off-axis mass analyzer, means that background levels are very low. Figure 6.6 shows a schematic of a quadrupole-based ICP-MS that utilizes a 90° ion optic design.[6,11]

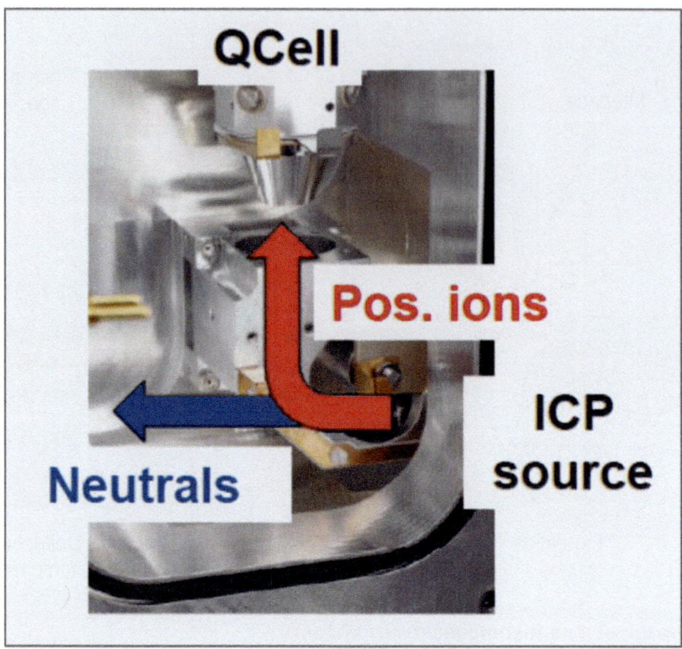

FIGURE 6.5 Right-angled cylinder lens showing how it deflects the positive ion beam 90° into the collision/ reaction cell, while allowing the neutral species to go straight through.

(Courtesy of Thermo Scientific)

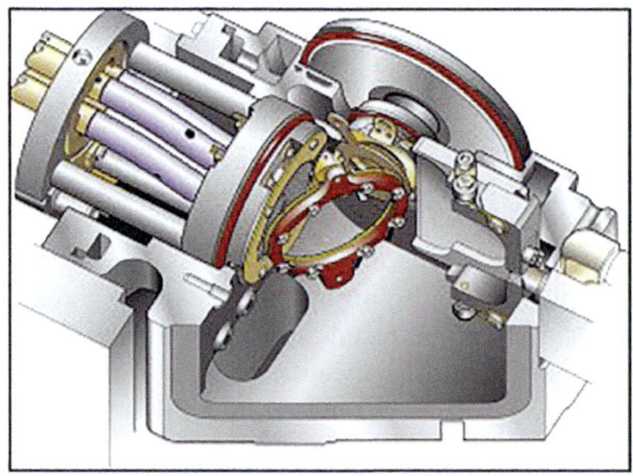

FIGURE 6.6 A 90° ion optic design used with curved fringe rods and an off-axis quadrupole mass analyzer.

(Courtesy of Analytik Jena)

The most novel commercial development in ion optic design utilizes a quadrupole ion deflector (QID), which utilizes a miniaturized quadrupole. This novel filtering technology bends the ion beam 90°, focusing ions of a specified mass into the mass-separation device, while discarding all neutral species, photons and particulates into the turbo pump. The major benefit of removing these nonionic species means they won't be deposited on component surfaces in the mass-analyzer region, which significantly minimizes drift and ensures good signal stability, even when running the most challenging sample matrices. This QID approach to ion filtering is shown in Figure 6.7.

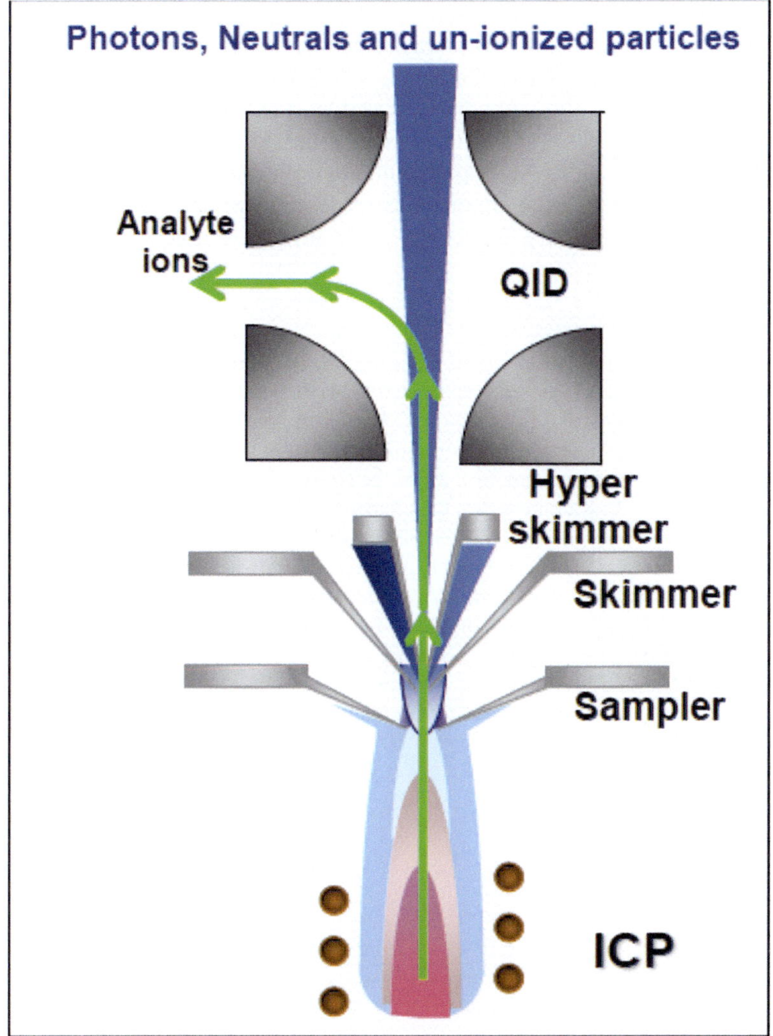

FIGURE 6.7 A novel commercial development in ion optic design utilizes a quadrupole ion deflector (QID). (Courtesy of PerkinElmer Inc.).

It is also worth emphasizing that a number of ICP-MS systems offer what is called a high-sensitivity option. All these designs work slightly differently but share similar components. By using a combination of slightly different cone geometry, higher vacuum at the interface, one or more extraction lenses and slightly modified ion optic design, they offer up to ten times the sensitivity of a traditional interface. However, in some systems, this increased sensitivity sometimes comes with slightly worse stability and an increase in background levels, particularly for samples with a heavy matrix. To get around this, these kinds of samples typically need to be diluted before analysis—which has somewhat limited their applicability to real-world samples with high dissolved solids.[12] However, they have found a use in non-liquid-based applications in which high sensitivity is crucial—for example, in the analysis of small areas/spots on the surface of a geological specimen using laser ablation ICP-MS. For this application, the instrument must offer high sensitivity, because a single laser pulse is often used to ablate very small amounts of the sample, which is then swept into the ICP-MS for analysis.

The importance of the ion-focusing system cannot be overemphasized, because it has a direct bearing on the number of ions that find their way to the mass analyzer. In addition to affecting background levels and instrument response across the entire mass range, it has a huge impact on both long- and short-term signal stability, especially in real-world samples. However, there are many different ways of achieving this. It is almost irrelevant whether the design of the ion optics is based on a multicomponent lens system, an RF-multipole guide or a right-angled deflection approach using an ion mirror or a quadrupole. The most important consideration when evaluating any ion-lens system is not the actual design but its ability to perform well with real sample matrices.

FURTHER READING

1. J. A. Olivares and R. S. Houk, *Analytical Chemistry*, **58**, 20, 1986.
2. D. J. Douglas and J. B. French, *Spectrochimica Acta*, **41B**(3), 197–201, 1986.
3. S. D. Tanner, L. M. Cousins, and D. J. Douglas, *Applied Spectroscopy*, **48**, 1367–1372, 1994.
4. D. Potter, *American Lab*, Agilent Technologies, July, 1994.
5. I. Kalinitchenko, *Ion Optical System for a Mass Spectrometer*, Patent Number 750860, 1996.
6. S. Elliott, M. Plantz, and L. Kalinitchenko, Oral Paper 1360–8, *Pittsburgh Conference*, Orlando, FL, 2003.
7. P. Turner, *Paper at 2nd International Conference on Plasma Source Mass Spec*, Durham, UK, 1990.
8. S. D. Tanner, D. J. Douglas, and J. B. French, *Applied Spectroscopy*, **48**, 1373–1378, 1994.
9. Y. Kishi, *Agilent Technologies Application Journal*, August, 1997.
10. E. R. Denoyer, D. Jacques, E. Debrah, and S. D. Tanner, *Atomic Spectroscopy*, **16**(1), 1–6, 1995.
11. I. Kalinitchenko, *Mass Spectrometer Including a Quadrupole Mass Analyzer Arrangement*, Patent Applied for—WO 01/91159 A1.
12. B. C. Gibson, *Paper at Surrey International Conference on ICP-MS*, London, UK, 1994.

7 Mass Analyzers
Quadrupole Technology

Chapters 7–10 deal with the heart of the ICP-MS system—the mass separation device. Sometimes called the mass analyzer, it is the region of the ICP mass spectrometer that separates the ions according to their mass-to-charge ratio. This selection process is achieved in a number of different ways, depending on the mass-separation device, but they all have one common goal, which is to separate the ions of interest from all other non-analyte, matrix, solvent and argon-based ions. Quadrupole mass filters are described in this chapter, followed by magnetic sector systems in Chapter 8, time-of-flight mass spectrometers in Chapter 9 and finally collision/reaction cell/interface technology in Chapter 10.

Although ICP-MS was commercialized in 1983, the first ten years of its development utilized a traditional quadrupole mass analyzer to separate the ions of interest. These worked exceptionally well for most applications, but proved to have limitations when determining difficult elements or dealing with more complex sample matrices. This led to the development of alternative mass-separation devices that allowed ICP-MS to be used for applications that were previously beyond the capabilities of quadrupole-based technology. Before we discuss these different mass spectrometers in greater detail, let us take a look at the proximity of the mass analyzer in relation to the ion optics and detector. Figure 7.1 shows this in greater detail.

As can be seen, the mass analyzer is positioned between the ion optics and detector, and it is maintained at a vacuum of approximately 10^{-6} torr with an additional turbomolecular pump to the one that is used for the lens chamber. Assuming the ions are emerging from the ion optics at the optimum kinetic energy, they are ready to be separated according to their mass-to-charge ratio (m/z) by the mass analyzer. There are basically three different kinds of commercially available mass analyzers: quadrupole mass filters, double-focusing magnetic sectors, time-of-flight mass spectrometers. It should be noted that although collision/reaction cell/interface technology has been given its own chapter (Chapter 10) in this book, it is not utilized on its own to carry out mass separation. It is typically used in conjunction with a quadrupole (or quadrupoles) to reduce the impact of polyatomic

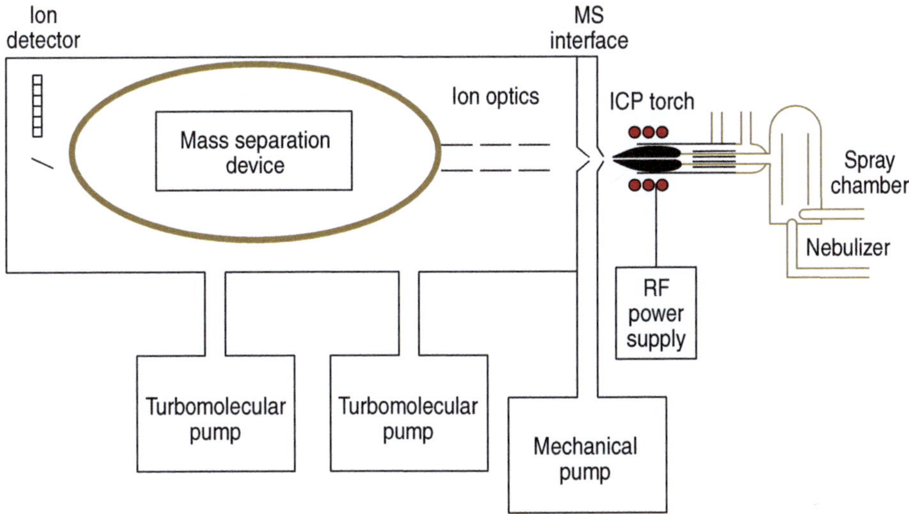

FIGURE 7.1 The mass separation device is positioned between the ion optics and the detector.

DOI: 10.1201/9781003187639-7

spectral interferences before the ions are passed into the mass analyzer for separation and detection. They are very powerful enhancements to the technique, but they are not considered primary separation devices on their own.

They all have their own strengths and weaknesses, which will be discussed in greater detail over the next four chapters (Chapters 7–10). Let us first begin with the most common type of mass-separation device used in ICP-MS—the quadrupole mass filter.

QUADRUPOLE TECHNOLOGY

Developed in the early 1980s for ICP-MS, quadrupole-based systems represent approximately 90% of all ICP mass spectrometers used today. This design was the first to be commercialized, and as a result, today's quadrupole ICP-MS technology is considered a very mature, routine trace element technique. A quadrupole usually consists of four cylindrical or hyperbolic metallic rods of the same length and diameter. Although in one design of collision/reaction cell, a quadrupole with flat sides is used (this "flatapole" design is described in greater detail in Chapter 10 on "Collision/Reaction Cells"). Quadrupole rods are typically made of stainless steel or molybdenum and sometimes coated with a ceramic coating for corrosion resistance. When used for ICP-MS, they are typically 15–25 cm in length, about 1 cm in diameter, and operate at a frequency of 2–3 MHz. Figure 7.2 shows a photograph of a quadrupole system mounted in its housing.

7.1 BASIC PRINCIPLES OF OPERATION

A quadrupole operates by placing both a direct current (DC) field and a time-dependent alternating current (AC) of radio frequency on opposite pairs of the four rods. By selecting the optimum AC/DC ratio on each pair of rods, ions of a selected mass are allowed to pass through the rods to the detector, whereas the others are unstable and ejected from the quadrupole. Figure 7.3 shows this in greater detail.

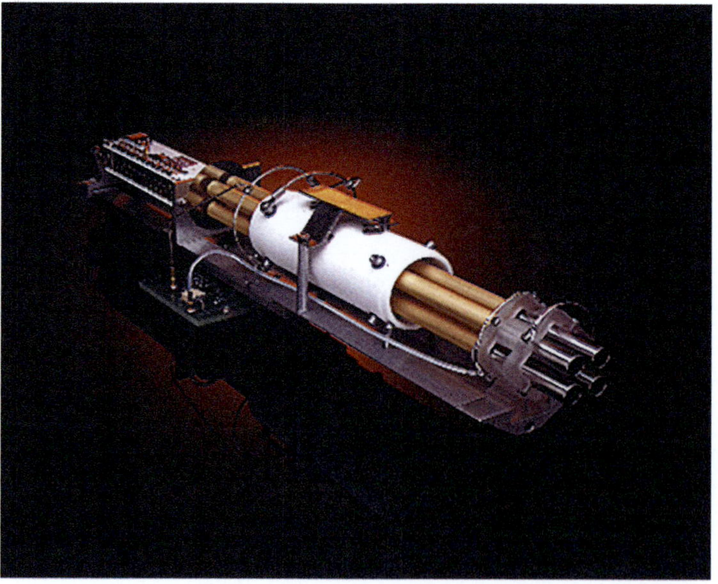

FIGURE 7.2 Photograph of a quadrupole system mounted in its housing.

In this simplified example, the analyte ion (black) and four other ions (gray) have arrived at the entrance to the four rods of the quadrupole. When a particular AC/DC potential is applied to the rods, the positive or negative bias on the rods will electrostatically steer the analyte ion of interest down the middle of the four rods to the end, where it will emerge and be converted to an electrical pulse by the detector. The other ions of different m/z values will be unstable, pass through the spaces between the rods, and be ejected from the quadrupole. This scanning process is then repeated for another analyte with a completely different mass-to-charge ratio until all the analytes in a multielement analysis have been measured. The process for the detection of one particular mass in a multielement run is represented in Figure 7.4.

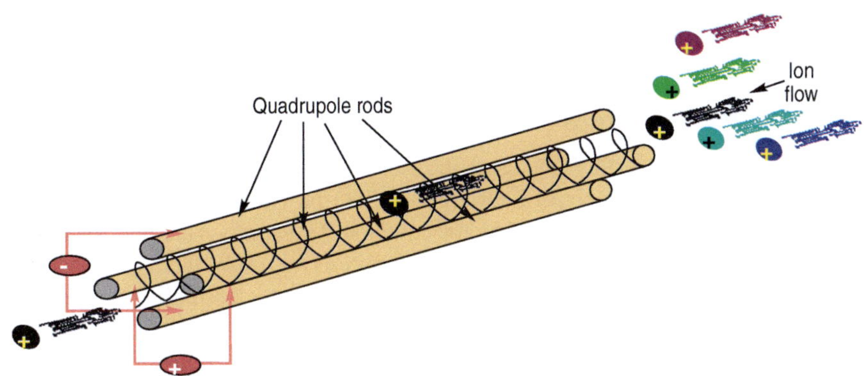

FIGURE 7.3 Schematic representation showing principles of mass separation using a quadrupole mass filter.

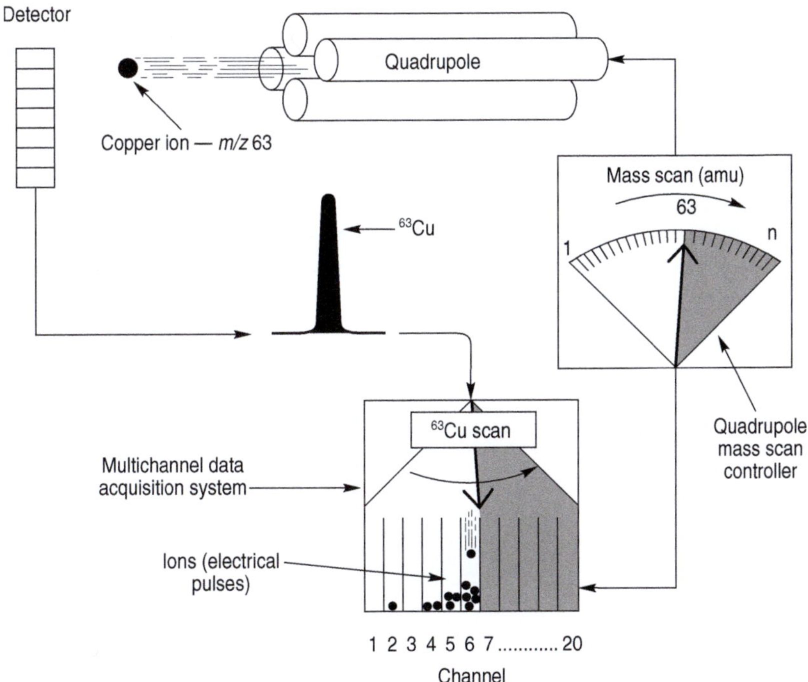

FIGURE 7.4 Profiles of different masses are built up using a multichannel data-acquisition system. (Courtesy of PerkinElmer Inc.)

It shows a $^{63}Cu^+$ ion emerging from the quadrupole and being converted to an electrical pulse by the detector. As the AC/DC voltage of the quadrupole—corresponding to $^{63}Cu^+$—is repeatedly scanned, the ions are stored and counted by a multichannel analyzer as electrical pulses. This multichannel data-acquisition system typically has 20 channels per mass. As the electrical pulses are counted in each channel, a profile of the mass is built up over the 20 channels, corresponding to the spectral peak of $^{63}Cu^+$. In a multielement run, repeated scans are made over the entire suite of analyte masses as opposed to just one mass represented in this example.

Quadrupole scan rates are typically on the order of 2,500–5,000 amu per second (depending on the commercial design) and can cover the entire mass range of 0–300 amu in about one-tenth of a second. However, real-world analytical speed and sample throughput is much slower than this, and in practice, 25 elements can be determined in duplicate with good precision in 1–2 min, depending on the analytical requirements.

7.2   QUADRUPOLE PERFORMANCE CRITERIA

There are two very important performance specifications of a mass analyzer that govern its ability to separate an analyte peak from a spectral interference. The first is the resolving power (R), which, in traditional mass spectrometry, is represented by the equation $R = m/\Delta m$, where m is the nominal mass at which the peak occurs and Δm is the mass difference between two resolved peaks.[1] However, for quadrupole technology, the term *resolution* is more commonly used and is normally defined as the width of a peak at 10% of its height. The second specification is abundance sensitivity, which is the signal contribution of the tail of an adjacent peak at one mass lower and one mass higher than the analyte peak.[2] Even though they are somewhat related and both define the quality of a quadrupole, the abundance sensitivity is probably the most critical. If a quadrupole has good resolution but poor abundance sensitivity, it will often prohibit the measurement of an ultratrace analyte peak next to a major interfering mass.

7.3   RESOLUTION

Let us now discuss this area in greater detail. The ability to separate different masses with a quadrupole is determined by a combination of factors, including: shape, diameter and length of the rods; frequency of quadrupole power supply; operating vacuum; applied RF/DC voltages; and the motion and kinetic energy of the ions entering and exiting the quadrupole. All these factors will have a direct impact on the stability of the ions as they travel down the middle of the rods, and therefore, the quadrupole's ability to separate ions with differing m/z values. This is represented in Figure 7.5, which shows a simplified version of the Mathieu mass stability plot of two separate masses (A and B) entering the quadrupole at the same time.[3]

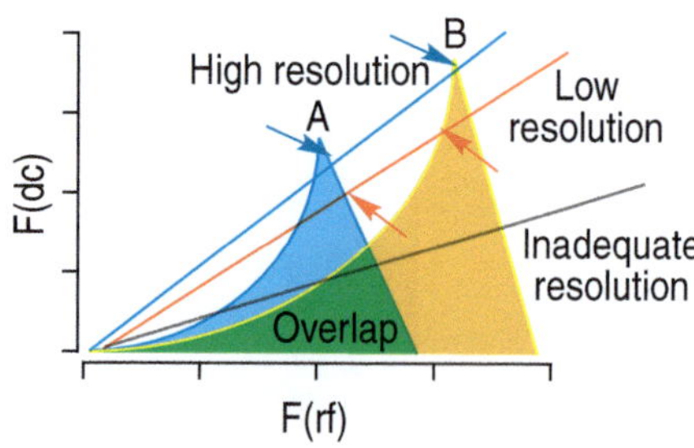

FIGURE 7.5   Simplified Mathieu stability diagram of a quadrupole mass filter, showing separation of two different masses A and B.

(From *Quadrupole Mass Spectrometry and Its Applications*: P. H. Dawson, Ed., Elsevier, Amsterdam, 1976; reissued by AIP Press, Woodbury, NY, 1995)

Any of the RF/DC conditions shown under the peak on the left will only allow mass A to pass through the quadrupole, whereas any combination of RF/DC voltages under the peak on the right will only allow mass B to pass through the quadrupole. If the slope of the RF/DC scan rate is steep, represented by the top line (high resolution), the spectral peaks will be narrow and masses A and B will be well separated. However, if the slope of the scan is shallow, represented by the middle line (low resolution), the spectral peaks will be wide and masses A and B will not be well separated. On the other hand, if the slope of the scan is too shallow, represented by the lower line (inadequate resolution), the peaks will overlap each other and both masses A and B will pass through the quadrupole without being separated. Theoretically, the resolution of a quadrupole mass filter can be varied between 0.3 and 3.0 amu, but is normally kept at 0.7–1.0 amu for most applications. However, improved resolution is always accompanied by a sacrifice in sensitivity, as seen in Figure 7.6, which shows a comparison of the same mass at resolutions of 3.0, 1.0 and 0.3 amu.

It can be seen that the peak height at 3.0 amu is much larger than that at 0.3 amu, but as expected, it is also much wider. This would prohibit using a resolution of 3.0 amu with spectrally complex samples. Conversely, the peak width at 0.3 amu is very narrow, but the sensitivity is low. For this reason, a compromise between peak width and sensitivity normally has to be reached, depending on the application. This can clearly be seen in Figure 7.7, which shows a spectral overlay of two copper isotopes—^{63}Cu$^+$ and ^{65}Cu$^+$—at resolution settings of 0.70 and 0.50 amu. In practice, the quadrupole is normally operated at a resolution of 0.7–1.0 amu for the majority of applications.

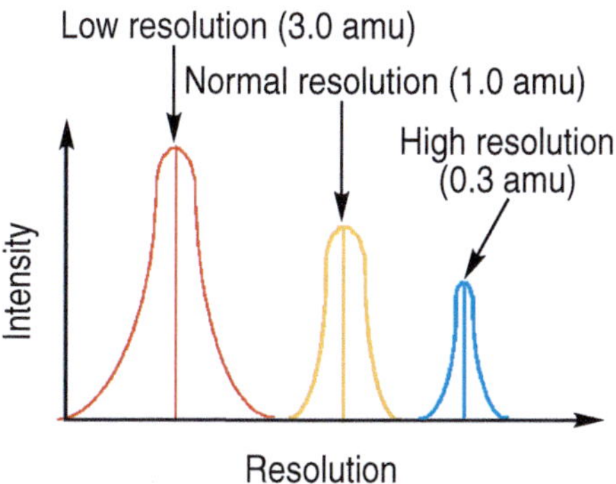

FIGURE 7.6   Sensitivity comparison of a quadrupole operated at 3.0, 1.0 and 0.3 amu resolutions.

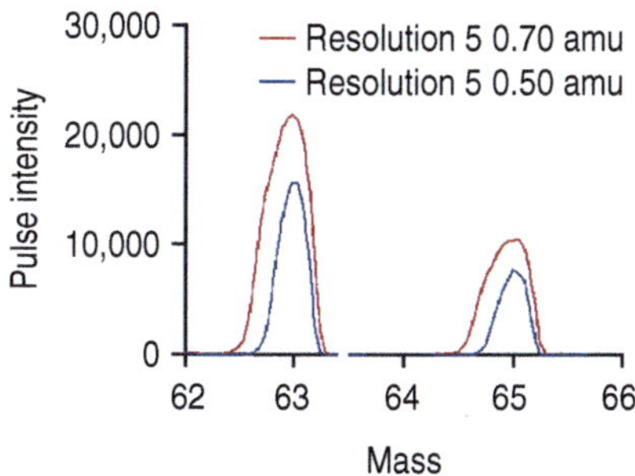

FIGURE 7.7   Sensitivity comparison of two copper isotopes—^{63}Cu$^+$ and ^{65}Cu$^+$—at resolution settings of 0.70 and 0.50 amu.

It is worth mentioning that most quadrupoles are operated in the first stability region, where resolving power is typically on the order of 500–700. If the quadrupole is operated in the second or third stability regions, resolving powers of 4,000[4] and 9,000[5], respectively, have been achieved. However, improving resolution using this approach has resulted in a significant loss of signal. Although there are ways of improving sensitivity, other problems have been encountered. As a result, to date there are no commercial quadrupole instruments available that use higher stability regions.

Some instruments can vary the peak width "on the fly", which means that the resolution can be changed between 3.0 and 0.3 amu for every analyte in a multielement run. Although this appears to offer some benefits, in reality they are few and far between, and for the vast majority of applications it is adequate to use the same resolution setting for every analyte. Even though quadrupoles can be operated at a higher resolution (in the first stability region), up to now the slight improvement has not been shown to be of practical benefit for most routine applications.

7.4 ABUNDANCE SENSITIVITY

It can be seen in Figure 7.7 that the tail of the spectral peaks drops off more rapidly at the high-mass end of the peak compared to the low-mass end. The overall peak shape, particularly its low-mass and high-mass tail, is determined by the abundance sensitivity of the quadrupole, which is impacted by a combination of factors, including the design of the rods, frequency of the power supply and operating vacuum.[6] Even though they are all important, probably the biggest impact on abundance sensitivity is the motion and kinetic energy of the ions as they enter and exit the quadrupole. If the Mathieu stability plot in Figure 7.5 is examined, it can be seen that the stability boundaries of each mass are less defined (not so sharp) on the low-mass side compared to the high-mass side.[3] As a result, the characteristic of ion motion at the low-mass boundary is different from that at the high-mass boundary and is therefore reflected in poorer abundance sensitivity at the low-mass side compared to the high-mass side. The velocity, and therefore the kinetic energy, of the ions entering the quadrupole will affect the ion motion and, as a result, will have a direct impact on the abundance sensitivity. For that reason, factors that affect the kinetic energy of the ions, such as high plasma potential and the use of lenses to accelerate the ion beam, could have a negative effect on the instrument's abundance sensitivity.[7]

These are the fundamental reasons why the peak shape is not symmetrical with a quadrupole and explains why there is always a pronounced shoulder at the low-mass side of the peak compared to the high-mass side—as represented in Figure 7.8, which shows the theoretical peak shape of a nominal mass M.

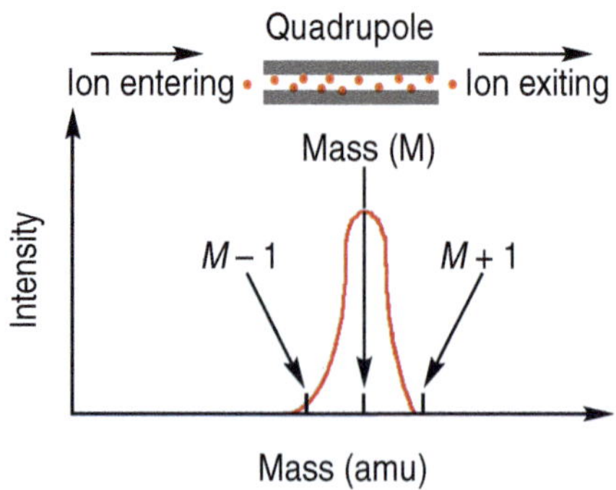

FIGURE 7.8 Ions entering the quadrupole are slowed down by the filtering process and produce peaks with a pronounced tail or shoulder at the low-mass end ($M - 1$) compared to the high-mass end ($M + 1$).

It can be seen that the shape of the peak at one mass lower ($M - 1$) is slightly different from the other side of the peak at one mass higher ($M + 1$) than the mass M. For this reason, the abundance-sensitivity specification for all quadrupoles is always worse on the low-mass side than the high-mass side and is typically 1×10^{-6} at $M - 1$ and 1×10^{-7} at $M + 1$. In other words, an interfering peak of 1 million counts per second (mcps) at $M - 1$ would produce a background of 1 cps at M, whereas it would take an interference of 10 million cps at $M + 1$ to produce a background of 1 cps at M.

7.5 BENEFIT OF GOOD ABUNDANCE SENSITIVITY

An example of the importance of abundance sensitivity is shown in Figure 7.9. Figure 7.9a is a spectral scan of 50 ppm of the doubly charged europium ion, $^{151}Eu^{++}$, at 75.5 amu (a doubly charged ion is one with two positive charges, as opposed to a normal singly charged positive ion, and exhibits an m/z peak at half its mass). It can be seen that the intensity of the peak is so great that its tail overlaps the adjacent mass at 75 amu, which is the only available mass for the determination of arsenic. This is highlighted in Figure 7.9b, which shows an expanded view of the tail of the $^{151}Eu^{++}$ together with a scan of 1 ppb of As at mass 75. It can be seen very clearly that the $^{75}As^+$ signal lies on the sloping tail of the $^{151}Eu^{++}$ peak. Measurement on a sloping background similar to this would result in a significant degradation in the arsenic detection limit, particularly as the element is monoisotopic and no alternative mass is available. In this particular example, a slightly higher resolution setting was also used (0.5 amu instead of 0.7 amu) to enhance the separation of the arsenic peak from the europium peak, but nevertheless still emphasizes the importance of good abundance sensitivity in ICP-MS.

There are many different designs of quadrupole used in ICP-MS, all made from different materials with varied dimensions, shape and physical characteristics. In addition, they are all maintained at a slightly different vacuum chamber pressure and operate at different frequencies. Theoretically, these hyperbolic rods should generate a better hyperbolic (elliptical) field than cylindrical rods, resulting in higher transmission of ions at higher resolution. It also tells us that a higher operating frequency means a higher rate of oscillation—and therefore separation—of the ions as they travel

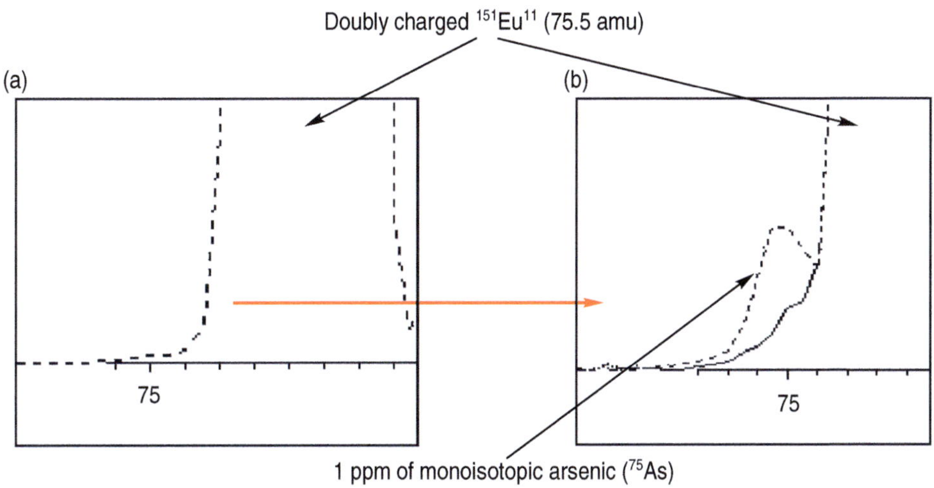

FIGURE 7.9 A low abundance-sensitivity specification is critical to minimize spectral interferences, as shown by (a), which represents a spectral scan of 50 ppb of $^{151}Eu^{++}$ at 75.5 amu, and (b), which shows how the tail of the $^{151}Eu^{++}$ elevates the spectral background of 1 ppb of As at mass 75.

(Courtesy of PerkinElmer Inc.)

down the quadrupole. Finally, it is very well accepted that a higher vacuum produces fewer collisions between gas molecules and ions, resulting in a narrower spread in kinetic energy of the ions, and therefore, a reduction in the tail at the low-mass side of a peak. Given all these theoretical differences, in reality the practical capabilities of most modern quadrupoles used in ICP-MS are very similar. However, there are some subtle differences in each instrument's measurement protocol and the software's approach to peak quantitation. This will be discussed in greater detail in Chapter 12 on "Peak Measurement Protocol".

FURTHER READING

1. F. Adams, R. Gijbels, and R. Van Grieken, *Inorganic Mass Spectrometry*, John Wiley and Sons, New York, 1988.
2. E. Montasser, ed., *Inductively Coupled Plasma Mass Spectrometry*, Wiley-VCH, Berlin, 1998.
3. P. H. Dawson, ed., *Quadrupole Mass Spectrometry and Its Applications*, Elsevier, Amsterdam, 1976; reissued by AIP Press, Woodbury, NY, 1995.
4. Z. Du, T. N. Olney, and D. J. Douglas, *Journal of American Society of Mass Spectrometry*, **8**, 1230–1236, 1997.
5. P. H. Dawson and Y. Binqi, *International Journal of Mass Spectrometry*, Ion Proc., **56**, 25–32, 1984.
6. D. Potter, *Agilent Technologies Application Note*, 228–349, January, 1996.
7. E. R. Denoyer, D. Jacques, E. Debrah, and S. D. Tanner, *Atomic Spectroscopy*, **16**(1), 1–6, 1995.

8 Mass Analyzers
Double-Focusing Magnetic-Sector Technology

Although quadrupole mass analyzers represent approximately 80–90% of all ICP-MS systems installed worldwide, limitations in their resolving power have led to the development of high-resolution spectrometers based on the double-focusing magnetic-sector design. In Chapter 8, we take a detailed look at the fundamental principles of this very powerful mass-separation device, which has found its niche in solving challenging application problems that require excellent detection capability, exceptional resolving power and very high precision. However, it should be emphasized that there are now many variations of magnet-sector mass spectrometers for elemental analysis, including Glow Discharge Mass Spectrometry (GDMS), Isotope Ratio Mass Spectrometry (IRMS), Thermal Ionization Mass Spectrometry (TIMS) and Noble Gas Mass Spectrometry (NGMS). These approaches are beyond the scope of this chapter, but if interested, you will find vendors of these types of instruments mentioned in Chapter 28 on Useful Contact Information.

As discussed in Chapter 7, a quadrupole-based ICP-MS system typically offers a resolution of 0.7–1.0 amu. This is quite adequate for most routine applications but has proved to be inadequate for many elements that are prone to argon-, solvent- and/or sample-based spectral interferences. These limitations in quadrupoles drove researchers in the direction of traditional high-resolution magnetic-sector technology to improve quantitation by resolving the analyte mass away from the spectral interference.[1] These ICP-MS instruments that were first commercialized in the late 1980s offered resolving power of up to 10,000, compared to that of a quadrupole, which for most applications is in the order of ~500. This dramatic improvement in resolving power allowed difficult elements such as Fe, K, As, V and Cr to be determined with relative ease, even in complex sample matrices.

8.1 MAGNETIC-SECTOR MASS SPECTROSCOPY: A HISTORICAL PERSPECTIVE

Mass spectrometers, using separation based on velocity focusing[2,3] and magnetic deflection,[4,5] were first developed over 80 years ago, primarily to investigate isotopic abundances and calculate atomic weights. Even though these designs were combined into one instrument in the 1930s to improve both sensitivity and resolving power,[6,7] they were still considered rather bulky and expensive to build. For that reason, in the late 1930s and 1940s, magnetic field technology, and in particular the small radius sector design of Nier,[8] became the preferred method of mass separation. Because Nier was a physicist, most of the early work carried out with this design was used for isotope studies in the disciplines of earth and planetary sciences. However, it was the oil industry that accelerated the commercialization of MS, because of its demand for the fast and reliable analysis of complex hydrocarbons in oil refineries.

Once scanning magnetic-sector technology became the most accepted approach for high-resolution mass separation in the 1940s, the challenges that lay ahead for mass spectroscopists were in the design of the ionization source—especially as the technique was being used more and more for the analysis of solids. The gas-discharge ion source that was developed for gases and high-vapor-pressure liquids proved to be inadequate for most solid materials. For this reason, one of the first successful methods of ionizing solids was carried out using the hot anode method,[9] where the previously dissolved material was deposited onto a strip of platinum foil and evaporated by passing an electric current through it. Unfortunately, although there were variations of this approach that all worked reasonably well, the main drawback of a thermal evaporation technique was selective ionization. In other words, because of the different volatilities of the elements, it could not be guaranteed that the ion beam properly represented the compositional integrity of the sample.

It was finally the work carried out by Dempster in 1946,[10] using a vacuum spark discharge and a high-frequency, high-voltage spark, that led researchers to believe that it could be applied to sample electrodes and used as a general-purpose source for the analysis of solids. The breakthrough came in 1954 with the development of the first modern spark source mass spectrometer (SSMS) based on the Mattauch-Herzog mass spectrometer design, which separated the ions in the same flat plane, so they could be detected by a linear detector such as a photographic plate.[11] Using this design, Hannay and Ahearn showed that it was possible to determine sub-ppm impurity levels directly in a solid material.[12]

Over the years, as a result of a demand for more stable ionization sources, lower detection capability and higher precision, researchers were led in the direction of other techniques such as secondary-ion mass spectrometry (SIMS),[13] ion microprobe mass spectrometry (IMMS)[14] and laser ionization mass spectrometry (LIMS).[15] Although they are considered somewhat complementary to SSMS, they all had their own strengths and weaknesses, depending on the analytical objectives for the solid material being analyzed. However, it should be emphasized that these techniques were predominantly used for microanalysis because only a very small area of the sample is vaporized. This meant that it could only provide meaningful analytical data of the bulk material if the sample was sufficiently homogeneous. For that reason, other ionization sources that sampled a much larger area, such as the glow discharge, became a lot more practical for the bulk analysis of solids by MS.[16]

8.2   USE OF MAGNETIC-SECTOR TECHNOLOGY FOR ICP-MS

Even though magnetic-sector technology was the most common mass-separation device for the analysis of inorganic compounds using traditional ion sources, it lost out to quadrupole technology when ICP-MS was first developed in the early 1980s. However, it was not until the mid- to late 1980s that the analytical community realized that quadrupole ICP-MS suffered from serious limitations in its ability to resolve troublesome polyatomic spectral interferences. In addition, because quadrupoles were scanning devices, the technique was not suitable for applications that required high precision, such as isotope-ratio measurements. As a result, analytical chemists began to look at double-focusing magnetic-sector technology to alleviate these kinds of limitations.

Initially, this technology was found to be unsuitable as a separation device for an ICP because of the high voltage required to accelerate the ions into the mass analyzer. This high potential at the interface region dramatically changed the energy of the ions entering the mass spectrometer, and therefore made it very difficult to steer the ions through the ion optics and still maintain a narrow spread of ion kinetic energies. For this reason, basic changes had to be made to the ion acceleration mechanism for magnetic-sector technology to be successfully used as a separation device for ICP-MS. This was a significant challenge when magnetic-sector systems were first developed in the late 1980s.

However, by the early 1990s, this problem was solved by moving the high-voltage components away from the plasma and positioning the interface closer to the mass spectrometer. Modern instrumentation has typically been based on two different approaches, the "standard" and "reverse" Nier–Johnson geometry. Both these designs, which use the same basic principles, consist of two analyzers: a traditional electromagnet and an electrostatic analyzer (ESA). In the standard (sometimes called forward) design, the ESA is positioned before the magnet, and in the reverse design it is positioned after the magnet. A schematic of the reverse Nier–Johnson spectrometer is shown in Figure 8.1.

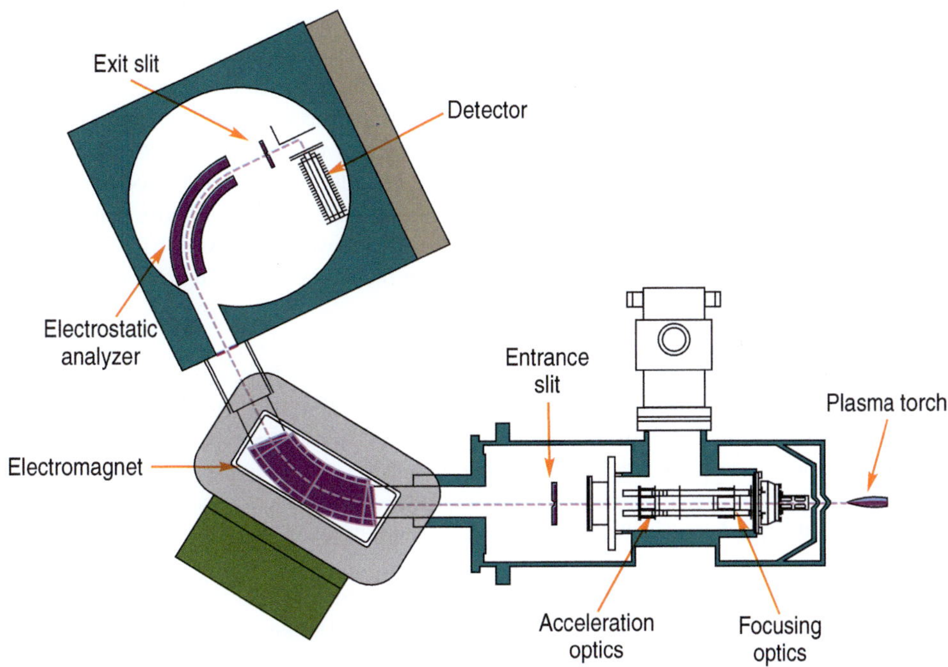

FIGURE 8.1   Schematic of a reverse Nier–Johnson double-focusing magnetic-sector mass spectrometer. (from U. Geismann and U. Greb, *Fresnius' Journal of Analytical Chemistry,* **350**, 186–193, 1994).

8.3   PRINCIPLES OF OPERATION OF MAGNETIC-SECTOR TECHNOLOGY

The original concept of magnetic-sector technology was to scan over a large mass range by varying the magnetic field over time with a fixed acceleration voltage. During a small window in time, which was dependent on the resolution chosen, ions of a particular mass-to-charge ratio were swept past the exit slit to produce the characteristic flat-top peaks. As the resolution of a magnetic-sector instrument is independent of mass, ion signals, particularly at low mass, are far apart. Unfortunately, this results in a relatively long time being spent scanning and settling the magnet. This was not such a major problem for qualitative analysis or mass spectral fingerprinting of unknown compounds, but proved to be impractical for rapid trace element analysis, where you had to scan to individual masses, slow down, settle the magnet, stop, take measurements and then scan to the next mass.

However, by using the double-focusing approach, the ions are sampled from the plasma in a conventional manner and then accelerated in the ion-optic region to a few kilovolts (kV) before they enter the mass analyzer. The magnetic field, which is dispersive with respect to ion energy and mass, then focuses all the ions with diverging angles of motion from the entrance slit. The ESA, which is only dispersive with respect to ion energy, then focuses all the ions onto the exit slit, where the detector is positioned. If the energy dispersion of the magnet and ESA are equal in magnitude but opposite in direction, they will focus both ion angles (first focusing) and ion energies (second or double focusing) when combined together. Changing the electric field in the opposite direction during the cycle time of the magnet (in terms of the mass passing the exit slit) has the effect of "freezing" the mass for detection. Then, as soon as a certain magnetic field strength is passed, the electric field is set to its original value and the next mass is "frozen". The voltage is varied on a per-mass basis, allowing the operator to scan only the mass peaks of interest rather than the full mass range.[17,18]

It should be pointed out that although this approach represents enormous time savings over traditional magnet-scanning technology, it is still slower than quadrupole-based instruments. The inherent problem lies in the fact that a quadrupole can be electronically scanned faster than a magnet. Typical speeds for a full mass scan (0–250 amu) of a magnet are on the order of 200 ms compared to approximately 50 ms for a quadrupole. In addition, it takes much longer for a magnet to slow down, settle and stop to take measurements—typically, 20 ms compared to < 1 ms for a quadrupole. So, even though in practice the electric scan dramatically reduces the overall analysis time, modern double-focusing magnetic-sector ICP-MS systems are still slower than state-of-the-art quadrupole instruments, which make them less than ideal for fast, high-throughput multielement applications or the characterization of rapid transient peaks.

8.4 RESOLVING POWER

As mentioned previously, most commercial magnetic-sector ICP-MS systems offer up to 10,000 resolving power (10% valley definition), which is high enough to resolve the majority of spectral interferences. It is worth emphasizing that resolving power is represented by the equation $R = m/\Delta m$, where m is the nominal mass at which the peak occurs and Δm is the mass difference between two resolved peaks.[19] In a quadrupole, the resolution is selected by changing the ratio of the RF and DC voltages on the quadrupole rods.

However, because a double-focusing magnetic-sector instrument involves focusing ion angles and ion energies, mass resolution is achieved by using two mechanical slits—one at the entrance to the mass spectrometer and another at the exit, prior to the detector. Varying resolution is achieved by scanning the magnetic field under different entrance and exit slit width conditions. Similar to optical systems, low resolution is achieved by using wide slits, whereas high resolution is achieved with narrow slits. Varying the width of both the entrance and exit slits effectively changes the operating resolution.

However, it should be emphasized that, similar to optical spectrometry, as the resolution is increased, the transmission decreases. So even though extremely high resolution is available, detection limits will be compromised under these conditions. This can be seen in Figure 8.2, which shows a plot of resolution against ion transmission.

It can be seen that a resolving power of 400 produces 100% transmission, but at a resolving power of 10,000, only ~2% is achievable. This dramatic loss in sensitivity could be an issue if low detection limits are required in spectrally complex samples that require the highest possible resolution. However, spectral demands of this nature are not very common. Table 8.1 shows the resolution required to resolve fairly common polyatomic interferences from a selected group of elemental isotopes, together with the achievable ion transmission.

Figure 8.3 is a comparison between a quadrupole and a magnetic-sector instrument of one of the most common polyatomic interferences—$^{40}Ar^{16}O^+$ on $^{56}Fe^+$, which requires a resolution of 2504 to separate the peaks. Figure 8.3a shows a spectral scan of $^{56}Fe^+$ using a quadrupole instrument. What it does not show is the massive polyatomic interference $^{40}Ar^{16}O^+$ (produced by oxygen ions from the

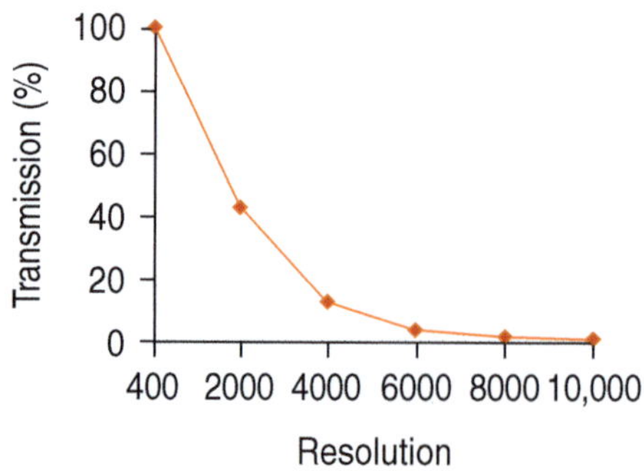

FIGURE 8.2   Ion transmission with a magnetic-sector instrument decreases as the resolution increases.

TABLE 8.1

Resolution Required to Resolve Some Common Polyatomic Interferences from a Selected Group of Isotopes

Isotope	Matrix	Interference	Resolution	Transmission (%)
$^{39}K^+$	H_2O	$^{38}ArH^+$	5570	6
$^{40}Ca^+$	H_2O	$^{40}Ar^+$	199,800	0
$^{44}Ca^+$	HNO_3	$^{14}N^{14}N\ ^{16}O^+$	970	80
$^{56}Fe^+$	H_2O	$^{40}Ar^{16}O^+$	2504	18
$^{31}P^+$	H_2O	$^{15}N^{16}O^+$	1460	53
$^{34}S^+$	H_2O	$^{16}O^{18}O^+$	1300	65
$^{75}As^+$	HCl	$^{40}Ar^{35}Cl^+$	7725	2
$^{51}V^+$	HCl	$^{35}Cl^{16}O^+$	2572	18
$^{64}Zn^+$	H_2SO_4	$^{32}S^{16}O^{16}O^+$	1950	42
$^{24}Mg^+$	Organics	$^{12}C^{12}C^+$	1600	50
$^{52}Cr^+$	Organics	$^{40}Ar^{12}C^+$	2370	20
$^{55}Mn^+$	HNO_3	$^{40}Ar^{15}N^+$	2300	20

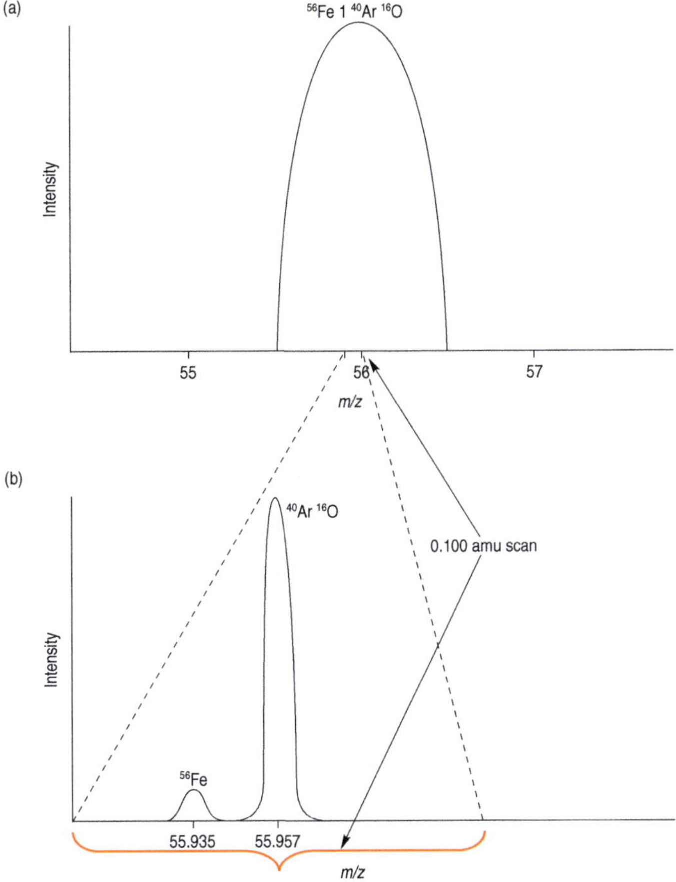

FIGURE 8.3 Comparison of resolution between (a) a quadrupole and (b) a magnetic-sector instrument for the polyatomic interference of $^{40}Ar^{16}O^+$ on $^{56}Fe^+$.

(From U. Greb and L. Rottman, *Labor Praxis*, August 1994)

water combining with argon ions from the plasma) completely overlapping the $^{56}Fe^+$. It shows very clearly that these two masses are not resolvable with a quadrupole. If that same spectral scan is carried out on a magnetic-sector-type instrument, the result is the scan shown in Figure 8.3b.[20] It should be pointed out that in order to see the spectral scan on the same scale, it is necessary to examine a much smaller range. For this reason, a 0.100 amu window was taken, as indicated by the dotted lines.

8.5 OTHER BENEFITS OF MAGNETIC-SECTOR INSTRUMENTATION

Besides high resolving power, another attractive feature of magnetic-sector instruments is their very high sensitivity combined with extremely low background levels. High ion transmission in low-resolution mode translates into sensitivity specifications of up to 1 billion counts per second (1 Bcps) per ppm (parts per million), whereas background levels resulting from extremely low dark current noise are typically 0.1–0.2 counts per second (cps). This compares to typical sensitivity levels of 100 million cps (Mcps) and background levels of 2–5 cps for a quadrupole technology, although the newer instruments are now capable of generating in excess of 200 Mcps and up to 500 Mcps when optimized for high sensitivity, with background (BG) levels in the order of 1cps. However, because background levels are usually significantly lower on a magnetic-sector system, detection limits, especially for high-mass elements such as uranium where high resolution is generally not required, are typically five to ten times better than a quadrupole-based instrument.

Besides good detection capability, another of the recognized benefits of the magnetic-sector approach is its ability to quantitate with excellent precision. Measurement of the characteristically flat-topped spectral peaks translates directly into high-precision data. As a result, in the low-resolution mode, relative standard deviation (RSD) values of 0.01–0.05% are fairly common, which makes it an ideal approach for carrying out high-precision isotope-ratio work.[21] Although precision is usually degraded as resolution is increased, modern instrumentation with high-speed electronics and low-mass bias are still capable of precision values of < 0.1% RSD in medium- or high-resolution mode.[22]

The demand for ultrahigh-precision data, particularly in the field of geochemistry, has led to the development of instruments dedicated to isotope-ratio analysis. These are based on the double-focusing magnetic-sector design, but instead of using just one detector, these instruments use multiple detectors. Often referred to as multi-collector systems, they offer the capability of detecting and measuring multiple ion signals at exactly the same time. As a result of this simultaneous-measurement approach, they are recognized as producing extremely low isotope-ratio precision.[23]

8.6 SIMULTANEOUS-MEASUREMENT APPROACH USING ONE DETECTOR

The scanning limitations of a single detector magnetic-sector instrument have been addressed by the development of spectrometers based on Mattauch–Herzog geometry and simultaneous measurement of the ions using linear plane array detectors.[24, 25, 26, 27] The early designs were limited in their applicability to the entire mass spectrum and the use of an ICP excitation source, but they eventually led to the development of a commercially available instrument in 2010.[28] This particular instrument is made up of an entrance slit, electrostatic analyzer, energy slit and a permanent magnet. In this design, the magnetic field focuses all of the ions onto a flat linear focal plane, without having to adjust the voltage of the mass analyzer or the strength of the magnetic field. In this way, no scanning is involved and, as a result, all the ions generated can be collected on a solid-state ion detector for the simultaneous detection of all the elemental masses in the sample, similar to using a CCD (charge-coupled detector) or CID (charge-injection detector) for optical emission work. The principles of this design are exemplified in Figure 8.4. The ion detector has over 4,800 channels, which is enough to record all 210 isotopes of the 75 elements that can be determined by ICP-MS, using an average of 20 channels per isotope. Every channel is a combination of two detector arrays with different signal amplifiers. In this way, it is possible to cover six orders of dynamic range plus an additional three orders by optimizing the measurement times. More details will be given about this type of detector in Chapter 11 on "Ion Detectors".

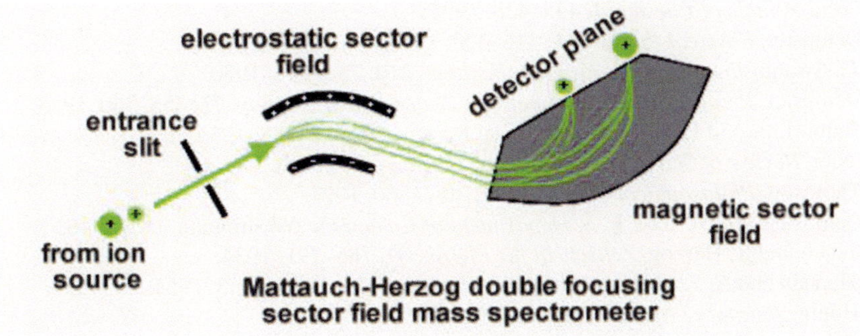

FIGURE 8.4 The principles of the Mattauch–Herzog double-focusing magnet-sector mass spectrometer with a simultaneous detector plane.

The major benefit of this approach is that no scanning of the magnet is required. For this reason, it is ideal for any applications that benefit from a rapid simultaneous measurement of the analyte masses. Some of these applications include:

- Analysis that requires high precision, such as isotope-ratio/dilution studies, or the real-time measurement of internal standards
- Multielement analysis of a fast transient event such as laser ablation or chromatographic separation techniques coupled with ICP-MS
- Applications where rapid speed of analysis is important, such as in high-throughput environmental contract labs
- Because the complete mass spectrum is collected every time, any mass can be interrogated after the analysis to check for any unexpected interferences

However, it should be emphasized that this type of magnetic-sector mass spectrometer does not offer the high resolving power of a traditional Nier–Johnson scanning system, which is typically on the order of 10,000. The resolution of this design of Mattauch–Herzog mass spectrometer is only slightly better than a quadrupole system, which limits its use for the analysis of spectrally complex sample matrices.

8.7　FINAL THOUGHTS

There is no question that magnetic-sector ICP-MS instruments are no longer novel analytical techniques. They have proved themselves to be a valuable addition to the trace element toolbox, particularly for challenging applications that require good detection capability, exceptional resolving power and very high precision. Even though they are not trying to compete with quadrupole instruments when it comes to rapid, high-sample-throughput applications, the measurement speed of modern systems, particularly the newer Mattauch–Herzog simultaneous-detection approach, have opened up the technique to brand-new application areas, which were traditionally beyond the scope of earlier designs of this technology.

FURTHER READING

1. N. Bradshaw, E. F. H. Hall, and N. E. Sanderson, *Journal of Analytical Atomic Spectrometry*, **4**, 801–803, 1989.
2. F. W. Aston, *Philosophical Magazine*, **38**, 707, 1919.

3. J. L. Costa, *Annals of Physics*, **4**, 425–428, 1925.

4. A. J. Dempster, *Physical Review*, **11**, 316–320, 1918.

5. W. F. G. Swann, *Journal of the Franklin Institute*, **210**, 751–756, 1930.

6. A. J. Dempster, *Proceedings of the American Philosophical Society*, **75**, 755–760, 1935.

7. K. T. Bainbridge and E. B. Jordan, *Physical Review*, **50**, 282–288, 1936.

8. A. O. Nier, *Review of Scientific Instruments*, **11**, 252–256, 1940.

9. G. P. Thomson, *Philosophical Magazine*, **42**, 857–860, 1921.

10. A. J. Dempster, *MDDC 370, U. S. Department of Commerce*, Washington, DC, 1946.

11. J. Mattauch and R. Herzog, *Zeitschrift für Physik*, **89**, 786–791, 1934.

12. N. B. Hannay and A. J. Ahearn, *Analytical Chemistry*, **26**, 1056–1062, 1954.

13. R. E. Honig, *Journal of Applied Physics*, **29**, 549–553, 1958.

14. R. Castaing and G. Slodzian, *Journal of Microscopy*, **1**, 394–401, 1962.

15. R. E. Honig and J. R. Wolston, *Applied Physics Letters*, **2**, 138–143, 1963.

16. J. W. Coburn, *Review of Scientific Instruments*, **41**, 1219–1223, 1970.

17. R. Hutton, A. Walsh, D. Milton, and J. Cantle, *ChemSA*, **17**, 213–215, 1991.

18. U. Geismann and U. Greb, *Fresnius' Journal of Analytical Chemistry*, **350**, 186–193, 1994.

19. F. Adams, R. Gijbels, and R. Van Grieken, *Inorganic Mass Spectrometry*, John Wiley and Sons, New York, 1988.

20. U. Greb and L. Rottman, *Labor Praxis*, Würzburg, August 1994.

21. F. Vanhaecke, L. Moens, R. Dams, and R. Taylor, *Analytical Chemistry*, **68**, 565–571, 1996.

22. M. Hamester, D. Wiederin, J. Willis, W. Keri, and C. B. Douthitt, *Fresnius' Journal of Analytical Chemistry*, **364**, 495–497, 1999.

23. J. Walder and P. A. Freeman, *Journal of Analytical Atomic Spectrometry*, **7**, 571–575, 1992.

24. J. H. Barnes, R. P. Sperline, M. B. Denton, C. J. Barinaga, D. W. Koppenaal, E. T. Young, and G. M. Hieftje, *Analytical Chemistry*, **74**(20), 5327–5332, 2002.

25. J. H. Barnes, G. D. Schilling, R. P. Sperline, M. B. Denton, E. T. Young, C. J. Barinaga, D. W. Koppenaal, and G. M. Hieftje, *Analytical Chemistry*, **76**(9), 2531–2536, 2004.

26. G. D. Schilling, F. J. Andrade, J. H. Barnes, R. P. Sperline, M. B. Denton, C. J. Barinaga, D. W. Koppenaal, and G. M. Hieftje, *Analytical Chemistry*, **78**(13), 4319–4325, 2006.

27. G. D. Schilling, S. J. Ray, A. A. Rubinshtein, J. A. Felton, R. P. Sperline, M. B. Denton, C. J. Barinaga, D. W. Koppenaal, and G. M. Hieftje, *Analytical Chemistry*, **81**(13), 5467–5473, 2009.

28. *A New Era in Mass Spectrometry: Spectro Analytical Instruments*, www.spectro.com/pages/e/p010402tab_overview.htm

9 Mass Analyzers
Time-of-Flight Technology

Let us turn our attention to time-of-flight (TOF) technology. Although the first TOF mass spectrometer was first described in the literature in the late 1940s,[1] it has taken over 50 years to adapt it for use with a commercial ICP mass spectrometer. The interest in TOF ICP-MS instrumentation has come about because of its ability to sample all ions generated in the plasma at exactly the same time, which is ideally suited for multielement determinations of rapid transient signals, high-precision isotope-ratio analysis, fast data acquisition for high throughput workloads and the rapid semi-quantitative fingerprinting of unknown samples.

9.1 BASIC PRINCIPLES OF TOF TECHNOLOGY

The simultaneous nature of sampling ions in TOF offers distinct advantages over traditional scanning (sequential) quadrupole technology for ICP-MS applications in which large amounts of data need to be captured in a short span of time. To understand the benefits of this mass-separation device, let us first take a look at its fundamental principles. All TOF mass spectrometers are based on the same principle: the kinetic energy (KE) of an ion is directly proportional to its mass (m) and velocity (V). This can be represented by the following equation:

$$KE = 1/2mV^2$$

Therefore, if a population of ions—all with different masses—is given the same KE by an accelerating voltage (U), the velocities of the ions will all be different, depending on their masses. This principle is then used to separate ions of different mass-to-charge ratios (m/e) in the time (t) domain over a fixed flight path distance (D), represented by the following equation:

$$m/e = \frac{2Ut^2}{D^2}$$

This is schematically shown in Figure 9.1, which shows three ions of different mass-to-charge ratios being accelerated into a "flight tube" and arriving at the detector at different times. It can be seen that, depending on their velocities, the lightest ion arrives first, followed by the medium mass ion, and finally the heaviest one. Using flight tubes of 1 m length, even the heaviest ions typically take less than 50 μs to reach the detector. This translates into approximately 20,000 mass spectra per second—three orders of magnitude faster than the sequential scanning mode of a quadrupole system.

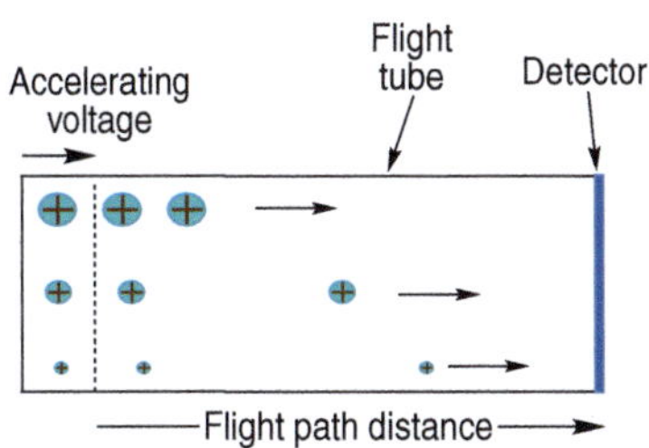

FIGURE 9.1 Principles of ion detection using time-of-flight technology, showing separation of three different masses in the time domain.

9.2   COMMERCIAL DESIGNS

Even though this process sounds fairly straightforward, it is not a trivial task to sample the ions in a simultaneous manner from a continuous source of ions being generated in the plasma discharge. There are basically two different sampling approaches used—the orthogonal design,[2] in which the flight tube is positioned at right angles to the sampled ion beam, and the axial design,[3] in which the flight tube is along the same axis as the ion beam.

In both designs, all ions that contribute to the mass spectrum are sampled through the interface cones, but instead of being focused into the mass filter in the conventional way, packets (groups) of ions are electrostatically injected into the flight tube at exactly the same time. With the orthogonal approach, an accelerating potential is applied at right angles to the continuous ion beam from the plasma source. The ion beam is then "chopped" by using a pulsed voltage supply coupled to the orthogonal accelerator to provide repetitive voltage "slices" at a frequency of a few kilohertz. The "sliced" packets of ions, which are typically tall and thin in cross section (in the vertical plane), are then allowed to "drift" into the flight tube, where the ions are temporally resolved according to their differing velocities. This is shown schematically in Figure 9.2.[4]

In the orthogonal approach, an accelerating potential is applied axially (in the same axis) to the incoming ion beam as it enters the extraction region. Because the ions are in the same plane as the detector, the beam has to be modulated using an electrode grid to repel the "gated" packet of ions into the flight tube. This kind of modulation generates an ion packet that is long and thin in cross section (in the horizontal plane). The different masses are then resolved in the time domain in a similar manner to the orthogonal design. An on-axis TOF system is schematically shown in Figure 9.3.[4]

Figures 9.2 and 9.3 offer a rather simplified explanation of the TOF principles of operation. In practice, there are many complex ion-focusing components in a TOF mass analyzer that ensure that the maximum number of analyte ions reach the detector, and also that undesired photons, neutral species and interferences are ejected from the ion beam. Some of these components are seen in Figure 9.4, which shows a more detailed view of a commercial orthogonal TOF ICP-MS system.

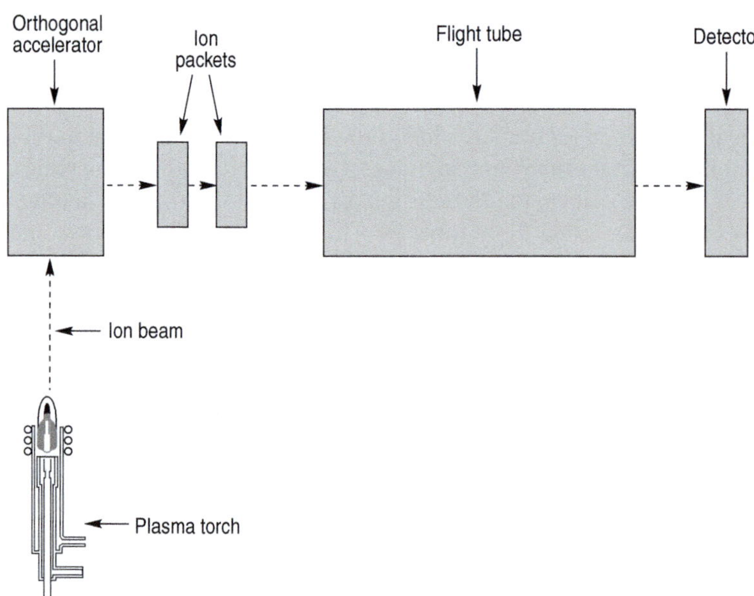

FIGURE 9.2   A schematic of an orthogonal acceleration TOF analyzer.

(From Technical Note: 001–0877–00, GBC Scientific, February 1998)

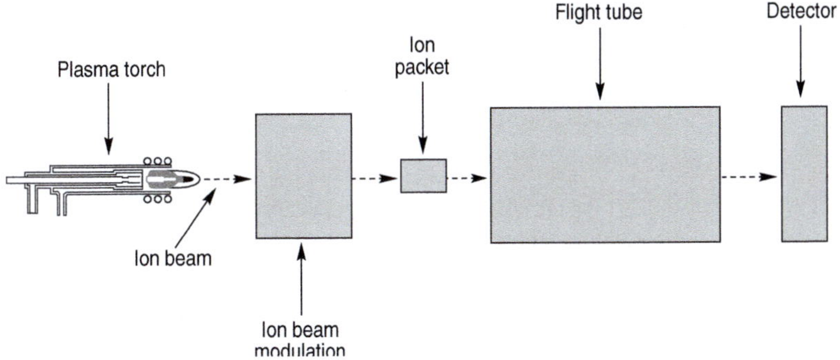

FIGURE 9.3 A schematic of an axial acceleration TOF analyzer.

(From Technical Note: 001–0877–00, GBC Scientific, February 1998)

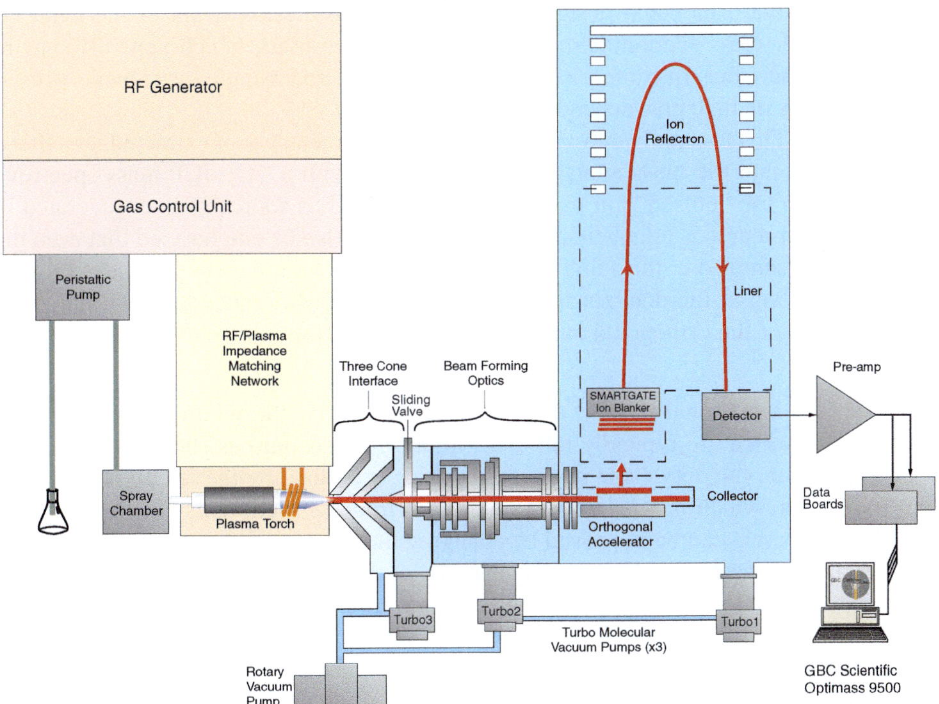

FIGURE 9.4 A more detailed view of an orthogonal (right-angled) TOF ICP-MS system, showing some of the ion-steering components.

(Courtesy of GBC Scientific Equipment Pty Ltd.)

It can be seen in this design that an orthogonal accelerator is used to inject packets of ions at right angles from the ion beam emerging from the MS interface. They are then directed toward an ion blanker, where unwanted ions are rejected from the flight path by deflection plates. The packets of ions are then directed into an ion reflectron where they do a U-turn and are deflected back 180°, where they are detected by a channel electron multiplier or discrete dynode detector. The reflectron, which is a type of ion mirror, functions as an energy-compensation device so that different ions of the same mass arrive at the detector at the same time.

9.3 DIFFERENCES BETWEEN ORTHOGONAL AND ON-AXIS TOF

Although there are real benefits of using TOF over quadrupole technology for some ICP-MS applications, there are also subtle differences in the capabilities of each type of TOF design. Some of these differences include:

- **Sensitivity:** The axial approach tends to produce higher ion transmission because the steering components are in the same plane as the ion-generation system (plasma) and the detector. This means that the direction and magnitude of greatest energy dispersion is along the axis of the flight tube. In addition, when ions are extracted orthogonally, the energy dispersion can produce angular divergence of the ion beam, resulting in poor transmission efficiency. However, on the basis of current evidence, the sensitivity of both TOF designs is generally an order of magnitude lower than the latest commercial quadrupole instruments.
- **Background levels:** The axial design tends to generate higher background levels because neutral species and photons stand a better chance of reaching the detector. This results in background levels on the order of 20–50 cps—approximately one order of magnitude higher than the orthogonal design. However, because the ion beam in the axial design has a smaller cross section, a smaller detector can be used, which generally has better noise characteristics. In comparison, most commercial quadrupole instruments offer background levels of 1–10 cps, depending on the design.
- **Duty cycle:** This is usually defined as the fraction (percentage) of extracted ions that actually make it into the mass analyzer. Unfortunately, with a TOF-ICP mass spectrometer, which has to use "pulsed" ion packets from a continuous source of ions generated in the plasma, this process is relatively inefficient. It should also be emphasized that even though the ions are sampled at the same time, detection is not simultaneous because of different masses arriving at the detector at different times. The difference between the sampling mechanisms of the orthogonal and axial TOF designs translates into subtle differences in their duty cycles.

  - With the orthogonal design, duty cycle is defined by the width of the extracted ion packets, which are typically tall and thin in cross section, as shown in Figure 9.2. In comparison, the duty cycle of an axial design is defined by the length of the extracted ion packet, which is typically wide and thin in cross section, as shown in Figure 9.3. The duty cycle can be improved by changing the cross-sectional area of the ion packet, but, depending on the design, it is generally at the expense of resolution. However, this is not a major issue, because TOF instruments are generally not used for high-resolution ICP-MS applications. In practice, the duty cycles for both orthogonal and axial designs are on the order of 15–20%.

- **Resolution:** The resolution of the orthogonal approach is slightly better because of its two-stage extraction/acceleration mechanism. Because a pulse of voltage pushes the ions from the extraction area into the acceleration region, the major energy dispersion lies along the axis of ion generation. For this reason, the energy spread is relatively small in the direction of extraction compared to spread with the axial approach, resulting in better resolution. However, the resolving power of commercial TOF-ICP-MS systems is typically on the order of 500–2,000, depending on the mass region, which makes them inadequate to resolve many of the problematic polyatomic species encountered in ICP-MS. In comparison, commercial high-resolution systems based on the double-focusing magnetic-sector design offer resolving power up to 10,000, whereas commercial quadrupoles can typically achieve 400–500.
- **Mass bias:** This is also known as mass discrimination and is the degree to which ion-transport efficiency varies with mass. All instruments show some degree of mass bias,

which is usually compensated for by measuring the difference between the theoretical and observed ratio of two different isotopes of the same element. In TOF, the velocity (energy) of the initial ion beam will affect the instrument's mass-bias characteristics. In theory, it should be less with the axial design because the extracted ion packets do not have any velocity in a direction perpendicular to the axis of the flight tube, which could potentially impact their transport efficiency.

9.4 BENEFITS OF TOF TECHNOLOGY FOR ICP-MS

It should be emphasized that these performance differences between the two designs are subtle and should not detract from the overall benefits of the TOF approach for ICP-MS. As mentioned earlier, a scanning device such as a quadrupole can only detect one mass at a time, which means that there is always a compromise between number of elements, detection limits, precision and the overall measurement time. However, with the TOF approach, the ions are sampled at the same moment in time, which means that multielement data can be collected with no significant deterioration in quality. The ability of a TOF system to capture a full mass spectrum, approximately three orders of magnitude faster than a quadrupole, translates into three major benefits—multielement determinations in a fast transient peak, improved precision, especially for isotope-ratio techniques, and rapid data acquisition for carrying out qualitative or semi-quantitative scans. Let us look at these in greater detail.

9.5 RAPID TRANSIENT PEAK ANALYSIS

Probably, the most exciting potential for TOF-ICP-MS is in the multielement analysis of a rapid transient signal generated by sampling accessories such as laser ablation (LA),[5] electrothermal vaporization (ETV)[6] and flow injection systems.[7] Even though a scanning quadrupole can be used for this type of analysis, it struggles to produce high-quality multielement data when the transient peak lasts only a few seconds. The simultaneous nature of TOF instrumentation makes it ideally suited for this type of analysis, because the entire mass range can be collected in less than 50 μs. In particular, when used with an ETV system, the high acquisition speed of TOF can help to reduce matrix-based spectral overlaps by resolving them from the analyte masses in the temperature domain.[6] There is no question that TOF technology is ideally suited (probably more than any other design of ICP mass spectrometer) for the analysis of transient peaks.

9.6 IMPROVED PRECISION

To better understand how TOF technology can help improve precision in ICP-MS, it is important to know the major sources of instability. The most common source of noise in ICP-MS is the flicker noise associated with the sample-introduction process (peristaltic pump pulsations, nebulization mechanisms, plasma fluctuations, etc.) and the shot noise derived from photons, electrons and ions hitting the detector. Shot noise is based on counting statistics and is directly proportional to the square root of the signal. It therefore follows that as the signal intensity increases, the shot noise has less of an impact on the precision (% RSD) of the signal. This means that at high ion counts, the most dominant source of imprecision in ICP-MS is derived from the flicker noise generated in the sample-introduction area.

One of the most effective ways to reduce instability produced by flicker noise is to use a technique called internal standardization, where the analyte signal is compared and ratioed to the signal of an internal standard element (usually of similar mass or ionization characteristics) that is spiked into the sample. Even though a quadrupole-based system can do an adequate job of compensating for these signal fluctuations, it is ultimately limited by its inability to measure the internal standard at precisely the same time as the analyte isotope. So, in order to compensate for sample introduction and plasma-based noise, and to achieve high precision, the analyte and internal standard isotopes need to be sampled and measured simultaneously. For this reason, the design of a TOF mass analyzer is perfect for simultaneous internal standardization required for high-precision work. It therefore follows that TOF is also well suited for high-precision isotope-ratio analysis, where its simultaneous nature of measurement is capable of achieving precision values close to the theoretical limits of counting statistics. Also, unlike a scanning-quadrupole-based system, it can measure ratios for as many isotopes or isotopic pairs as needed—all with excellent precision.[8]

9.7 RAPID DATA ACQUISITION

As with a scanning ICP-OES system, the speed of a quadrupole ICP mass spectrometer is limited by its scanning rate. To determine ten elements in duplicate with good precision and detection limits, an integration time of three seconds per mass is normally required. When overhead scanning and settling times are added for each mass and replicate, this translates into approximately two minutes per sample. With a TOF system, the same analysis would take significantly less time, because all the data are captured simultaneously. In fact, detection-limit levels in a TOF instrument are typically achieved within a 10–30 s integration time, which translates into a five- to ten-fold improvement in data-acquisition time over a quadrupole instrument. The added benefit of a TOF instrument is that the speed of analysis is not impacted by the number of analytes being determined: it would not matter if the method contained 10 or 70 elements—the measurement time would be virtually the same. However, there is one point that must be stressed. A large portion of the overall analysis time is taken up for flushing an old sample out of and pumping a new sample into the sample-introduction system. This can be as much as 2 min per sample for real-world matrices. So, when this is taken into account, the difference between the sample throughput of a quadrupole and a TOF-ICP mass spectrometer is not so evident.

Another benefit of the fast acquisition time is that qualitative or semi-quantitative analysis is relatively seamless compared to scanning-quadrupole technology, because every multielement scan contains data for every mass. This also makes spectral identification much easier by comparing the spectral fingerprint of unknown samples against a known reference standard. This is particularly useful for forensic work, where the evidence is often an extremely small sample.

Commercial TOF-ICP-MS instruments have been available since the late 1990s, mainly being used for the multielement characterization of fast transient signals and high-precision isotope-ratio studies. However, two recent novel developments of TOF technology are widening its applicability and extending its real-world analytical capabilities.

9.8 HIGH-SPEED MULTI-ELEMENTAL IMAGING USING LASER ABLATION COUPLED WITH TOF-ICP-MS

A more recent example of the benefits of TOF technology for the characterizing rapid transients is in the field of high-speed multielement imaging using laser ablation. Since its first appearance in the mid-1980s, laser ablation inductively coupled plasma mass spectrometry (LA-ICP-MS) has established itself as a routine method for the quantitation of trace elements in solid samples. In recent times, LA-ICP-MS has further become an important tool for elemental imaging applied in geological, biological and medical research studies. To date, the most common imaging approach is to run the laser in continuous-scan mode, which involves firing the laser continuously with a specific repetition rate (usually 1 to 10 Hz) while slowly moving the stage with the mounted sample underneath the laser beam. The image is then constructed by ablating parallel lines on the sample surface. The ablated aerosol is washed out of the airtight ablation chamber (cell) to the ICP with a continuous flow of inert gas.

The washout times for conventional ablation cells are on the order of 0.5 to 30 seconds. After ionization in the ICP, the ablation signal is measured in a time-resolved manner, most commonly using a quadrupole or a sector field mass spectrometer. These instruments operate sequentially, resulting in the measurement of only one isotope at a time. For the analysis of multiple isotopes/ elements, sequential peak-hopping must be performed, which, depending on the number of elements of interest, can become very restrictive and time consuming. The conventional approach to laser-ablation imaging is associated with drawbacks on spatial resolution and speed of analysis, which is usually limited to 1–2 pixels per second. The recent advent of fast-washout ablation cells and high-speed TOF-ICP-MS allowed some of these limitations to be overcome by enabling spot-resolved imaging.

In order to capitalize on the benefits of fast-washout ablation cells, a mass spectrometer capable of extremely fast and simultaneous multielement data acquisition is required, and in particular its ability to handle rapid transient signals from these cells. The TOF-ICP-MS is well suited for the shot-resolved imaging approach due to its simultaneous and fast multielement analysis, because the entire mass spectrum is measured simultaneously with each ion package extraction. Different isotopes from this package are then separated based on their time of flight traversing the region from extraction to detector, which is directly related to their mass-to-charge ratio. Thus, in contrast to conventional sequential mass spectrometers, it is not necessary to pre-define or to limit a range of isotopes for analysis. Moreover, the TOF-ICP-MS can record complete mass spectra at a rate of 30 μs, which amounts to a maximum of ~33,000 spectra per second. This allows short transient signals from single laser pulses to be temporally resolved with sufficient sampling density (e.g., up to ~1,000 spectra per 30 ms pulse). In practice, however, multiple spectra are integrated to improve counting statistics and to simplify data processing. Due to this simultaneous measurement of complete mass spectra, TOF-ICP-MS can have a higher total ion utilization than sequential mass spectrometers, and as a result, a greater proportion of the ablated sample is actually used for analysis. This work has been presented in a recent publication by Bussweiler and coworkers.[9]

9.9 LASER ABLATION LASER IONIZATION TIME-OF-FLIGHT MASS SPECTROMETRY

Laser ablation, laser ionization coupled with TOF mass spectrometry (LALI-TOF-MS) is a brand-new development that offers virtually the entire periodic table of the elements at mass spec detection capability. It has broad applicability for any type of lab that is looking to avoid sample-digestion procedures to carry out multielement analysis directly on solid samples.

LALI-TOF-MS utilizes dual-laser technology to first extract and ionize material in two discrete steps—first, ablation of the sample and then ionization of the analyte elements. The first step uses a focused laser beam to ablate material from a solid sample surface using the Nd:YAG laser with an adjustable wavelength. A short delay (< 1 μs) allows dispersion of plasma-generated ions before a second Nd:YAG laser is triggered for ionization.

By ionizing elements across a wide range of ionization energies, this technique serves as a highly efficient ion source and replaces the Ar plasma of ICP-MS instruments. Once ionized, an ion funnel collects and focuses the ions into a quadrupole ion deflector (QID) and turns the ion beam 90 degrees through an einzel lens stack into a reflectron time-ff-Flight mass spectrometer, which completes the mass separation and detection. More detailed information about this very exciting technology can be found in Chapter 23 on "Other AS Techniques".

9.10 FINAL THOUGHTS

There is no question that TOF-ICP-MS, with its rapid, simultaneous mode of measurement, excels at multielement applications that generate fast transient signals, such as laser ablation or speciation studies using chromatographic techniques. In addition, it offers excellent precision, for isotope-ratio techniques, and also has the potential for rapid data acquisition when carrying out multielement semi-quantitative analysis. And when coupled with novel excitation/ionization sources, it has the potential to be utilized for the determination of trace elements directly on solid samples. So although TOF mass-separation devices are a fairly recent commercial development compared to quadrupole technology, it definitely should be considered as an option if the application demands it.

FURTHER READING

1. A. E. Cameron and D. F. Eggers, *The Review of Scientific Instruments*, **19**(9), 605–608, 1948.
2. D. P. Myers, G. Li, P. Yang, and G. M. Hieftje, *Journal of American Society of Mass Spectrometry*, **5**, 1008–1016, 1994.
3. D. P. Myers, *12th Asilomar Conference on Mass Spectrometry*, Pacific Grove, CA, September 20–24, 1996.
4. Technical Note: 001–0877–00, *GBC Scientific*, February 1998.
5. P. Mahoney, G. Li, and G. M. Hieftje, *Journal of American Society of Mass Spectrometry*, **11**, 401–406, 1996.
6. Technical Note: 001–0876–00, *GBC Scientific*, February 1998.
7. R. E. Sturgeon, J. W. H. Lam, and A. Saint, *Journal of Analytical Atomic Spectrometry*, **15**, 607–616, 2000.
8. F. Vanhaecke, L. Moens, R. Dams, L. Allen, and S. Georgitis, *Analytical Chemistry*, **71**, 3297–3303, 1999.
9. Y. Bussweiler, O. Borovinskaya, M. Tanner, Laser Ablation and Inductively Coupled Plasma Time-of-Flight Mass Spectrometry—A Powerful Combination for High-Speed Multi-Elemental Imaging on the Micrometer-Scale, *Spectroscopy Magazine*, **32**(5), 14–22, 2017.

10 Mass Analyzers
Collision/Reaction Cell and Interface Technology

The detection capability for some elements using traditional quadrupole mass-analyzer technology is severely compromised because of the formation of polyatomic spectral interferences generated by a combination of argon, solvent and matrix-derived ions. Although there are ways to minimize these interferences, including correction equations, cool plasma technology and matrix separation, they cannot be completely eliminated. However, a novel approach using collision/reaction cell and interface technology has been developed that significantly reduces the formation of many of these harmful species before they enter the mass analyzer. Chapter 10 takes a detailed look at this very powerful technique and the exciting potential it has to offer and how it might be beneficial for the analysis of difficult and varied sample matrices.

There are a small number of elements that are recognized as having poor detection limits by ICP-MS. These are predominantly elements that suffer from major spectral interferences generated by ions derived from the plasma gas, matrix components or the solvent/acid used in sample preparation. Examples of these interferences include the following:

- $^{40}Ar^{16}O^+$ in the determination of $^{56}Fe^+$
- $^{38}ArH^+$ in the determination of $^{39}K^+$
- $^{40}Ar^+$ in the determination of $^{40}Ca^+$
- $^{40}Ar^{40}Ar^+$ in the determination of $^{80}Se^+$
- $^{40}Ar^{35}Cl^+$ in the determination of $^{75}As^+$
- $^{40}Ar^{12}C^+$ in the determination of $^{52}Cr^+$
- $^{35}Cl^{16}O^+$ in the determination of $^{51}V^+$

The cold/cool plasma approach, which uses a lower temperature to reduce the formation of the argon-based interferences, has been a very effective way to get around some of these problems.[1] However, this approach can sometimes be difficult to optimize, is only suitable for a few of the interferences, is susceptible to more severe matrix effects and it can be time consuming to change back and forth between normal and cool plasma conditions. These limitations and the desire to improve performance have led to the commercialization of collision/reaction cells (CRC) and collision/reaction interfaces (CRI) (note: CRIs have more recently been referred to as the integrated Collision Reaction Cell (iCRC) design). Designs for CRC and CRI were based on the early work of Rowan and Houk, who used Xe and CH_4 in the late 1980s to reduce the formation of ArO^+ and Ar_2^+ species in the determination of Fe and Se with a modified tandem mass spectrometer.[2] This research was investigated further by Koppenaal and coworkers in 1994, who carried out studies using an ion trap for the determination of Fe, V, As and Se in a 2%-hydrochloric acid matrix.[3] However, it was not until 1996 that studies describing the coupling of a collision/reaction cell with a traditional quadrupole ICP mass spectrometer were published. Eiden and coworkers experimented using hydrogen as a collision gas,[4] whereas Turner and coworkers based their investigations on using helium gas.[5] These studies and the work of other groups at the time[6,7] proved to be the basis for modern collision and reaction cells and interfaces that are commercially available today.

Let's take a look at the fundamental principles of collision/reaction cells and interfaces.

DOI: 10.1201/9781003187639-10

10.1 BASIC PRINCIPLES OF COLLISION/REACTION CELLS

With all collision/reaction cells, ions enter the interface in the normal manner, and then are directed into a collision/reaction cell positioned prior to the analyzer quadrupole. A collision/reaction gas (e.g., helium, hydrogen, ammonia or oxygen, depending on the design) is then bled via an inlet aperture into the cell containing a multipole (a quadrupole, hexapole or octapole), usually operated in the RF-only mode. The RF-only field does not separate the masses as a traditional quadrupole mass analyzer does, but instead, it has the effect of focusing the ions, which then collide and react with molecules of the collision/reaction gas. By a number of different ion–molecule collision and reaction mechanisms, polyatomic interfering ions such as $^{40}Ar^+$, $^{40}Ar^{16}O^+$ and $^{38}ArH^+$ will either be converted to harmless noninterfering species, or the analyte will be converted to another ion that is not interfered with. This process is exemplified by the following equation, which shows the use of hydrogen as a reaction gas to reduce the $^{40}Ar^+$ interference in the determination of $^{40}Ca^+$.

$$H_2 + {}^{40}Ar^+ = Ar + H_2^+$$
$$H_2 + {}^{40}Ca^+ = {}^{40}Ca^+ + H_2 \text{ (no reaction)}$$

It can be seen that the hydrogen molecule interacts with the argon interference to form atomic argon and the harmless H_2^+ ion. However, there is no interaction between the hydrogen and the calcium. As a result, the $^{40}Ca^+$ ions, free of the argon interference, emerge from the collision/reaction cell through the exit aperture, where they are directed towards the quadrupole analyzer for normal mass separation. Other gases are better suited to reduce the $^{40}Ar^+$ interference, but this process at least demonstrates the principles of the reaction mechanisms in a collision/reaction cell. The layout of a typical collision/reaction cell within the instrument is shown in Figure 10.1.

The previous equation is an example of an ion–molecule reaction using the process of charge transfer. By the transfer of a positive charge from the argon ion to the hydrogen molecule, an innocuous neutral Ar atom is formed, which is invisible to the mass analyzer. There are many other reaction and collisional mechanisms that can take place in the cell, depending on the nature of the analyte ion, the interfering species, the reaction/collision gas and the type of multipole used. Other possible mechanisms that can occur in the cell, in addition to charge transfer, include the following:

- *Proton transfer*—The interfering polyatomic species gives up a proton, which is then transferred to the reaction gas molecule to form a neutral atom.
- *Hydrogen-atom transfer*—A hydrogen atom is transferred to the interfering ion, which is converted to an ion at one mass higher.
- *Molecular association reactions*—An interfering ion associates with a neutral species (atom or molecule) to form a molecular ion.
- *Collisional fragmentation*—The polyatomic ion is broken apart or fragmented by the process of multiple collisions with the gaseous atoms.
- *Collisional retardation*—The gas atoms/molecules undergo multiple collisions with the polyatomic interfering ion in order to retard or lower its kinetic energy. Because the interfering ion has a larger cross-sectional area than the analyte ion, it undergoes more collisions, and as a result, can be separated or discriminated from the analyte ion based on kinetic energy differences.
- *Collisional focusing*—Analyte ions lose energy as they collide with the gaseous molecules, and depending on the molecular weight of the gas, will either enhance ion transmission as the ions migrate towards the central axis of the cell, or decrease sensitivity if ion scattering takes place.

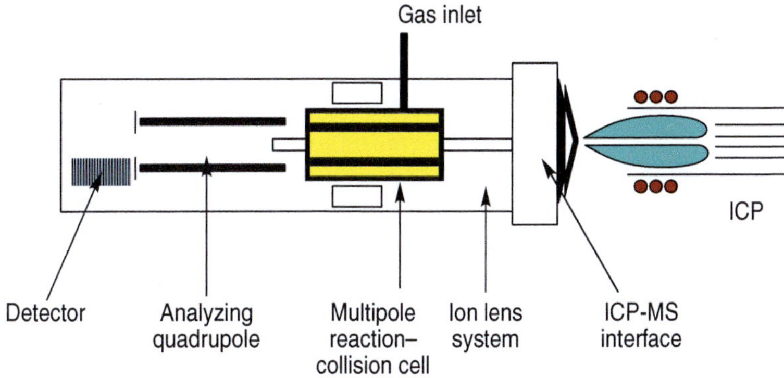

FIGURE 10.1 Layout of a typical collision/reaction cell instrument.

The collision/reaction interface, which will be discussed later in this chapter, uses a slightly different principle to remove the interfering ions. It does not use a pressurized cell before the mass analyzer, but instead, injects a reaction/collision gas directly into the aperture of the interface skimmer cone. The injection of the collision/reaction into this region of the ion beam produces collisions between the argon gas and the injected gas molecules, and as a result, argon-based polyatomic interferences are destroyed or removed before they are extracted into the ion optics.

10.2 DIFFERENT COLLISION/REACTION CELL APPROACHES

All these possible interactions between ions and molecules indicate that many complex secondary reactions and collisions can take place, which generate undesirable interfering species. If these species are not eliminated or rejected, they could potentially lead to additional spectral interferences. There are basically two different approaches used to reduce the formation of polyatomic interferences and discriminate the products of these unwanted side reactions from the analyte ion. They are:

- **Collisional mechanisms** using non- or low-reactive gases and kinetic energy discrimination (KED).[8]
- **Reaction mechanisms** using highly reactive gases and discrimination by selective band-pass mass filtering.[9]

The major differences between the two approaches are how the gaseous molecules interact with the interfering species and what type of multipole is used in the cell. These dictate whether it is an ion–molecule collision or reaction mechanism taking place. Let us take a closer look at each process, because there are distinct differences in the way the interference is rejected and separated from the analyte ion.

10.3 COLLISIONAL MECHANISMS USING NONREACTIVE GASES AND KINETIC ENERGY DISCRIMINATION

The collisional-mechanisms approach was adapted from collision-induced dissociation (CID) technology, which was first used in the early to mid-1990s in the study of organic molecules using tandem mass spectrometry. The basic principle relies on using a nonreactive gas in a hexapole collision cell to stimulate ion–molecule collisions. The more collision-induced daughter species that are generated, the better the chance of identifying the structure of the parent molecule.[10,11] However, this very desirable CID characteristic for identifying and quantifying biomolecules was a disadvantage in inorganic mass spectrometry, where uncontrolled secondary reactions are generally something to be avoided. The limitation restricted the use of hexapole-based collision cells in ICP-MS to inert gases such as helium or low-reactivity gases such as hydrogen, because of the potential to form undesirable reaction by-products, which could spectrally interfere with other analytes. Unfortunately, higher-order multipoles have little control over these secondary reactions because their stability boundaries are very diffuse and not well defined like a quadrupole. As a result, they do not provide adequate mass-separation capabilities to suppress the unwanted secondary reactions. Thus, the need is to rely mainly on collisional mechanisms and a process called *kinetic energy discrimination* (KED) to distinguish the interfering ions and by-product species from the analyte ions. So, what is KED?

KED relies on the principle of separating ions depending on their different ion energies. As ions enter the interface region, they all have differing kinetic energies based on the ionization process in the plasma and their mass-to-charge ratio. When the analyte and plasma/matrix-based interfering ions (sometimes referred to as *precursor ions*) enter the pressurized cell, they will undergo multiple collisions with the collision gas. Because the collisional cross-sectional area of the precursor ions and other collision-induced by-product ions are usually larger than the analyte ion, they will undergo more collisions. This has the effect of lowering the kinetic energy of these interfering species compared to the analyte ion. If the collision cell rod offset potential is set slightly more negative than the mass filter potential, the polyatomic ions with lower kinetic energy are rejected or discriminated by the potential energy barrier at the cell exit. On the other hand, the analyte ions, which have a higher kinetic energy, are transmitted to the mass analyzer. Figure 10.2 shows the principles of KED.

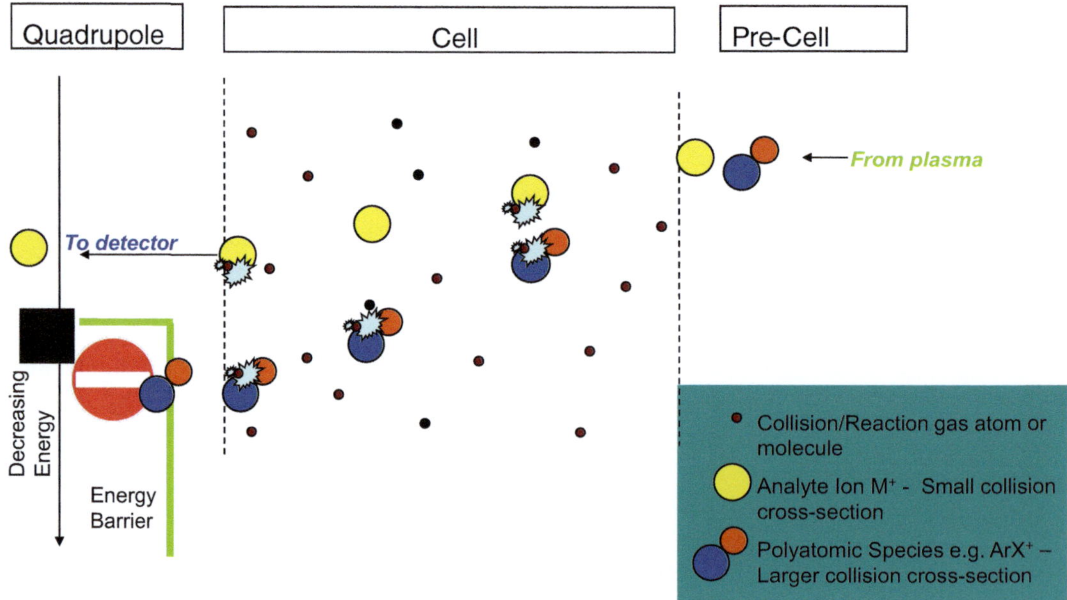

FIGURE 10.2   Principles of kinetic energy discrimination.

(Courtesy of Thermo Scientific)

In addition to the kinetic energy generated by the ionization process, the spread of ion energies will also be dictated by the efficiency of the RF-grounding mechanism (refer to Chapter 5 on the "Interface Region"). Therefore, for KED to work properly, the ion energy spread of ions generated in the plasma must be as narrow as possible to ensure that there is very little overlap between the analyte and the polyatomic interfering ion as they enter the mass analyzer. This means that it is absolutely critical that the RF-grounding mechanism work efficiently to guarantee a low potential at the interface. If this is not the case, and there is a secondary discharge between the plasma and the interface, it will increase the ion energy spread of ions entering the collision cell and make it extremely difficult to separate the polyatomic interfering ion from the analyte of interest based on their kinetic energy difference. The relevance of having a narrow spread of ion energies is shown in Figure 10.3.

It can be seen that the ion energy spread of the analyte ion (gray peak) and a polyatomic ion (black peak) is very similar as they enter the collision cell. This allows the collision process and KED system to easily separate the ions as they exit the cell. If the ion energy spread is larger, there would be more of an overlap as the ions enter the mass analyzer, and therefore compromise the detection limit for that analyte.

KED using helium as the collision gas works very well when the interfering polyatomic ion is physically larger than the analyte ion. This is exemplified in Figure 10.4, which shows helium flow optimization plots for six elements in 1:10 diluted seawater.[12]

It can be seen that the signal intensity for the analytes—Cr, V, Co, Ni, Cu and As—are all at a maximum, whereas their respective matrix, argon and solvent-based polyatomic interferences are all at a minimum, at a similar helium flow rate of 5–6 mL/min. All the analytes show very good sensitivity because the KED process has allowed the analytes to be efficiently separated from their respective polyatomic interfering ions. The additional benefit of using helium is that it is inert, and even if it is not being used in an interference-reduction mode, it can have a beneficial effect on the other elements in a multielement run by increasing sensitivity via the process of collisional focusing. This makes the use of helium and KED very useful for both quantitative and semi-quantitative multielement analysis using one set of tuning conditions.

For the KED process to work efficiently, there must be a distinct difference between the kinetic energy of the analyte compared to the interfering ion. In most cases where the

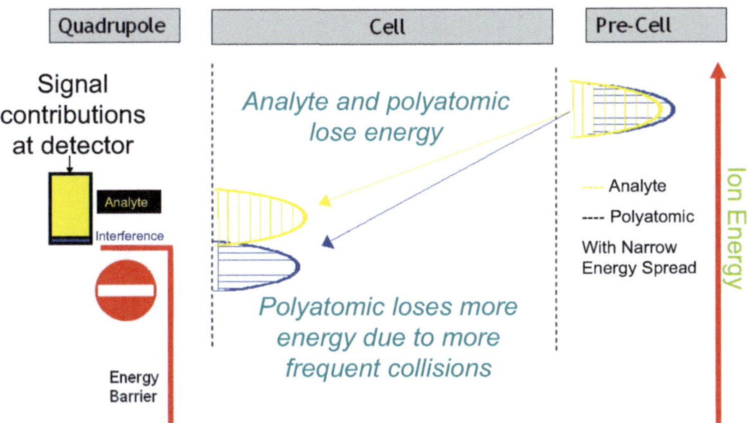

FIGURE 10.3　For optimum separation of analyte ion from the interfering species, it is important to have a narrow spread of ion energies entering the collision cell.

(Courtesy of Thermo Scientific)

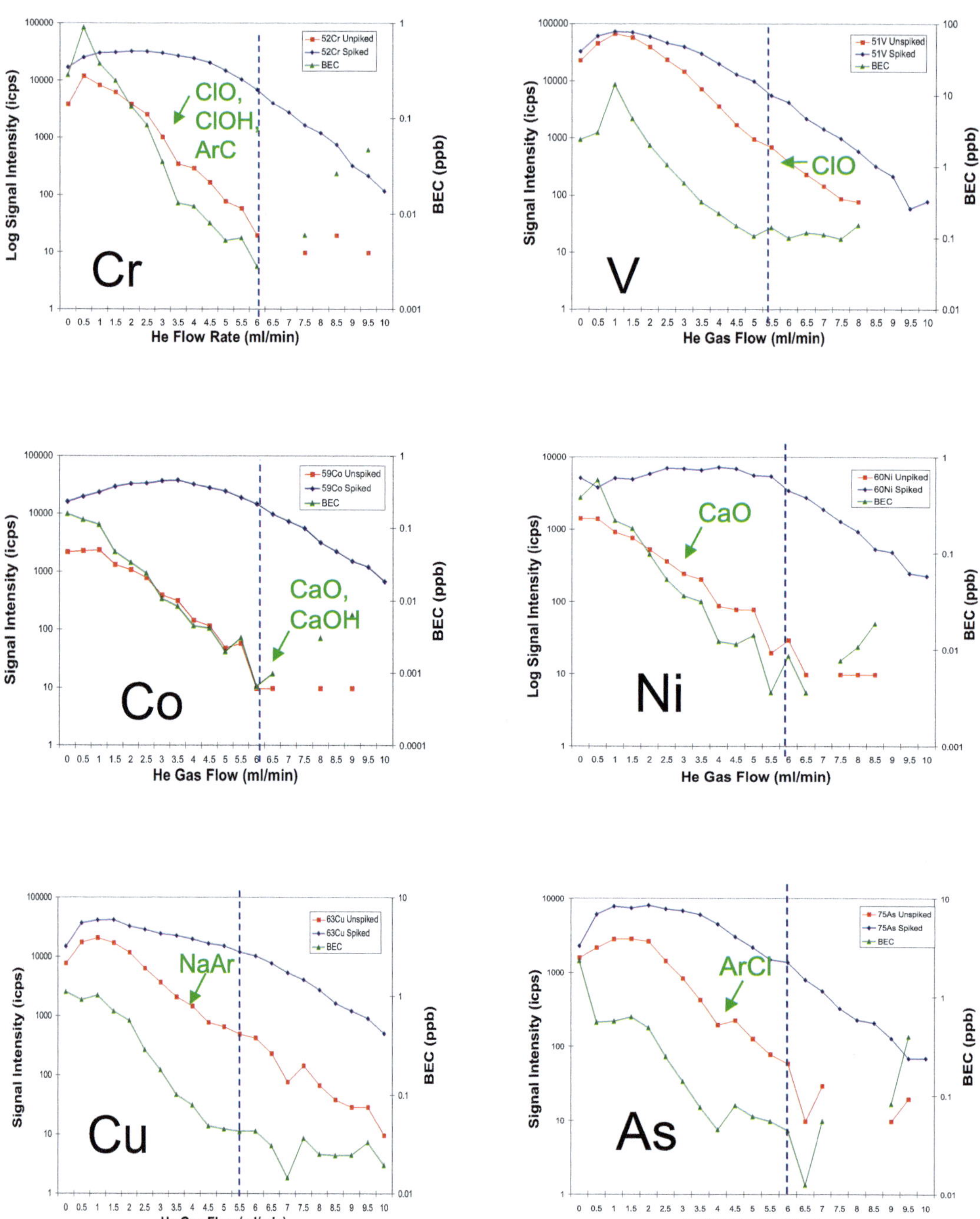

FIGURE 10.4   Helium cell gas-flow optimization plots for Cr, V, Co, Ni, Cu and As in 1:10 diluted seawater, showing that all polyatomic interfering ions are reduced to an acceptable level under one set of cell gas-flow conditions.

(Courtesy of Thermo Scientific)

polyatomic ion is large, this is not a problem. However, in many cases where the interfering and analyte ions have a similar physical size (cross section), the process requires an extremely large number of collisions, which will have an impact not only on the attenuation of the interfering ion, but also on the analyte ion. This means that for some situations, especially if the requirement is for ultra-trace detection capability, the collisional process with KED is not enough to reduce the interference to acceptable levels. That is why most collision cells also have basic reaction-mechanism capabilities, which allows low-reactivity gases such as hydrogen and, in some cases, small amounts of highly reactive gases such as ammonia or oxygen mixed with helium to be used.[13] But it should be pointed out that even though this initiates a basic reaction mechanism, rejection of the reaction by-product ions is still handled through the process of KED, which in some applications might not offer the most efficient way of reducing the interfering species. Using the collision cell as a basic reaction cell with low-reactivity gases is described later in this chapter.

One further point to keep in mind is that a KED-based cell relies on interactions of the interfering ion with an inert or low-reactivity gas, such that it can be separated from the analyte based on their differences in kinetic energy. If the gas contains impurities such as organic compounds or water vapor, the impurity could be the dominant reaction or collision pathway as opposed to the predicted collision/reaction with bulk gas. In addition, other unexpected ion–molecule reactions can readily occur if there are chemical impurities in the gas. This could also pose a secondary problem because of the formation of unexpected cluster ions such as metal oxide and hydroxide species, which have the potential to interfere with other analyte ions. Fortunately, many of these new ions formed in the cell as a result of reactions with the impurities have low energy and are adequately handled by KED. However, depending on the level of the impurity, some of the ions formed have higher energies and are therefore too high to be attenuated by the KED process, which could negatively impact the performance of the interference-reduction process. For this reason, it is strongly advised that the highest-purity collision/reaction gases be used. If this is not an option, it is recommended that a gas-purifier (getter) system be placed in the gas line to cleanse the collision/reaction gas of impurities, such as H_2O, O_2, CO_2, CO or hydrocarbons. If you want to learn more about this process, Yamada and coworkers published a very interesting paper describing the effects of cell–gas impurities and KED in an octapole-based collision cell.[14]

Let us now go on to discuss the other major way interference rejection in a collision/reaction cell uses highly reactive gases and mass (bandpass) filtering discrimination.

10.4 REACTION MECHANISMS WITH HIGHLY REACTIVE GASES AND DISCRIMINATION BY SELECTIVE BANDPASS MASS FILTERING

Another way of rejecting polyatomic interfering ions and the products of secondary reactions/collisions is to discriminate them by mass. As mentioned previously, higher-order multipoles cannot be used for efficient mass discrimination because the stability boundaries are diffuse, and sequential secondary reactions cannot be easily intercepted. The only way this can be done is to utilize a quadrupole (instead of a hexapole or octapole) inside the reaction/collision cell, and use it as a selective bandpass (mass) filter. There are a number of commercial designs using this approach, so let's take a look at them in greater detail in order to better understand how they work and how they differ.

10.5 DYNAMIC REACTION CELL

The first commercial instrument to use this approach was called *dynamic reaction cell* (DRC) *technology*.[15] Similar in appearance to the hexapole and octapole collision/reaction cells, the dynamic reaction cell is a pressurized multipole positioned prior to the analyzer quadrupole. However, this is where the similarity ends. In DRC technology, a quadrupole is used instead of a hexapole or octapole. A highly reactive gas such as ammonia, oxygen or methane is bled into the cell, which is a catalyst for ion–molecule chemistry to take place. By a number of different reaction mechanisms, the gaseous molecules react with the interfering ions to convert them into either an innocuous species different from the analyte mass or a harmless neutral species. The analyte mass then emerges from the dynamic reaction cell free of its interference and is steered into the analyzer quadrupole for conventional mass separation.

The advantage of using a quadrupole in the reaction cell is that the stability regions are much better defined than higher-order multipoles, so it is relatively straightforward to operate the quadrupole inside the reaction cell as a mass or bandpass filter and not just as an ion-focusing guide. Therefore, by careful optimization of the quadrupole electrical fields, unwanted reactions between the gas and the sample matrix or solvent, which could potentially lead to new interferences, are prevented. It means that every time an analyte and interfering ions enter the dynamic reaction cell, the bandpass of the quadrupole can be optimized for that specific problem and then changed on the fly for the next one. This is shown schematically in Figure 10.5, where an analyte ion $^{56}Fe^+$ and an isobaric interference $^{40}Ar^{16}O^+$ enter the dynamic reaction cell. As can be seen, the reaction gas NH_3 picks up a positive charge from the $^{40}Ar^{16}O^+$ ion to form atomic oxygen, argon and a positive NH_3 ion (this is known as a "charge-transfer reaction"). There is no reaction between the $^{56}Fe^+$ and the NH_3, as predicted by thermodynamic reaction kinetics. The quadrupole's electrical field is then set to allow the transmission of the analyte ion $^{56}Fe^+$ to the analyzer quadrupole, free of the problematic isobaric interference, $^{40}Ar^{16}O^+$. In addition, the NH_3^+ is prevented from reacting further to produce a new interfering ion.

The practical benefit of using highly reactive gases is that they increase the number of ion–molecule reactions taking place inside the cell, which results in a faster, more efficient removal of the interfering species. Of course, they will also generate more side reactions, which, if not prevented, will lead to new polyatomic ions being formed and could possibly interfere with other

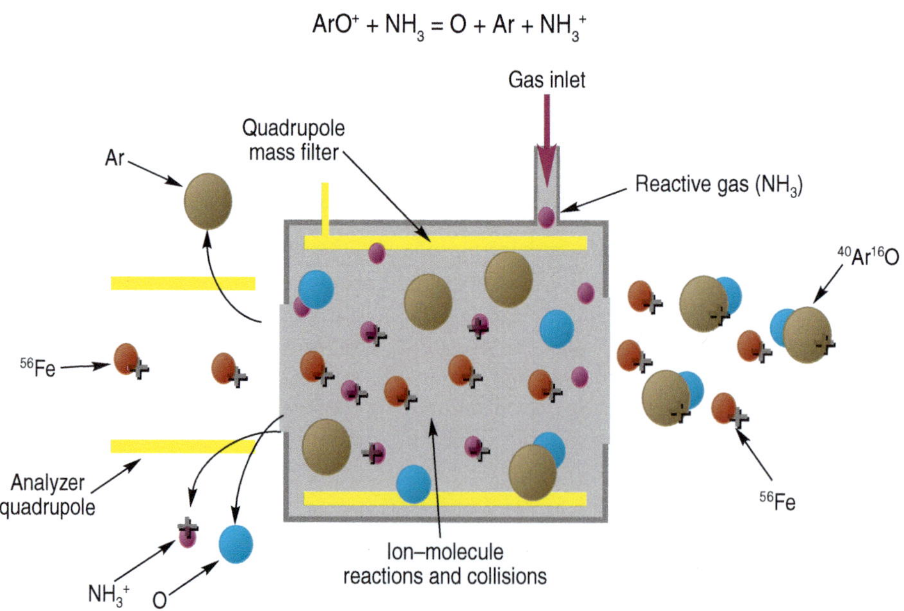

FIGURE 10.5   Elimination of the $^{40}Ar^{16}O^+$ interference with a dynamic reaction cell.[20]

analyte masses. However, the quadrupole reaction cell is well characterized by well-defined stability boundaries. So, by careful selection of bandpass parameters, ions outside the mass/charge (m/z) stability boundaries are efficiently and rapidly ejected from the cell. It means that additional reaction chemistries, which could potentially lead to new interferences, are successfully interrupted. In addition, the bandpass of the reaction-cell quadrupole can be swept in concert with the bandpass of the quadrupole mass analyzer. This allows a dynamic bandpass to be defined for the reaction cell so that the analyte ion can be efficiently transferred to the analyzer quadrupole. The overall benefit is that within the reaction cell, the most efficient thermodynamic reaction chemistries can be used to minimize the formation of plasma- and matrix-based polyatomic interferences, in addition to simultaneously suppressing the formation of further reaction by-product ions.

The process described can be exemplified by the elimination of $^{40}Ar^+$ by NH_3 gas in the determination of $^{40}Ca^+$. The reaction between NH_3 gas and the $^{40}Ar^+$ interference, which is predominantly charge transfer/exchange, occurs because the ionization potential of NH_3 (10.2 eV) is low compared to that of Ar (15.8 eV). This makes the reaction extremely exothermic and fast. However, as the ionization potential of Ca (6.1 eV) is significantly less than that of NH_3, the reaction, which is endothermic, is not allowed to proceed.[15] This can be seen in greater detail in Figure 10.6.

Of course, other secondary reactions are probably taking place, which you would suspect with such a reactive gas as ammonia, but by careful selection of the cell quadrupole electrical fields, the optimum bandpass only allows the analyte ion to be transported to the analyzer quadrupole, free of the interfering species. This highly efficient reaction mechanism and selection process translates into a dramatic reduction of the spectral background at mass 40, which is shown graphically in Figure 10.7. It can be seen that at the optimum NH_3 flow, a reduction in the $^{40}Ar^+$ background signal of about eight orders of magnitude is achieved, resulting in a detection limit of approximately 0.1 ppt for $^{40}Ca^+$.

One final thing to point out is that when highly reactive gases are used, the purity of the gas is not so critical because the impurity is almost insignificant in determining the ion–molecule reaction

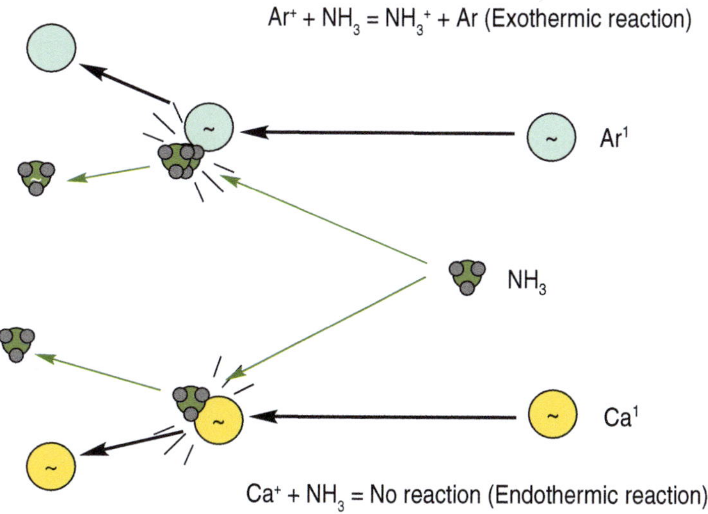

FIGURE 10.6 The reaction between NH_3 and Ar^+ is exothermic and fast, whereas there is no reaction between NH_3 and Ca^+ in the dynamic reaction cell.[20]

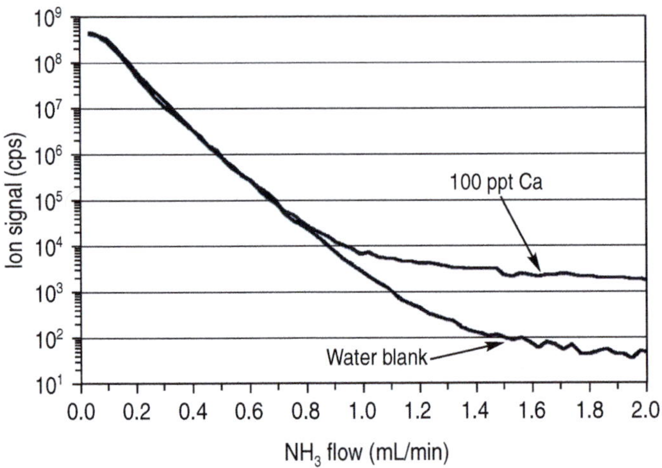

FIGURE 10.7 A reduction of eight orders of magnitude in the $^{40}Ar^+$ background signal is achievable with the dynamic reaction cell, resulting in a 0.1 ppt detection limit for $^{40}Ca^+$.[20]

mechanism. On the other hand, with collision and low-reactivity gases that contain impurities, such as carbon dioxide, hydrocarbons or water vapor, the impurity could be the dominant reaction pathway as opposed to the predicted collision/reaction with the bulk gas. In addition, the formation of unexpected by-product ions or other interfering species, which have the potential to interfere with other analyte ions in a KED-based collision cell, are not such a serious problem with the DRC system because of its ability to intercept and stop these side reactions using the bandpass mass-filtering discrimination process.

These observations were, in fact, made by Hattendorf and Günter, who attempted to quantify the differences between KED and bandpass tuning with regard to suppression of interferences generated in a collision/reaction cell for a group of mainly monoisotopic element (Sc, Y, La, Th) oxides.[16] They observed that when the collision/reaction gas contains impurities such as water vapor or ammonia, a broad range of additional interferences are produced in the cell. Depending on the relative mass of the precursor ions (ions that are formed in the plasma) compared to the by-product ions formed in the cell, there will be significant differences in the way these interferences are suppressed. They concluded that unless the mass (or energy) differences between the precursor and by-product ions are large, there will be significant overlap of kinetic energy distribution, making it very difficult to separate them, which limits the effectiveness of KED to suppress the cell-generated ions. On the other hand, bandpass tuning can tolerate a much smaller difference in mass between the precursor and by-product interfering ions because of its ability to set the optimum mass/charge cutoff at the point where these interfering ions are rejected. In addition, they found that the bandpass tuning method can use a heavier or denser collision gas if desired, without suffering a loss of sensitivity due to scattering observed with the KED method. The overall conclusion of their study was that "under optimized conditions, the bandpass tuning approach provides superior analytical performance because it retains a significantly higher elemental sensitivity and provides more efficient suppression of cell-generated oxide ions, when compared to kinetic energy discrimination".

10.6 LOW-MASS CUTOFF COLLISION/REACTION CELL

A variation on bandpass filtering uses slightly different control of the filtering process. By operating the cell in the RF-only mode, the quadrupole's stability boundaries can be tuned to cut off

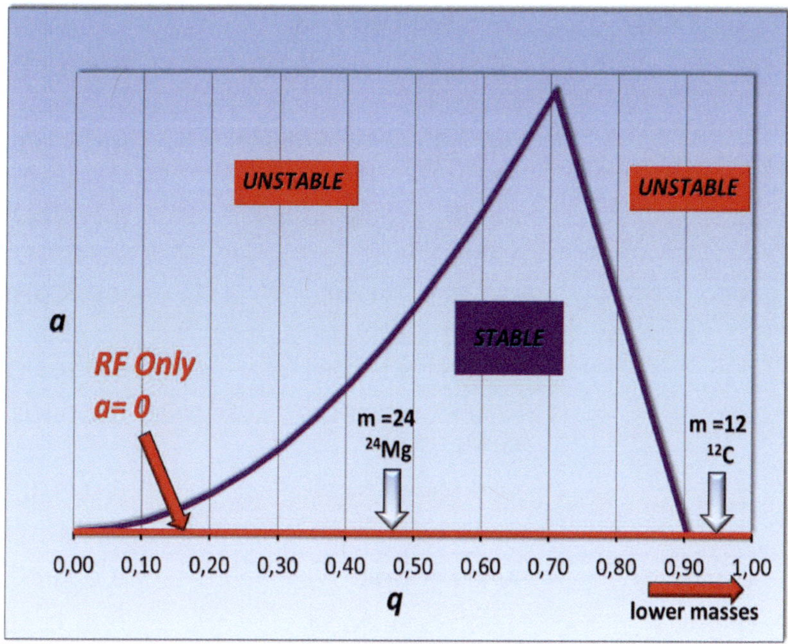

FIGURE 10.8   Basic principles of a low-mass cutoff collision/reaction cell.

(Courtesy of Thermo Scientific)

low masses where the majority of the interferences occur.[17] This stops many of the problematic argon-, matrix- and solvent-based ions from entering the collision/reaction cell, therefore reducing the likelihood of creating new precursor ions in the cell, which have the potential to negatively impact the determination of the analytes. The basic principles of this technique are shown in Figure 10.8, which shows a typical Mathieu stability plot for a quadrupole. It can be seen with a fixed DC electrical field (a) of zero (red horizontal line) and an RF electrical field scan (q) of 0–0.9 (purple triangle); all masses above the red line and inside the purple triangle are stable and will pass through the quadrupole rods, while all masses outside the triangle will be unstable and will be ejected out of the electrical field. So, for example, a q-value of 0.47 might be equivalent to 24 amu, which would mean ^{24}Mg is stable and allowed to be transmitted, while a q-value of 0.95 will be equivalent to mass 12, which would mean ^{12}C is unstable and be rejected. And if ^{12}C is not transmitted, the resulting $^{12}C^{12}C$ polyatomic ion at mass 24 would not be present to interfere with ^{24}Mg.

The benefit of this approach is seen in Figure 10.9, which shows a list of analytes in the far left-hand column, and the suggested cutoff mass in the next column. By using this cutoff mass, the potential polyatomic interferents in the third column won't be created, because the majority of the precursor ions shown in the final column will be unstable and won't be present to react and combine with the other matrix- or solvent-based ions in the cell. For some elements, this can translate into a four- to five-fold improvement in detection capability over a traditional collision cell, particularly elements such as $^{31}P^+$ and $^{34}S^+$, which are prone to interferences from the $^{16}O^+$ Ion.

The major difference of this technology is that it doesn't use a traditional quadrupole with spherical or elliptical-shaped rods. It uses a flatapole, which has beveled, or straight edges, as opposed to round edges. The benefit is that the transmission efficiency tends to be similar to higher-order multipoles, thus allowing more ions through. The difference between the fields of a flatapole and quadrupole are shown in Figure 10.10.

Analyte	Cut-Off Mass	Potential Interferent	Precursors
^{45}Sc	29	$^{13}C^{16}O_2$, $^{12}C^{16}O_2H$, ^{44}CaH, $^{32}S^{12}CH$, $^{32}S^{13}C$, $^{33}S^{12}C$	H, C, O, **S, Ca**
^{47}Ti	32	$^{31}P^{16}O$, ^{46}CaH, $^{35}Cl^{12}C$, $^{32}S^{14}NH$, $^{33}S^{14}N$	H, C, N, O, P, **S, Cl, Ca**
^{49}Ti	33	$^{31}P^{18}O$, ^{48}CaH, $^{35}Cl^{14}N$, $^{37}Cl^{12}C$, $^{32}S^{16}OH$, $^{33}S^{16}O$	H, C, N, O, P, S, **Cl, Ca**
^{50}Ti	34	$^{34}S^{16}O$, $^{32}S^{18}O$, $^{35}Cl^{14}NH$, $^{37}Cl^{12}CH$	H, C, N, O, S, **Cl**
^{51}V	35	$^{35}Cl^{16}O$, $^{37}Cl^{14}N$, $^{34}S^{16}OH$	H, O, N, S, **Cl**
^{52}Cr	36	$^{36}Ar^{16}O$, $^{40}Ar^{12}C$, $^{35}Cl^{16}OH$, $^{37}Cl^{14}NH$, $^{34}S^{18}O$	H, C, O, N, S, **Cl, Ar**
^{55}Mn	39	$^{37}Cl^{18}O$, $^{23}Na^{32}S$, $^{23}Na^{31}PH$	H, O, Na, P, S, Cl, **Ar**
^{56}Fe	39	$^{40}Ar^{16}O$, $^{40}Ca^{16}O$	O, **Ar, Ca**
^{57}Fe	40	$^{40}Ar^{16}OH$, $^{40}Ca^{16}OH$	H, O, Ar, Ca
^{58}Ni	41	$^{40}Ar^{18}O$, $^{40}Ca^{18}O$, $^{23}Na^{35}Cl$	O, Na, Cl, Ar, Ca
^{59}Co	42	$^{40}Ar^{18}OH$, $^{43}Ca^{16}O$, $^{23}Na^{35}ClH$	H, O, Na, Cl, Ar, Ca
^{60}Ni	43	$^{44}Ca^{16}O$, $^{23}Na^{37}Cl$	O, Na, Cl, Ca
^{61}Ni	44	$^{44}Ca^{16}OH$, $^{38}Ar^{23}Na$, $^{23}Na^{37}ClH$	H, O, Na, Cl, Ca
^{63}Cu	45	$^{40}Ar^{23}Na$, $^{12}C^{15}O^{35}Cl$, $^{12}C^{14}N^{37}Cl$, $^{31}P^{32}S$, $^{31}P^{16}O_2$	C, N, O, Na, P, S, Cl
^{64}Zn	46	$^{32}S^{16}O_2$, $^{32}S_2$, $^{36}Ar^{12}C^{16}O$, $^{38}Ar^{12}C^{14}N$, $^{48}Ca^{16}O$	C, N, O, S, Ar, Ca
^{65}Cu	47	$^{32}S^{16}O^2H$, $^{32}S^2H$, $^{14}N^{15}O^{35}Cl$, $^{48}Ca^{16}OH$	H, N, O, S, Cl, Ca
^{66}Zn	47	$^{34}S^{16}O$, $^{32}S^{34}S$, ^{33}S, ^{48}C, ^{18}O	O, C, S
^{67}Zn	47	$^{32}S^{34}SH$, $^{33}S_2H$, $^{48}Ca^{18}OH$, $^{14}N^{16}O^{37}Cl$, $^{35}Cl^{16}O_2$	H, N, O, S, Cl, Ca
^{68}Zn	47	$^{32}S^{18}O_2$, $^{34}S_2$	O, S
^{69}Ga	47	$^{32}S^{18}O_2H$, $^{34}S_2H$, $^{37}Cl^{16}O_2$	H, O, S, Cl
^{70}Zn	47	$^{34}S^{18}O_2$, $^{35}Cl_2$	O, S, Cl
^{75}As	47	$^{40}Ar^{34}SH$, $^{40}Ar^{35}Cl$, $^{40}Ca^{35}Cl$, $^{37}Cl_2H$	H, S, Cl, Ca, Ar
^{77}Se	47	$^{40}Ar^{37}Cl$, $^{40}Ca^{37}Cl$	Cl, Ca, Ar
^{78}Se	47	$^{40}Ar^{38}Ar$	Ar
^{80}Se	47	$^{40}Ar_2$, $^{40}Ca_2$, $^{40}Ar^{40}Ca$, $^{32}S_2\,^{16}O$, $^{32}S^{16}O_3$	O, S, Ar, Ca

FIGURE 10.9 A list of elements with potential interferences that would benefit from the low-mass cutoff approach.

(Note: Elements in red are above the low-mass cutoff and could still contribute to a polyatomic interference)[19]

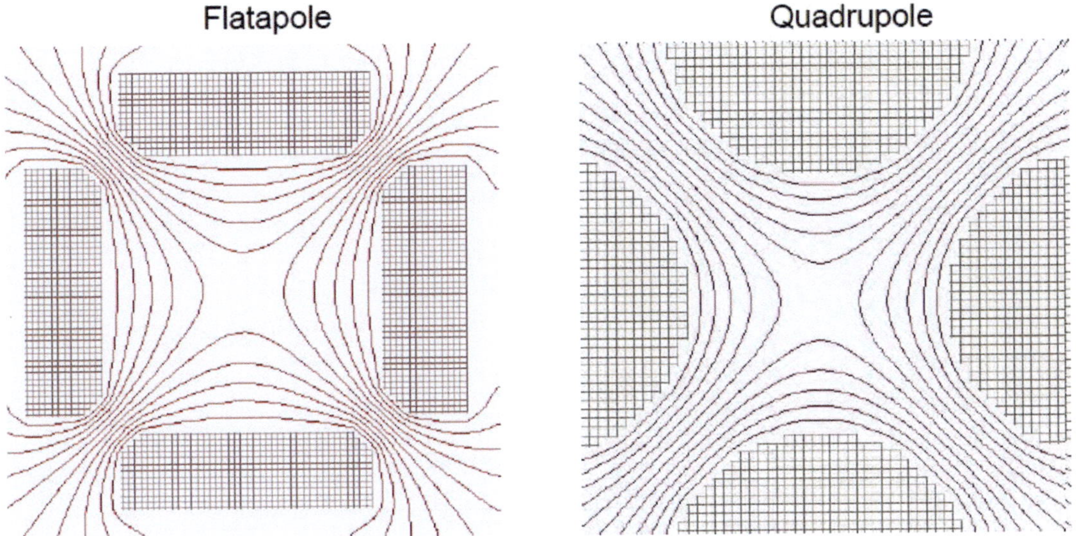

FIGURE 10.10   The difference between the electric fields of a flatapole compared to a quadrupole.[19]

10.7   USING REACTION MECHANISMS IN A COLLISION CELL

After almost 20 years of solving real-world-application problems, the practical capabilities of both hexapole and octapole collision cells using KED are fairly well understood. It is clear that the majority of applications are being driven by the demand for routine multielement analysis of well-characterized matrices, where a rapid sample turnaround is required. This technique has also been promoted by the vendors as a fast, semi-quantitative tool for unknown samples. The fact that it requires very little method development, just one collision gas and one set of tuning conditions, makes it very attractive for these kinds of applications.[18]

However, it is well recognized that this approach will not work well for many of the more complex interfering species, especially if the analyte is at ultra-trace levels. For example, the collision mode using helium is not the best choice for quantifying selenium using its two major isotopes ($^{80}Se^+$, $^{78}Se^+$) because it requires a large number of collisions to separate the argon dimers ($^{40}Ar_2^+$ and $^{40}Ar^{38}Ar^+$) from the analyte ions using kinetic energy discrimination. As a result of the inefficient interference-reduction process, the signal intensity for the analyte ion is also suppressed. For this reason, it is well accepted that the best approach is to use a reaction gas in order to initiate some kind of charge-transfer mechanism. This can be seen in Figure 10.11, which shows the calibration for $^{78}Se^+$ using hydrogen as the reaction gas to minimize the impact of the argon dimer, $^{40}Ar^{38}Ar^+$. On the left is the 0, 1.0, 2.5, 5.0 and 10.0 ppb calibration using helium and on the right is the same calibration using hydrogen as the reaction gas. It can be seen very clearly that the signal intensity for the calibration standards using hydrogen gas is approximately five times higher than the calibration standards using helium, producing a selenium detection limit of 15 times lower (1 ppt compared to 15 ppt); this is also seen in the table following the calibration graphs.[19]

Likewise, helium has very little effect on reducing the $^{40}Ar^+$ interference in the determination of $^{40}Ca^+$. So, when using a collision cell with helium, the quantification of Ca must be carried out using the $^{44}Ca^+$ isotope, which is about 50 times less sensitive than $^{40}Ca^+$. For this reason, in order to achieve the lowest detection limits for calcium, a low-reactivity gas, such as pure hydrogen, is the better option.[22] By initiating an ion–molecule reaction, it allows the most sensitive calcium

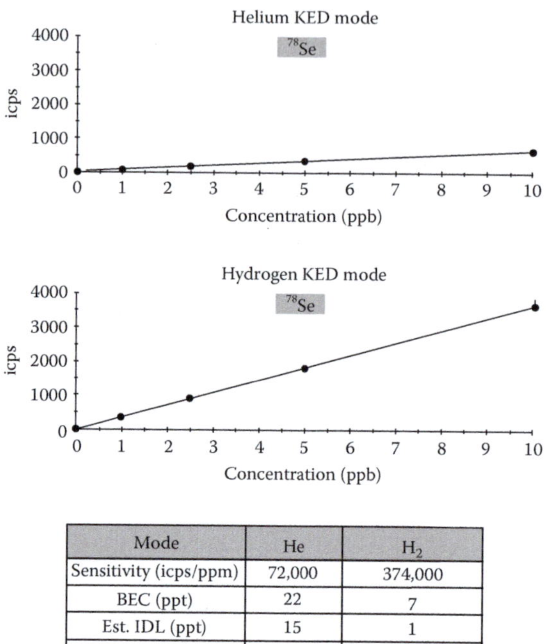

Mode	He	H$_2$
Sensitivity (icps/ppm)	72,000	374,000
BEC (ppt)	22	7
Est. IDL (ppt)	15	1
RSD % at 1 ppb level	8	1

FIGURE 10.11   Comparison of $^{78}Se^+$ calibration plots and detection limits using the collision mode with helium (left) and reaction mode with hydrogen (right).[19]

isotope at 40 amu to be used for quantitation. In fact, even though the use of hydrogen significantly improves the detection limit for calcium, the best interference reduction is achieved using a mixture of ammonia and helium.

Another example of the benefits of using more reactive gases, such as an ammonia-and-helium mixture over pure helium or hydrogen gas, is in the determination of vanadium in a high-concentration chloride matrix. The collision mode using helium works reasonably well on the reduction of the $^{35}Cl^{16}O^+$ interference at 51 amu. However, when 1% NH_3 in helium is used, the interference is dramatically reduced by the process of charge/electron transfer. This allows the most abundant isotope, $^{51}V^+$, to be used for the quantitation of vanadium in matrices such as seawater or hydrochloric acid. Vanadium-detection capability in a chloride matrix is improved by a factor of 50–100 times using the reaction chemistry of NH_3 in helium compared to pure helium in the collision mode.[19]

In addition to ammonia–helium mixtures, oxygen is sometimes the best reaction gas to use, because it offers the possibility of either moving the analyte ion to a region of the mass spectrum where the interfering ion does not pose a problem or moving the interfering species away from the analyte ion by forming an oxygen-derived polyatomic ion 16 amu higher. An example of changing the mass of the interfering ion is in the determination of $^{114}Cd^+$ in the presence of high concentrations of molybdenum. In the plasma, the molybdenum forms a very stable oxide species $^{98}Mo^{16}O^+$ at 114 amu, which interferes with the major isotope of cadmium, also at mass 114. By using pure oxygen as the reaction gas, the $^{98}Mo^{16}O^+$ interference is converted to the $^{98}Mo^{16}O^{16}O^+$ complex at 16 amu higher than the analyte ion, allowing the $^{114}Cd^+$ isotope to be used for quantitation.[19]

The benefits of using reaction mechanisms for the determination of calcium, vanadium and cadmium are seen in Figure 10.12, which shows (a) a 0–300-ppt calibration plot for $^{40}Ca^+$ using hydrogen gas, (b) a 0–500-ppt calibration plot for $^{51}V^+$ in 2% hydrochloric acid using 1% NH_3 in helium and (c) a 0–10-ppb calibration plot for $^{114}Cd^+$ in a molybdenum matrix using pure oxygen. Background equivalent concentration (BEC) values of 4 ppt, 2 ppt and 3 ppt were achieved for calcium, vanadium and cadmium, respectively.[19]

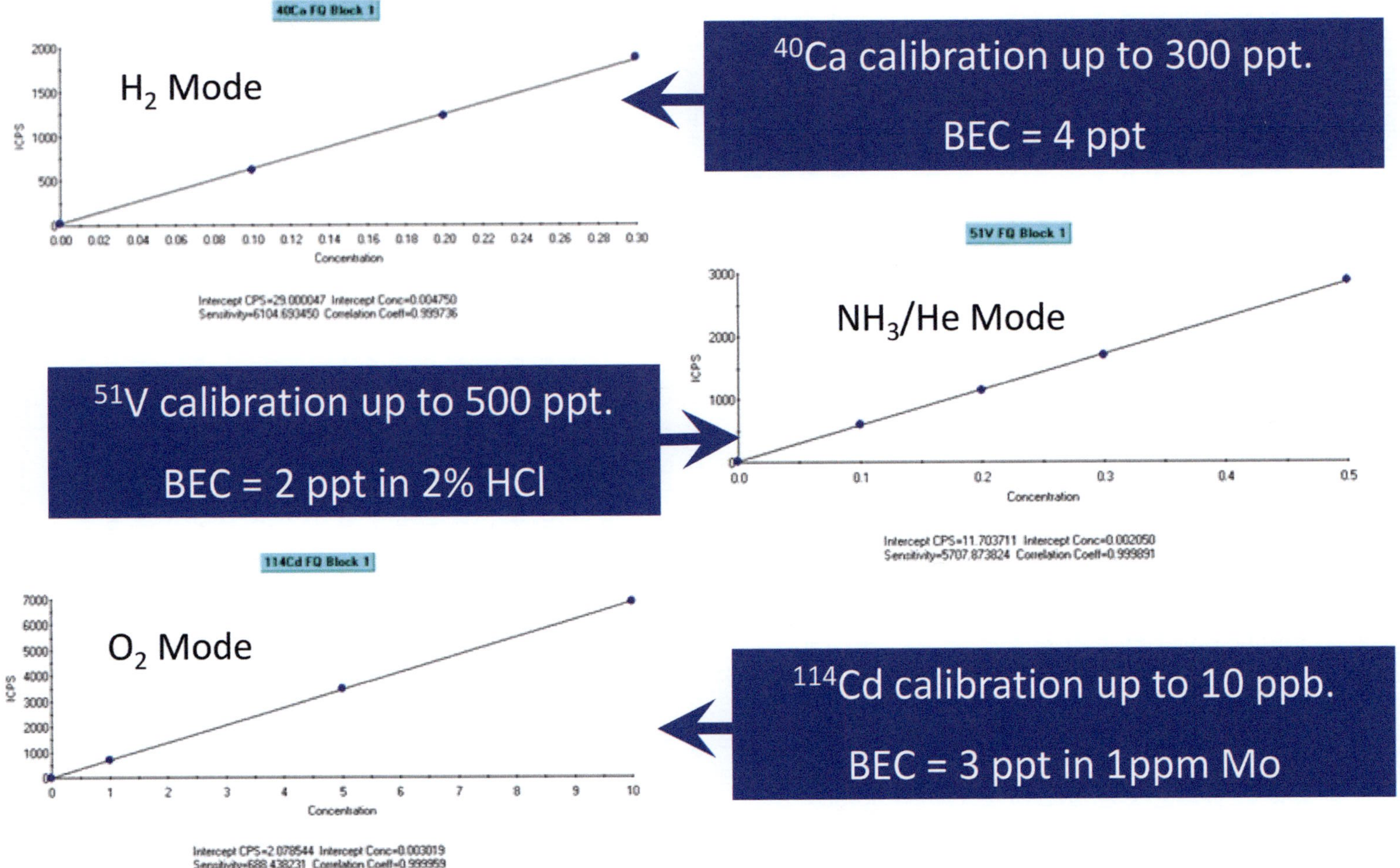

FIGURE 10.12   Calibration plots for (1) $^{40}Ca^+$ using hydrogen, (2) $^{51}V^+$ in 2% HCl using 1% NH_3 in helium and (3) $^{114}Cd^+$ in high concentrations of molybdenum using oxygen.[19]

However, it is important to point out that even though a higher-order multipole collision cell with KED can use low-reactivity gases, it usually requires significantly more interactions than a reaction cell that uses a highly reactive gas.[23] Take, for example, the reduction of the $^{40}Ar^+$ interferences in the determination of $^{40}Ca^+$. In a collision cell, even though the kinetic energy of the $^{40}Ar^+$ ion will be reduced by reactive collisions with molecules of hydrogen gas, the $^{40}Ca^+$ will also lose kinetic energy, because it, too, will collide with the reaction gas. Now, even though these interactions are basically nonreactive with respect to $^{40}Ca^+$, it will experience the same number of collisions because it has a similar cross-sectional area as the $^{40}Ar^+$ ion. So, in order to achieve many orders of interference reduction with a low-reactivity gas such as hydrogen, a high number of collisions are required. This means that in addition to the interference being suppressed, the analyte will also be affected to a similar extent. As a result, the energy distribution of both the interfering species and the analyte ion at the cell exit will be very close, if not overlapping, resulting in compromised interference-reduction capabilities compared to a reaction cell that uses highly reactive gases and discrimination by mass filtering. This compromise translates into a detection limit for $^{40}Ca^+$ with a KED-based collision cell using hydrogen being approximately 50–100 times worse than a dynamic reaction cell using pure ammonia.[20, 21]

10.8 THE UNIVERSAL CELL

A recent development is the commercialization of an instrument that can be used in both the collision mode and reaction mode. More commonly known as the Universal Cell, this innovative design can be utilized as both a simple collision cell using KED and a dynamic reaction using a pure reaction gas.[22] When operating in the collision-cell mode, the universal cell works on the principle that the interfering ion is physically larger than the analyte ion. If both ions are allowed to pass through the cell containing the inert gas molecules, the interfering ion will collide more frequently with the inert gas atoms than the analyte ion due to its larger size. This results in a greater loss of kinetic energy by the interferent compared to the analyte ion. An energy barrier is then placed at the exit of the cell, so that the higher-energy analyte ions are allowed to pass through it, while the lower-energy interferences are not. As mentioned earlier in this chapter, this process is commonly referred to as kinetic energy discrimination (KED).

When the universal cell is operating in the reaction mode, it takes advantage of whether the interaction of the analyte and interfering species with the reaction gas is either exothermic (fast) or endothermic (slow). The reactivity will be based on the relative ionization potential of the analyte and interfering species compared to the reaction gas. Typically, interfering ions tend to react exothermally with a reactive gas like ammonia, while analyte ions react endothermically. This means if both interferent ions and analyte ions enter the cell, the interferent ions will be reacting with the reactive gas and will be converted to a new species with a different mass, while the analyte ions will be unaffected and pass through the reaction cell into the filtering quadrupole, free of the interference.

But it should be emphasized that while a reactive gas efficiently removes interferences, it is also capable of creating new interferences if not properly controlled. For that reason, an optimized reaction cell (coupled with a single mass analyzer), requires the use of a scanning quadrupole to prevent these new interferences from forming through the creation of a unit-resolution bandpass filter, to allow only a single mass to pass through.

The real benefit of the "universal" collision/reaction-cell approach is that it can be used in both the collision-cell and the reaction-cell modes. This means the operator has the flexibility of operating the system in three different modes all in the same multielement method—the standard mode for elements where interferences are not present; in collision mode for removal of minor interference; and in dynamic-reaction mode for the most severe polyatomic spectral interferences.

Let us now go on to discuss a slightly different approach to reduce interferences using a collision/reaction cell (CRC)—the collision/reaction interface (CRI).

10.9   THE COLLISION/REACTION INTERFACE

Unlike collision and reaction cells, the collision/reaction interface does not use a pressurized multipole-based cell before the mass analyzer. Instead, it injects a reaction/collision gas (typically He, or H_2) at relatively high flow rates (100–150 mL/min) into the plasma through the aperture of the interface skimmer cone, where the plasma density is high.[23] This increases the rate of interactions between the introduced gas and interfering ions, giving improved attenuation of interfering ions. In addition, the reaction/collision gas is supplied directly to the plasma, which means that the plasma electrons are still available to assist in attenuating the interfering ions through electron–ion recombination. The presence of plasma electrons also significantly reduces the generation of secondary by-products ions produced from the interference-attenuation process. The overall result is that most argon-based polyatomic interferences are destroyed or removed before they are extracted into the ion optics. Figure 10.13 shows the basic principles of the CRI.

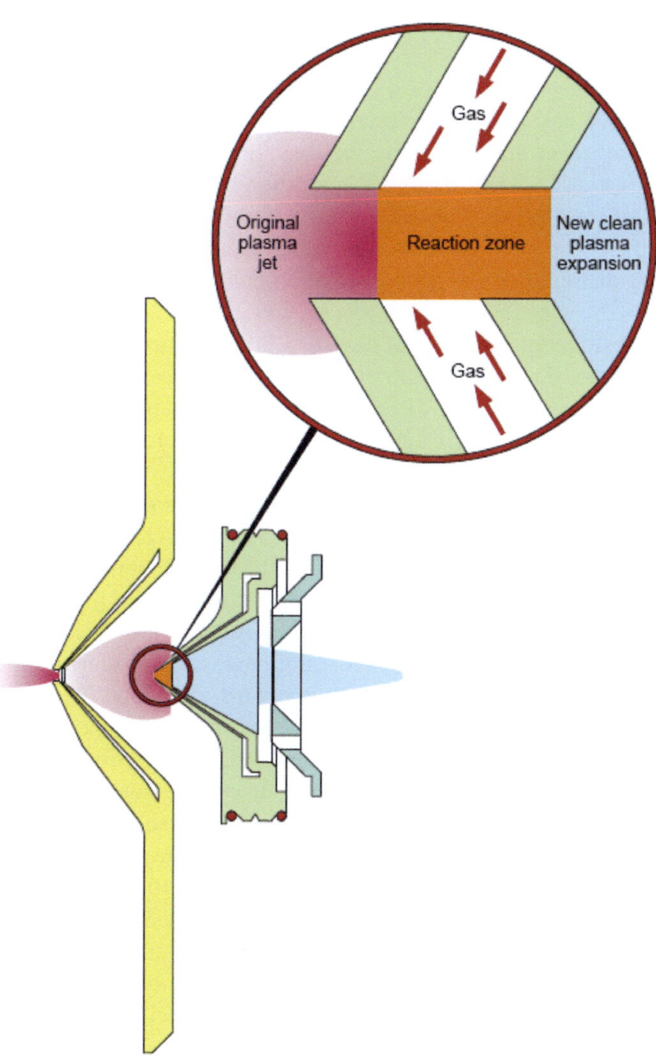

FIGURE 10.13   Principles of the collision/reaction interface (CRI).

(Courtesy of Analytic Jena)

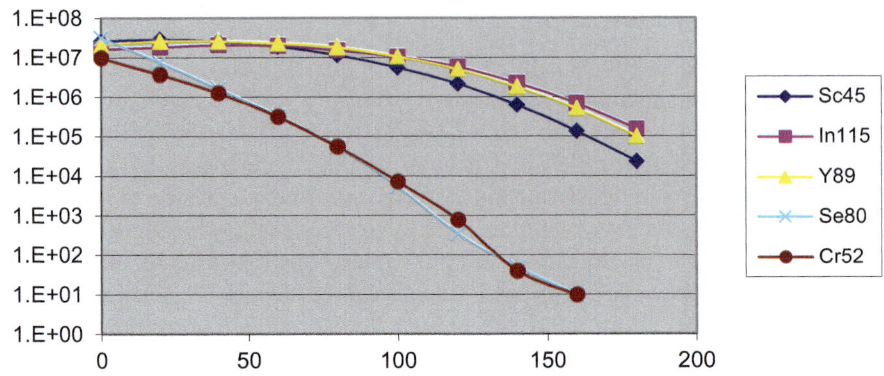

FIGURE 10.14 Optimization of hydrogen-gas flow rate (mLs/min), showing three internal standard (Sc, In, Y) signals while monitoring the interferences $^{40}Ar^{40}Ar$ (^{80}Se) and $^{40}Ar^{12}C$ (^{52}Cr).[24]

The limitations of the earlier designs restricted their use for real-world samples because there appeared to be no way to effectively focus the ions, and therefore there was very little control over the collision process. So, even though the addition of a collision/reaction gas helped reduce plasma-based spectral interferences, it did virtually nothing for matrix-induced spectral interferences. In addition, there appeared to be no way to carry out KED in the interface region and, as a result, it made it very difficult to take advantage of collisional mechanisms using an inert gas such as helium.

However, in the most recent commercial design, all the reaction/collision processes are actually taking place inside the tip of the skimmer and not between the sampler and skimmer cone, as with earlier designs. Because of this subtle difference, simple, loosely bonded polyatomic species can receive sufficient energy through collisional (vibrational and rotational) excitation mechanisms to bring about the dissociation of the interference, whereas the analyte ions simply lose energy as they collide with the gas molecules. And where the collisional impact is not suitable for interference reduction, as in the removal of the argon dimer ($^{40}Ar_2^+$) in the determination of $^{80}Se^+$, or the elimination of the $^{40}Ar^{12}C^+$ interference in the determination of $^{52}Cr^+$, a low-reactivity gas such as hydrogen can be used to initiate an ion–molecule reaction.

This can be seen in Figure 10.14, which shows the reduction of the $^{40}Ar_2^+$ and $^{40}Ar^{12}C^+$ interfering ions using hydrogen as the collision/reaction gas, in the determination of ^{80}Se and ^{52}Cr respectively. The sensitivies of three internal standard elements, ^{45}Sc, ^{89}Y, ^{115}In, were monitored at the same time as monitoring the two interfering ions. It should be noted that there is no Se or Cr in this solution, so the signals at mass 80 and 52 amu are contributions from the $^{40}Ar^{40}Ar$ and $^{40}Ar^{12}C$ polyatomic ions respectively. It can be seen very clearly that there is a sharper decrease of the interferent signals as compared to those of the internal standards, showing evidence of the removal of the ArAr$^+$ and ArC$^+$ polyatomic interferences with increasing H_2 flow rate. The optimization plot shows that at a flow rate of 140 mL/min, signals of the interfering ions decrease by six orders of magnitude while that of the Sc, Y and In are only reduced by two orders of magnitude.[24]

The CRI looks to be a very interesting concept, which appears to offer a relatively straightforward, non-cell-based solution to minimizing plasma- and matrix-based spectral interferences in ICP-MS. Each year more and more challenging applications appear in the public domain, showing the capabilities of CRI systems. If there is an interest in this approach, it is worth checking out the vendor application notes, which show the capabilities of the CRI in many different, real-world matrices.[21]

10.10 DETECTION-LIMIT COMPARISON OF SINGLE-QUADRUPOLE CRC SYSTEMS

In general, highly reactive gases are recognized as being more efficient at reducing the interference and generating better signal-to-noise ratio than either an inert gas such as helium, or low-reactivity gases such as hydrogen or mixtures of ammonia and helium. However, it's very difficult to make a detection-limit comparison with this technology because detection capability tends to be application-specific. In other words, depending on the interference-reduction capability of the collision/reaction cell interface/device, it might offer better detection capability than another instrument in one particular sample matrix, but offer inferior detection limits in another. In addition, vendors' instrument detection limits are typically carried out in simple aqueous solutions, which is unlikely to show a significant difference in the performance of the different collision/reaction cell/interface technology. The only way the interference-reduction capability of a particular device can be truly evaluated is in a real sample containing matrix and solvent components. For that reason, please refer to the previous cited references, which are a selection of technical/data sheets and application literature showing performance characteristics of the different collision/reaction cell/interface approaches.

10.11 TRIPLE-QUADRUPOLE SYSTEMS

It should be noted that, historically, single-multipole-based collision/reaction cells dominated the commercial landscape, but in the late 2000s, multiple-multipole technology was developed, which placed an additional quadrupole prior to the collision/reaction-cell multipole and the analyzer quadrupole. This first quadrupole acts as a simple mass filter to allow only the analyte masses to enter the cell, while rejecting all other masses. With all non-analyte, plasma and sample matrix ions excluded from the cell, sensitivity and interference removal efficiency is significantly improved compared to traditional collision/reaction cell technology coupled with a single-quadrupole mass analyzer.

This very exciting collision/reaction technology is known as a "triple-quadrupole" collision/reaction cell[18]—a name derived from the LC-MS-MS technique, where three quadrupoles are used to separate, detect and confirm the presence of organic molecules such as proteins and peptides in biological samples. However, the term is not technically correct in this configuration used for ICP-MS. Even though the first quadrupole (Q1) is a mass filter, and the second quadrupole (Q2) is the analyzer quadrupole, the middle multipole device is actually an octapole or a hexapole collision/reaction cell. This means that the cell is passive and cannot be used as a conventional

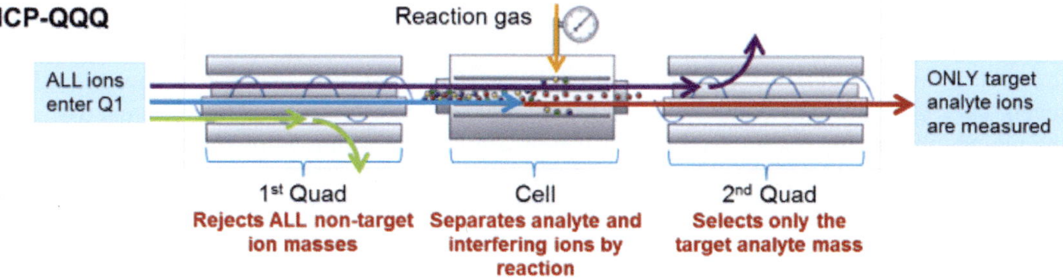

FIGURE 10.15 The fundamental principles of "triple-quadrupole" technology used in ICP-MS.

(Courtesy of Agilent Technologies)

bandpass filter, like a single-quadrupole-based dynamic-reaction cell or low-mass cutoff device. Reactive gases can be used to initiate ion–molecule chemistry in the octopole cell, but it is then used just to pass all the product ions formed into the analyzer quadrupole (Q2), which is used to separate and select the mass or masses of interest. The principles of this technology are shown in Figure 10.15.[25]

The capability and flexibility of the "triple-quad" collision/reaction cell has shown enormous potential. There are a number of different ways it can be utilized depending on the severity of the analytical problem. The two most common modes of analysis are:

- M/S Mode
- MS/MS Mode

10.12   M/S MODE

In its most basic configuration, the instrument can be utilized in the single M/S mode, where Q1 acts as a simple ion guide, allowing all ions through to the collision/reaction cell, similar to a traditional single-quad ICP-MS system that uses an octopole/hexapole-based collision cell. It can also be used in single M/S mode where Q1 acts as a bandpass filter, allowing a "window" of masses through, above/below the Q2 mass range selected by the user. In this mode, it functions in a similar way to a single quad, with a "scanning" bandpass filter-type collision/reaction cell, except masses outside the bandpass window are rejected before they can enter the cell.

10.13   MS/MS MODE

The instrument can also be used in the MS/MS mode, where the first quadrupole is operated with a 1-amu fixed bandpass window, allowing only the target ions to enter the collision/reaction cell. This process can be implemented in two different ways:

- On-Mass Mode
- Mass-Shift Mode

10.14   ON-MASS MS/MS MODE

In this configuration, Q1 and Q2 are both set to the target mass. Q1 allows only the precursor-ion mass to enter the cell (analyte and on-mass polyatomic interfering ions). The octopole/hexapole collision/reaction cell then separates the analyte ion from the interferences using the reaction chemistry of a reactive gas, while Q2 measures the analyte ion at the target mass after the on-mass interferences have been removed by reactions in the cell.

An example of this is in the removal of sulfur-based interferences using ammonia (NH_3) gas in the determination of vanadium in the presence of a sulfuric acid matrix. The major isotope of vanadium is $^{51}V^+$. However, In the presence of high concentrations of H_2SO_4, the interfering ions $^{33}S^{18}O^+$ and $^{34}S^{16}OH^+$ will overlap the $^{51}V^+$ ion. By using NH_3 as the reaction gas, which reacts very quickly with S-based polyatomic interferences, but is virtually unreactive with the vanadium ion, the $^{33}S^{18}O^+$ and $^{34}S^{16}OH^+$ ions are removed, thus allowing the $^{51}V^+$ ion to be detected, free of any interferences. So, in this example, Q1 and Q2 would be set at mass 51 to take advantage of the reactive properties of ammonia to effectively remove the SO^+/SOH^+ interferences. No new analyte- or matrix-based NH_3 cluster ions can be created at mass 51, as no other ions are able to enter the cell. A detection limit in the order of 13 ppt is achievable for the determination of vanadium in percentage levels of sulfuric acid as demonstrated by the calibration seen in Figure 10.16. This detection capability is similar to what is achievable in an aqueous solution.

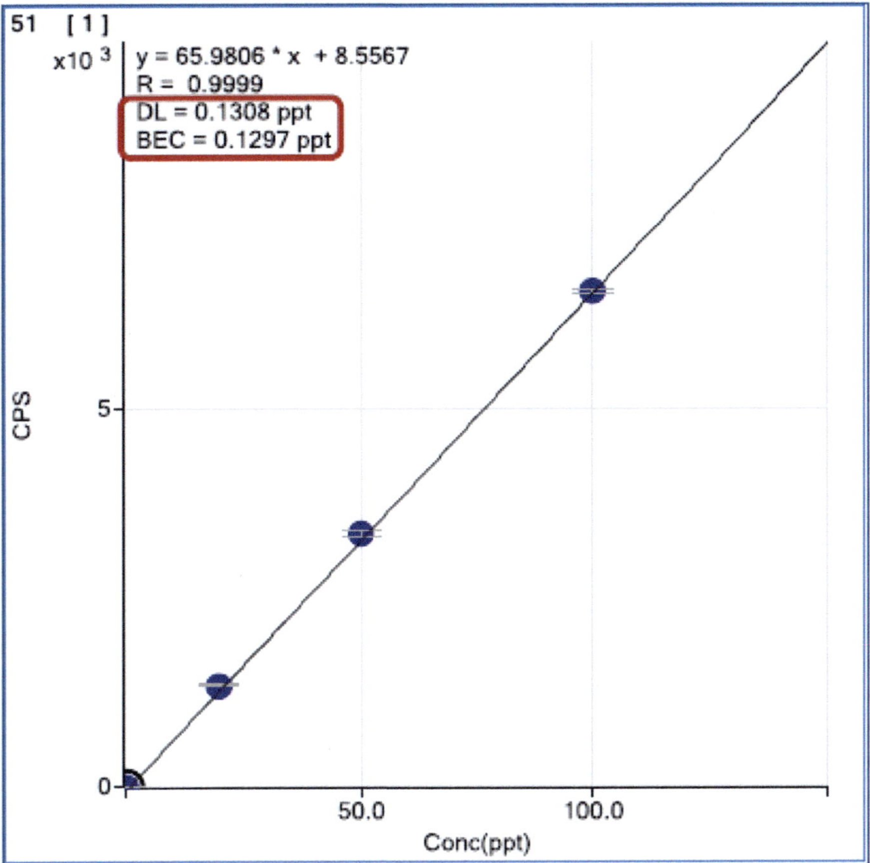

FIGURE 10.16   Detection capability of V in a sulfuric acid matrix using the on-mass MS/MS mode is in the order of 13 ppt as shown in this calibration plot.[25]

10.15   MASS-SHIFT MS/MS MODE

In this configuration, Q1 and Q2 are set to different masses. Similar to the "on-mass mode", Q1 is set to the precursor-ion mass (analyte and on-mass polyatomic interfering ions), controlling the ions that enter octapole collision/reaction cells. However, in this mode, Q2 is then set to the mass of a target reaction product ion containing the original analyte. Mass-shift mode is typically used when the analyte ion is reactive, while the interfering ions are unreactive with a particular collision/reaction-cell gas. The basic principles of this approach are shown in Figure 10.17.

An example of this is in the determination of arsenic in the presence of transition metal oxides. Because As is monoisotopic, it can only be determined at 75 amu, which is overlapped by the $^{59}Co^{16}O$ polyatomic interference. However, by reacting ^{75}As with oxygen gas in the cell, the $^{75}As^{16}O$ species at 91 amu can be used for quantitation. This is achieved by setting a band-pass to only allow ions of mass 75 amu to pass through Q1. All ions at 75 amu will then enter the octapole collision/reaction cell where they will interact with the oxygen-cell gas. The ^{75}As will react with O_2 to form $^{75}As^{16}O$ species at 91 amu, while the $^{59}Co^{16}O$ at mass 75 amu will be unreactive. Q2 is then set at 91 amu to only allow the $^{75}As^{16}O$ ion through to be detected and quantified.

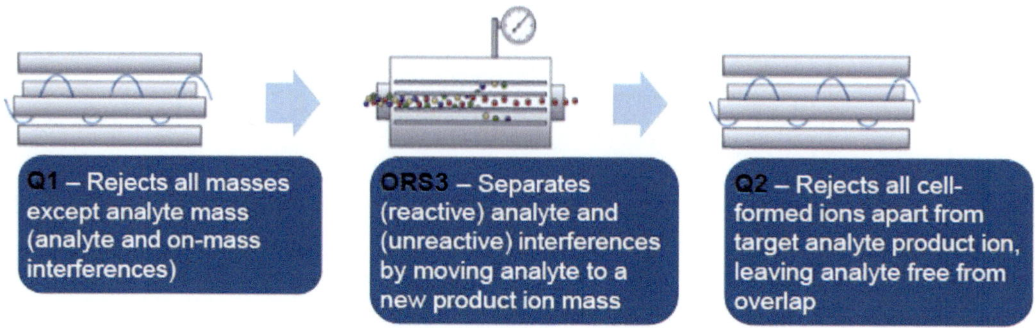

FIGURE 10.17   The basic principles of the mass-shift MS/MS mode.[25]

For more advanced applications, this technology offers other modes of interference reduction, including:

- **Precursor-Ion Scan,** where Q2 is set to the target ion mass, while Q1 scans over a user-defined mass range to select the precursor ions that enter the cell and react with the collision/reaction gas. An example of this is the monitoring of the by-product $^{14}NH_4$ ion in the determination of Hg, using NH_3 as a reaction gas for a multielement method. In this example, NH_3 is the optimum reaction gas for the majority of the other elements, but by measuring the $^{14}NH_4$ ion, Hg can also be determined in the same suite.
- **Product-Ion Scan,** where Q1 is set to allow only the target precursor-ion mass to enter the cell, while Q2 scans to measure all the product ions formed in the cell, including controlled cluster-ion analysis. An example of this is the use of NH_3 gas to create cluster ions of an analyte like titanium. By allowing only ^{48}Ti through Q1, only titanium cluster complex ions are formed in the cell, and not other potentially interfering transition metal cluster ions, as with a traditional collision/reaction cell.
- **Neutral-Gain Scan,** where Q1 and Q2 are scanned together, with a user-defined mass difference. This mode allows monitoring of the product ions from a particular transition for all ions in the Q1 scan range. An example of this is in the determination of titanium using oxygen as the collision/reaction gas. By only scanning Q1 at the masses for titanium (^{46}Ti, ^{47}Ti, ^{48}Ti, ^{49}Ti, ^{50}Ti) and Q2 at 16 amu or higher ($^{46}Ti^{16}O$, $^{47}Ti^{16}O$, $^{48}Ti^{16}O$, $^{49}Ti^{16}O$, $^{50}Ti^{16}O$), the isotopic abundances of the titanium oxide ions can be unambiguously measured, without the presence of other transition metals in the mass spectrum.

There is no question that the potential of this "triple-quad" approach to reducing interferences using a collision/reaction-cell analysis is a truly very exciting addition. However, it is probably more suited for research-type applications, or in an academic environment, where non-routine investigations are being carried out. In my opinion, it will be competing with the double-focusing, magnetic-sector technology as a problem-solving tool and for the analysis of more complex sample matrices. The traditional, single-quadrupole ICP-MS instrumentation will still represent the vast majority of instruments sold into the marketplace, for carrying out high-throughput, routine applications. Although, for companies that are purchasing a second instrument, a triple quad probably represents a good investment. These two references highlight more details about the analytical capabilities of commercially available triple-quad ICP-MS systems.[25, 26]

10.16 MULTI-QUAD SYSTEMS

In a traditional triple-quadrupole design, the collision/reaction cell (Q2) is passive and is placed between the first transmission analyzer quadrupole (Q1) and the final quadrupole (Q3). This approach, popular in organic mass spectrometry where parent ions are subjected to collision-induced dissociation in the cell, was commercially adopted for ICP-MS in 2012. This design allows the analyst to select only the nominal mass of interest in the first analyzer quadrupole, thereby eliminating other masses from the ion beam so that only the analyte and interfering ions remain. The interference is then removed through the use of reaction chemistry, and sometimes KED, in the reaction/collision (Q2) cell, which allows the analyte to be separated from the interference in the final analyzer quadrupole. One limitation of this approach is that the cell is passive and, as such, offers no control over any potential side reactions between the reaction by-products and the gas itself or impurities in the gas.

In a multi-quadrupole design, the passive cell of traditional triple-quadrupole designs is replaced with an active cell, which possesses mass-filtering capabilities. Such cells can operate in either passive (triple-quad) or active (multi-quad) modes, and offer the ability to reject the reaction by-products before they can potentially generate any new interferences. Figure 10.18 shows the layout configuration of the four quadrupoles.[27]

This design allows the ion beam to be cleaned up in the first quadrupole (Q0), followed by mass selection in the first transmission analyzer quadrupole (Q1), thereby removing all interferences that are not at the desired nominal mass. After this, the analyte and any interferences at the same nominal mass are introduced into the active quadrupole cell (Q2) where the dynamic bandpass mass-tuning capability of the cell effectively eliminates reaction by-products before they have a chance to form new interferences. Thereafter, the interfering and analyte ions are efficiently separated in the final transmission analyzer quadrupole (Q3) in a manner that depends upon the mode of operation (MS/MS or mass shift). In this way, it not only controls the ions that enter the cell, but also controls the reactions, ensuring that by-product ions are not formed.

The fundamental differences between a conventional triple-quad system and multi-quadrupole technology can be exemplified by highlighting the determination of chromium in an organic solvent, which is particularly difficult using a passive reaction/collision cell.[28]

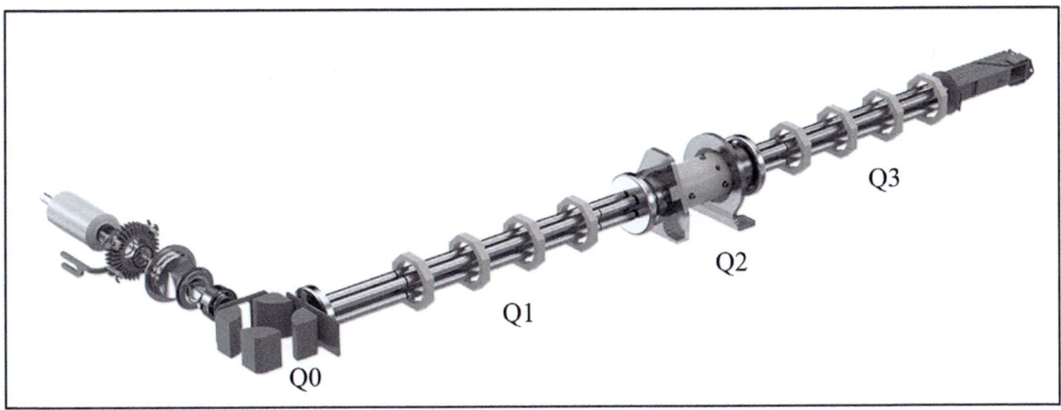

FIGURE 10.18 Layout of the four quadrupoles in the multi-quadrupole ICP-MS technology. (Courtesy of PerkinElmer Inc.)[27]

10.17 DIFFERENCE BETWEEN A TRIPLE AND MULTI QUAD SYSTEM

Detection limits for chromium tend to be poor in organic matrices due to the $^{40}Ar^{12}C$ polyatomic interference, which forms as a result of the plasma gas and organic solvent. Figure 10.19 shows a product-ion scan of 1% methanol (MeOH), where Q1 was set to mass 52 and Q3 was set to scan the mass region from 4–100 amu with an NH_3 gas flow of 1 mL/min. Pure ammonia was used for this reaction, since it is known to produce highly efficient and reproducible reactions. In order to duplicate a passive collision cell (triple-quad mode), an RPq value of 0.05 was used. Four peaks can be clearly observed at masses 18, 19, 35 and 52 (m/z 52 ca. 4,500 cps), which are due to $^{14}NH_4{}^+$, $^{15}NH_4{}^+$, $^{14}NH_4{}^{14}NH_3{}^+$ and $^{14}NH_4({}^{14}NH_3)_2{}^+$ ions respectively. The presence of these ions was confirmed by peaks at mass 18 and 19 whose abundance has the same relationship as the two isotopes of nitrogen (shown in Table 10.1) as was determined using the formula shown in the following equation.

$$\text{Calculated abundance of } {}^{15}N = ({}^{15}N \text{ cps})/({}^{14}N \text{ cps} + {}^{15}N \text{ cps})$$

Also contributing to the peak at mass 52 are Cr impurities in the MeOH, as a result of the MeOH HPLC gradient grade being used ($\geq$ 99.8% HiPerSolv CHROMANORM®, VWR Chemicals BDH).

As is evidenced in Figure 10.20, it can be clearly seen that additional species have formed from the NH_3 in the cell, even though only mass 52 was allowed through Q1 and the $^{40}Ar^{12}C$ interference removed via the reaction mode. However, once dynamic bandpass mass tuning was applied by increasing the RPq (shown in Figure 10.20), it can be clearly seen that all peaks which had been generated from the reaction gas are completely eliminated, leaving behind a clean and significantly

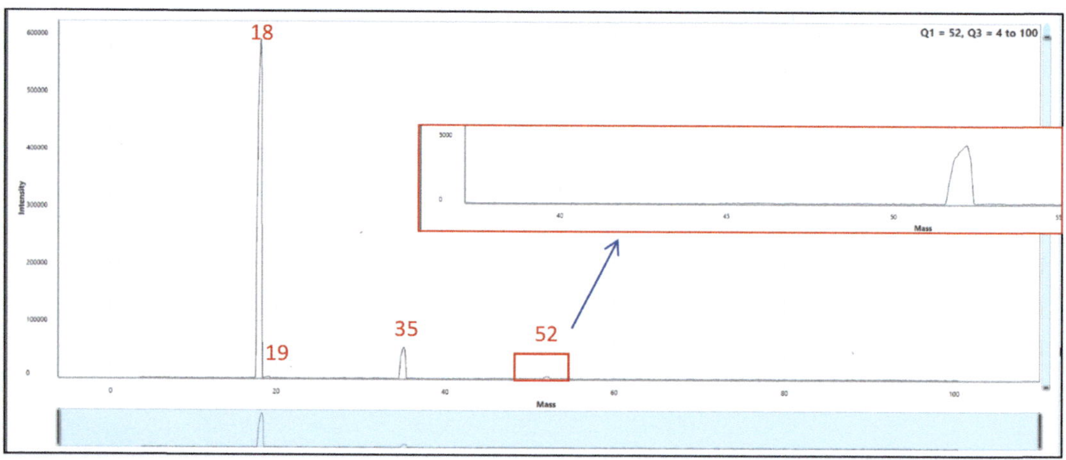

FIGURE 10.19 Product-ion scan with Q1 set to mass 52, Q3 set to scan masses 4–100 in "triple-quad mode" and NH_3 gas flow of 1 mL/min.[28]

TABLE 10.1

Calculated Relative Abundance of Peaks at Mass 18 and 19 Equating to $^{14}N^1H_4{}^+$, $^{15}N^1H_4{}^+$ [28]

Isotope	cps	Theoretical relative abundance	Calculated relative abundance
^{14}N	591019	99.6337%	99.6174%
^{15}N	2270	0.3663%	0.3826%

smaller peak (ca. 500 cps) at mass 52, which equates to approximately 5 ppt Cr impurity, which could be due to the fact that MeOH is traditionally stored in stainless steel containers.

By plotting the background equivalent concentrations (BECs) against the RPq, the benefits of an active cell are apparent, as exemplified in Figure 20.21. Operating in "triple-quad mode" at an RPq of 0.05, the BEC was around 966.7 ppt. By using the power of dynamic bandpass mass tuning, a significant lowering of the BEC down to 5 ppt can be clearly observed.

The previous example highlights why, despite the effective way that triple-quad designs deal with interferences, it is challenging for these systems to reach sub-1 ppt BECs for some elements using hot plasma conditions. However, with the use of multi-quadrupole technology, BECs of < 1 ppt in a hot plasma can easily be achieved.

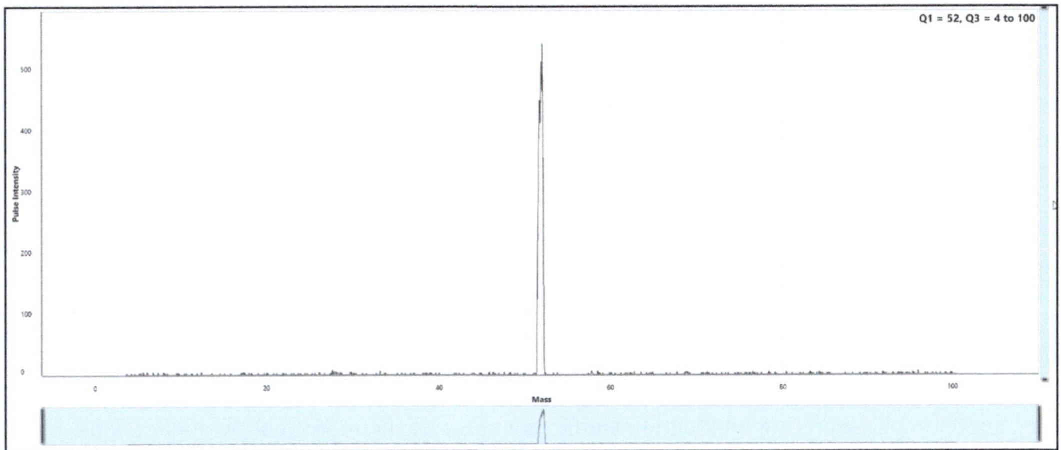

FIGURE 10.20   Product-ion scan with Q1 set to mass 52, Q3 set to scan masses 4–100 with an RPq of 0.7 "multi-quad mode" and an NH_3 gas flow of 1 mL/min.[28]

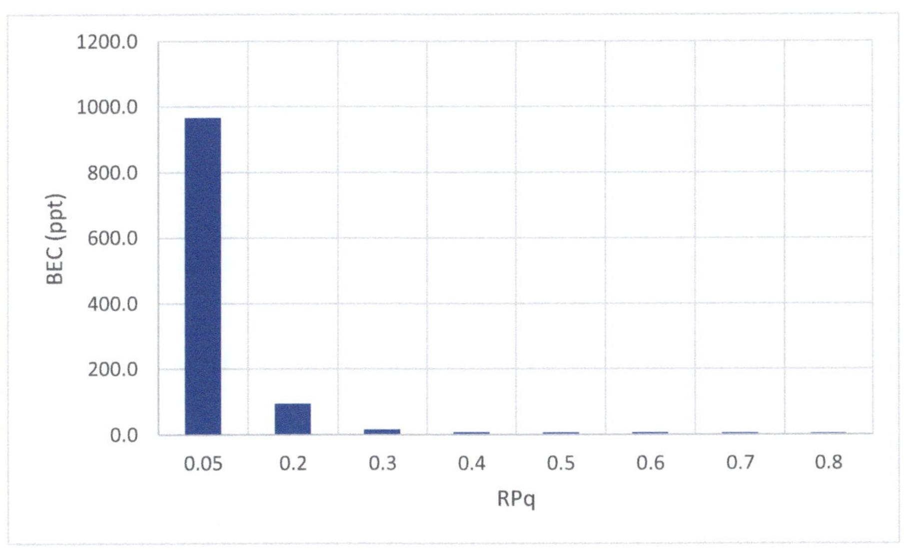

FIGURE 10.21   Lowering of BEC (ppt) for Cr with increasing RPq.[28]

10.18　FINAL THOUGHTS

There is no question that collision/reaction cells and interfaces have given a new lease on life to quadrupole mass analyzers used in ICP-MS. They have enhanced its performance and flexibility, and most definitely opened up the technique to more demanding applications that were previously beyond its capabilities. This is most definitely the case with the new triple-quad system, which will be competing with the magnetic-sector, high-resolution systems for the more difficult, research-type applications. It should also be noted that when I wrote my ICP-MS textbook in 2014, there was only one vendor of triple-quad ICP-MS instrumentation. Today there is a second vendor, with a near certainty that other vendors will be offering the technology in the near future.[27]

However, it must be emphasized that when assessing collision/reaction-cell technology, it is critical that you fully understand the capabilities of the different approaches, especially how they match up to your application objectives. The KED-based collision cell using an inert gas such as helium is probably better suited to doing multielement analysis in a routine environment. However, you have to be aware that its detection capability is compromised for some elements depending on the type of samples being analyzed. By using ion–molecule reactions as opposed to collisions, detection limits for many of these elements can be improved quite significantly. Of course, if two or even three different gases have to be used, the convenience of using one gas goes away.

On the other hand, using highly reactive gases with discrimination by mass filtering appears to offer the best performance and the most flexibility of all the different commercial approaches. By careful matching of the reaction gas with the analyte ion and polyatomic interference, extremely low detection limits can be achieved by ICP-MS, even for many of the notoriously difficult elements. It should be emphasized that selection of the optimum reaction gas and selection of the best quadrupole bandpass parameters can sometimes translate into quite lengthy method development, especially if there is very little application data available. However, most vendors do a very good job of generating application studies for some of the more routine applications. If you analyze out-of-the-ordinary or complex samples, you might initially need to spend the time to develop an analytical method that is both robust and routine. If you are uncertain, which is the best approach for your application, you should consider investing in a system that offers both collision-cell and reaction-cell capability in the same instrument. The simple collision-cell approach using helium could be used for your routine samples, while the more powerful and dynamic reaction cell could be utilized for the more difficult sample matrices. Even if you don't need both, it will at least give you peace of mind that you have the flexibility to tackle the most demanding applications if needed. And finally, if you are investing in a second instrument, a triple-quad system might be the best option.

So, when evaluating this technique, particularly for its capabilities with different sample types, pay attention not only to what the technique can do for your application problem but also to what it cannot do, which is equally important. In other words, make sure you evaluate its capabilities on the basis of all your present and future analytical requirements. This is particularly important, because it might be suitable for reducing interferences on your current panel of analytes, but when you are required to expand the elemental suite, based on future regulations, make sure it is capable of mitigating potential interferences on other metals. For that reason, when assessing vendor-generated data, make sure the performance is achievable in your laboratory today as well as on all future elemental and sample workload demands.

FURTHER READING

1. K. Sakata and K. Kawabata, *Spectrochimica Acta*, **49B**, 1027–1032, 1994.
2. J. T. Rowan and R. S. Houk, *Applied Spectroscopy*, **43**, 976–980, 1989.
3. D. W. Koppenaal, C. J. Barinaga, and M. R. Smith, *Journal of Applied Analytical Chemistry*, **9**, 1053–1058, 1994.

4. G. C. Eiden, C. J. Barinaga, and D. W. Koppenaal, *Journal of Applied Analytical Chemistry*, **11**, 317–322, 1996.

5. P. Turner, T. Merren, J. Speakman, and C. Haines, *Plasma Source Mass Spectrometry: Developments and Applications*, Royal Society of Chemistry, London, pp. 28–34, 1996, ISBN 0-85404-727-1.

6. D. J. Douglas and J. B. French, *Journal of American Society of Mass Spectrometry*, **3**, 398, 1992.

7. B. A. Thomson, D. J. Douglas, J. J. Corr, J. W. Hager, and C. A. Joliffe, *Analytical Chemistry*, **67**, 1696–1704, 1995.

8. *X Series ICP-MS: Enhanced Collision Cell Technology CCT*, Thermo Scientific Product Specifications, July 2004, www.thermo.com/eThermo/CMA/PDFs/Articles/articlesFile_24138.pdf.

9. E. R. Denoyer, S. D. Tanner, and U. Voellkopf, *Spectroscopy*, **14**, 2, 1999.

10. H. H. Willard, L. L. Merritt, J. A. Dean, and F. A. Settle, *Instrumental Methods of Analysis*, Spectroscopic and Spectrometric Techniques (Chapters 5–13), Wadsworth Publishing Co., Belmont, CA, pp. 465–507, 1988.

11. E. De Hoffman, J. Charette, and V. Stroobant, *Mass Spectrometry, Principles and Applications*, John Wiley and Sons, Paris, France, 1996.

12. *Analysis of Ultra-trace Levels of Elements in Seawaters Using 3rd-Generation Collision Cell Technology*, Thermo Scientific Product Application Note: 40718, April 2007, www.thermo.com/eThermo/CMA/PDFs/Articles/articlesFile_26161.pdf.

13. J. Takahashi, *Determination of Impurities in Semiconductor Grade Hydrochloric Acid Using the Agilent 7500 cs ICP-MS*, Agilent Technologies Application Note 5989–4348EN, January 2006,

14. N. Yamada, J. Takahashi, and K. Sakata, *Journal of Analytical Atomic Spectrometry*, **17**, 1213–1222, 2002.

15. S. D. Tanner and V. I. Baranov, *Atomic Spectroscopy*, **20**(2), 45–52, 1999.

16. B. Hattendorf and D. Günter, *Journal of Analytical Atomic Spectrometry*, **19**, 600–606, 2004.

17. L. Rottmann, G. Jung, T. Vincent, and J. Wills, *Collision/Reaction Cell for ICP-MS—a New Concept for an Improved Removal of Low Masses*, Thermo Fisher Scientific, www.thermoscientific.com/content/dam/tfs/ATG/CMD/CMD%20Documents/QCell-Poster-IMSC-2012-final.pdf

18. *Agilent Technologies 7900 ICP-MS Product Overview*, www.agilent.com/en/products/icp-ms/icp-ms-systems/7900-icp-ms

19. F. Keenan and W. Spence, *Theory and Applications of Collision/Reaction Cells: How Collision and Reaction Cells Work for Interference Removal in ICP-MS*, Thermo Fisher Scientific Web-based Presentation, 2007, http://breeze.thermo.com/crcs.

20. S. D. Tanner, V. I. Baranov, and D. R. Bandura, *Spectrochimica Acta*, **57B**(9), 1361–1452, 2002.

21. K. Kawabata, Y. Kishi, and R. Thomas, *Spectroscopy*, **18**(1), 16–31, 2003.

22. *Overview of the NexION 300 ICP-MS Using the "Universal" Collision/Reaction Cell*, PerkinElmer Inc., www.perkinelmer.com/Catalog/Family/ID/NexION

23. *PlasmaQuant MS ICP-MS*, Analytik Jena, www.analytik-jena.us/products/elemental-analysis/icp-ms/plasmaquant-ms/

24. Strategies for Achieving the Lowest Possible Detection Limits in ICP-MS, October 01, 2019, *Spectroscopy Magazine*, R. Chemnitzer, Analytik Jena, www.spectroscopyonline.com/column-atomic-perspectives

25. *Product Overview of the Agilent Technologies 8900 Triple Quadrupole ICP-MS*, www.agilent.com/en/products/icp-ms/icp-ms-systems/8900-triple-quadrupole-icp-ms

26. Thermo Fisher Scientific, *iCAP TQ Triple Quad ICP-MS Landing Page*, www.thermofisher.com/order/catalog/product/731546

27. *NexION 5000 ICP-MS Product Technical Note*, www.perkinelmer.com/Product/nexion-5000-icp-ms-n8160010

28. E. Kroukamp and F. Abou Shakra, Multi-Quadrupole ICP-MS: Pushing Limits of Detection to the Next Decimal, *Spectroscopy Magazine*, **35**(9), 16–22, 2020, www.spectroscopyonline.com/view/multi-quadrupole-icp-ms-pushing-limits-of-detection-to-the-next-decimal.

11 Ion Detectors

Chapter 11 looks at the detection system—an important area of the mass spectrometer that detects and quantifies the number of ions emerging from the mass analyzer. The detector converts the ions into electrical pulses, which are then counted using its integrated measurement circuitry. The magnitude of the electrical pulses corresponds to the number of analyte ions present in the sample, which is then used for trace element quantitation by comparing the ion signal with known calibration or reference standards. In this section we will take a look at conventional dynode detection, which monitors discrete ions emerging from the mass-separation device in a sequential manner, in addition to describing the new breed of array detectors, which can monitor the entire mass spectrum simultaneously.

Since ICP-MS was first introduced in the early 1980s, a number of different ion-detection designs have been utilized, the most popular being electron multipliers for low-ion-count rates and Faraday collectors for high-count rates. Today, the majority of ICP-MS systems that are used for ultratrace analysis use detectors that are based on the active film or discrete-dynode electron multiplier. They are very sophisticated pieces of equipment and are very efficient at converting ion currents emerging from the mass analyzer into electrical signals. The location of the detector in relation to the mass analyzer is shown in Figure 11.1.

Before I go on to describe discrete-dynode detectors in greater detail, it is worth looking at two of the earlier designs—the channel electron multiplier (Channeltron®)[1] and the Faraday cup—to get a basic understanding of how the ICP-MS ion-detection process works.

11.1 CHANNEL ELECTRON MULTIPLIER

The operating principles of the channel electron multiplier are similar to a photomultiplier tube used in ICP-OES. However, instead of using individual dynodes to convert photons to electrons, the Channeltron is an open glass cone (coated with a semiconductor-type material) to generate electrons from ions that impinge on its surface. For the detection of positive ions, the front of the cone

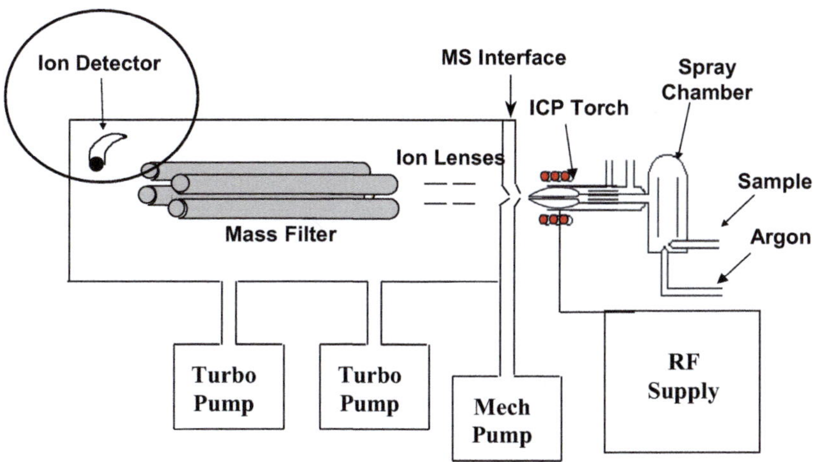

FIGURE 11.1 The location of the detector in relation to the mass analyzer.

DOI: 10.1201/9781003187639-11

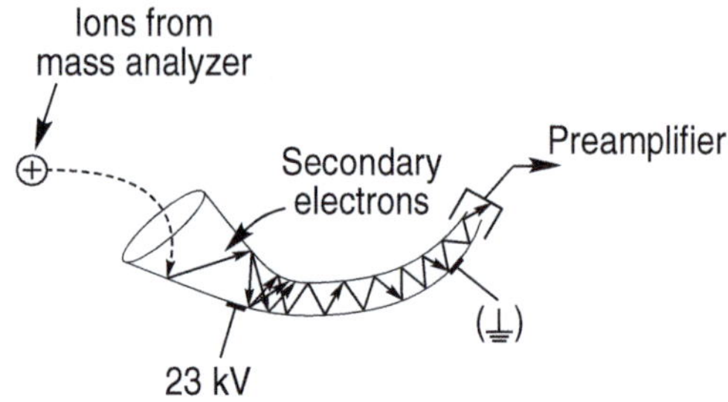

FIGURE 11.2 Basic principles of a channel electron multiplier.

(From Channeltron®, *Electron Multiplier Handbook for Mass Spectrometry Applications*, Galileo Electro-Optic Corp., 1991. Channeltron is a registered trademark of Galileo Corp)

is biased at a negative potential while the far end near the collector is kept at ground. When the ion emerges from the quadrupole mass analyzer, it is attracted to the high negative potential of the cone. When the ion hits this surface, one or more secondary electrons are formed. The potential gradient inside the tube varies based on position, so the secondary electrons move further down the tube. As these electrons strike new areas of the coating, more secondary electrons are emitted. This process is repeated many times. The result is a discrete pulse, which contains many millions of electrons generated from an ion that first hits the cone of the detector.[1] This process is shown simplistically in Figure 11.2.

This pulse is then sensed and detected by a very fast preamplifier. The output pulse from the pre-amplifier then goes to a digital discriminator and counting circuitry that only counts pulses above a certain threshold value. This threshold level needs to be high enough to discriminate against pulses caused by spurious emission inside the tube, from any stray photons from the plasma itself, or photons generated from fast-moving ions striking the quadrupole rods.

It is worth pointing out that the rate at which ions hit the detector is sometimes too high for the measurement circuitry to handle in an efficient manner. This is caused by ions arriving at the detector during the output pulse of the preceding ion and not being detected by the counting system. This "dead time", as it is known, is a fundamental limitation of the multiplier detector and is typically 30–50 ns (nanoseconds), depending on the detection system. Compensation in the measurement circuitry has to be made for this dead time to count the maximum number of ions hitting the detector.

11.2 FARADAY CUP

For some applications, where ultratrace detection limits are not required, the ion beam from the mass analyzer is directed into a simple metal electrode, or Faraday cup. With this approach, there is no control over the applied voltage (gain), so they can only be used for high ion currents. Their lower working range is on the order of 10^4 cps, which means that if they are to be used as the only detector, the sensitivity of the ICP mass spectrometer will be severely compromised. For this reason, they are normally used in conjunction with a Channeltron or discrete-dynode detector to extend the dynamic range of the instrument. An additional problem with the Faraday cup is that because of the time constant used in the DC amplification process to measure the ion current, they are limited to relatively low scan rates. This limitation makes them unsuitable for the fast scan rates required for traditional pulse counting used in ICP-MS, and also limits their ability to handle fast transient peaks.

The Faraday cup was never sensitive enough for quadrupole ICP-MS technology, because it was not suitable for very low ion-count rates. An attempt was made in the early 1990s to develop an ICP-MS system using a Faraday-cup detector for the environmental market, but its sensitivity was compromised, and as a result was considered more suitable for applications requiring ICP-OES trace-level detection capability. However, Faraday-cup technology is still utilized in some magnetic-sector instruments, particularly where high ion signals are encountered in the determination of high-precision isotope ratios, using a multi-collector detection system.

11.3 DISCRETE-DYNODE ELECTRON MULTIPLIER

These detectors, which are often called *active film multipliers*, work in a similar way to the Channeltron, but utilize discrete dynodes to carry out the electron multiplication.[2] Figure 11.3 illustrates the principles of operation of this device.

The detector is positioned off-axis to minimize the background noise from stray radiation and neutral species coming from the ion source. When an ion emerges from the quadrupole, it sweeps through a curved path before it strikes the first dynode. On striking the first dynode, it liberates secondary electrons. The electron-optic design of the dynode produces acceleration of these secondary electrons to the next dynode where they generate more electrons. This process is repeated at each dynode, generating a pulse of electrons that are finally captured by the multiplier collector or anode. Because of the materials used in the discrete-dynode detector and the difference in the way electrons are generated, it is typically 50–100% more sensitive than Channeltron technology and with much lower dead time (typically 5–10 ns).

Although most discrete-dynode detectors are very similar in the way they work, there are subtle differences in the way the measurement circuitry handles low and high ion-count rates. When ICP-MS was first commercialized, it could only handle up to five orders of dynamic range. However, when attempts were made to extend the dynamic range, certain problems were encountered. Before we discuss how modern detectors deal with this issue, let us first look at how it was addressed in earlier instrumentation.

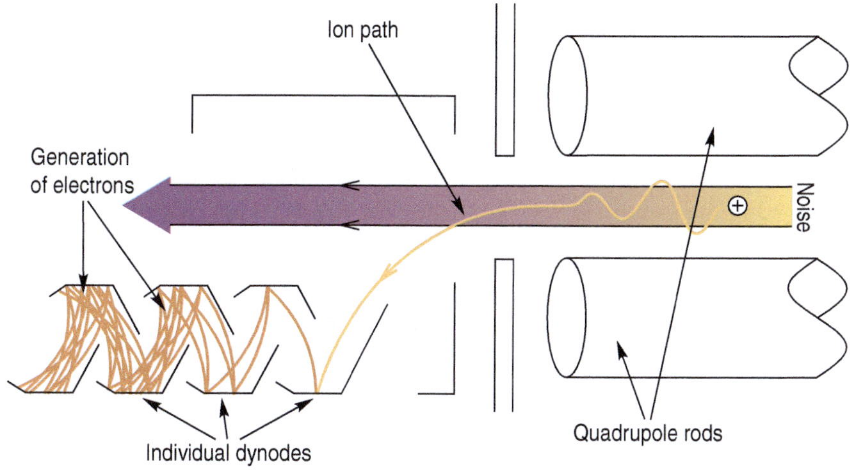

FIGURE 11.3 Schematic of a discrete-dynode electron multiplier.

(From K. Hunter, *Atomic Spectroscopy*, **15**[1], 17–20, 1994)

11.4   EXTENDING THE DYNAMIC RANGE

Traditionally, ICP-MS using the pulse-counting measurement is capable of about five orders of linear dynamic range. This means that ICP-MS calibration curves, generally speaking, are linear from detection limit up to a few hundred ppb. However, there are a number of ways of extending the dynamic range of ICP-MS another four to five orders of magnitude and working from sub-ppt levels up to hundreds of ppm. Here is a brief overview of some different approaches that have been used.

11.5   FILTERING THE ION BEAM

One of the very first approaches to extending the dynamic range in ICP-MS was to filter the ion beam. This was achieved by putting a non-optimum voltage on one of the ion lens components or the quadrupole itself, to limit the number of ions reaching the detector. This voltage offset, which was set on an individual-mass basis, acted as an energy filter to electronically screen the ion beam and reduce the subsequent ion signal to within a range covered by pulse-counting ion detection. The main disadvantage with this approach was that the operator had to have prior knowledge of the sample to know what voltage to apply to the high concentration masses.

11.6   USING TWO DETECTORS

Another technique that was used on some of the early ICP-MS instrumentation was to utilize two detectors, such as a channel electron multiplier and a Faraday cup, to extend the dynamic range. With this technique, two scans would be made. In the first scan it would measure the high concentration masses using the Faraday cup; in the second scan it would skip over the high concentration masses and carry out pulse counting of the low concentration masses with a channel electron multiplier. This worked reasonably well, but struggled with applications that required rapid switching between the two detectors, because the ion beam had to be physically deflected to select the optimum detector. Not only did this degrade the measurement duty cycle, but detector switching and stabilization times of several seconds also precluded fast-transient signal detection.

11.7   USING TWO SCANS WITH ONE DETECTOR

A more recent approach is to use just one detector to extend the dynamic range. This has typically been done by using the detector both in pulse and analog mode, so high and low concentrations can be determined in the same sample. There are basically three approaches to using this type of detection system: two of them involve carrying out two scans of the sample, whereas the third only requires one scan.

The first approach uses an electron multiplier operated in both digital and analog mode.[3] Digital counting provides the highest sensitivity, whereas operation in the analog mode (achieved by reducing the high voltage applied to the detector) is used to reduce the sensitivity of the detector, thus extending the concentration range for which ion signals can be measured. The system is implemented by scanning the spectrometer twice for each sample. The first scan, in which the detector is operated in the analog mode, provides signals for elements present at high concentrations. A second scan in which the detector voltage is switched to digital pulse-counting mode provides high-sensitivity detection for elements present at low levels. A major advantage of this technology is that the user does not need to know in advance whether to use analog or digital detection, because the system automatically scans all elements in both modes. However, one of the drawbacks is that two independent mass scans are required to gather data across an extended signal range. This not only results in degraded measurement efficiency and slower analyses, but it also means that the system is not ideally suited for fast-transient signal analysis, because mode switching is generally too slow.

An alternative way of extending the dynamic range is similar to the first approach, except that the first scan is used as an investigative tool to examine the sample spectrum before analysis.[4] This first pre-scan establishes the mass positions at which the analog and pulse modes will be used for subsequently collecting the spectral signal. The second analytical scan is then used for data collection, switching the detector back and forth rapidly between pulse and analog mode at each analytical mass.

Even though these approaches worked very well, their main disadvantage was that two separate scans are required to measure high and low levels. With conventional nebulization, this is not such a major problem except that it can impact sample throughput. However, it does become a concern when it comes to working with transient peaks found in laser sampling (LS), flow injection analysis (FIA) or electro thermal vaporization (ETV) ICP-MS. Because these transient peaks often only last a few seconds, all the available time must be spent measuring the masses of interest to get the best detection limits. When two scans have to be made, time is wasted collecting data, which is not contributing to the analytical signal.

11.8   USING ONE SCAN WITH ONE DETECTOR

The limitation of having to scan the sample twice led to the development of an improved design using a dual stage discrete-dynode detector.[5] This technology utilizes measurement circuitry that allows both high and low concentrations to be determined in one scan. This is achieved by measuring the ion signal as an analog signal at the midpoint dynode. When more than a threshold number

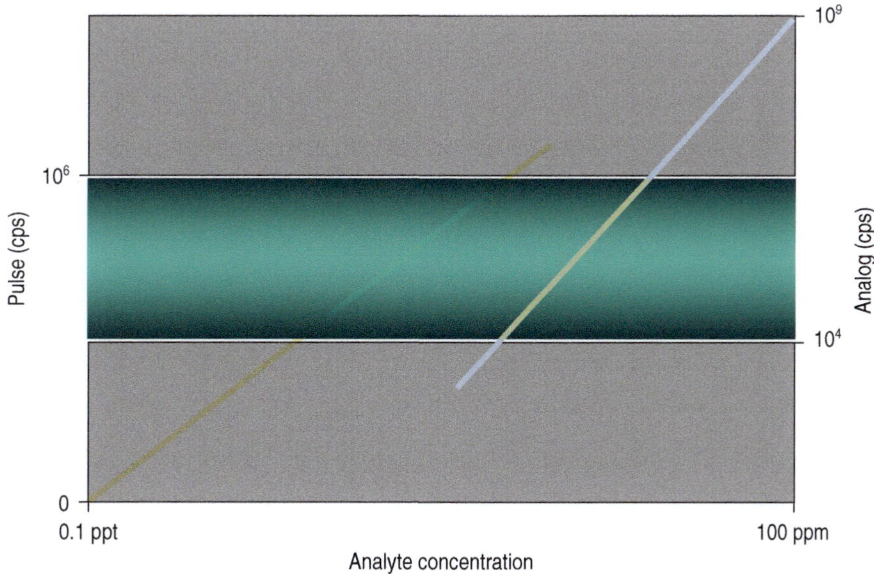

FIGURE 11.4   Dual-stage discrete-dynode detector measurement circuitry.

(From E. R. Denoyer, R. J. Thomas, and L. Cousins, *Spectroscopy,* **12**[2], 56–61, 1997. Covered by U.S. Patent Number 5,463,219)

of ions are detected, the signal is processed through the analog circuitry. When fewer than the threshold number of ions is detected, the signal cascades through the rest of the dynodes and is measured as a pulse signal in the conventional way. This process, which is shown in Figure 13.4, is completely automatic and means that both the analog and the pulse signals are collected simultaneously in one scan.[6]

The pulse-counting mode is typically linear from zero to about 10^6 cps, whereas the analog circuitry is suitable from 10^4 to 10^{10} cps. To normalize both ranges, a cross-calibration is carried out to cover concentration levels, which produces a pulse and an analog signal. This is possible because the analog and pulse outputs can be defined in identical terms of incoming pulse counts per second, based on knowing the voltage at the first analog stage, the output current and a conversion factor defined by the detection circuitry electronics. By carrying out a cross-calibration across the mass range, a dual-mode detector of this type is capable of achieving approximately nine to ten orders of dynamic range in one simultaneous scan. This can be seen in Figures 13.5 and 13.6. Figure 13.5 shows that the pulse-counting calibration curve (left-hand line) is linear up to 10^6 cps, whereas the analog calibration curve (right-hand line) is linear from 10^4 to 10^9 cps. Figure 13.6 shows that after cross-calibration, the two curves are normalized, which means the detector is suitable for concentration levels between 0.1 ppt and 100 ppm and above—typically nine to ten orders of magnitude for most elements.[5]

There are subtle variations of this type of detection system, but its major benefit is that it requires only one scan to determine both high and low concentrations. It, therefore, not only offers the potential to improve sample throughput, but also means that the maximum data can be collected on a transient signal that only lasts a few seconds. This is described in greater detail in Chapter 12, where we discuss different measurement protocols and peak integration routines.

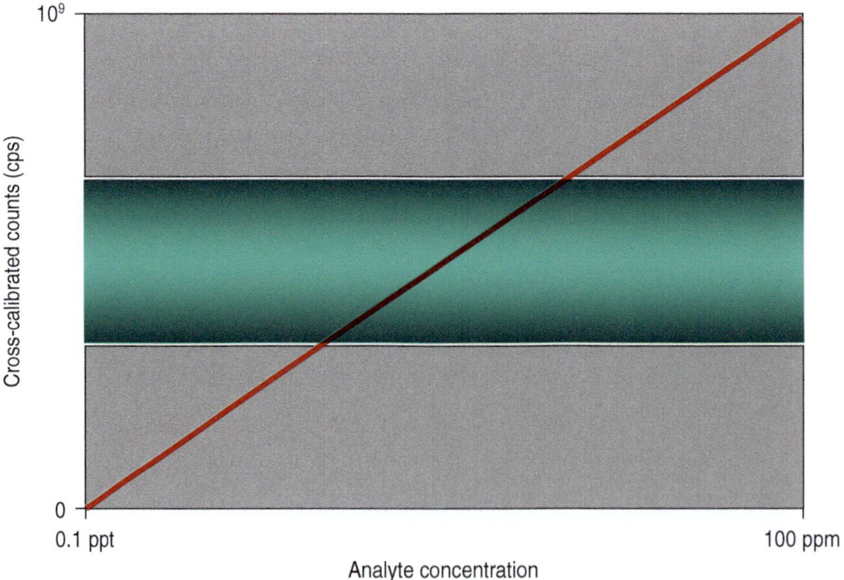

FIGURE 11.5 The pulse-counting mode covers up to 10^6 cps, while the analog circuitry is suitable from 10^4 to 10^9 cps, with a dual-mode discrete-dynode detector.

(From E. R. Denoyer, R. J. Thomas, and L. Cousins, *Spectroscopy,* **12**[2], 56–61, 1997. Covered by U.S. Patent Number 5,463,219)

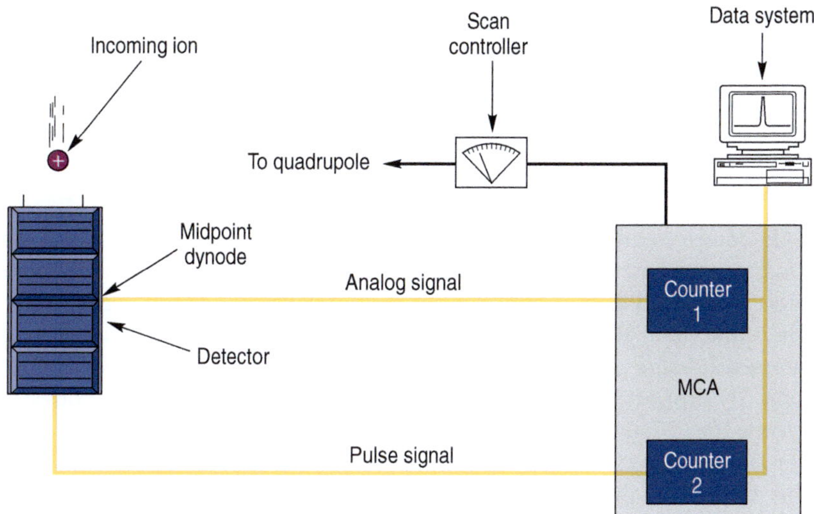

FIGURE 11.6   Using cross-calibration of the pulse and analog modes, quantitation from sub-ppt to high-ppm levels is possible.

(From E. R. Denoyer, R. J. Thomas, and L. Cousins, *Spectroscopy,* **12**[2], 56–61, 1997. Covered by U.S. Patent Number 5,463,219)

11.9   EXTENDING THE DYNAMIC RANGE USING PULSE-ONLY MODE

Another recent development in extending the dynamic range is to use the pulse-only signal. This is achieved by monitoring the ion flux at one of the first few dynodes of the detector (before extensive electron multiplication has taken place) and then attenuating the signal up to 10,000:1 by applying a control voltage. Electron pulses passed by the attenuation section are then amplified to yield pulse heights that are typical in normal pulse-counting applications.

There are basically three ways of implementing this technology based on the types of samples being analyzed. It can be run in conventional pulse-only mode for normal low-level work. It can also be run using an operator-selected attenuation factor if the higher level elements being determined are known and similar in concentration. If the samples are complete unknowns and have not been well characterized beforehand, a dynamic attenuation mode of operation is available. In this mode, an additional pre-measurement time is built into the quadrupole settling time to determine the optimum detector attenuation for the selected dwell times used.

This novel, pulse-only approach to extending the dynamic range looks to be a very interesting development, which does not have the limitation of having to calibrate where pulse and analog signals cross over. However, it does require a pre-analysis attenuation calibration to be carried out on a fairly frequent basis to determine the extent of signal attenuation required. The frequency of this calibration will vary depending on sample workload, but is expected to be on the order of once every four weeks.

However, it should be strongly emphasized that, irrespective of which extended range technology is used, if low and high concentrations of the same analyte are expected in a suite of samples, it is unrealistic to think you can accurately quantitate down at the low end and at the top end of the linear range with the same calibration graph. If you want to achieve accurate and precise data at or near the limit of quantitation, you must run a set of appropriate calibration standards to cover your low-level samples. In addition, if you are expecting high and low concentrations in the same suite of samples, you have to be absolutely sure that a high concentration sample has been thoroughly washed out from the spray chamber/nebulizer system, before a low-level sample is introduced. For this reason, caution must be taken when setting up the method with an autosampler, because if the read delay/integration times are not optimized for a suite of samples, erroneous results can be generated, which might necessitate a rerun under the manual supervision of the instrument operator.

11.10 SIMULTANEOUS ARRAY DETECTORS

Discrete-dynode detectors are designed to handle a sequential stream of separated ions emerging from the mass spectrometer. Similar to a photomultiplier tube that converts photons from an optical-emission signal into a pulse of electrons, these detectors cannot capture the entire mass spectrum at the same time. However, a new breed of ion detectors have recently been developed that are based on solid-state, direct-charge arrays, similar to CID/CCD technology used in ICP optical emission. By projecting all the separated ions from a mass-separation device onto a two-dimensional array, these detectors can view the entire mass spectrum simultaneously.[7] Designed specifically for the Mattauch–Herzog double-focusing magnetic-sector technology described in Chapter 8, this CMOS (complementary metal–oxide semiconductor) ion-sensitive device is a 12-cm-long array that covers the entire mass range simultaneously in 4,800 separate channels.[8] The basic principle of the detector is similar to that of a Faraday cup: when a charged ion arrives at a detector array, it is discharged by receiving an electron, which generates a signal. The detector is referred to as a direct-charge detector (DCD) because every ion arriving at the detector contributes to the signal. Furthermore, with 4,800 detector arrays covering the mass range (from 5 to 240 amu) every mass unit is covered on average by 20 separate channels, resulting in a true mass spectrum rather than a single point for each atomic mass unit. A photograph of this direct-charge detector is shown in Figure 11.7.

To cover linearity of up to eight orders of magnitude required in ICP-MS measurements, each detector channel incorporates separate high- and low-gain detector elements, allowing it to independently handle a wide range of signal levels in the basic integration cycle. The dynamic range can be further extended by optimizing the readout process. In the basic integration cycle, each channel is monitored by the electronics every 20 ms. If the signal integrated in this time interval nears the threshold of the channel, the integrated signal of that channel is automatically logged, the channel is reset and its measurement cycle repeated. This is repeated until the end of the defined measurement time, when all the collected data is integrated to produce the final signal. This means that the detector is always working within its linear response range and that longer integration times can

FIGURE 11.7 A direct-charge detector (DCD) used for simultaneous measurement of ions separated by a Mattauch–Herzog double-focusing magnetic-sector mass spectrometer.

be used without fear of detector saturation. For the benefits of simultaneous detection in ICP mass spectrometry, please refer to Chapter 8 on "Magnetic-Sector" technology and Chapter 12 on "Peak Measurement Protocol".

FURTHER READING

1. Channeltron®, *Electron Multiplier Handbook for Mass Spectrometry Applications*, Galileo Electro-Optic Corp., 1991 (Channeltron is a registered trademark of Galileo Corp).
2. K. Hunter, *Atomic Spectroscopy*, **15**(1), 17–20, 1994.
3. R. C. Hutton, A. N. Eaton, and R. M. Gosland, *Applied Spectroscopy*, **44**(2), 238–242, 1990.
4. Y. Kishi, *Agilent Technologies Application Journal*, **6**, 12–16, August 1997.
5. E. R. Denoyer, R. J. Thomas, and L. Cousins, *Spectroscopy*, **12**(2), 56–61, 1997. Covered by U.S. Patent Number 5,463,219.
6. J. Gray, R. Stresau, and K. Hunter, Ion Counting Beyond 10 GHz, *Poster Presentation Number 890–6P*, Pittsburgh Conference and Exposition, Orlando, FL, 2003.
7. G. D. Schilling, S. J. Ray, A. A. Rubinshtein, J. A. Felton, R. P. Sperline, M. B. Denton, C. J. Barinaga, D. W. Koppenaal, and G. M. Hieftje, *Analytical Chemistry*, **81**(13), 5467–5473, 2009.
8. *A New Era in Mass Spectrometry: Spectro Analytical Instruments*, www.spectro.com/pages/e/p010402tab_overview.htm.

12 Peak Measurement Protocol

With its multielement capability, superb detection limits, wide dynamic range and high sample throughput, ICP-MS is proving to be a compelling technique for more and more diverse application areas. However, it is very unlikely that two different application areas have identical analytical requirements. For example, environmental, pharmaceutical, cannabis and clinical contract/testing laboratories, although wanting reasonably low detection limits, are not really pushing the technique to its extreme detection capability. Their main requirement usually is high sample throughput, because the number of samples these laboratories can analyze in a day directly impacts their revenue. On the other hand, a semiconductor fabrication plant or a supplier of high-purity chemicals to the electronics industry is interested in the lowest detection limits the technique can offer, because of the contamination problems associated with manufacturing high-performance electronic devices. Chapter 12 looks at the many different measurement protocols associated with identifying and quantifying the analyte peak in ICP-MS and how they impact sample throughput and the quality of the data generated.

To meet such diverse application needs, modern ICP-MS instrumentation has to be very flexible if it is to keep up with the increasing demands of its users. Nowhere is this more important than in the area of peak integration and measurement protocol. The way the analytical signal is managed in ICP-MS has a direct impact on its multielement characteristics, isotopic capability, detection limits, dynamic range and sample throughput—the five major strengths that attracted the trace element community to the technique over 35 years ago. To understand signal management in greater detail and its implications on data quality, we will discuss how measurement protocol is optimized based on the application's analytical requirements, and its impact on both continuous signals generated by traditional nebulization devices and transient signals produced by alternative sample-introduction techniques such as automated sample delivery systems, chromatographic-separation studies and particle size measurements.

12.1 MEASUREMENT VARIABLES

There are many variables that affect the quality of the analytical signal in ICP-MS. The analytical requirements of the application will often dictate this, but there is no question that instrumental detection and measurement parameters can have a significant impact on the quality of data in ICP-MS. Some of the variables that can potentially impact the quality of the data, particularly when carrying out multielement analysis, are as follows:

- A continuous or transient signal
- The temporal length of the sampling event
- Volume of sample available
- Number of samples being analyzed
- Number of replicates per sample
- Number of elements being determined
- Detection limits required
- Precision/accuracy expected
- Dynamic range needed
- Integration time used
- Peak quantitation routines

Before we go on to discuss these in greater detail and how these parameters affect the data, it is important to remind ourselves how a scanning device like a quadrupole mass analyzer works. Although we will focus on quadrupole technology, the fundamental principles of measurement

DOI: 10.1201/9781003187639-12

protocol will be very similar for all types of mass spectrometers that use a sequential approach for multielement peak quantitation.

12.2 MEASUREMENT PROTOCOL

The principles of scanning with a quadrupole mass analyzer are shown in Figure 12.1. In this simplified example, the analyte ion in front (black) and four other ions have arrived at the entrance to the four rods of the quadrupole. When a particular RF/DC voltage is applied to each pair of

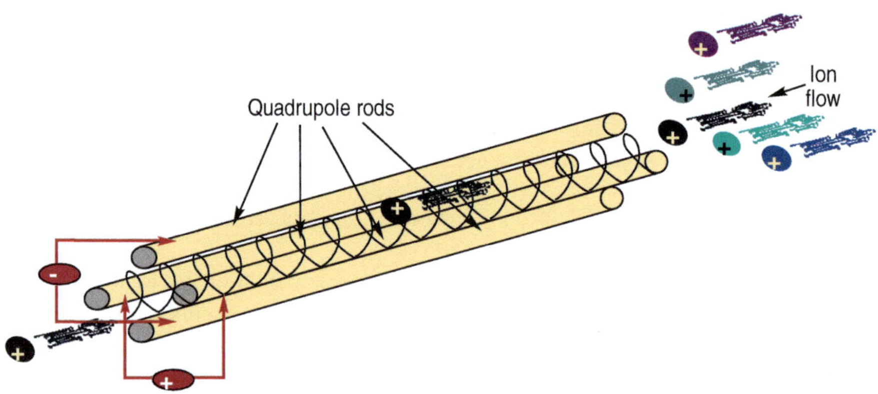

FIGURE 12.1   Principles of mass selection with a quadrupole mass filter.

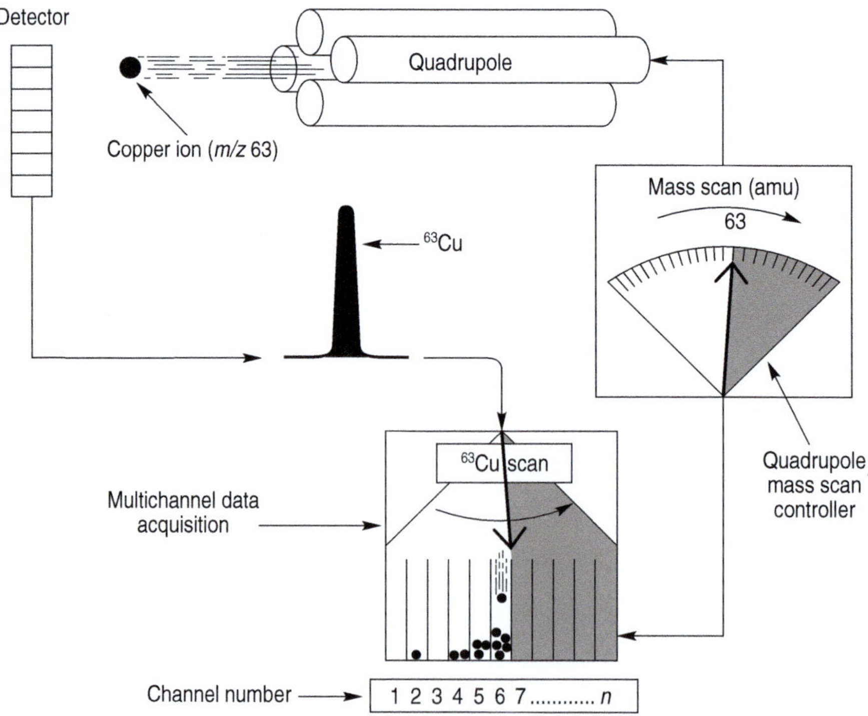

FIGURE 12.2   Detection and measurement protocol using a quadrupole mass analyzer.

(From *Integrated MCA Technology in the ELAN ICP-Mass Spectrometer*, Application Note TSMS-25, PerkinElmer Instruments, 1993)

rods, the positive or negative bias on the rods will electrostatically steer the analyte ion of interest down the middle of the four rods to the end, where it will emerge and be converted to an electrical pulse by the detector. The other ions of different mass-to-charge ratio will pass through the spaces between the rods and be ejected from the quadrupole. This scanning process is then repeated for another analyte at a completely different mass-to-charge ratio until all the analytes in a multielement analysis have been measured.

The process for the detection of one particular mass in a multielement run is represented in Figure 12.2. It shows a $^{63}Cu^+$ ion emerging from the quadrupole and being converted to an electrical

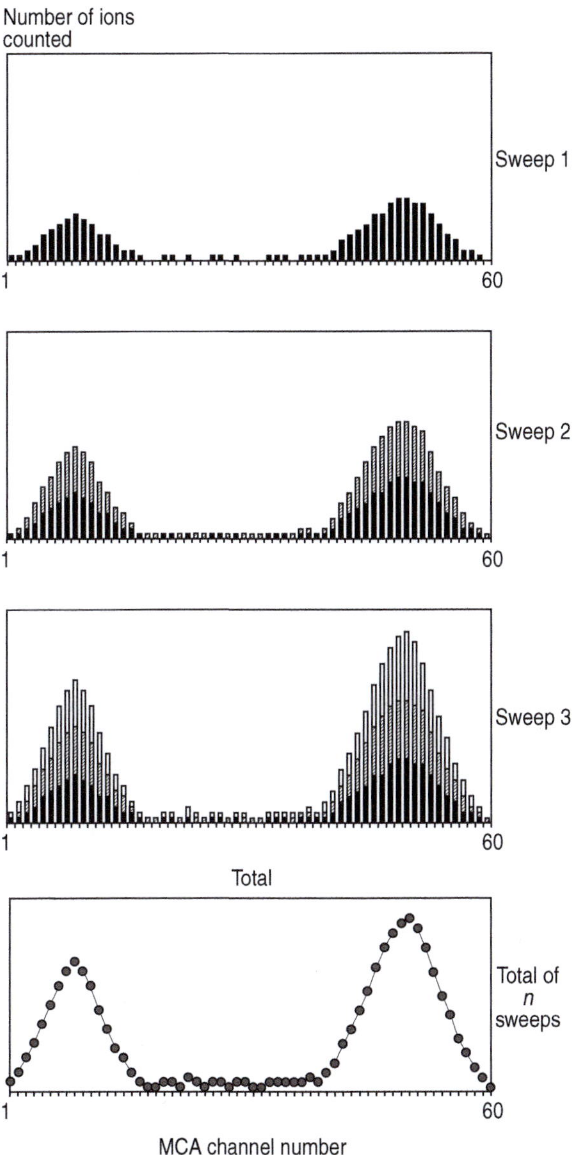

FIGURE 12.3 A profile of the peak is built up by continually sweeping the quadrupole across the mass spectrum.

(From *Integrated MCA Technology in the ELAN ICP-Mass Spectrometer*, Application Note TSMS-25, PerkinElmer Instruments, 1993)

pulse by the detector. As the optimum RF-to-DC ratio is applied for $^{63}Cu^+$ and repeatedly scanned, the ions as electrical pulses are stored and counted by a multichannel analyzer. This multichannel data-acquisition system typically has 20 channels per mass, and as the electrical pulses are counted in each channel, a profile of the mass is built up over the 20 channels, corresponding to the spectral peak of $^{63}Cu^+$. In a multielement run, repeated scans are made over the entire suite of analyte masses, as opposed to just one mass represented in this example.

The principles of multielement peak acquisition are shown in Figure 12.3. In this example, signal pulses for two masses are continually collected as the quadrupole is swept across the mass

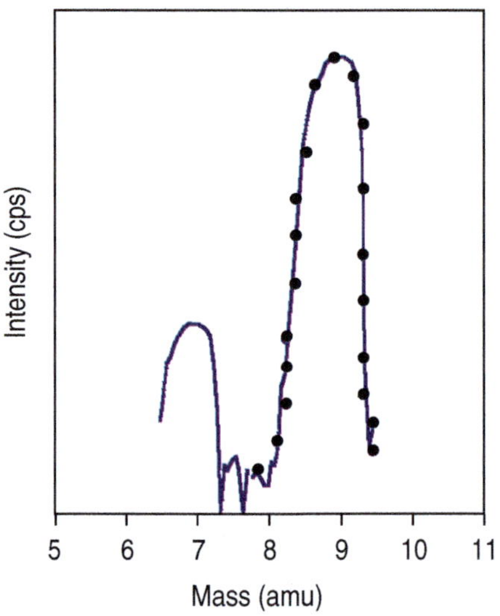

FIGURE 12.4   Multichannel ramp scanning approach using 20 channels per amu.

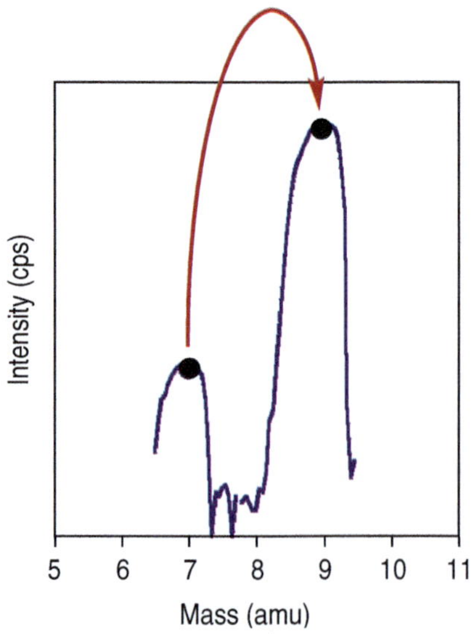

FIGURE 12.5   Peak-hopping approach.

spectrum, shown by sweeps 1–3. After a fixed number of sweeps (determined by the user), the total number of signal pulses in each channel is obtained, resulting in the final spectral peak.[1]

When it comes to quantifying an isotopic signal in ICP-MS, there are basically two approaches to consider. One is the multichannel ramp scanning approach, which uses a continuous smooth ramp of 1—n channels (where n is typically 20) per mass across the peak profile. This is shown in Figure 12.4.

Also, there is the peak-hopping approach, in which the quadrupole power supply is driven to a discrete position on the peak (normally the maximum point), allowed to settle and a measurement is taken for a fixed amount of time. This is represented in Figure 12.5.

The multipoint-scanning approach is best for accumulating spectral and peak-shape information when doing mass scans. It is normally used for doing mass calibration and resolution checks, and as a classical qualitative-method development tool to find out what elements are present in the sample and to assess their spectral implications on the masses of interest. Full peak profiling is not normally used for doing rapid quantitative analysis, because valuable analytical time is wasted taking data on the wings and valleys of the peak, where the signal-to-noise ratio is poorest.

When the best possible detection limits are required, the peak-hopping approach is best. It is important to understand that to get the full benefit of peak hopping, the best detection limits are achieved when single-point peak-hopping at the peak maximum is chosen. However, to carry out single-point peak-hopping, it is essential that the mass stability be good enough to reproducibly go to the same mass point every time. If good mass stability can be guaranteed (usually by thermostating the quadrupole power supply), measuring the signal at the peak maximum will always give the best detection limits for a given integration time. It is well documented that there is no benefit in spreading the chosen integration time over more than one measurement point per mass. If time is a major consideration in the analysis, then using multiple points is wasting valuable time on the wings and valleys of the peak, which contribute less to the analytical signal and more to the background noise. This is shown in Figure 12.6, which demonstrates the degradation in signal-to-background noise of 10 ppb Rh with an increase in the number of points per peak, spread over the same total integration time. Detection-limit improvement for a selected group of elements using 1 point/peak compared to 20 points/peak is shown in Figure 12.7.

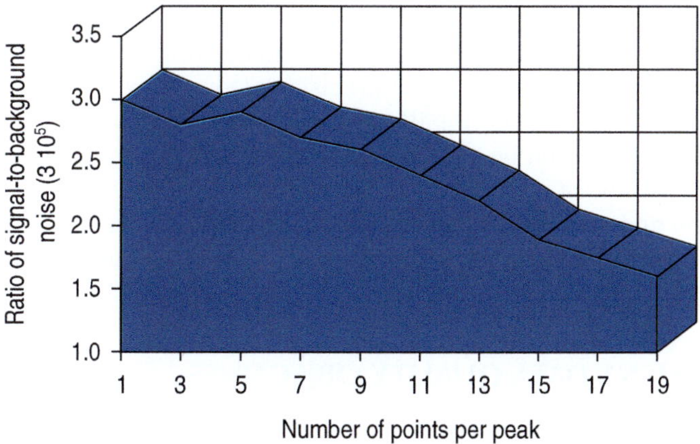

FIGURE 12.6 Signal-to-background noise degrades when more than one point, spread over the same integration time, is used for peak quantitation.

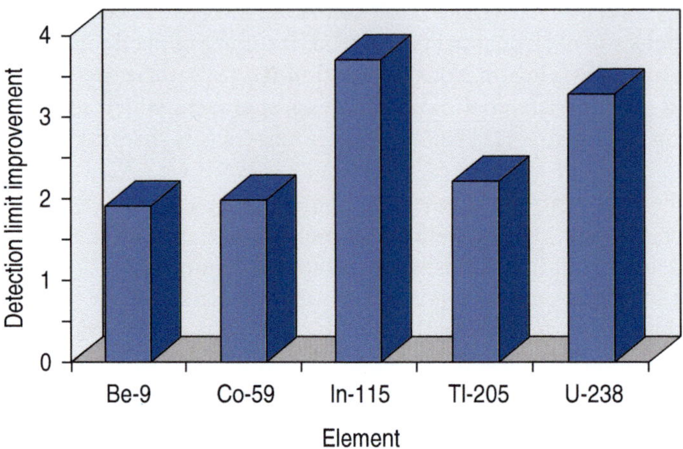

FIGURE 12.7   Detection-limit improvement using one point per peak compared to 20 points per peak over the mass range.

(From E. R. Denoyer, *Atomic Spectroscopy,* **13**[3], 93–98, 1992)

12.3   OPTIMIZATION OF MEASUREMENT PROTOCOL

Now that the fundamentals of the quadrupole-measuring electronics have been described, let us now go into more detail on the impact of optimizing the measurement protocol based on the requirements of the application. When multielement analysis is being carried out by ICP-MS, there are a number of decisions that need to be made. First, we need to know if we are dealing with a continuous signal from a nebulizer or a transient signal from a sampling accessory such as the laser ablation system of a chromatographic-separation device. If it is a transient event, how long will the signal last? Another question that needs to be addressed is how many elements are going to be determined? With a continuous signal, this is not such a major problem, but could be an issue if we are dealing with a transient signal that lasts only a few seconds. We also need to be aware of the level of detection capability required. This is a major consideration with a short laser pulse of a few seconds' duration, and even more critical when characterizing nanoparticles which only last for a few milliseconds. But it is also an issue with a continuous signal produced by a concentric nebulizer, where we might have to accept a compromise of detection limit based on the speed of analysis requirements or amount of sample available. What analytical precision is expected? If isotope ratio/dilution work is being done, how many ions do we have to count to guarantee good precision? Does increasing the integration time of the measurement help the precision? Finally, is there a time constraint on the analysis? A high-throughput laboratory might not be able to afford to use the optimum sampling time to get the ultimate in detection limit. In other words, what compromises need to be made between detection limit, precision and sample throughput? It is clear that before the measurement protocol can be optimized, the major analytical requirements of the application need to be defined. Let us look at this in greater detail.

12.4   MULTIELEMENT DATA QUALITY OBJECTIVES

Because multielement detection capability is probably the major reason why most laboratories invest in ICP-MS, it is important to understand the impact of measurement criteria on detection limits. We know that in a multielement analysis, the quadrupole's RF-to-DC ratio is "driven" or scanned to mass regions, which represent the elements of interest. The electronics are allowed to

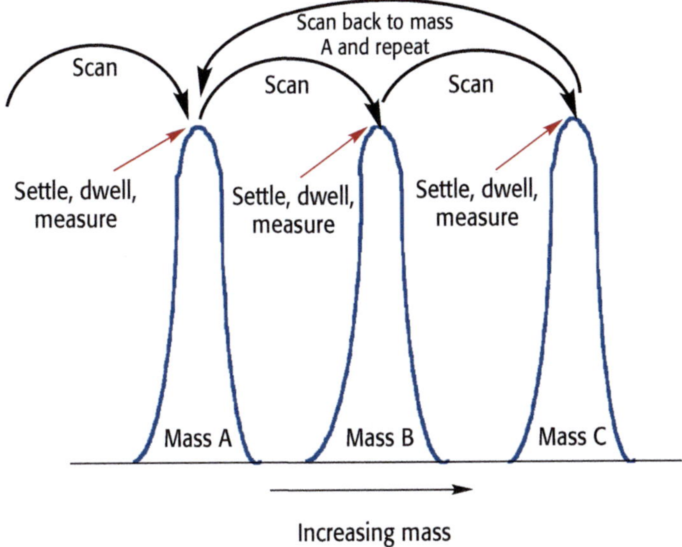

FIGURE 12.8 Multielement scanning and peak measurement protocol used in a quadrupole.

settle and then "sit" or dwell on the peak and take measurements for a fixed period of time. This is usually performed a number of times until the total integration time is fulfilled. For example, if a dwell time of 50 ms is selected for all masses and the total integration time is 1 s, then the quadrupole will carry out 20 complete sweeps per mass, per replicate. It will then repeat the same routine for as many replicates that have been built into the method. This is shown in a simplified manner in Figure 12.8, which displays the scanning protocol of a multielement scan of three different masses.

In this example, the quadrupole is scanned to mass A. The electronics are allowed to settle (settling time), left to dwell for a fixed period of time at one or multiple points on the peak (dwell time) and intensity measurements taken (based on the dwell time). The quadrupole is then scanned to masses B and C, and the measurement protocol is repeated. The complete multielement measurement cycle (sweep) is repeated as many times as needed to make up the total integration per peak. It should be emphasized that this is a generalization of the measurement routine—management of peak integration by the software will vary slightly based on different instrumentation.

It is clear from this that during a multielement analysis, there is a significant amount of time spent scanning and settling the quadrupole, which does not contribute to the quality of the analytical signal. Therefore, if the measurement routine is not optimized carefully, it can have a negative impact on data quality. The dwell time can usually be selected on an individual mass basis, but the scanning and settling times are normally fixed because they are a function of the quadrupole and detector electronics. For this reason, it is essential that the dwell time, which ultimately affects detection limit and precision, dominate the total measurement time, compared to the scanning and settling times. It follows, therefore, that the measurement duty cycle (percentage of actual measuring time compared to total integration time) is maximized when the quadrupole and detector electronics settling times are kept to an absolute minimum. This can be seen in Figure 12.9, which shows a plot of percent measurement duty cycle against dwell time for four different quadrupole settling times—0.2, 1.0, 3.0 and 5.0 ms for one replicate of a multielement scan of five masses, using one point per peak. In this example, the total integration time for each mass was 1 s, with the number of sweeps varying depending on the dwell time used. For this exercise, the duty-cycle percentage is defined by the following equation:

$$\frac{\text{Dwell time} \times \text{\# sweeps} \times \text{\# elements} \times \text{\# replicates}}{\left\{ \left(\text{Dwell time} \times \text{\# sweeps} \times \text{\# elements} \times \text{\# replicates} \right) + \left(\dfrac{\text{Scanning}}{\text{settling time}} \times \text{\# sweeps} \times \text{\# elements} \times \text{\# reps} \right) \right\}} \times 100$$

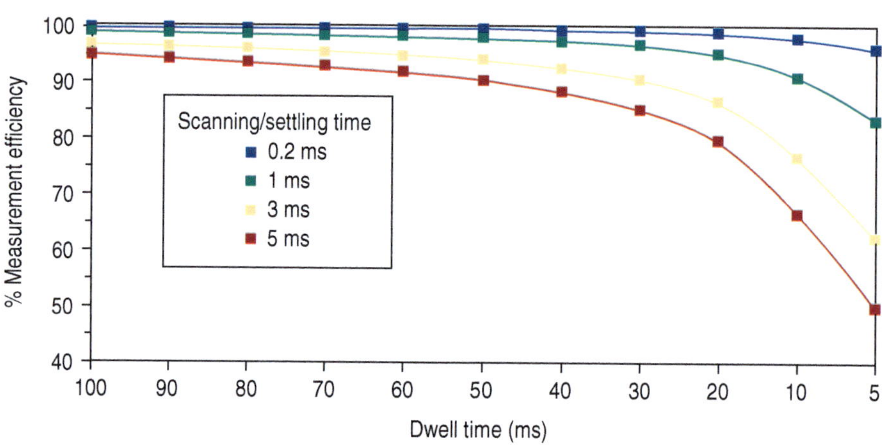

FIGURE 12.9   Measurement duty cycle as a function of dwell time with varying scanning/settling times.

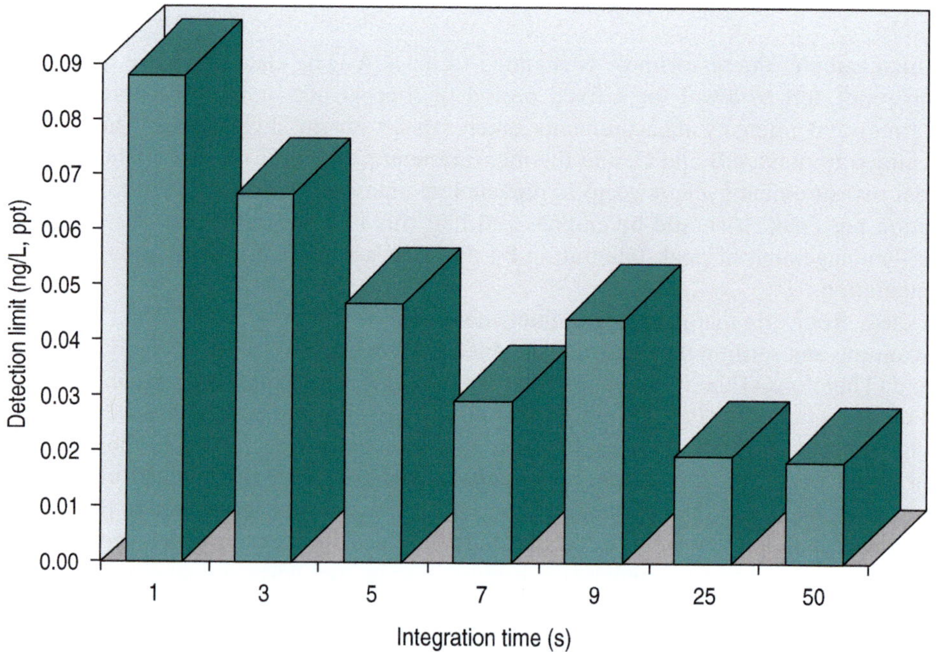

FIGURE 12.10   Plot of detection limit against integration time for $^{238}\text{U}^+$.

To achieve the highest duty cycle, the nonanalytical time must be kept to an absolute minimum. This leads to more time being spent counting ions and less time scanning, and settling, which do not contribute to the quality of the analytical signal. This becomes critically important when a rapid transient peak is being quantified, because the available measuring time is that much shorter.[3] It is therefore a good rule of thumb, when setting up your measurement protocol in ICP-MS, to avoid using multiple points per peak and long settling times, because it ultimately degrades the quality of the data for a given integration time.

It can also be seen in Figure 12.9 that shorter dwell times translate into a lower duty cycle. For this reason, for normal quantitative-analysis work, it is probably desirable to carry out multiple sweeps with longer dwell times (typically, 50 ms) to get the best detection limits. So, if an integration time of 1 s is used for each element, this would translate into 20 sweeps of 50 ms dwell time per mass. Although 1 s is long enough to achieve reasonably good detection limits, longer integration times generally have to be used to reach the lowest possible detection limits. This is shown in Figure 12.10, which shows detection-limit improvement as a function of integration time for $^{238}U^+$.

(COURTESY OF PERKINELMER INC.)

As would be expected, there is a fairly predictable improvement in the detection limit as the integration time is increased because more ions are being counted without an increase in the background noise. However, this only holds true up to the point where the pulse-counting detection system becomes saturated and no more ions can be counted. In the case of $^{238}U^+$, it can be seen that this happens at around 25 s, because there is no obvious improvement in D/L at a higher integration time. So from this data, we can say that there appears to be no real benefit in using longer than a seven-second integration time. When deciding the length of the integration time in ICP-MS, you have to weigh up the detection-limit improvement against the time taken to achieve that improvement. Is it worth spending 25 s measuring each mass to get a 0.02 ppt-detection limit, if 0.03 ppt can be achieved using a seven-second integration time? Alternatively, is it worth measuring for 7 s when 1 s will only degrade the performance by a factor of 3? It really depends on your data quality objectives.

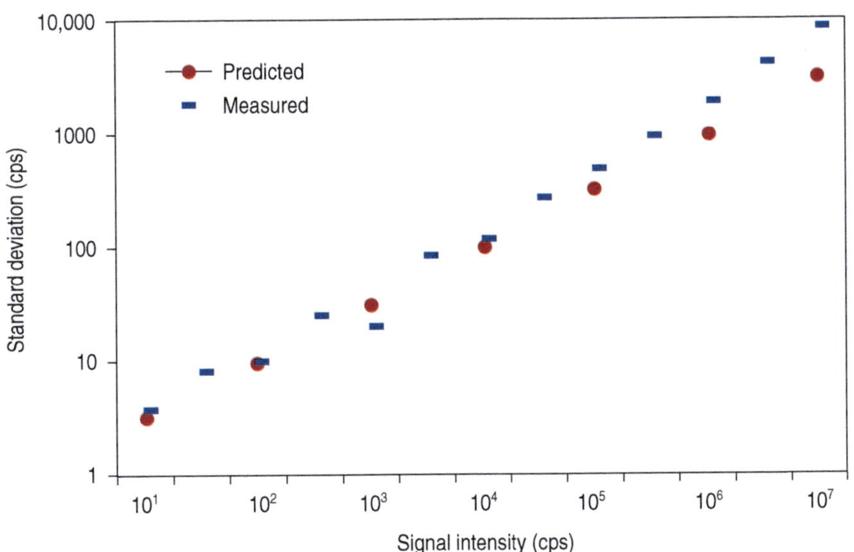

FIGURE 12.11 Comparison of measured standard deviation of a $^{208}Pb^+$ signal against that predicted by counting statistics.

(From E. R. Denoyer, *Atomic Spectroscopy*, **13**[3], 93–98, 1992)

For some applications like isotope dilution/ratio studies, high precision is also a very important data quality objective.[4] However, to understand what is realistically achievable, we have to be aware of the practical limitations of measuring a signal and counting ions in ICP-MS. Counting statistics tells us that the standard deviation of the ion signal is proportional to the square root of the signal. It follows, therefore, that the RSD or precision should improve with an increase in the number (N) of ions counted, as shown by the following equation:

$$\%\text{RSD} = \sqrt{\frac{N}{N}} \times 100$$

In practice, this holds up very well as can be seen in Figure 12.11. In this plot of standard deviation as a function of signal intensity for $^{208}\text{Pb}^+$, the black dots represent the theoretical relationship as predicted by counting statistics. It can be seen that the measured standard deviation (black bars) follows theory very well up to about 100,000 cps. At that point, additional sources of noise (e.g., sample-introduction pulsations/plasma fluctuations) dominate the signal, which lead to poorer standard deviation values.

So, based on counting statistics, it is logical to assume that the more ions are counted, the better the precision will be. To put this in perspective, it means that at least 1 million ions need to be counted to achieve an RSD of 0.1%. In practice, of course, these kinds of precision values are very difficult to achieve with a scanning-quadrupole system because of the additional sources of noise. If this information is combined with our knowledge of how the quadrupole is scanned, we begin to understand what is required to get the best precision. This is confirmed by the spectral scan in Figure 12.12, which shows the predicted precision at all 20 channels of a 5 ppb $^{208}\text{Pb}^+$ peak.[2]

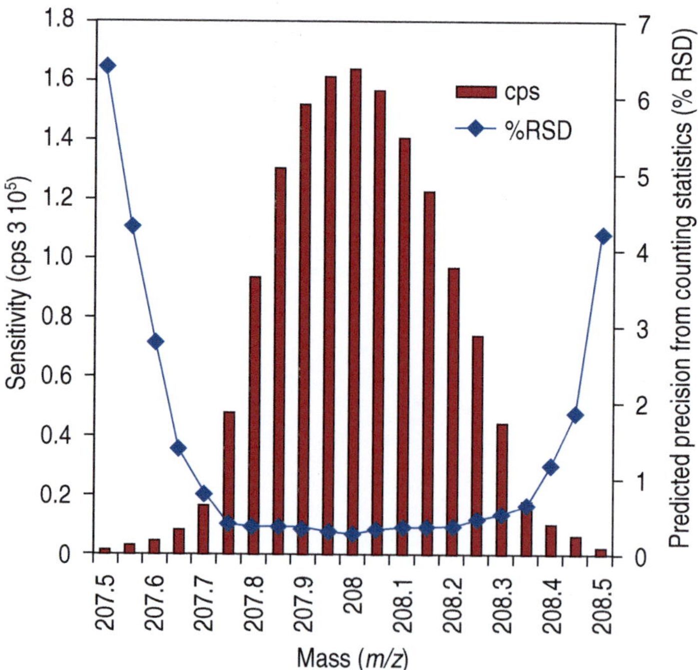

FIGURE 12.12   Comparison of % RSD with signal intensity across the mass profile of a $^{208}\text{Pb}^+$ peak.

(From E. R. Denoyer, *Atomic Spectroscopy,* **13**[3], 93–98, 1992)

TABLE 12.1

Precision of Pb Isotope-Ratio Measurement as a Function of Dwell Time Using a Total Integration Time of 5.5 s

Dwell time (ms)	% RSD $^{207}Pb^+/^{206}Pb^+$	% RSD $^{208}Pb^+/^{206}Pb^+$
2	0.40	0.36
5	0.38	0.36
10	0.23	0.22
25	0.24	0.25
50	0.38	0.33
100	0.41	0.38

Source: L. Halicz, Y. Erel, and A. Veron, *Atomic Spectroscopy,* **17**(5), 186–189, 1996.

This tells us that the best precision is obtained at the channels where the signal is highest, which as we can see are the ones at or near the center of the peak. For this reason, if good precision is a fundamental requirement of your data quality objectives, it is best to use single-point peak-hopping with integration times on the order of 5–10 s. On the other hand, if high-precision isotope ratio or isotope dilution work is being done, where analysts would like to achieve precision values approaching counting statistics, then much longer measuring times are required. That is why integration times on the order of 5–10 min are commonly used for determining isotope ratios involving environmental pollutants[5] or clinical metabolism studies.[6] For this type of analysis, when two or more isotopes are being measured and ratioed to each other, it follows that the more simultaneous the measurement, the better the precision becomes. Therefore, the ability to make the measurement as simultaneous as possible is considered more desirable than any other aspect of the measurement. This is supported by the fact that the best isotope ratio precision data is achieved with multicollector, magnetic-sector ICP-MS technology that carries out many isotopic measurements at the same time using multiple detectors.[7] Also, TOF technology, which simultaneously samples all the analyte ions in a slice of the ion beam, offers excellent precision, particularly when internal standardization measurement is also carried out in a simultaneous manner.[8] So, the best way to approximate simultaneous measurement with a rapid scanning device such as a quadrupole is to use shorter dwell times (but not too short that insufficient ions are counted) and keep the scanning/settling times to an absolute minimum, which results in more sweeps for a given measurement time. This can be seen in Table 12.1, which shows the precision of Pb isotope ratios at different dwell times carried out by researchers at the Geological Survey of Israel.[9] The data are based on nine replicates of a NIST SRM-981 (75 ppb Pb) solution, using 5.5 s integration time per isotope.

From these data, the researchers concluded that a dwell time of 10 or 25 ms offered the best isotope-ratio precision measurement (quadrupole settling time was fixed at 0.2 ms). They also found that they could achieve slightly better precision by using a 17.5 s integration time (700 sweeps at 25 ms dwell time), but felt the marginal improvement in precision for nine replicates was not worth spending the approximately three-and-a-half-times-longer analysis time. This can be seen in Table 12.2.

This work shows the benefit of being able to optimize the dwell time, settling time and the number of sweeps to get the best isotope-ratio precision data. It also helps to be working with relatively healthy ion signals for the three Pb isotopes, ^{206}Pb, ^{207}Pb and ^{208}Pb (24.1%, 22.1%, and 52.4% abundance, respectively). If the isotopic signals were dramatically different as in the measurement of two of the uranium isotopes, ^{235}U to ^{238}U, which are 0.72% and 99.2745% abundant, respectively, then

TABLE 12.2

Impact of Integration Time on the Overall Analysis Time for Pb Isotope Ratios

Dwell time (ms)	No. of sweeps	Integration time (s)/mass	% RSD $^{207}Pb^+/^{206}Pb^+$	% RSD $^{207}Pb^+/^{206}Pb^+$	Time for 9 reps (min/s)
25	220	5.5 s	0.24	0.25	2 m 29 s
25	500	12.5 s	0.21	0.19	6 m 12 s
25	700	17.5 s	0.20	0.17	8 m 29 s

Source: L. Halicz, Y. Erel, and A. Veron, *Atomic Spectroscopy,* **17**(5), 186–189, 1996.

the ability to optimize the measurement protocol for individual isotopes becomes of even greater importance to guarantee good precision data.[10]

12.5 DATA QUALITY OBJECTIVES FOR SINGLE-PARTICLE ICP-MS STUDIES

The ability to optimize the measurement protocol is even more critical when using ICP-MS to characterize nanomaterials using the single-particle mode of analysis. Engineered nanomaterials, as they are known, refer to the process of producing materials that contain particles < 100 nm in size. They often possess different properties compared to bulk materials of the same composition, making them of great interest to a broad spectrum of industrial and commercial applications. Unfortunately many of these nanomaterials, once they get into the environment, have proved to be harmful to humans. So in order to better understand the impact of nanoparticles on human health, several key properties need to be assessed, such as concentration, composition, particle size and shape. Recent studies have shown that ICP-MS is proving to be a critical tool to characterize nanoparticles using the single-particle technique.[11] However, these nanoparticles only exist for a few milliseconds, so the ability to measure them with good accuracy and precision is dependent on optimizing the dwell time, settling time and the speed of the data-acquisition electronics. If the measurement protocol is not handled correctly, it can significantly impact the quality of data collected.

12.6 FINAL THOUGHTS

It is clear that the analytical demands put on ICP-MS are probably higher than any other trace element technique, because it is continually being asked to solve a wide variety of application problems at increasingly lower levels. However, by optimizing the measurement protocol to fit the analytical requirement, ICP-MS has shown that it has the unique capability to carry out rapid trace element analysis, with superb detection limits and good precision on both continuous and transient signals, and still meet the most stringent data quality objectives.

FURTHER READING

1. *Integrated MCA Technology in the ELAN ICP-Mass Spectrometer*, Application Note TSMS-25, PerkinElmer Instruments, 1993.
2. E. R. Denoyer, *Atomic Spectroscopy*, **13**(3), 93–98, 1992.
3. E. R. Denoyer and Q. H. Lu, *Atomic Spectroscopy*, **14**(6), 162–169, 1993.
4. T. Catterick, H. Handley, and S. Merson, *Atomic Spectroscopy*, **16**(10), 229–234, 1995.
5. T. A. Hinners, E. M. Heithmar, T. M. Spittler, and J. M. Henshaw, *Analytical Chemistry*, **59**, 2658–2662, 1987.

6. M. Janghorbani, B. T. G. Ting, and N. E. Lynch, *Microchemica Acta*, **3**, 315–328, 1989.
7. J. Walder and P. A. Freeman, *Journal of Analytical Atomic Spectrometry*, **7**, 571–576, 1992.
8. F. Vanhaecke, L. Moens, R. Dams, L. Allen, and S. Georgitis, *Analytical Chemistry*, **71**, 3297–3303, 1999.
9. L. Halicz, Y. Erel, and A. Veron, *Atomic Spectroscopy*, **17**(5), 186–189, 1996.
10. D. R. Bandura and S. D. Tanner, *Atomic Spectroscopy*, **20**(2), 69–72, 1999.
11. D. M. Mitrano, E. K. Leshner, A. Bednar, J. Monserud, C. P. Hig Gins, and J. F. Ranville, Detection of Nanoparticulate Silver Using Single Particle Inductively Coupled Plasma Mass Spectrometry, *Environmental Toxicology and Chemistry*, **31**(1), 115–141, 2014.

13 Methods of Quantitation

There are many different ways to carry out trace element analysis by ICP-MS, depending on your data quality objectives. Such is the flexibility of the technique that it allows detection from sub-ppt up to high-ppm levels using a wide variety of calibration methods, from full quantitative and semi-quantitative analysis to one of the very powerful isotope-ratio techniques. Chapter 13 looks at the most common quantitation methods available in ICP-MS.

This ability of ICP-MS to carry out isotopic measurements allows the technique to carry out quantitation methods that are not available to any other trace element technique. They include the following:

- Quantitative analysis
- Semi-quantitative routines
- Isotope dilution
- Isotope ratio
- Internal standardization

Each of these techniques offers varying degrees of accuracy and precision; so, it is important to understand their strengths and weaknesses to know which one will best meet the data quality objectives of the analysis. In this chapter we will focus on using methods of quantitation for carrying out the analysis of liquids using continuous nebulization. However, even though the principles of calibration are similar, we covered the issues of quantitation of transient peaks in the previous chapter on "Peak Measurement Protocol" and will deal with specific sampling accessories in Chapter 18 on "Performance and Productivity Enhancement Techniques".

Anyway, let's first take a look at each of the methods of quantitation in greater detail.

13.1 QUANTITATIVE ANALYSIS

As in more trace element techniques such as AA and ICP-OES, quantitative analysis in ICP-MS is the fundamental tool used to determine analyte concentrations in unknown samples. In this mode of operation, the instrument is calibrated by measuring the intensity for all elements of interest in a number of known calibration standards that represent a range of concentrations likely to be encountered in your unknown samples. When the full range of calibration standards and blank have been run, the software creates a calibration curve of the measured intensity versus concentration for each element in the standard solutions. Once calibration data are acquired, the unknown samples are analyzed by plotting the intensity of the elements of interest against the respective calibration curves. The software then calculates the concentrations for the analytes in the unknown samples.

This type of calibration is often called external standardization and is usually used when there is very little difference between the matrix components in the standards and the samples. However, when it is difficult to closely match the matrix of the standards with the samples, external standardization can produce erroneous results, because matrix-induced interferences will change analyte sensitivity based on the amount of matrix present in the standards and samples. When this occurs, better accuracy is achieved by using the method of standard addition or a similar approach called addition calibration. Let us look at these three variations of quantitative analysis to see how they differ.

13.2 EXTERNAL STANDARDIZATION

As explained earlier, this involves measuring a blank solution followed by a set of standard solutions to create a calibration curve over the anticipated concentration range. Typically, a blank and three standards containing different analyte concentrations are run. Increasing the number of points on the calibration curve by increasing the number of standards may improve accuracy in circumstances where the calibration range is very broad. However, it is seldom necessary to run a calibration with more than five standards. After the standards have been measured, the unknown samples are analyzed and their analyte intensities read against the calibration curve. Over extended analysis times, it is common practice to update the calibration curve, either by recalibrating the instrument with a full set of standards or by running one midpoint standard. The following protocol summarizes a typical calibration using external standardization:

1. Blank >
2. Std. 1 >
3. Std. 2 >
4. Std. 3 >
5. Sample 1 >
6. Sample 2 >
7. Sample . . . n
8. Recalibrate
9. Sample $n + 1$, etc.

This can be seen more clearly in Figure 13.1, which shows a typical calibration curve using a blank and three standards of 2, 5 and 10 ppb. This calibration curve shows a simple *linear regression* but usually other modes of calibration are also available, like *weighted linear* to emphasize measurements at the low concentration region of the curve and *linear through zero*, where the linear regression is forced through zero. Whatever approach is used, it is critically important to select your range of standards, based on the expected concentration levels in your samples, if you want to ensure to optimum accuracy.

It should be emphasized that this graph represents a single-element calibration. However, because ICP-MS is usually used for multielement analysis, multielement standards are typically used to generate calibration data. For that reason, it is absolutely essential to use multielement standards that

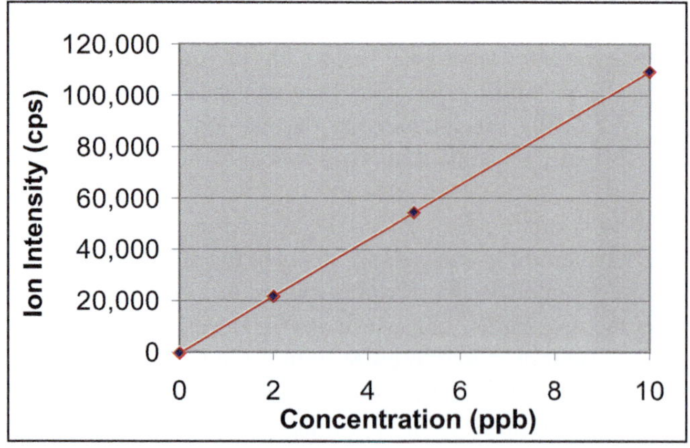

FIGURE 13.1 A simple linear regression calibration curve.

have been manufactured specifically for ICP-MS. Single-element AA standards are not suitable, because they usually have only been certified for the analyte element and not for any others. The purity of the standard cannot be guaranteed for any other element and as a result cannot be used to make up multielement standards for use with ICP-MS. For the same reason, ICP-OES multielement standards are not advisable either, because they are only certified for a group of elements and could contain other elements at higher levels, which will affect the ICP-MS multielement calibration.

13.3 STANDARD ADDITIONS

This mode of calibration provides an effective way to minimize sample-specific matrix effects by spiking samples with known concentrations of analytes.[1,2] In standard addition calibration, the intensity of a blank solution is first measured. Next, the sample solution is "spiked" with known concentrations of each element to be determined. The instrument measures the response for the spiked samples and creates a calibration curve for each element for which a spike has been added. The calibration curve is a plot of the blank subtracted intensity of each spiked element against its concentration value. After creating the calibration curve, the un-spiked sample solutions are then analyzed and compared to the calibration curve. Based on the slope of the calibration curve and where it intercepts the x-axis, the instrument software determines the un-spiked concentration of the analytes in the unknown samples. This can be seen in Figure 13.2, which shows a calibration of the sample intensity and the sample spiked with 2 and 5 ppb of the analyte. The concentration of sample is where the calibration line intercepts the negative side of the x-axis.

The following protocol summarizes a typical calibration using the method of standard additions:

1. Blank >
2. Spiked Sample 1 (Spike Conc. 1) >
3. Spiked Sample 1 (Spike Conc. 2) >
4. Un-spiked Sample 1 >
5. Blank >
6. Spiked Sample 2 (Spike Conc. 1) >
7. Spiked Sample 2 (Spike Conc. 2) >
8. Un-spiked Sample 2 >
9. Blank >
10. Etc.

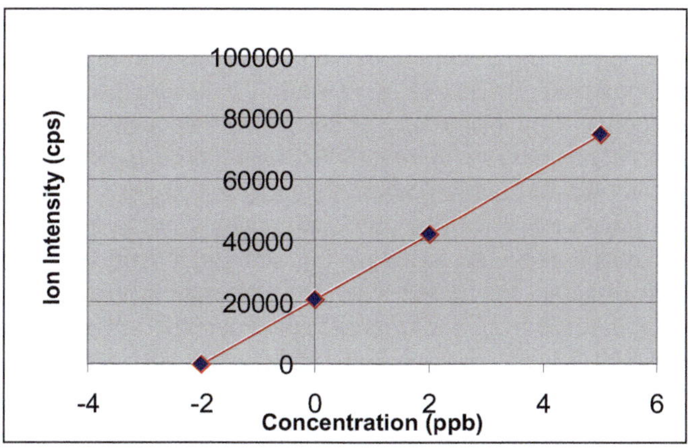

FIGURE 13.2 A typical "method of additions" calibration curve.

13.4 ADDITION CALIBRATION

Unfortunately, with the method of standard additions, each and every sample has to be spiked with all the analytes of interest, which becomes extremely labor-intensive when many samples have to be analyzed. For this reason, a variation of standard additions called *addition calibration* is more widely used in ICP-MS. However, this method can only be used when all the samples have a similar matrix. It uses the same principle as standard additions, but only the first (or representative) sample is spiked with known concentrations of analytes, and then analyzes the rest of the sample batch against the calibration, assuming all samples have a similar matrix to the first one. The following protocol summarizes a typical calibration using the method of addition calibration:

1. Blank >
2. Spiked Sample 1 (Spike Conc. 1) >
3. Spiked Sample 1 (Spike Conc. 2) >
4. Un-spiked Sample 1 >
5. Un-spiked Sample 2 >
6. Un-spiked Sample 3 >
7. Etc.

13.5 SEMIQUANTITATIVE ANALYSIS

If your data quality objectives for accuracy and precision are less stringent, ICP-MS offers a very rapid semiquantitative mode of analysis. This technique enables you to automatically determine the concentrations of up to 75 elements in an unknown sample, without the need for calibration standards.[3,4] This is an approach that could be extremely useful for initially screening samples, before quantitative analysis is carried out.

There are slight variations in the way different instruments approach semiquantitative analysis, but the general principle is to measure the entire mass spectrum, without specifying individual elements or masses. It relies on the principle that each element's natural isotopic abundance is fixed. By measuring the intensity of all their isotopes, correcting for common spectral interferences, including molecular, polyatomic and isobaric species and applying heuristic, knowledge-driven routines in combination with numerical calculations, a positive or negative confirmation can be made for each element present in the sample. Then, by comparing the corrected intensities against a stored isotopic response table, a good semiquantitative approximation of the sample components can be made.

Semiquant, as it is often called, is an excellent approach to rapidly characterizing unknown samples. Once the sample has been characterized, you can choose to either update the response table with your own standard solutions to improve analytical accuracy, or switch to the quantitative analysis mode to focus on specific elements and determine their concentrations with even greater accuracy and precision. Whereas a semiquantitative determination can be performed without using a series of standards, the use of a small number of standards is highly recommended for improved accuracy across the full mass range. Unlike traditional quantitative analysis, in which you analyze standards for all the elements you want to determine, semiquant calibration is achieved using just a few elements distributed across the mass range. This calibration process, shown more clearly in Figure 13.3, is used to update the reference response curve data that correlates measured ion intensities to the concentrations of elements in a solution. During calibration, this response data is adjusted to account for changes in the instrument's sensitivity due to variations in the sample matrix.

This process is often called semiquantitative analysis using external calibration, and, like traditional quantitative analysis using external standardization, works extremely well for samples that have a similar matrix. However, if you are analyzing samples containing widely different

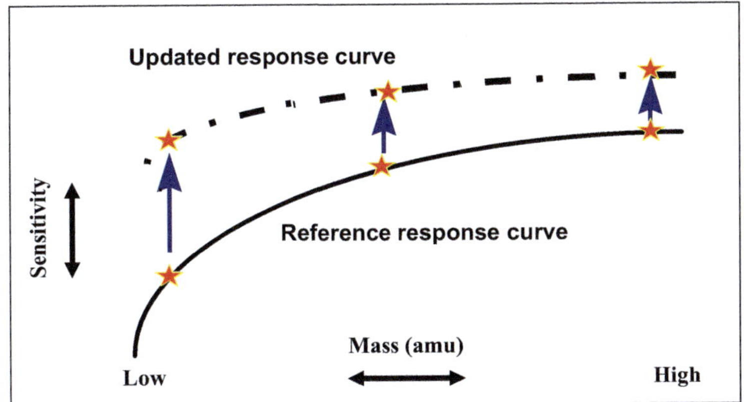

FIGURE 13.3 In semiquantitative analysis, a small group of elements are used to update the reference response curve to improve the accuracy as the sample matrix changes.

concentrations of matrix components, external calibration does not work very well because of matrix-induced suppression effects on the analyte signal. If this is the case, semiquant using a variation of standard addition calibration should be used. Similar to standard addition calibration used in quantitative analysis, this procedure involves adding known quantities of specific elements to every unknown sample before measurement. The major difference with semiquant is that the elements you add must not already be present in significant quantities in the unknown samples, because they are being used to update the stored reference response curve. As with external calibration, the semiquant software then adjusts the stored response data for all remaining analytes relative to the calibration elements. This procedure works very well, but tends to be very labor-intensive because the calibration standards have to be added to every unknown sample.

The other factor to be wary of with semiquantitative analysis is the spectral complexity of unknown samples. If you have a spectrally rich sample and are not making any compensation for spectral overlaps close to the analyte peaks, it could possibly give you a false positive for that element. Therefore, you have to be very cautious when reporting semiquantitative results on completely unknown samples. They should be characterized first, especially with respect to the types of spectral interferences generated by the plasma gas, the matrix and the solvents/acids/chemicals used for sample preparation. Collision/reaction cells/interfaces can help in the reduction of some of these interferences, but extreme care should be taken, as these devices are known to have no effect on some polyatomic interferences, and in some cases can increase the spectral complexity by generating other interfering complexes.

13.6 ISOTOPE DILUTION

Although quantitative and semiquantitative analysis methods are suitable for the majority of applications, there are other calibration methods available, depending on your analytical requirements. For example, if your application requires even greater accuracy and precision, the "isotope dilution" technique may offer some benefits. Isotope dilution is an absolute means of quantitation based on altering the natural abundance of two isotopes of an element by adding a known amount of one of the isotopes, and is considered one of the most accurate and precise approaches to elemental analysis.[5,6,7,8]

For this reason, a prerequisite of isotope dilution is that the element must have at least two stable isotopes. The principle works by spiking a known weight of an enriched stable isotope into your sample solution. By knowing the natural abundance of the two isotopes being measured, the abundance of the spiked enriched isotopes, the weight of the spike and the weight of the sample, the original trace element concentration can be determined by using the following equation:

$$C = \frac{\left[\text{Aspike} - \left(R \times \text{Bspike}\right)\right] \times \text{Wspike}}{\left[R \times \left(\text{Bsample} - \text{Asample}\right)\right] \times \text{Wsample}}$$

where
  C = Concentration of trace element
  Aspike = % of higher abundance isotope in spiked enriched isotope
  Bspike = % of lower abundance isotope in spiked enriched isotope
  Wspike = Weight of spiked enriched isotope
  R = Ratio of the % of higher abundance isotope to lower abundance isotope in the spiked sample
  Bsample = % of higher natural abundance isotope in sample
  Asample = % of lower natural abundance isotope in sample
  Wsample = Weight of sample

This might sound complicated, but in practice, it is relatively straightforward. This is illustrated in Figure 13.4, which shows an isotope dilution method for the determination of copper in a 250 mg sample of orchard leaves, using the two copper isotopes ^{63}Cu and ^{65}Cu.

In the bar graph on top, it can be seen that the natural abundance of the two isotopes are 69.09% and 30. 91% respectively, for ^{63}Cu and ^{65}Cu. In the middle graph, it shows that 4 μg of an enriched isotope of 100% ^{65}Cu (and 0 % ^{63}Cu) is spiked into the sample, which now produces a spiked sample containing 71.4% of ^{65}Cu and 28.6% of ^{63}Cu, as seen in the bottom plot.[9] If we plug these data into the preceding equation, we get:

$$C = \frac{\left[100 - \left(71.4/28.6 \times 0\right)\right] \times 4 \ \mu\text{g}}{\left[\left(71.4/28.6 \times 69.09\right) - 30.91\right] \times 0.25 \ \text{g}}$$
$$C = 400/35.45 = 11.3 \ \mu\text{g/g}$$

The major benefit of the isotope dilution technique is that it provides measurements that are extremely accurate because you are measuring the concentration of the isotopes in the same solution as your unknown sample, and not in a separate external calibration solution. In addition, because it is a ratioing technique, loss of solution during the sample preparation stage has no influence on the accuracy of the result. The technique is also extremely precise, because, using a simultaneous detection system such as a magnetic-sector multicollector or a simultaneous ion sampling device such as a TOF-ICP-MS, the results are based on measuring the two isotopes' solution at the same instant in time, which compensates for imprecision of the signal due to sources of sample introduction related noise, such as plasma instability, peristaltic pump pulsations and nebulization fluctuations. Even when using a scanning mass analyzer such as a quadrupole, the measurement protocol can be optimized to scan very rapidly between the two isotopes and achieve very good precision. However, isotope dilution has some limitations, which makes it suitable only for certain applications. These limitations include the following:

- The element you are determining must have more than one isotope, because calculations are based on the ratio of one isotope to another isotope of the same element; this makes it unsuitable for approximately 15 elements that can be determined by ICP-MS.
- It requires certified enriched isotopic standards, which can be very expensive, especially those that are significantly different from the normal isotopic abundance of the element.

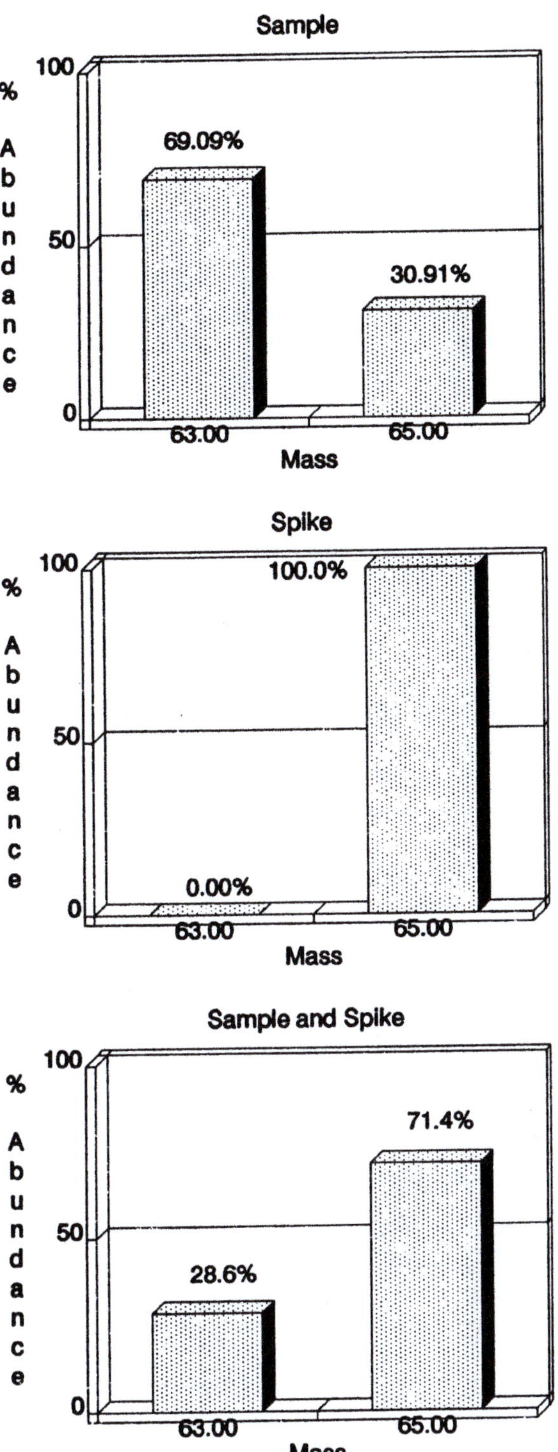

FIGURE 13.4   Quantitation of trace levels of copper in a sample of SRM orchard leaves using isotope dilution methodology.

(From *Multi-elemental Isotope Dilution Using the Elan* ICP-MS *Elemental Analyzer,* ICP-MS *Technical Summary TSMS-1,* PerkinElmer Instruments, 1985)

- It compensates for interferences due to signal enhancement or suppression, but does not compensate for spectral interferences. For this reason, an external blank solution must always be run.

13.7 ISOTOPE RATIOS

The ability of ICP-MS to determine individual isotopes also makes it suitable for another isotopic measurement technique called "isotope-ratio" analysis. The ratio of two or more isotopes in a sample can be used to generate very useful information, including an indication of the age of a geological formation, a better understanding of animal metabolism, and to help identify sources of environmental contamination.[10,11,12,13,14] Similar to isotope dilution, isotope-ratio analysis uses the principle of measuring the exact ratio of two isotopes of an element in the sample. With this approach, the isotope of interest is typically compared to a reference isotope of the same element. For example, you might want to compare the concentration of ^{204}Pb to that of ^{206}Pb. Alternatively, the requirement might be to compare one isotope to all remaining reference isotopes of an element, e.g., the ratio of ^{204}Pb to ^{206}Pb, ^{207}Pb and ^{208}Pb. The ratio is then expressed in the following manner:

$$\text{Isotope ratio} = \frac{\text{Intensity of isotope of interest}}{\text{Intensity of reference isotope}}$$

As this ratio can be calculated from within a single sample measurement, classic external calibration is not normally required. However, if there is a large difference between the concentrations of the two isotopes, it is recommended to run a standard of known isotopic composition. This is done to verify that the higher concentration isotope is not suppressing the signal of the lower concentration isotope and biasing the results. This effect, called mass discrimination, is less of a problem if the isotopes are relatively close in concentration, e.g., ^{107}Ag to ^{109}Ag, which are 51.839% and 48.161% abundant, respectively. However, it can be an issue if there is a significant difference in their concentration values, e.g., ^{235}U to ^{238}U, which are 0.72% and 99.275% abundant, respectively. Mass discrimination effects can be reduced by running an external reference standard of known isotopic concentration, comparing the isotope ratio with the theoretical value, and then mathematically compensating for the difference. The principles of isotope-ratio analysis and how to achieve optimum precision values are explained in greater detail in Chapter 12 on "Peak Measurement Protocol".

13.8 INTERNAL STANDARDIZATION

Another method of standardization commonly employed in ICP-MS is called *internal standardization*. It is not considered an absolute calibration technique, but instead is used to correct for changes in analyte sensitivity caused by variations in the concentration and type of matrix components found in the sample. An internal standard is a non-analyte isotope that is added to the blank solution, standards and samples before analysis. It should also be noted that the chosen internal standard element must not be present in the sample matrix. It is typical to add three or four internal standard elements to the samples to cover the analyte elements of interest. The software adjusts the analyte concentration in the unknown samples by comparing the intensity values of the internal standard intensities in the unknown sample to those in the calibration standards.

The implementation of internal standardization varies according to the analytical technique that is being used. For quantitative analysis, the internal standard elements are selected on the basis of the similarity of their ionization characteristics to the analyte elements. Each internal standard is bracketed with a group of analytes. The software then assumes that the intensities of all elements within a group are affected in a similar manner by the matrix. Changes in the ratios of the internal standard intensities are then used to correct the analyte concentrations in the unknown samples.

For semiquantitative analysis that uses a stored response table, the purpose of the internal standard is similar, but a little different in implementation to quantitative analysis. A semiquant internal standard is used to continuously compensate for instrument drift or matrix-induced suppression over a defined mass range. If a single internal standard is used, all the masses selected for the determination are updated by the same amount based on the intensity of the internal standard. If more than one internal standard is used, which is recommended for measurements over a wide mass range, the software interpolates the intensity values based on the distance in mass between the analyte and the nearest internal standard element.

It is worth emphasizing that if you do not want to compare your intensity values to a calibration graph, most instruments allow you to report raw data. This enables you to analyze your data using external data-processing routines, to selectively apply a minimum set of ICP-MS data processing methods, or to just view the raw data file before reprocessing it. The availability of raw data is primarily intended for use in non-routine applications such as chromatographyseparation techniques and laser sampling devices that produce a time-resolved transient peak or by users whose sample set requires data processing using algorithms other than those supplied by the instrument software.

FURTHER READING

1. D. Beauchemin, J. W. McLaren, A. P. Mykytiuk, and S. S. Berman, *Analytical Chemistry*, **59**, 778–783, 1987.
2. E. Pruszkowski, K. Neubauer, and R. Thomas, *Atomic Spectroscopy*, **19**(4), 111–115, 1998.
3. M. Broadhead, R. Broadhead, and J. W. Hager, *Atomic Spectroscopy*, **11**(6), 205–209, 1990.
4. E. Denoyer, *Journal of Analytical Atomic Spectrometry*, **7**, 1187–1193, 1992.
5. J. W. McLaren, D. Beauchemin, and S. S. Berman, *Analytical Chemistry*, **59**, 610–615, 1987.
6. H. Longerich, *Atomic Spectroscopy*, **10**(4), 112–115, 1989.
7. A. Stroh, *Atomic Spectroscopy*, **14**(5), 141–143, 1993.
8. T. Catterick, H. Handley, and S. Merson, *Atomic Spectroscopy*, **16**(10), 229–234, 1995.
9. *Multi-elemental Isotope Dilution Using the Elan ICP-MS Elemental Analyzer, ICP-MS Technical Summary TSMS-1*, PerkinElmer Instruments, 1985.
10. B. T. G. Ting and M. Janghorbani, *Analytical Chemistry*, **58**, 1334–1343, 1986.
11. M. Janghorbani, B. T. G. Ting, and N. E. Lynch, *Microchemica Acta*, **3**, 315–328, 1989.
12. T. A. Hinners, E. M. Heithmar, T. M. Spittler, and J. M. Henshaw, *Analytical Chemistry*, **59**, 2658–2662, 1987.
13. L. Halicz, Y. Erel, and A. Veron, *Atomic Spectroscopy*, **17**(5), 186–189, 1996.
14. M. Chaudhary-Webb, D. C. Paschal, W. C. Elliott, H. P. Hopkins, A. M. Ghazi, B. C. Ting, and I. Romieu, *Atomic Spectroscopy*, **19**(5), 156–166, 1998.

14 Review of ICP-MS Interferences

Now that we have covered the fundamental principles of ICP-MS and its measurement and calibration routines, Chapter 14 focuses on the most common interferences found in ICP-MS and the methods that are used to compensate for them. Although interferences are reasonably well understood in ICP-MS, it can often be difficult and time consuming to compensate for them, particularly in complex sample matrices. Prior knowledge of the interferences associated with a particular set of samples will often dictate the sample preparation steps and the instrumental methodology used to analyze them.

Interferences in ICP-MS are generally classified into three major groups: spectral, matrix and physical. Each of them has the potential to be problematic in its own right, but modern instrumentation and good software combined with optimized analytical methodologies has minimized their negative impact on trace element determinations by ICP-MS. Let us look at these interferences in greater detail and describe the different approaches used to compensate for them.

14.1 SPECTRAL INTERFERENCES

Spectral overlaps are probably the most serious types of interferences seen in ICP-MS. The most common are known as a polyatomic or molecular spectral interference and are produced by the combination of two or more atomic ions. They are caused by a variety of factors, but are usually associated with the plasma/nebulizer gas used, matrix components in the solvent/sample, other elements in the sample or entrained oxygen/nitrogen from the surrounding air. For example, in the argon plasma, spectral overlaps caused by argon ions and combinations of argon ions with other species are very common. The most abundant isotope of argon is at mass 40, which dramatically interferes with the most abundant isotope of calcium at mass 40, whereas the combination of argon and oxygen in an aqueous sample generates the $^{40}Ar^{16}O^+$ interference, which has a significant impact on the major isotope of Fe at mass 56. The complexity of these kinds of spectral problems can be seen in Figure 14.1, which shows a mass spectrum of deionized water from mass 40 to mass 90.

In addition, argon can form polyatomic interferences with elements found in the acids used to dissolve the sample. For example, in a hydrochloric acid medium, $^{40}Ar^+$ combines with the most abundant chlorine isotope at 35 amu to form $^{40}Ar^{35}Cl^+$, which interferes with the only isotope of arsenic at mass 75, whereas in an organic solvent matrix, argon and carbon combine to form $^{40}Ar^{12}C^+$, which interferes with $^{52}Cr^+$, the most abundant isotope of chromium. Sometimes, matrix or solvent ions combine to form spectral interferences of their own. A good example is in a sample that contains sulfuric acid. The dominant sulfur isotope, $^{32}S^+$, combines with two oxygen ions to form a $^{32}S^{16}O^{16}O^+$ molecular ion, which interferes with the major isotope of Zn at mass 64. In the analysis of samples containing high concentrations of sodium, such as seawater, the most abundant isotope of Cu at mass 63 cannot be used because of interference from the $^{40}Ar^{23}Na^+$ molecular ion. There are many more examples of these kinds of polyatomic and molecular interferences, which have been comprehensively reviewed.[1] Table 14.1 represents some of the most common matrix–solvent spectral interferences seen in ICP-MS.

DOI: 10.1201/9781003187639-14

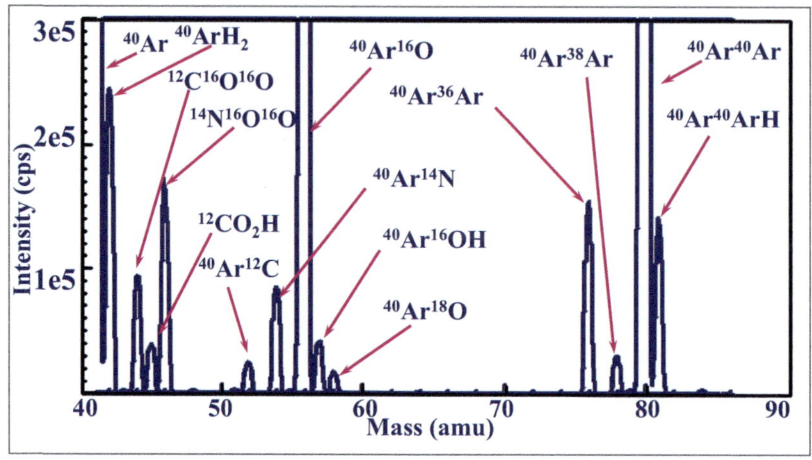

FIGURE 14.1   ICP mass spectrum of deionized water from mass 40 to mass 90.

TABLE 14.1

Some Common Plasma/Matrix/Solvent-Related Polyatomic Spectral Interferences Seen in ICP-MS

Element/isotope	Matrix/solvent	Interference
^{39}K$^+$	H_2O	^{38}ArH$^+$
^{40}Ca$^+$	H_2O	^{40}Ar$^+$
^{56}Fe$^+$	H_2O	^{40}Ar^{16}O$^+$
^{80}Se$^+$	H_2O	^{40}Ar^{40}Ar$^+$
^{51}V$^+$	HCl	^{35}Cl^{16}O$^+$
^{75}As$^+$	HCl	^{40}Ar^{35}Cl$^+$
^{28}Si$^+$	HNO_3	^{14}N^{14}N$^+$
^{44}Ca$^+$	HNO_3	^{14}N^{14}N ^{16}O$^+$
^{55}Mn$^+$	HNO_3	^{40}Ar^{15}N$^+$
^{48}Ti$^+$	H_2SO_4	^{32}S^{16}O$^+$
^{52}C$^+$r	H_2SO_4	^{34}S^{18}O$^+$
^{64}Zn$^+$	H_2SO_4	^{32}S^{16}O^{16}O$^+$
^{63}Cu$^+$	H_3PO_4	^{31}P^{16}O^{16}O$^+$
^{24}Mg$^+$	Organics	^{12}C^{12}C$^+$
^{52}Cr$^+$	Organics	^{40}Ar^{12}C$^+$
^{65}Cu$^+$	Minerals	^{48}Ca^{16}OH$^+$
^{64}Zn$^+$	Minerals	^{48}Ca^{16}O$^+$
^{63}Cu$^+$	Seawater	^{40}Ar^{23}Na$^+$

14.2   OXIDES, HYDROXIDES, HYDRIDES AND DOUBLY CHARGED SPECIES

Another type of spectral interference is produced by elements in the sample combining with H^+, ^{16}O$^+$ or ^{16}OH$^+$ (either from water or air) to form molecular hydrides (+ H^+), oxides (+ ^{16}O$^+$) and hydroxides (+ ^{16}OH$^+$), which occur at 1, 16 and 17 mass units, respectively, higher than the element's mass.[2]

TABLE 14.2

Some Elements That Readily Form Oxides, Hydroxides, Hydrides and Doubly Charged Species in the Plasma, Together with the Analytes Affected by the Interference

Oxide, hydroxide, hydride, doubly charged species	Analyte affected by interference
$^{40}Ca^{16}O^+$	$^{56}Fe^+$
$^{48}Ti^{16}O^+$	$^{64}Zn^+$
$^{98}Mo^{16}O^+$	$^{114}Cd^+$
$^{138}Ba^{16}O^+$	$^{154}Sm^+$, $^{154}Gd^+$
$^{139}La^{16}O^+$	$^{155}Gd^+$
$^{140}Ce^{16}O^+$	$^{156}Gd^+$, $^{156}Dy^+$
$^{40}Ca^{16}OH^+$	$^{57}Fe^+$
$^{31}P^{18}O^{16}OH^+$	$^{66}Zn^+$
$^{79}BrH^+$	$^{80}Se^+$
$^{31}P^{16}O_2H^+$	$^{64}Zn^+$
$^{138}Ba^{2+}$	$^{69}Ga^+$
$^{139}La^{2+}$	$^{69}Ga^+$
$^{140}Ce^{2+}$	$^{70}Ge^+$, $^{70}Zn^+$

These interferences are typically produced in the cooler zones of the plasma, immediately before the interface region. They are usually more serious when rare earth or refractory-type elements are present in the sample, because many of them readily form molecular species (particularly oxides), which create spectral overlap problems on other elements in the same group. If the oxide species is mainly derived from entrained air around the plasma, it can be reduced by either using an elongated outer tube to the torch or a metal shield between the plasma and the RF coil.

Associated with oxide-based spectral overlaps are doubly charged spectral interferences. These are species that are formed when an ion is generated with a double positive charge as opposed to a normal single charge and produces an isotopic peak at half its mass. Similar to the formation of oxides, the level of doubly charged species is related to the ionization conditions in the plasma and can usually be minimized by careful optimization of the nebulizer gas flow, RF power and sampling position within the plasma. It can also be impacted by the severity of the secondary discharge present at the interface,[3] which was described in greater detail in Chapter 5. Table 14.2 shows a selected group of elements that readily form oxides, hydroxides, hydrides and doubly charged species, together with the analytes affected by them.

14.3 ISOBARIC INTERFERENCES

The final classification of spectral interferences is called isobaric overlaps, produced mainly by different isotopes of other elements in the sample creating spectral interferences at the same mass as the analyte. For example, vanadium has two isotopes at 50 and 51 amu. However, mass 50 is the only practical isotope to use in the presence of a chloride matrix because of the large contribution from the $^{16}O^{35}Cl^+$ interference at mass 51. Unfortunately, mass 50 amu, which is only 0.25% abundant, also coincides with isotopes of titanium and chromium, which are 5.4% and 4.3% abundant, respectively. This makes the determination of vanadium in the presence of titanium and chromium very difficult unless mathematical corrections are made. Figure 14.2 shows all the possible naturally occurring isobaric spectral overlaps in ICP-MS.[4]

Relative Abundance of the Natural Isotopes

Isotope		%		%		%
1	H	99.985				
2	H	0.015				
3			He	0.000137		
4			He	99.999863		
5						
6					Li	7.5
7					Li	92.5
8						
9	Be	100				
10			B	19.9		
11			B	80.1		
12					C	98.90
13					C	1.10
14	N	99.643				
15	N	0.366				
16			O	99.762		
17			O	0.038		
18			O	0.200		
19					F	100
20	Ne	90.48				
21	Ne	0.27				
22	Ne	9.25				
23			Na	100		
24					Mg	78.99
25					Mg	10.00
26					Mg	11.01
27	Al	100				
28			Si	92.23		
29			Si	4.67		
30			Si	3.10		
31					P	100
32	S	95.02				
33	S	0.75				
34	S	4.21				
35			Cl	75.77		
36	S	0.02			Ar	0.337
37			Cl	24.23		
38					Ar	0.063
39	K	93.2581				
40	K	0.0117	Ca	96.941	Ar	99.600
41	K	6.7302				
42			Ca	0.647		
43			Ca	0.135		
44			Ca	2.086		
45					Sc	100
46	Ti	8.0	Ca	0.004		
47	Ti	7.3				
48	Ti	73.8	Ca	0.187		
49	Ti	5.5				
50	Ti	5.4	V	0.250	Cr	4.345
51			V	99.750		
52					Cr	83.789
53					Cr	9.501
54	Fe	5.8			Cr	2.365
55			Mn	100		
56	Fe	91.72				
57	Fe	2.2				
58	Fe	0.28			Ni	68.077
59			Co	100		
60					Ni	26.223

Isotope		%		%		%
61					Ni	1.140
62					Ni	3.634
63	Cu	69.17				
64			Zn	48.6	Ni	0.926
65	Cu	30.83				
66			Zn	27.9		
67			Zn	4.1		
68			Zn	18.8		
69					Ga	60.108
70	Ge	21.23	Zn	0.6		
71					Ga	39.892
72	Ge	27.66				
73	Ge	7.73				
74	Ge	35.94	Se	0.89		
75					As	100
76	Ge	7.44	Se	9.36		
77			Se	7.63		
78	Kr	0.35	Se	23.78		
79					Br	50.69
80	Kr	2.25	Se	49.61		
81					Br	49.31
82	Kr	11.6	Se	8.73		
83	Kr	11.5				
84	Kr	57.0	Sr	0.56		
85					Rb	72.165
86	Kr	17.3	Sr	9.86		
87			Sr	7.00	Rb	27.835
88			Sr	82.58		
89					Y	100
90	Zr	51.45				
91	Zr	11.22				
92	Zr	17.15	Mo	14.84		
93					Nb	100
94	Zr	17.38	Mo	9.25		
95			Mo	15.92		
96	Zr	2.80	Mo	16.68	Ru	5.52
97			Mo	9.55		
98			Mo	24.13	Ru	1.88
99					Ru	12.7
100			Mo	9.63	Ru	12.6
101					Ru	17.0
102	Pd	1.02			Ru	31.6
103			Rh	100		
104	Pd	11.14			Ru	18.7
105	Pd	22.33				
106	Pd	27.33	Cd	1.25		
107					Ag	51.839
108	Pd	26.46	Cd	0.89		
109					Ag	48.161
110	Pd	11.72	Cd	12.49		
111			Cd	12.80		
112	Sn	0.97	Cd	24.13		
113			Cd	12.22	In	4.3
114	Sn	0.65	Cd	28.73		
115	Sn	0.34			In	95.7
116	Sn	14.53	Cd	7.49		
117	Sn	7.68				
118	Sn	24.23				
119	Sn	8.59				
120	Sn	32.59	Te	0.096		

Isotope		%		%		%
121					Sb	57.36
122	Sn	4.63	Te	2.603		
123			Te	0.908	Sb	42.64
124	Sn	5.79	Te	4.816	Xe	0.10
125			Te	7.139		
126			Te	18.95	Xe	0.09
127	I	100				
128			Te	31.69	Xe	1.91
129					Xe	26.4
130	Ba	0.106	Te	33.80	Xe	4.1
131					Xe	21.2
132	Ba	0.101			Xe	26.9
133			Cs	100		
134	Ba	2.417			Xe	10.4
135	Ba	6.592				
136	Ba	7.854	Ce	0.19	Xe	8.9
137	Ba	11.23				
138	Ba	71.70	Ce	0.25	La	0.0902
139					La	99.9098
140			Ce	88.48		
141					Pr	100
142	Nd	27.13	Ce	11.08		
143	Nd	12.18				
144	Nd	23.80	Sm	3.1		
145	Nd	8.30				
146	Nd	17.19				
147			Sm	15.0		
148	Nd	5.76	Sm	11.3		
149			Sm	13.8		
150	Nd	5.64	Sm	7.4		
151					Eu	47.8
152	Gd	0.20	Sm	26.7		
153					Eu	52.2
154	Gd	2.18	Sm	22.7		
155	Gd	14.80				
156	Gd	20.47	Dy	0.06		
157	Gd	15.65				
158	Gd	24.84	Dy	0.10		
159					Tb	100
160	Gd	21.86	Dy	2.34		
161			Dy	18.9		
162	Er	0.14	Dy	25.5		
163			Dy	24.9		
164	Er	1.61	Dy	28.2		
165					Ho	100
166	Er	33.6				
167	Er	22.95				
168	Er	26.8	Yb	0.13		
169					Tm	100
170	Er	14.9	Yb	3.05		
171			Yb	14.3		
172			Yb	21.9		
173			Yb	16.12		
174			Yb	31.8	Hf	0.162
175	Lu	97.41				
176	Lu	2.59	Yb	12.7	Hf	5.206
177					Hf	18.606
178					Hf	27.297
179					Hf	13.629
180	Ta	0.012	W	0.13	Hf	35.100

Isotope		%		%		%
181	Ta	99.988				
182			W	26.3		
183			W	14.3		
184	Os	0.02	W	30.67		
185					Re	37.40
186	Os	1.58	W	28.6		
187	Os	1.6			Re	62.60
188	Os	13.3				
189	Os	16.1				
190	Os	26.4			Pt	0.01
191			Ir	37.3		
192	Os	41.0			Pt	0.79
193			Ir	62.7		
194					Pt	32.9
195					Pt	33.8
196	Hg	0.15			Pt	25.3
197			Au	100		
198	Hg	9.97			Pt	7.2
199	Hg	16.87				
200	Hg	23.10				
201	Hg	13.18				
202	Hg	29.86				
203					Tl	29.524
204	Hg	6.87	Pb	1.4		
205					Tl	70.476
206			Pb	24.1		
207			Pb	22.1		
208			Pb	52.4		
209	Bi	100				
210						
211						
212						
213						
214						
215						
216						
217						
218						
219						
220						
221						
222						
223						
224						
225						
226						
227						
228						
229						
230						
231	Pa	100				
232	Th	100				
233						
234	U	0.0055				
235	U	0.7200				
236						
237						
238	U	99.2745				

"Isotopic Compositions of the Elements 1989" Pure Appl. Chem., Vol. 63, No. 7, pp. 991-1002, 1991. © 1991 IUPAC

FIGURE 14.2   Relative isotopic abundances of the naturally occurring elements, showing all the potential isobaric interferences.

(From UIPAC Isotopic Composition of the Elements, *Pure and Applied Chemistry* **75**[6], 683–799, 2003)

14.4 WAYS TO COMPENSATE FOR SPECTRAL INTERFERENCES

Let us now look at the different approaches used to compensate for spectral interferences. One of the very first ways used to get around severe matrix-derived spectral interferences was to remove the matrix somehow. In the early days, this involved precipitating the matrix with a complexing agent and then filtering off the precipitate. However, more recently, this has been carried out by automated matrix-removal/analyte-preconcentration techniques using chromatography-type equipment. In fact, this is the preferred method for carrying out trace metal determinations in seawater, because of the matrix and spectral problems associated with such high concentrations of sodium and chloride ions.[5]

14.5 MATHEMATICAL CORRECTION EQUATIONS

Another method that has been successfully used to compensate for isobaric interferences and some less severe polyatomic overlaps (when no alternative isotopes are available for quantitation) is to use mathematical interference correction equations. Similar to interelement corrections (IECs) in ICP-OES, this method works on the principle of measuring the intensity of the interfering isotope or interfering species at another mass, which is ideally free of any interferences. A correction is then applied by knowing the ratio of the intensity of the interfering species at the analyte mass to its intensity at the alternate mass. Let us look at a "real-world" example to exemplify this type of correction. The most sensitive isotope for cadmium is at mass 114. However, there is also a minor isotope of tin at mass 114. This means that if there is any tin in the sample, quantitation using $^{114}Cd^+$ can only be carried out if a correction is made for $^{114}Sn^+$. Fortunately, Sn has a total of ten isotopes, which means that probably at least one of them is going to be free of a spectral interference. Therefore, by measuring the intensity of Sn at one of its most abundant isotopes (typically, $^{118}Sn^+$) and ratioing it to $^{114}Sn^+$, a correction is made in the method software in the following manner:

$$\text{Total counts at mass } 114 = {}^{114}Cd^+ + {}^{114}Sn^+$$
$$\text{Therefore, } {}^{114}Cd^+ = \text{total counts at mass } 114 - {}^{114}Sn^+$$

To find out the contribution from $^{114}Sn^+$, it is measured at the interference-free isotope of $^{118}Sn^+$ and a correction of the ratio of $^{114}Sn^+/^{118}Sn^+$ is applied, which means $^{114}Cd^+$ = Counts at mass $114 - (^{114}Sn^+/^{118}Sn^+) \times (^{118}Sn^+)$.

Now, the ratio $(^{114}Sn^+/^{118}Sn^+)$ is the ratio of the natural abundances of these two isotopes (065%/24.23%) and is always constant.

$$\text{Therefore, } {}^{114}Cd^+ = \text{mass } 114 - (0.65\%/24.23\%) \times (^{118}Sn^+)$$
$$\text{or } {}^{114}Cd^+ = \text{mass } 114 - (0.0268) \times (^{118}Sn^+)$$

An interference correction for $^{114}Cd^+$ would then be entered in the software as

$$- (0.0268) \times (^{118}Sn^+)$$

This is a relatively simple example, but explains the basic principles of the process. In practice, especially in spectrally complex samples, corrections often have to be made to the isotope being used for the correction in addition to the analyte mass, which makes the mathematical equation far more complex.

This approach can also be used for some less severe polyatomic-type spectral interferences. For example, in the determination of V at mass 51 in diluted brine (typically 1,000 ppm NaCl), there is a substantial spectral interference from $^{35}C^{16}O^+$ at mass 51. By measuring the intensity of the $^{37}C^{16}O^+$ at mass 53, which is free of any interference, a correction can be applied in a similar way to the previous example.

14.6　COOL/COLD PLASMA TECHNOLOGY

If the intensity of the interference is large, and analyte intensity is extremely low, mathematical equations are not ideally suited as a correction method. For that reason, alternative approaches have to be considered to compensate for the interference. One such approach, which has helped to reduce some of the severe polyatomic overlaps, is to use cold/cool plasma conditions. This technology, which was reported in the literature in the late 1980s, uses a low-temperature plasma to minimize the formation of certain argon-based polyatomic species.[6] Under normal plasma conditions (typically 1,000–1,400 W RF power, 15–20 L/min coolant flow and 0.8–1.0 L/min of nebulizer gas flow), argon ions combine with matrix and solvent components to generate problematic spectral interferences such as $^{38}ArH^+$, $^{40}Ar^+$ and $^{40}Ar^{16}O^+$, which impact the detection limits of a small number of elements including K, Ca and Fe. By using cool plasma conditions (e.g. 500–800 W RF power, 10 L/min coolant flow and 1.5–1.8 L/min nebulizer gas flow), the ionization conditions in the plasma are changed so that many of these interferences are dramatically reduced. The result is that detection limits for this group of elements are significantly enhanced.[7] An example of this improvement is shown in Figure 14.3. It shows a spectral scan of 100 ppt of $^{56}Fe^+$ (its most sensitive isotope) using cool plasma conditions. It can be clearly seen that there is virtually no contribution from $^{40}Ar^{16}O^+$, as indicated by the extremely low background for deionized water, resulting in single-figure ppt detection limits for iron. Under normal plasma conditions, the $^{40}Ar^{16}O^+$ intensity is so large that it would completely overlap the $^{56}Fe^+$ peak.[8]

Unfortunately, even though the use of cool plasma conditions is recognized as being a very useful tool for the determination of a small group of elements, its limitations are well documented.[9] A summary of the limitations of cool-plasma technology includes the following:

- As a result of less energy being available in a cool plasma, elements that form a strong bond with one of the matrix/solvent ions cannot be easily decomposed and, as a result, their detection limits are compromised.
- Elements with high ionization potentials cannot be ionized, because there is much less energy compared to a normal, high-temperature plasma.
- Elemental sensitivity is severely affected by the sample matrix; so, cool plasma often requires the use of standard additions or matrix matching to achieve satisfactory results.
- When carrying out multielement analysis, normal plasma conditions must also be used. This necessitates the need for stabilization times on the order of 3 min to change from a normal to a cool plasma, which degrades productivity and results in higher sample consumption.

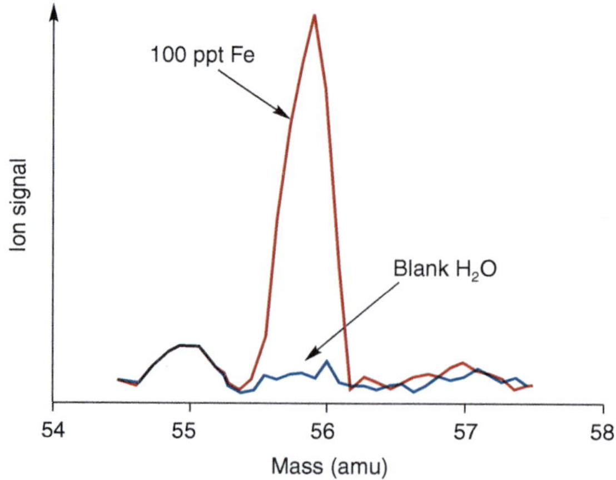

FIGURE 14.3　Spectral scan of 100 ppt ^{56}Fe and deionized water using cool plasma conditions.

(From S. D. Tanner, M. Paul, S. A. Beres, and E. R. Denoyer, *Atomic Spectroscopy*, **16**[1], 16, 1995)

For this reason, it is not ideally suited for the analysis of complex samples, but it does offer real detection-limit improvement for elements with low ionization potential, such as sodium and lithium, which benefit from the ionization conditions of the cooler plasma. However, recent advances in solid-state, free-running RF generators appear to have enhanced the capability of cold-plasma technology. Researchers have reported that by using the combination of highly reactive gases in a collision/reaction cell with cold plasma conditions, efficient reduction of some polyatomic interferences was achieved. This new approach allowed ultra-trace levels of a suite of elements to be accurately measured in high-purity hydrochloric acid using both cold- and normal-plasma operating conditions.[10]

14.7 COLLISION/REACTION CELLS

The earlier limitations of cool-plasma technology have led to the development of collision/reaction cells and interfaces, which utilize ion–molecule collisions and reactions to cleanse the ion beam of harmful polyatomic and molecular interferences before they enter the mass analyzer. Quadrupole mass analyzers fitted with these devices are showing enormous potential to eliminate many spectral interferences, allowing the use of the most sensitive elemental isotopes that were previously unavailable for quantitation.

A full description and review of collision/reaction cell and interface technology and how they handle spectral interferences was given in Chapter 10. However, it's worth noting that one of the unique features of using reaction chemistry in a collision/reaction cell is that it can be used in the mass-shift mode, which is particularly beneficial if a spectral interference is encountered with an analyte which is monoisotopic (only one mass is available for quantitation). An example of this is that when high levels of chloride are present in the sample (whether it's from the matrix itself or from hydrochloric acid used in the sample digestion procedure), it can be problematic to quantitate arsenic and vanadium at low levels, because of polyatomic spectral interferences $^{40}Ar^{75}Cl$ and $^{35}Cl^{16}O$, which interfere with the major isotopes of ^{75}As and ^{51}V respectively. A collision cell with KED can reduce these interferences to a certain level, but if the requirement is for ultra trace determinations of these analytes, reaction chemistry is the better option. Pure ammonia works extremely well to reduce the $^{35}Cl^{16}O$ interference when determining trace levels of vanadium, while oxygen gas is best suited for the determination of low levels of arsenic because only one mass (^{75}As) is available for quantitation. When using oxygen to determine arsenic, reaction chemistry works by implementing the mass-shift mode. In this approach, the arsenic reacts rapidly with oxygen to form $^{75}As^{16}O$ at mass 91, 16 amu away from the $^{40}Ar^{35}Cl$ interference at mass 75. Because $^{40}Ar^{35}Cl$ does not react with oxygen, $^{75}As^{16}O$ is measured free of interferences at 91 amu. Figure 14.4 exemplifies this,

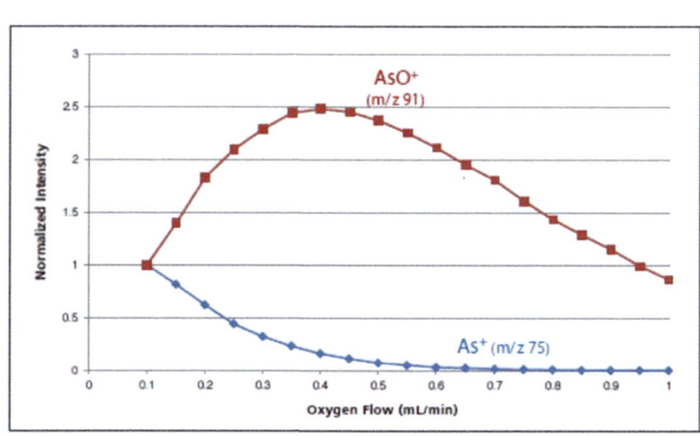

FIGURE 14.4 Conversion of ^{75}As to $^{75}As^{16}O$ as a function of oxygen flow.

showing the conversion of ^{75}As to ^{75}As^{16}O as a function of oxygen flow. As the O$_2$ flow increases, the signal for ^{75}As decreases (blue plot), while the signal for ^{75}As^{16}O increases (red plot), demonstrating complete conversion.[10]

14.8 HIGH-RESOLUTION MASS ANALYZERS

The best and probably most efficient way to remove spectral overlaps is to resolve them using a high-resolution mass spectrometer.[11] Over the past ten years, this approach, particularly double-focusing magnetic-sector mass analyzers, has proved to be invaluable for separating many of the problematic polyatomic and molecular interferences seen in ICP-MS, without the need to use cool-plasma conditions or collision/reaction cells. This can be seen in Figure 14.5, which shows a spectral peak for 10 ppb of ^{75}As$^+$ resolved from the ^{40}Ar^{35}Cl$^+$ interference in a 1% hydrochloric acid matrix, using a resolution setting of 5,000.[12]

Although their resolving capability is far more powerful than quadrupole-based instruments, there is a sacrifice in sensitivity if extremely high resolution is used, which can often translate into a degradation in detection capability for some elements, compared to other spectral interference correction approaches. A full review of magnetic-sector technology for ICP-MS is given in Chapter 11.

14.9 MATRIX INTERFERENCES

Let us now look at the other class of interference in ICP-MS—suppression of the signal by the matrix itself. There are basically three types of matrix-induced interferences. The first and simplest to overcome is often called a *sample transport effect* and is a physical suppression of the analyte signal, brought on by the level of dissolved solids or acid concentration in the sample. It is caused by the sample's impact on droplet formation in the nebulizer or droplet-size selection in the spray chamber. In the case of organic matrices, it is usually caused by variations in the pumping rate of

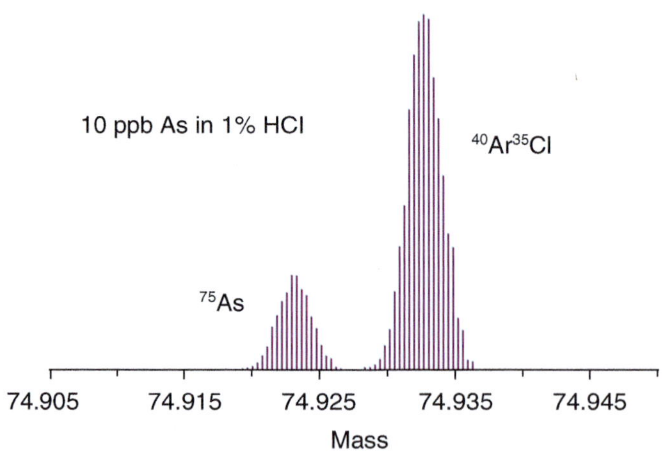

FIGURE 14.5 Separation of ^{75}As$^+$ from ^{40}Ar^{35}Cl$^+$ using high resolving power (5,000) of a double-focusing magnetic-sector instrument.

(From W. Tittes, N. Jakubowski, and D. Stuewer, Poster Presentation at Winter Conference on Plasma Spectrochemistry, San Diego, 1994)

solvents with different viscosities. The second type of matrix suppression is caused when the sample affects the ionization conditions of the plasma discharge. This results in the signal being suppressed by varying amounts, depending on the concentration of the matrix components. This type of interference is exemplified when different concentrations of acids are aspirated into cool plasma. The ionization conditions in the plasma are so fragile that higher concentrations of acid result in severe suppression of the analyte signal. This can be seen very clearly in Figure 14.6, which shows sensitivity for a selected group of elements in varying concentrations of nitric acid in a cool plasma.[9]

14.10   COMPENSATION USING INTERNAL STANDARDIZATION

The classic way to compensate for a physical interference is to use internal standardization (IS). With this method of correction, a small group of elements (usually at the ppb level) are spiked into the samples, calibration standards and blank to correct for any variations in the response of the elements caused by the matrix. As the intensities of the internal standards change, the element responses are updated every time a sample is analyzed. The following criteria are typically used for selecting an internal standard:

- It is not present in the sample.
- The sample matrix or analyte elements do not spectrally interfere with it.
- It does not spectrally interfere with the analyte masses.
- It should not be an element that is considered an environmental contaminant.
- It is usually grouped with analyte elements of a similar mass range. For example, a low-mass internal standard is grouped with the low-mass analyte elements, and so on, up the mass range.
- It should be of a similar ionization potential to the group of analyte elements, so it behaves in a similar manner in the plasma.

Some of the most common elements/masses reported to be good candidates for internal standards include ^{9}Be, ^{45}Sc, ^{59}Co, ^{74}Ge, ^{89}Y, ^{103}Rh, ^{115}In, ^{169}Tm, ^{175}Lu, ^{187}Re and ^{232}Th. An internal standard is

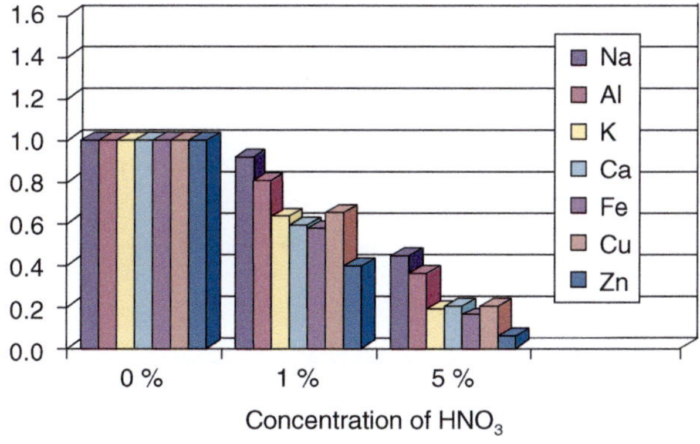

FIGURE 14.6   Matrix suppression caused by increasing concentrations of HNO$_3$, using cool-plasma conditions (RF power: 800 W, nebulizer gas: 1.5 L/min).

(From J. M. Collard, K. Kawabata, Y. Kishi, and R. Thomas, *Micro*, January 2002)

also used to compensate for long-term signal drift as a result of matrix components slowly blocking the sampler and skimmer-cone orifices. Even though total dissolved solids are usually kept below 0.2% in ICP-MS, this can still produce instability of the analyte signal over time with some sample matrices. It should also be emphasized that the difference in intensities of the internal standard elements across the mass range will indicate the flatness of the mass response curve. The flatter the mass response curve (i.e., less mass discrimination), the easier it is to compensate for matrix-based suppression effects using IS.

14.11   SPACE CHARGE-INDUCED MATRIX INTERFERENCES

Many of the early researchers reported that the magnitude of signal suppression in ICP-MS increased with decreasing atomic mass of the analyte ion.[12] More recently, it has been suggested that the major cause of this kind of suppression is the result of poor transmission of ions through the ion optics due to matrix-induced space charge effects.[13] This has the effect of defocusing the ion beam, which leads to poor sensitivity and detection limits, especially when trace levels of low-mass elements are being determined in the presence of large concentrations of high-mass matrices. Unless any compensation is made, the high-mass matrix element will dominate the ion beam, pushing the lighter elements out of the way.[14] This can be seen in Figure 14.7, which shows the classic space charge effects of a uranium (major isotope $^{238}U^+$) matrix on the determination of $^7Li^+$, $^9Be^+$, $^{24}Mg^+$, $^{55}Mn^+$, $^{85}Rb^+$, $^{115}In^+$, $^{133}Cs^+$, $^{205}Tl^+$ and $^{208}Pb^+$. It can clearly be seen that the suppression of the low-mass elements such as Li and Be is significantly higher than with the high-mass elements such as Tl and Pb in the presence of 1,000 ppm uranium.

There are a number of ways to compensate for space charge matrix suppression in ICP-MS. IS has been used but unfortunately does not address the fundamental cause of the problem. The most common approach used to alleviate or at least reduce space charge effects is to apply voltages to individual lens components of the ion optics. This is achieved in a number of different ways, but irrespective of the design of the ion-focusing system, its main function is to reduce matrix-based suppression effects by steering as many of the analyte ions through to the mass analyzer while rejecting the maximum number of matrix ions. For more details on space charge effects and different designs of ion optics, refer to Chapter 9 on the ion-focusing system.

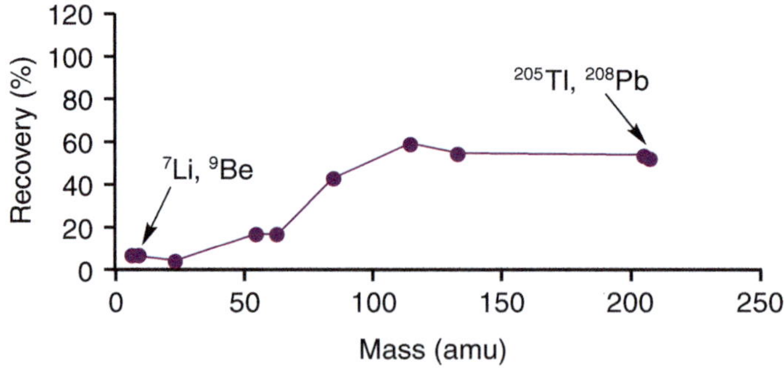

FIGURE 14.7   Space charge matrix suppression caused by 1,000 ppm uranium is significantly higher on low-mass elements such as Li and Be than it is with the high-mass elements such as Tl and Pb.

(From S. D. Tanner, *Journal of Analytical Atomic Spectrometry*, **10**, 905, 1995)

FURTHER READING

1. M. A. Vaughan and G. Horlick, *Applied Spectroscopy*, **41**(4), 523–530, 1987.
2. S. N. Tan and G. Horlick, *Applied Spectroscopy*, **40**(4), 445–4453, 1986.
3. D. J. Douglas and J. B. French, *Spectrochimica Acta*, **41B** (3), 197–200, 1986.
4. UIPAC Isotopic Composition of the Elements, *Pure and Applied Chemistry*, **75**(6), 683–799, 2003.
5. S. N. Willie, Y. Iida, and J. W. McLaren, *Atomic Spectroscopy*, **19**(3), 67–73, 1998.
6. S. J. Jiang, R. S. Houk, and M. A. Stevens, *Analytical Chemistry*, **60**, 1217–1226, 1988.
7. K. Sakata and K. Kawabata, *Spectrochimica Acta*, **49B**, 1027–1033, 1994.
8. S. D. Tanner, M. Paul, S. A. Beres, and E. R. Denoyer, *Atomic Spectroscopy*, **16**(1), 16–22, 1995.
9. J. M. Collard, K. Kawabata, Y. Kishi, and R. Thomas, *Micro*, 54–56, January 2002.
10. K. Neubauer, The Use of Ion-Molecule Chemistry Combined with Optimized Plasma Conditions to Meet SEMI Tier C Guidelines for Elemental Impurities in Semiconductor Grade Hydrochloric Acid by ICP-MS, *Spectroscopy*, **32**(10), 22–30, October 2017.
11. R. Hutton, A. Walsh, D. Milton, and J. Cantle, *ChemSA*, **17**, 213–215, 1992.
12. W. Tittes, N. Jakubowski, and D. Stuewer, *Poster Presentation at Winter Conference on Plasma Spectrochemistry*, San Diego, 1994.
13. J. A. Olivares and R. S Houk, *Analytical Chemistry*, **58**, 20–23, 1986.
14. S. D. Tanner, D. J. Douglas, and J. B. French, *Applied Spectroscopy*, **48**, 1373–1281, 1994.
15. S. D. Tanner, *Journal of Analytical Atomic Spectrometry*, **10**, 905–910, 1995.

15 Routine Maintenance

The components of an ICP mass spectrometer are generally more complex than other atomic-spectroscopic techniques, and as a result, more time is required to carry out routine maintenance to ensure that the instrument is performing to the best of its ability. Some tasks involve a simple visual inspection of a part, whereas others involve cleaning or changing components on a regular basis. However, routine maintenance is such a critical part of owning an ICP-MS system that it can impact both the performance and the lifetime of the instrument, particularly with complex samples. Chapter 15 covers this topic in greater detail.

The fundamental principle of inductively coupled plasma mass spectrometry (ICP-MS), which gives the technique its unequalled isotopic selectivity and sensitivity, also unfortunately contributes to some of its weaknesses. The fact that the sample "flows into" the spectrometer and is not "passed by it" at right angles, such as flame AA and radial inductively coupled plasma optical emission spectroscopy (ICP-OES), means that the potential for thermal problems, corrosion, chemical attack, blockage, matrix deposits and drift is much higher than with the other atomic spectrometry (AS) techniques. However, being fully aware of this fact and carrying out regular inspection of instrumental components can reduce and sometimes eliminate many of these potential problem areas. There is no question that a laboratory which initiates a routine maintenance schedule stands a much better chance of having an instrument ready and available for analysis whenever it is needed, compared to a laboratory that basically ignores these issues and assumes the instrument will look after itself.

Let us now look at the areas of the instrument that a user needs to pay attention to. I will not go into great detail but just give a brief overview of what is important, so you can compare it with maintenance procedures of trace element techniques you are more familiar with. These areas should be very similar with all commercial ICP-MS systems, but depending on the design of the instrument and the types of samples being analyzed, the regularity of changing or cleaning components might be slightly different (particularly if the instrument is being used for laser ablation work). The main areas that require inspection and maintenance on a routine or semi-routine basis include the following:

- Sample-introduction system
- Plasma torch
- Interface region
- Ion optics
- Roughing pumps
- Air/water filters

Other areas of the instrument require less attention, but nevertheless, the user should also be aware of maintenance procedures required to maximize their lifetime. They will be discussed at the end of this section.

15.1 SAMPLE-INTRODUCTION SYSTEM

The sample-introduction system, comprising the peristaltic/pneumatic pump, nebulizer, spray chamber and drain system, takes the initial abuse from the sample matrix and, as a result, is an area of the ICP mass spectrometer that requires a great deal of attention. Brennan et al. published a very informative article on how best to maintain the sample-introduction system.[1] The principles of the sample-introduction area have been described in great detail in Chapter 3, so let us now examine what kind of routine maintenance it requires.

15.2 PERISTALTIC PUMP TUBING

If the instrument uses a peristaltic pump, the sample is pumped at about 1 mL/min into the nebulizer. The constant motion and pressure of the pump rollers on the pump tubing, which is typically made from a polymer-based material, ensure a continuous flow of liquid to the nebulizer. However, over time, this constant pressure of the rollers on the pump tubing has the tendency to stretch it, which changes its internal diameter, and therefore, the amount of sample being delivered to the nebulizer. The impact could be a change in the analyte intensity, and therefore, a degradation in short-term stability.

Therefore, the condition of the pump tubing should be examined every few days, particularly if your laboratory has a high-sample workload or if extremely corrosive solutions are being analyzed. The peristaltic pump tubing is probably one of the most neglected areas, so it is absolutely essential that it be a part of your routine maintenance schedule. Here are some suggested tips to reduce pump-tubing-based problems:

- Manually stretch the new tubing before use.
- Maintain the proper tension on tubing.
- Ensure tubing is placed correctly in channel of the peristaltic pump.
- Periodically check flow of sample delivery, and throw away tubing if in doubt.
- Replace tubing if there is any sign of wear; do not wait until it breaks.
- With high-sample workload, change tubing every day or every other day.
- Release pressure on pump tubing when instrument is not in use.
- Pump and capillary tubing can be a source of contamination.
- Pump tubing is a consumable item—keep a large supply of it on hand.

A very useful tool to diagnose any problems associated with the peristaltic pump tubing (or the nebulizer) is a digital thermoelectric flow meter. By inserting this device in the sample line, you always know the actual rate of sample uptake to your nebulizer. This enhances the day-to-day reproducibility of your results and reduces the need to repeat measurements due to a blocked nebulizer, worn pump tubing or incorrect clamping of the pump tube. In addition, the borosilicate glass sample path ensures that there is no memory effect or sample contamination. A commercially available digital thermoelectric flow meter is shown in Figure 15.1.

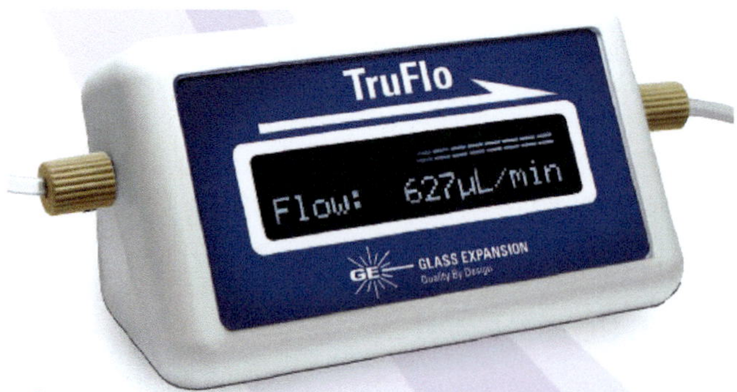

FIGURE 15.1 A commercially available digital thermoelectric flow meter to diagnose problems associated with peristaltic pump tubing and nebulizer blockages.

(Courtesy of Glass Expansion)

15.3 NEBULIZERS

The frequency of nebulizer maintenance will primarily depend on the types of samples being analyzed and the design of the nebulizer being used. For example, in a cross-flow nebulizer, the argon gas is directed at right angles to the sample capillary tip, in contrast to the concentric nebulizer, where the gas flow is parallel to the capillary. This can be seen in Figures 15.2 and 15.3, which show schematics of a concentric and cross-flow nebulizer, respectively.

The larger diameter of the liquid capillary and longer distance between the liquid and gas tips of the cross-flow design make it far more tolerant to dissolved solids and suspended particles in the sample than the concentric design. On the other hand, aerosol generation of a cross-flow nebulizer is far less efficient than a concentric nebulizer, and therefore it produces droplets of less optimum size than that required for the ionization process. As a result, concentric nebulizers generally produce higher sensitivity and slightly better precision than the cross-flow design, but are more prone to clogging.

So, the choice of which nebulizer to use is usually based on the types of samples being aspirated and the data quality objectives of the analysis. However, whichever type is being used, attention should be paid to the tip of the nebulizer to ensure it is not getting blocked. Sometimes, microscopic particles can build up on the tip of the nebulizer without the operator noticing, which, over time, can cause a loss of sensitivity, imprecision and poor long-term stability. In addition, O-rings and the sample capillary can be affected by the corrosive solutions being aspirated, which can also degrade

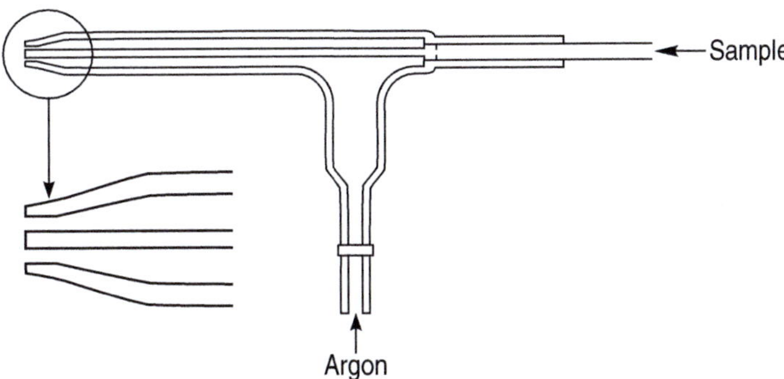

FIGURE 15.2   Schematic of a concentric nebulizer.

(Courtesy of Meinhard Glass Products)

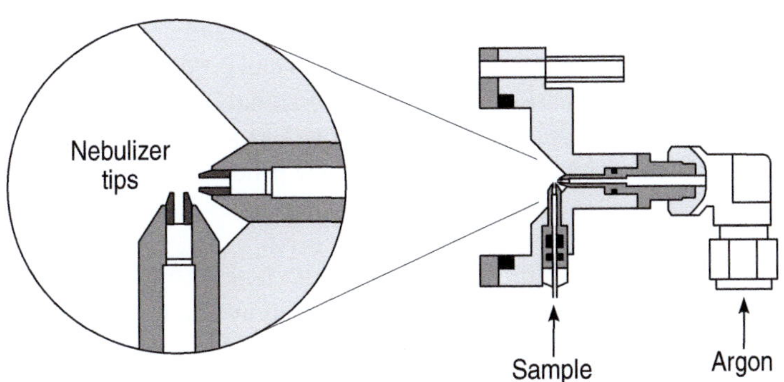

FIGURE 15.3   Schematic of a cross-flow nebulizer.

(Courtesy of PerkinElmer Inc.)

performance. For these reasons, the nebulizer should always be a part of the regular maintenance schedule. Some of the most common things to check include the following:

- Visually check the nebulizer aerosol by aspirating water—a blocked nebulizer will usually result in an erratic spray pattern with lots of large droplets.
- Remove blockage by either using backpressure from argon line or dissolving the material by immersing the nebulizer in an appropriate acid or solvent—an ultrasonic bath can sometimes be used to aid dissolution, but check with manufacturer first in case it is not recommended. (Note: Never stick any wires down the end of the nebulizer, because it could do permanent damage.)
- Ensure that the nebulizer is securely seated in spray chamber end cap.
- Check all O-rings for damage or wear.
- Ensure the sample capillary is inserted correctly into the sample line of the nebulizer.
- Nebulizer should be inspected every one to two weeks, depending on the workload.

The digital thermoelectric flow meter described earlier is also very useful to diagnose problems with the nebulizer, even if you are using a self-aspirating nebulizer because you are concerned about imprecision from the pulsing of a peristaltic pump. By placing the device in-line, you always know what your sample uptake is and can take immediate corrective action if there is any change. You can also record your sample flow in order to check that you are using the same flow from day to day.

If the flow meter indicates a blocked nebulizer tip, there are also nebulizer-cleaning devices offered by most of the third-party consumables/accessories companies. Traditionally, if particulate matter from the sample lodges itself in the end of the nebulizer, cleaning wires or ultrasonic baths were the only way to remove the obstruction, which often resulted in permanent damage. These new cleaning devices are designed to efficiently deliver a pressurized cleanser through the nebulizer capillary to safely dislodge particle buildup and thoroughly clean the nebulizer, without fear of damage.

15.4 SPRAY CHAMBER

By far the most common design of spray chamber used in commercial ICP-MS instrumentation is the double-pass design, which selects the small droplets by directing the aerosol into a central tube. The larger droplets emerge from the tube, and, by gravity, exit the spray chamber via a drain tube. The liquid in the drain tube is kept at positive pressure (usually by way of a loop), which forces the small droplets back between the outer wall and the central tube; they emerge from the spray chamber into the sample injector of the plasma torch. Scott double-pass spray chambers come in a variety of shapes, sizes and materials, but are generally considered the most rugged design for routine use. Figure 15.4 shows a double-pass spray chamber (made of a polymer material) coupled to a cross-flow nebulizer.

The most important maintenance with regard to the spray chamber is to make sure that the drain is functioning properly. A malfunctioning or leaking drain can produce a change in the spray chamber backpressure, producing fluctuations in the analyte signal, resulting in erratic and imprecise data. Less frequent problems can result from degradation of O-rings between the spray chamber and sample injector of the plasma torch. Typical maintenance procedures regarding the spray chamber include the following:

- Make sure the drain tube fits tightly and there are no leaks.
- Ensure the waste solution is being pumped from the spray chamber into the drain properly.
- If a drain loop is being used, make sure the level of liquid in the drain tube is constant.
- Check O-ring or ball joint between spray chamber exit tube and torch sample injector—ensure the connection is snug.

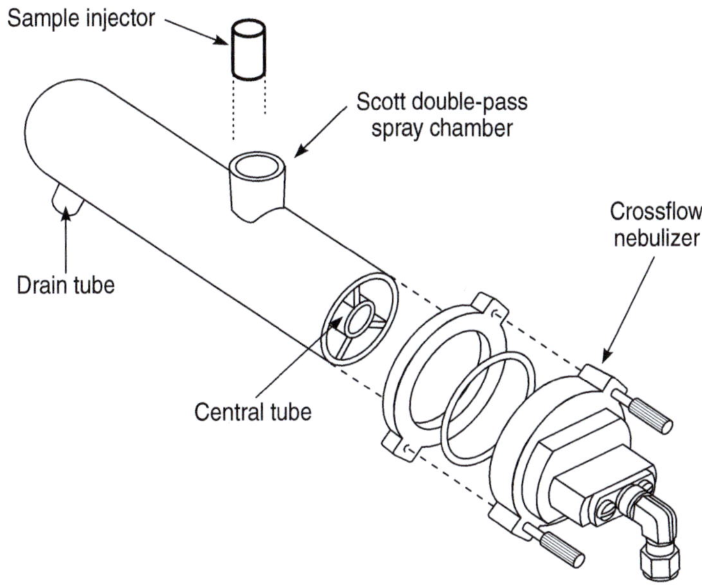

FIGURE 15.4   A double-pass spray chamber coupled to a cross-flow nebulizer.

(Courtesy of PerkinElmer Inc.)

- The spray chamber can be a source of contamination with some matrices/analytes, so flush thoroughly between samples.
- Empty spray chamber of liquid when instrument is not in use.
- Spray chamber and drain maintenance should be inspected every one to two weeks, depending on workload.

15.5   PLASMA TORCH

Not only are the plasma torch and sample injector exposed to the sample matrix and solvent, but they also have to sustain the analytical plasma at approximately 10,000 K. This combination makes for a very hostile environment and therefore is an area of the system that requires regular inspection and maintenance. A plasma torch positioned in the RF coil is shown in Figure 15.5.

As a result, one of the main problems is staining and discoloration of the outer tube of the quartz torch because of heat and the corrosiveness of the liquid sample. If the problem is serious enough, it has the potential to cause electrical arcing. Another potential problem area is blockage of the sample injector due to matrix components in the sample. As the aerosol exits the sample injector, desolvation takes place, and the sample changes from small liquid droplets to minute solid particles prior to entering the base of the plasma. Unfortunately, with some sample matrices, these particles can deposit themselves on the tip of the sample injector over time, leading to possible clogging and drift. In fact, this can be a potentially serious problem when aspirating organic solvents, because carbon deposits can rapidly build up on the sample injector and cones unless a small amount of oxygen is added to the nebulizer gas flow. Some torches also use metal plates or shields to reduce the secondary discharge between the plasma and the interface. These are consumable items, because of the intense heat and the effect of the RF field on the shield. A shield in poor condition can affect instrument performance, so the user should always be aware of this and replace it when necessary.

FIGURE 15.5   A plasma torch mounted in the torch box.

(Courtesy of Analytik Jena)

Some useful maintenance tips with regard to the torch area include the following:

- Look for discoloration or deposits on the outer tube of the quartz torch. Remove material by soaking the torch in appropriate acid or solvent if required.
- Check torch for thermal deformation. A nonconcentric torch can cause loss of signal.
- Check sample injector for blockages. If the injector is demountable, remove the material by immersing it in an appropriate acid or solvent if required (if the torch is one piece, soak the entire torch in the acid).
- Ensure that the torch is positioned in the center of the load coil and at the correct distance from the interface cone when replacing the torch assembly.
- If the coil has been removed for any reason, make sure the gap between the turns is correct as per recommendations in the operator's manual.
- Inspect any O-rings or ball joints for wear or corrosion. Replace if necessary.
- If a shield or plate is used to ground the coil, ensure it is always in good condition; otherwise, replace when necessary.
- The torch should be inspected every one to two weeks, depending on workload.

15.6   INTERFACE REGION

As the name suggests, the interface is the region of the ICP mass spectrometer where the plasma discharge at atmospheric pressure is "coupled" to the mass spectrometer at 10^{-6} torr by way of two interface cones—a sampler and skimmer. This coupling of a high-temperature ionization source such as an ICP to the metallic interface of the mass spectrometer imposes demands on this region of the instrument that are unique to this AS technique. When this is combined with matrix, solvent and analyte ions together with particulates and neutral species being directed at high velocity at the interface cones, an extremely harsh environment is the result. The most common types of problems associated with the interface are blocking or corrosion of the sampler cone and, to a lesser extent, the skimmer cone. A schematic of the interface cones showing potential areas of blockage are shown in Figure 15.6.

A blockage is not always obvious, because often the buildup of material on the cone or corrosion around the orifice can take a long time to reveal itself. For that reason, the sampler and skimmer interface cones have to be inspected and cleaned on a regular basis. The frequency will often depend on the types of samples being analyzed and also the design of the ICP mass spectrometer. Neufeld and coworkers wrote an excellent article on maintaining the cones and interface area to

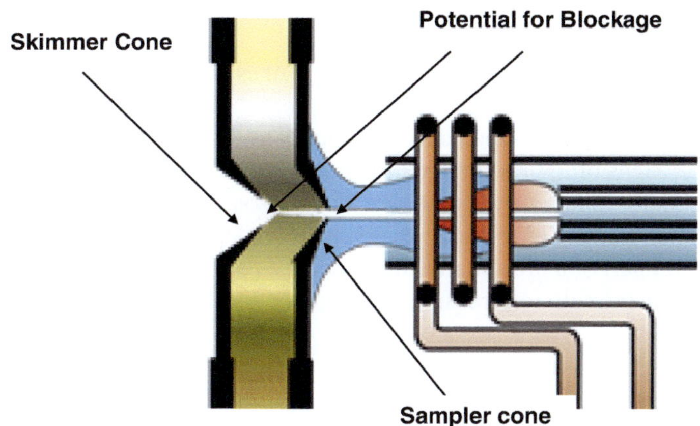

FIGURE 15.6 A schematic of the interface cones showing potential areas of blockage.

maximize productivity and uptime.[2] For example, it is well documented that a secondary discharge at the interface can prematurely discolor and degrade the sampler cone, especially when complex matrices are being analyzed or if the instrument is being used for high sample throughput.

Besides the cones, the metal interface housing itself is also exposed to the high-temperature plasma. Therefore, it needs to be cooled by a recirculating water system, usually containing some kind of antifreeze or corrosion inhibitor or by a continuous supply of mains water. Recirculating systems are probably more widely used because the temperature of the interface can be controlled much better. There is no real routine maintenance involved with the interface housing, except maybe to check the quality of the coolant from time to time, to make sure there is no corrosion of the interface cooling system. If for any reason the interface gets too hot, there are usually built-in safety interlocks that will turn the plasma off. Some useful hints to prolong the lifetime of the interface and cones include the following:

- Check that both sampler and skimmer cone are clean and free of sample deposits. The typical frequency is weekly, but will depend on sample type and workload.
- If necessary, remove and clean cones using the manufacturer's recommendations. Typical approaches include immersion in a beaker of weak acid or detergent placed in a hot-water or ultrasonic bath. Abrasion with fine wire wool or a coarse polishing compound has also been used.
- Never stick any wire into the orifice; it could do permanent damage.
- Nickel cones will degrade rapidly with harsh sample matrices. Use platinum cones for highly corrosive solutions and organic solvents.
- Periodically check cone-orifice diameter and shape with a magnifying glass (10–20× magnification). An irregular-shaped orifice will affect instrument performance.
- Thoroughly dry cones before installing them back in the instrument because water/solvent could be pulled back into the mass spectrometer.
- Check coolant in recirculating system for signs of interface corrosion such as copper or aluminum salts (or predominant metal of interface).

15.7 ION OPTICS

The ion optic system is usually positioned just behind or close to the skimmer cone to take advantage of the maximum number of ions entering the mass spectrometer. There are many different commercial designs and layouts, but they all have one attribute in common, and that is to transport the maximum number of analyte ions while allowing the minimum number of matrix ions through to the mass analyzer.

The ion-focusing system is not traditionally thought of as a component that needs frequent inspection, but because of its proximity to the interface region, it can accumulate minute particulates and neutral species that over time can dislodge, find their way into the mass analyzer and affect instrument performance. Signs of a dirty or contaminated ion optic system are poor stability or a need to gradually increase lens voltages over time. For that reason, no matter what design of ion optics is used, inspection and cleaning every three to six months (depending on workload and sample type) should be an integral part of a preventative maintenance plan. Some useful maintenance tips for the ion optics to ensure maximum ion transmission and good stability include the following:

- Look for sensitivity loss over time, especially in complex matrices.
- If sensitivity is still low after cleaning the sample-introduction system, torch and interface cones, it could indicate that the ion lens system is becoming dirty.
- Try retuning or re-optimizing the lens voltages.
- If voltages are significantly different (usually higher than previous settings), it probably means lens components are getting dirty.
- When the lens voltages become unacceptably high, the ion-lens system will probably need replacing or cleaning. Use recommended procedures outlined in the operator's manual.
- Depending on the design of the ion optics, some single-lens systems are considered consumables and are discarded after a period of time, whereas multicomponent lens systems are usually cleaned using abrasive papers or polishing compounds, and rinsed with water and an organic solvent.
- If cleaning ion optics, make sure they are thoroughly dry, because water or solvent could be sucked back into the mass spectrometer.
- Gloves are usually recommended when reinstalling an ion optic system because of the possibility of contamination.
- Do not forget to inspect or replace O-rings or seals when replacing ion optics.
- Depending on instrument workload, you should expect to see some deterioration in the performance of the ion lens system after three to four months of use. This is a good approximation of when it should be inspected and cleaned or replaced if necessary.
- With some instruments, you will need to break the vacuum to get to the ion optic region. Even though vacuum can be re-established very quickly, this should be a consideration when carrying out your own ion-lens cleaning procedures.

15.8 ROUGHING PUMPS

Typically, two roughing pumps are used in commercial instruments. One pump is used on the interface region, and the other is used as a backup to the turbomolecular pumps on the main vacuum chamber. They are usually oil-based rotary or diffusion pumps, where the oil needs to be changed on a regular basis, depending on the instrument usage. The oil in the interface pump will need changing more often than the one on the main vacuum chamber because it is pumping for a longer period. A good indication of when the oil needs to be changed is the color in the "viewing glass". If it appears dark brown, there is a good chance that heat has degraded its lubricating properties, and it needs to be changed. With the roughing pump on the interface, the oil should be changed every one to two months, and with the main vacuum chamber pump, it should be changed every three to six months. These times are only approximations and will vary depending on the sample workload and the time the instrument is actually running. Some important tips when changing the roughing pump oil:

- Do not forget to turn the instrument and the vacuum off. If the oil is being changed from "cold", it might be useful to run the instrument for 10–15 min beforehand to get the oil to flow better.

- Drain the oil into a suitable vessel; caution, the oil might be very hot if the instrument has been running all day.
- Fill the oil to the required level in the "viewing glass".
- Check for any loose hose connections.
- Replace oil filter if necessary.
- Turn the instrument back on. Check for any oil leaks around filling cap, and tighten if necessary.

15.9   AIR FILTERS

Most of the electronic components, especially the ones in the RF generator, are air-cooled. Therefore, the air filters should be checked, cleaned or replaced on a fairly regular basis. Although this is not carried out as routinely as the sample-introduction system, a typical time frame to inspect the air filters is every three to six months, depending on the workload and instrument usage.

15.10   OTHER COMPONENTS TO BE PERIODICALLY CHECKED

It is also important to emphasize that other components of the ICP mass spectrometer have a finite lifetime, and will need to be replaced or at least inspected from time to time. These components are not considered a part of the routine maintenance schedule, and usually require a service engineer (or at least an experienced user) to clean or to change them. These areas to be cleaned are described in the following text.

15.11   THE DETECTOR

Depending on the usage and levels of ion signals measured on a routine basis, the electron multiplier should last about 12 months. A sign of a failing detector is a rapid decrease in the "gain" setting, despite attempts to increase the detector voltage. The lifetime of a detector can be increased by avoiding measurements at masses that produce extremely high ion signals, such as those associated with the argon gas, solvent or acid used to dissolve the sample (e.g., hydrogen, oxygen and nitrogen) or any mass associated with the matrix itself. It is important to emphasize that the detector should be replaced by an experienced person wearing gloves, to reduce the possibility of contamination from grease or organic/water vapor from the operator's hands. It is advisable that a spare detector be purchased with the instrument.

15.12   TURBOMOLECULAR PUMPS

Most of the instruments running today use two turbomolecular pumps to create the operating vacuum for the main mass analyzer/detector chamber and the ion optic region. However, some of the newer instruments use a single, twin-throated turbo pump. The lifetime of turbo pumps, in general, is dependent on a number of factors, including the pumping capacity of the pump (usually expressed as L/s), the size (or volume) of the vacuum chamber to be pumped, the orifice diameter of the interface cones (in mm) and the time the instrument is running. Although some instruments still use the same turbo pumps after five to ten years of operation, the normal lifetime of a pump in an instrument that has a reasonably high sample workload is on the order of three to four years. This is an approximation and will obviously vary depending on the make and design of the pump (especially the type of bearings used). As the turbomolecular pump is one of the most expensive components of an ICP-MS system, this should be factored into the overall running costs of the instrument over its operating lifetime.

It is worth pointing out that although the turbo pump is not generally included in routine maintenance, most instruments use a "Penning" (or similar) gauge to monitor the vacuum in the main chamber. Unfortunately, this gauge can become dirty over time and lose its ability to measure the correct pressure. The frequency of this is almost impossible to predict but is closely related to types and numbers of samples analyzed. A sudden drop in pressure or fluctuations in the signal are two of the most common indications of a dirty Penning gauge. When this happens, the gauge must be removed and cleaned. This should be performed by an experienced operator or service engineer, because removing the gauge, cleaning it, maintaining the correct electrode geometry and reinstalling it correctly into the instrument is a fairly complicated procedure. It is further complicated by the fact that a Penning gauge is operated at high voltage.

15.13   MASS ANALYZER AND COLLISION/REACTION CELL

Under normal circumstances, there is no need for the operator to be concerned about routine maintenance of the mass analyzer or collision/reaction cell. With modern turbomolecular pumping systems, it is highly unlikely there will be any pump contamination problems associated with the quadrupole, magnetic sector, or TOF mass analyzer. And very few sample matrix components ever make it into the mass spectrometer region, which dramatically reduces the frequency of routine maintenance tasks. This certainly was not the case with some of the early instruments that used oil-based diffusion pumps, because many researchers found that the quadrupole and pre-filters were contaminated by oil vapors from the pumps. Today, it is fairly common for turbomolecular-based mass analyzers to require no maintenance of the analyzer or the collision/reaction cell quadrupole rods over the lifetime of the instrument, other than an inspection carried out by a service engineer on an annual basis. However, in extreme cases, particularly with older instruments, removal and cleaning of the quadrupole assembly might be required to get acceptable peak resolution and abundance sensitivity performance.

15.14   FINAL THOUGHTS

The overriding message I would like to leave you with on this subject is that routine maintenance cannot be overemphasized in ICP-MS. Even though it might be considered a mundane and time-consuming chore, it can have a significant impact on the uptime of your instrument. Read the routine-maintenance section of the operator's manual and understand what is required. It is essential that time be scheduled on a weekly, monthly and quarterly basis for preventative maintenance on your instrument. In addition, you should budget for an annual preventative-maintenance contract under which the service engineer checks out all the important instrumental components and systems on a regular basis to make sure they are all working correctly. This might not be as critical if you work in an academic environment, where the instrument might be down for extended periods, but in my opinion, it is absolutely critical if you work in a commercial laboratory, which is using the instrument to generate revenue. There is no question that spending the time to keep your ICP mass spectrometer in good working order can mean the difference between owning an instrument whose performance could be slowly degrading without your knowledge or one that is always working in "peak" condition.

However, I must give credit to the instrument designers and accessories suppliers in making today's instrumentation extremely easy to maintain and keep clean. The technique will be 40 years old when this book is published, and its commercial success has been built on acceptance by the analytical community for applying the technique to truly routine, real-world analysis. The only areas that need cleaning on a regular basis are the sample-introduction system and interface cones, depending on usage and sample matrices being aspirated. And many of today's instruments have alarms, which can be set to remind the operator when it's time for the few preventative-maintenance tasks that are required, such as oil changes and tubing replacement. Some systems will even display how many hours various components have been used and when they might need attention. So even though routine maintenance is very important, instrument vendors have put a great deal of time and effort into making this as straightforward and seamless as possible. Please refer to the websites of all the instrument vendors, and sample-introduction and accessories companies in Chapter 28 on useful contact information for troubleshooting hints and tips.

FURTHER READING

1. L. Neufeld, ICP-MS Interface Cones: Maintaining the Critical Interface between the Mass Spectrometer and the Plasma Discharge to Optimize Performance and Maximize Instrument Productivity, *Spectroscopy Magazine*, **34**(7), 12–17, 2019.
2. R. Brennan, J. Dulude, R. Thomas, Approaches to Maximize Performance and Reduce the Frequency of Routine Maintenance in ICP-MS, *Spectroscopy Magazine*, **30**(10), 12–25, 2015.

16 Sampling and Sample-Preparation Techniques

Sample preparation is extremely important in ICP-MS, particularly with regard to how it can impact contamination. If you have been using flame AA or ICP-OES, you will probably have to rethink your sample-preparation procedures for ICP-MS. Chapter 16 gives an overview of the fundamental principles and commercially available approaches of sampling and sample-digestion procedures for trace element analysis, as well as focusing on optimized sample-preparation approaches for plasma spectrochemical techniques. In particular, it takes a closer look at the major causes of contamination and analyte loss in ICP-MS and how they affect both the analysis and the method-development process.

There are many factors that influence the ability to get the correct result with any trace element technique. Unfortunately, with ICP-MS, the problem is magnified even more because of its extremely high sensitivity. So, in order to ensure that the data reported is an accurate reflection of the sample in its natural state, the analyst must not only be aware of all the potential sources of contamination, but also the many reasons why analyte loss is a problem in ICP-MS. Figure 15.1 shows the major factors that can impact the analytical result. Let's take a look at each of those procedures in greater detail.

16.1 COLLECTING THE SAMPLE

Collecting the sample and maintaining its integrity is a science of its own and is beyond the scope of this book. However, it is worth taking a big picture look, in order to understand its importance in the overall scheme of collecting, preparing and analyzing a sample for analysis. The object of sampling is to collect a portion of the material that is small enough in size to be conveniently transported and handled, and at the same time accurately represents the bulk material being sampled. Depending on the sampling requirements and the type of matrices being analyzed, there are basically three main types of sampling procedures.

- **Random sampling** is the most basic type of sampling and only represents the composition of the bulk material at the time and place it was sampled. If the composition of the material is known to vary with time, individual samples collected at suitable intervals and analyzed separately can reflect the extent, frequency and duration of these variations.
- **Composite sampling** is when a number of samples are collected at the same point, but at different times and mixed together before being analyzed.
- **Integrated sampling** is achieved by mixing together a number of samples that have been collected simultaneously from different points.

It is not the intent of this chapter to discuss which type of sampling is the most effective, but it must be emphasized that unless the correct sampling or subsampling procedure is used, the analytical data generated by the ICP-MS instrumentation may be seriously flawed, because it may not represent the original bulk material. If the sample is a liquid, it is also important to collect the sample in clean containers that have been thoroughly washed. In addition, if the sample is to be kept for a long period of time before analysis, it is essential that the analytes stay in solution in a preservative such as a dilute acid (this will also help stop the analytes from being absorbed into the walls of the container). It is also important to keep the samples as cool as possible (preferably refrigerated) to avoid losses through evaporation. This is especially important if mercury is being measured, because mercury ions could be reduced to elemental mercury if the solutions are left for extended periods of

DOI: 10.1201/9781003187639-16

time at room temperature before being analyzed. If this happens, the mercury can be absorbed into the walls of the containers or lost when the cap is removed. Kratochvil and Taylor give an excellent review of the importance of sampling for chemical analysis.[1]

16.2   PREPARING THE SAMPLE

If the sample is not in a convenient form to be dissolved, it has to be ground to a smaller particle size, mainly to improve the homogeneity of the original sample taken and make it more representative when taking a subsample. The ideal particle size will vary depending on the sample, but the sample is typically ground to pass through a fine-mesh sieve (0.1 mm^2 mesh). This uniform particle size ensures that the particles in the test portion are the same size as the particles in the rest of the ground sample. Another reason for grinding the sample into small uniform particles is that they are easier to dissolve.

The process of grinding a sample with mortar and pestle or ball mill and passing it through a metallic sieve can be time consuming and a major cause of contamination. This can occur from the remains of a previous sample that had been prepared earlier or from materials used in the manufacture of the grinding or sieving equipment. For example, sieves, which are made from stainless steel, bronze or nickel, can also introduce metallic contamination into the sample. In order to minimize some of these problems, plastic/polymer sieves are often used. However, still remaining is the problem of contamination from the grinding equipment. For this reason, with traditional sampling, it is usual to discard the first portion of the sample or even to use different grinding and sieving equipment for different kinds of samples.

16.3   CRYOGENIC GRINDING

Over the past few years, a new type of sample-grinding technique has become available. These grinders are based on the principle of cryogenically freezing the sample in liquid nitrogen. At this kind of temperature (−196° C), many materials become very brittle and a result can be ground to a fine powder ready for a dissolution procedure. The technology incorporates an insulated tub into which liquid nitrogen is poured. The grinding mechanism is a magnetic-coil assembly suspended in the liquid-nitrogen bath. Cooling materials to temperatures approaching −200° C make samples extremely brittle, so they can be pulverized quickly by impact milling. This ability allows difficult-to-process samples (bone, rocks, polymers, metals, food and pharmaceutical capsules) to be more efficiently processed before analysis. The sample is placed in a closed grinding vial and thoroughly cooled before grinding by the magnetic coil shuttling the impactor rapidly back and forth, pulverizing the sample against the end plugs of the vial, as shown in Figure 16.1. The type of vial and impactor for grinding samples is selected to reduce any potential cross contamination of metals. For example, for certain sample types, and the analyte levels being measured, it might be better to use polymer vials and polymer-encased impactors instead of metal ones to reduce the possibility of metal contamination.

Note: It should also be emphasized that if any trace element spiked additions need to be made to comply with the validation protocols, they are carried out before the digestion/sample-preparation step to ensure that there is no loss of analyte or contamination from the microwave-dissolution procedure. This could also indicate if any matrix suppression or enhancement effects are occurring from the dissolution acids/chemicals.

16.4   SAMPLE DISSOLUTION

Let's now take a closer look at how the digestion technique can be optimized for different sample types. If the sample under investigation is in a liquid form, it can be analyzed by direct aspiration or perhaps by simply diluting in an aqueous or organic solvent (see Chapter 3 on sample-introduction

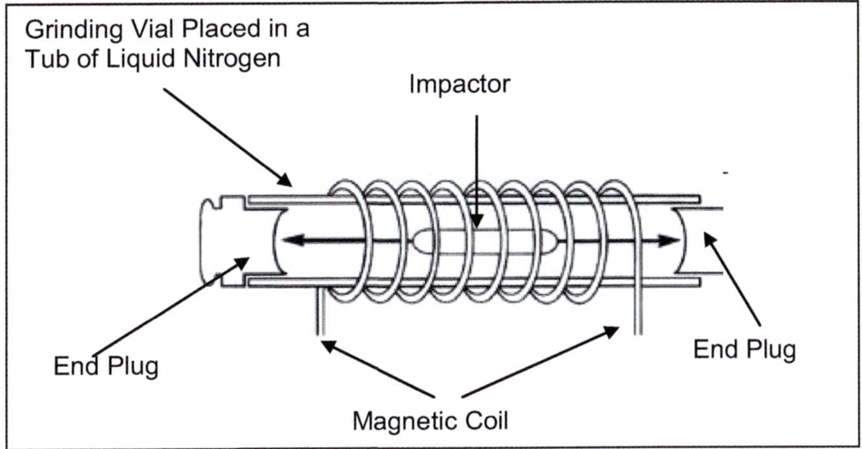

FIGURE 16.1   Schematic of a cryogenic grinding system.

(Courtesy of Spex Sample Prep, an Antylia company).

techniques). However, if the sample is a solid, it will have to be brought into solution either via a hot-plate dissolution technique using concentrated mineral acids, or a closed-vessel, microwave digestion procedure. Let's take a closer look at why sample dissolution is necessary.

16.5   REASONS FOR DISSOLVING SAMPLES

Sample dissolution using acid-digestion techniques can add a significant amount of time to the overall analytical procedure. For that reason, it's important to fully understand the benefits of working with a solution:

- Solid-sampling analytical techniques are notoriously prone to sampling inhomogeneity—taking multiple portions of the solid material under testing and dissolving them represents the best option for working with a with a homogeneous sample.
- Solution-based analytical techniques need a homogenous sample, which is representative of the sample matrix under testing—taking a one-off solid sample such as a tablet formulation, a cookie or gummy bear can produce erroneous results, because it may not be truly representative of the batch of samples.
- With notoriously inhomogeneous and varied botanical samples, it is important that the portion of the digested solid sample being presented to the instrument is representative of the bulk sample. For that reason, it is critical that enough of the bulk material is sampled, so that increments can be taken at random locations throughout the bulk sample.
- Measurements take a finite amount of time where the signal must stay constant—dissolving the sample and obtaining a clear solution is the best way to achieve signal stability.

16.6   DIGESTED SAMPLE WEIGHTS

It's also important to understand that the sample weight and final volume will be dictated by the expected contamination levels and total-dissolved-solids (TDS) limitations of the instrumental technique being used. Typical sample weights for ICP-MS are in the order of 0.2–0.5 g, depending on the type of sample and mineral acid used. However, it's fair to say that when using microwave digestion procedures, the dilution factor used in the sample-preparation step will ultimately have an impact on the ability of the technique to detect the expected contamination levels. So, it is inevitable that there will have to be a certain level of compromise with the dissolution of the sample, based on the level of acceptable TDS for the analytical technique, particularly if ICP-MS is being used, to ensure that the analyte is measurable above the limit of quantitation (LOQ) of the instrument.

16.7 MICROWAVE DIGESTION CONSIDERATIONS

Many standardized methods call for the use of closed-vessel microwave digestion in order to completely destroy and solubilize the sample matrix. Microwave digestion systems are commonly used for trace elemental analysis studies in a multitude of application areas to get the samples into solution, because they are easy to use and can rapidly process many samples at a time, which makes them ideally suited for high sample throughput environments.

16.8 WHY USE MICROWAVE DIGESTION?

So let's remind ourselves why closed or pressurized microwave digestion offers the best way to get samples into solution:

- Dissolution temperatures above boiling point of the solvent can be achieved.
- The oxidation potential of reagents is higher at elevated temperatures, which means digestion is faster and more complete.
- Under these conditions, concentrated nitric acid and/or hydrochloric can be used for the majority of materials.
- Microwave-dissolution conditions and parameters can be reproduced from one sample to the next.
- They are safer for laboratory personnel, as there is less need to handle hot acids.
- Samples can be dissolved very rapidly.
- The digestion process can be fully automated.
- High sample throughput can be achieved.
- Less hazardous fumes in the laboratory.

Typically, 0.2–0.5 g of sample is weighed and placed into a plastic vessel along with the appropriate acids, which are then sealed with a tight fitting cap to create a pressurized environment. Once samples are digested, which takes 10–30 minutes, depending on the matrix, the resulting liquid is then transferred to a volumetric flask and made up to the required volume using high-purity water. Note: in its heavy metals testing regulations, the state of California requires a minimum of 0.5 g of sample (dried to constant weight) to be digested, which is on the high side for a microwave digestion procedure, so care should be taken to ensure the microwave digestion equipment can handle the increase in pressure from this amount of sample (refer to section later in this chapter about commercial systems).

16.9 CHOICE OF ACIDS

The choice of acids used for the preparation of digested samples is also important. Typically concentrated nitric and/or hydrochloric acids are used in various concentrations, depending on the sample type. The presence of hydrochloric acid is useful for stabilization of many elements including mercury, but can sometimes produce insoluble chlorides, particularly if there is any silver in the sample. The presence of chloride can also be detrimental when ICP-MS is the chosen technique, as the chloride ions combine with other ions in the sample matrix and the argon plasma to generate polyatomic spectral interferences. An example of this is the formation of the $^{40}Ar^{35}Cl$ polyatomic ion in the determination of ^{75}As, and $^{35}Cl^{16}O$ in the determination of ^{51}V. These polyatomic interferences can usually be removed by the use of collision- or reaction-cell technology (CRC) if the

ICP-MS system offers that capability. However, CRC can slow the analysis down because stabilization times have to be built into a multielement method to determine analytes that require both cell and non-cell conditions.

Nitric acid and hydrogen peroxide are often used for the dissolution of organic matrices, as they are both strong oxidizing agents that effectively destroy the organic matter. In some cases, hydrofluoric acid (HF) may need to be used to dissolve certain silicate- or titanium dioxide-based materials, particularly if fillers or excipients have been used in a final tablet formulation, or lotions or creams are being analyzed. In cases where HF is required, specialized-plastic (PTFE) sample-introduction components need to be used, including the use of buffering agents, like boric acid, to dissolve insoluble fluorides and neutralize excess HF. It should be emphasized that HF is a highly corrosive acid, and extreme caution should be taken whenever it is used.[2]

However, it's important to understand that the more complex the sample preparation, the longer the analytical procedure will become, which will have a negative impact on the overall analysis time, particularly in a lab with a high-sample workload. In addition, the sample-preparation steps could potentially have an effect on the overall TDS levels, so it is important to consider this when looking at the preparation of these samples.

16.10 COMMERCIAL MICROWAVE TECHNOLOGY

There are a number of different microwave digestion technologies on the market. Depending on the types and variety of samples being digested and the degree of automation, they have their own strengths and weaknesses. However, if a suitable digestion procedure has been optimized for a particular sample type, the digestion chemistry should be applicable to any type of system. For example, many botanical-related matrices can be digested by taking 0.2–0.5 g of sample, adding 10 mL of acid mixture (9 mL HNO_3 and 1 mL of HCl) and placing it in the microwave digestion vessel. Depending on the sample and program, it can be fully dissolved to a clear/colorless liquid in approximately 15–20 minutes.[3,4]

So let's take a closer look the major types of microwave digestion technology on the market, and specifically outline the differences. The optimum choice will often depend on the workload and sample diversity of the lab carrying out the analysis.[5,6] This section will describe the basic principles of the major types of microwave digestion technology, and offer suggestions as to which might be the best approach, based on sample matrix, digestion efficiency, sample throughput, productivity and overall cost of analysis.

16.11 DIGESTION STRATEGIES

Obtaining analytical data required to ensure quality products starts with the crucial step of preparing the sample for analysis. Reducing handling steps, eliminating outside contamination and minimizing reagent blank contribution are all necessary for good sample preparation. It is well recognized that closed-vessel microwave digestion offers the best approach for getting your samples into solution for analysis by ICP-OES or ICP-MS. However, there are basically four different commercially available designs.

- Sequential systems
- Rotor-based systems
- Single reaction chamber technology
- Single cavity mode technology

So how do you go about selecting the optimum technology for your sample workload? What types of mineral acids will be best suited for your elements of interest, and what temperature and pressure will be required for the digestion process of your sample matrices? It's only when you have a good understanding of these issues that you can begin to look more closely at the pros and cons of the different commercially available microwave technology. So first, let's take a closer look at the fundamental principles of microwave digestion.

16.12 FUNDAMENTAL PRINCIPLES OF MICROWAVE DIGESTION TECHNOLOGY

First, it should be emphasized that sample digestion should be carried out using reagents compatible with ICP-OES and ICP-MS instrumentation. For example, the chemical/physical properties and concentration of the mineral acids used and how they affect the sample-introduction nebulization processes and the potential matrix suppression effects in the plasma should be taken into consideration. It is therefore well recognized that the most plasma spectrochemical-friendly reagents are typically strong oxidizing agents such as nitric acid (HNO_3) and hydrogen peroxide (H_2O_2), which are extremely efficient, but tend to generate large amounts of carbon dioxide (CO_2) and various oxides of nitrogen (NO_x) when they react with the samples. The microwave system and its components will therefore not only need to accommodate the high temperature required to digest all the different organic sample types, but also be able to handle the subsequent increase in pressure produced by the generation of large volumes of these gases. For some samples, the addition of small amounts of hydrochloric acid will also help to stabilize some elements, particularly mercury (Hg) and the platinum-group elements. However, it should be noted that if ICP-MS is being used as the analytical technique, the $^{40}Ar^{35}Cl$ polyatomic species could potentially interfere with monoisotopic arsenic (As) at 75 atomic mass units (amu). This can be alleviated using a collision/reaction cell (CRC), but it is important to be aware of this, so that the optimum instrumental conditions are used.

16.13 SEQUENTIAL SYSTEMS

The microwave cavity used in this technology produces an extremely homogeneous-focused microwave field, which enables the system to digest samples faster and with greater reproducibility than batch-style systems. However, it carries out the digestion in a sequential manner, one sample type at a time, which obviously limits their sample throughput capabilities. To carry out multi-sample analysis, they can be automated to move samples into the cavity automatically when the digestion of the previous sample has been completed. Another limitation of this approach is that because of their inherent design and the sample vails used, they operate at a much lower pressure, which means the achievable temperature is much lower than with other microwave approaches.

16.14 ROTOR-BASED TECHNOLOGY

With rotor-based technology, microwaves are directed onto vessels containing the sample and the digestion reagents, which are placed in a rotating carousel. The digestion process is accomplished by raising the pressure and temperature through microwave irradiation, as the carousel is rotating. This increase in temperature and pressure, together with the optimum reagent, increases both the speed of thermal decomposition of the sample and the solubility of metals in solution. Digestion conditions inside the vessels are monitored using fiber-optic technology, infrared temperature sensors and/or internal pressure controls to ensure the samples are digesting in a safe and controlled manner. A typical rotor-based microwave digestion system is shown in Figure 16.2.

Rotor-based systems work extremely well for similar matrices, by batching all the samples together that react in the same way. By carrying out the digestion process using the same microwave power/temperature/pressure conditions, it will ensure similar digestion quality in all positions. To increase throughput, different-sized carousels can be used, depending on the sample workload. However, when many different sample matrices have to be digested, productivity could

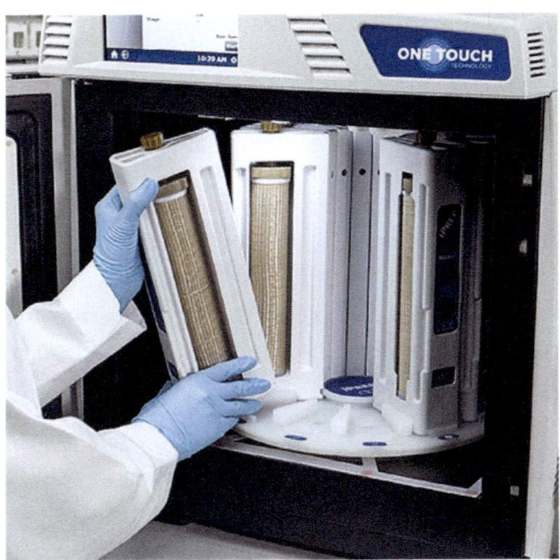

FIGURE 16.2 A sample carousel being loaded into a rotor-based microwave digestion system. (Courtesy of CEM Corp)

be sacrificed, because each sample type has to be batched together, which unfortunately limits the ability to digest completely different samples together in the same sample run.

However, recent developments with vessel design, sensor technology, temperature measurement/control as well as improved software algorithms have meant that power levels can be optimized so each individual sample vessel reaches a similar temperature. This means that different matrices can be mixed in the same batch to achieve an efficient digestion. This method of control is now becoming standard with rotor-based systems and has enhanced the capability of this technology from earlier designs.[7,8]

Traditional rotor-based systems can achieve relatively high pressures of around 100 bar, but cannot really go much higher because of their design. For that reason, the strength of any rotor-based system is the ability to safely vent the excess pressure caused by the buildup of CO_2 and NO_X during the digestion process. There are basically three different approaches used to carry out this process. They are:

- Burst disk
- Vent and reseal
- Self-regulating

Let's take a closer look at how they work.

BURST DISK: This is the simplest method, which employs a burst disk in the cap designed to fail in an overpressure situation, instantly releasing all pressure in the vessel. There is no danger of liquid or fumes escaping from the microwave, so it is completely safe for people working in the area around the microwave. However, when this happens, the sample contents are usually lost and the run has to be manually stopped. The result is that cleanup of the cavity is typically required, depending on the extent of the spill. There is also a strong possibility that corrosion of internal components will occur if high-concentration mineral acids are used for sample digestion.

VENT AND RESEAL: This type of technology eliminates vessel failure in the case of an out-of-control exothermic reaction. In the vent-and-reseal method, the vessel cap is held in place by a dome-shaped spring. In the case of overpressure due to a highly exothermic reaction, the spring is flattened, allowing the cap to lift up, slightly releasing excess pressure. Immediately, the excess pressure is released, the spring reseals the vessel, the digestion continues and the microwave program continues to completion with no loss of sample.

Self-regulating: This third approach is typically used in high-throughput rotors, which were developed to address the needs of labs that process larger sample volumes on a routine basis. Self-regulating vessels are very easy to assemble/disassemble and rely on the Teflon sealing plug inside the cap deforming to release pressure. Their compact design results in more moderate temperature and pressure capabilities, but allows for a large number of vessels to fit onto the rotor. These characteristics make self-regulating rotors ideal for high-workload labs with relatively straightforward applications such as clinical, environmental, food, etc. Most recently, the self-regulating approach has been applied to high-pressure rotors designed specifically for more challenging sample types. Since self-regulating vessels are designed to vent, incomplete digestions may be observed, given that the pressure loss does not allow the required temperatures necessary for complete digestion to be achieved.

16.15 SINGLE REACTION CHAMBER TECHNOLOGY

Let's take a more detailed look at the principles of single reaction chamber (SRC) technology, and how it differs from the rotor-based system.[9] Instead of a rotor with discrete sample vessel, the samples are put into vials with loose-fitting caps, which are sitting in a rack that is lowered into a larger vessel containing a base load of acidified water. It's this baseload that absorbs the microwave energy and transfers it to the vial. This allows every vial to react independently within the base load and ensures that all samples can reach a maximum pressure of up to 200 bar and achieve a digestion temperature of around 300 °C. This means that no batching of samples is necessary, and any combination of sample type and acid chemistry can be run simultaneously in the same chamber. In addition, this typically allows SRC technology to use higher sample weights than the other approaches.

This high-pressure/high-temperature capability is particularly useful for botanical-type samples, because of their high carbon content. At lower digestion temperatures, it is well recognized that the oxidizing conditions are not aggressive enough to completely digest many organic-based samples. As result, they often have a cloudiness to them, which indicates the carbonaceous material is not completely in solution. This is problematic, not only for the buildup of carbon on the interface cones, but also the negative impact of carbon-based polyatomic interferences on many analyte elements by ICP-MS.

A schematic of the SRC is shown on the left in Figure 16.3, while an actual photograph of sample vials being lowered into the chamber is shown on the right.

As previously mentioned, loose-fitting caps are used to seal the vials. This is possible because they are pre-pressurized with 40 bar of nitrogen prior to the start of the microwave program, which acts as a gas cap and keeps all the vials independently closed. As the pressure builds, equilibrium is achieved both inside and outside the vial. As a result, a variety of vial types, including disposable glass, quartz and Teflon or another material in any combination, can be used.

To exemplify that 40 bar of nitrogen is sufficient to seal the vials, an experiment was carried out where vials containing 110 ppm of mercury were placed right next to blank solutions in a 15-rack sample holder. In other words, every alternate sample vial in the rack was either 110 ppm or 0 ppm mercury. It is well recognized that mercury is highly volatile, particularly when heated, and would contaminate any surrounding vessels if not capped tightly. However, it can be seen in Figure 16.4 that the measurement of every alternate blank sample by ICP-MS is actually at the limit of detection for the technique, which is a clear indication that SRC technology using pressurized nitrogen gas as a sealant eliminates the potential of cross contamination in the sample chamber.

Every lab's trace element analysis workload and sample-digestion requirements are different. Rotor-based systems work extremely well, but their main limitation has traditionally been that they require batching of similar matrices and chemistries, because control of the power and therefore temperature is based on the reaction of one vessel at a time. By batching similar samples, under-digestion of some samples can be minimized due to the pressure and temperature required by

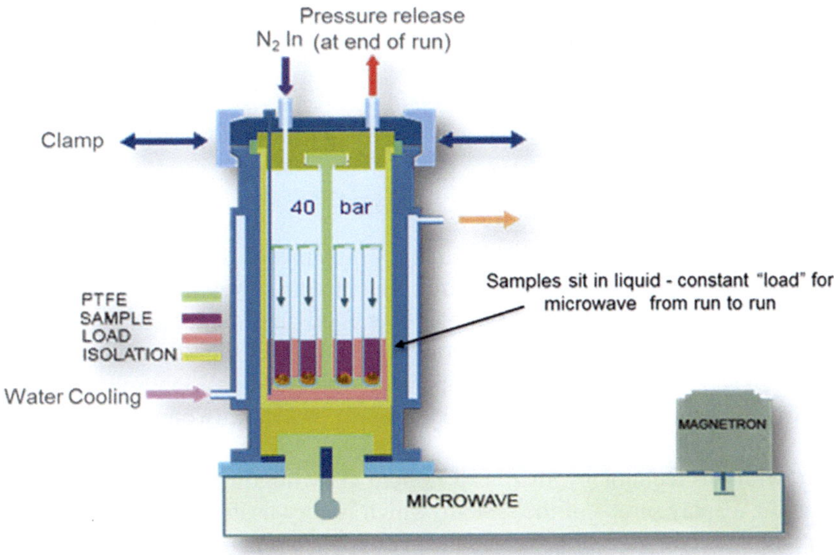

FIGURE 16.3 Single reaction chamber microwave digestion technology.
(Courtesy of Milestone Inc.)

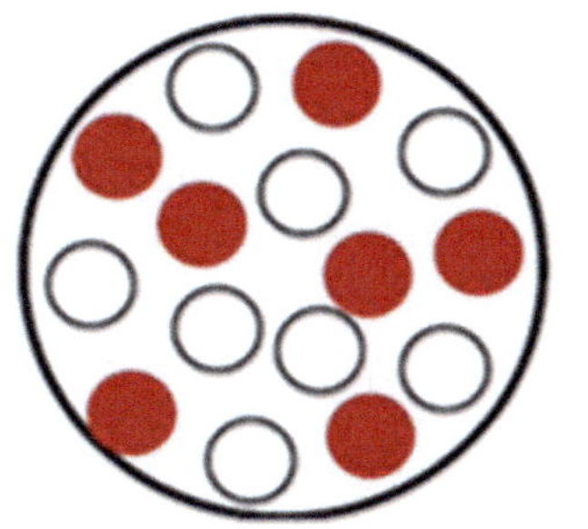

(15 Position) White vials: blanks.
Red vials: 110 ppm Hg

	Hg (ppt)
Bk position 01	0.020
Bk position 03	0.032
Bk position 05	0.001
Bk position 07	0.002
Bk position 09	-0.003
Bk position 11	-0.007
Bk position 13	-0.007
Bk position 15	-0.006

FIGURE 16.4 Study of sealing capability of nitrogen gas shows no sign of contamination from mercury.
(Courtesy of Milestone Inc.)

others. In addition, with rotor-based systems, the vessels are typically made from PTFE-type materials, which put limits on the achievable maximum pressure and therefore the available digestion temperature.

It's important to emphasize that rotor-based technology is perfectly adequate for digesting most routine-type samples. However, if the lab knows it will be handling an extremely wide variety of different sample types, especially if they are very difficult to get into solution, SRC technology offers some tangible benefits in being able to digest many different matrices (and using larger weights) at the same time, particularly in a high sample throughput environment. Unfortunately, the price tag of an SRC system might be prohibitive for many smaller testing labs that don't have the sample diversity and throughput to justify the higher cost.

16.16   SINGLE CAVITY MODE

A slightly modified approach to single reaction chamber technology has recently been commercialized. It is called single cavity mode (SCM), which provides a fully automated sequential alternative to traditional rotor-based microwave digestion.[10]

16.17   PRINCIPLES OF SINGLE CAVITY MODE

This technology uses a focused single-mode cavity, which means there is more microwave density to heat a solution compared to samples that are immersed in a water bath. This allows for faster heating and cooling, which means that samples like foods and biological samples can be completely digested and cooled in just over five minutes. However, it should be noted that even though every sample is under control of pressure, temperature and power as opposed to single sensors for a batch-type SRC system, it is still a sequential approach, where only one sample is being digested at a time.

In addition, with this technology, there is a camera allowing the operator to watch the digestion in real time on the display screen of the system, which is very useful for methods development and visual confirmation of total digestion. This is made possible with 60-mL high-purity quartz vessels, and brightly illuminated with LED lighting. In addition, a TFM liner is available for samples requiring HF and a snap-on cap to provide the initial sealing of the vessel so it can be safely transported from the fume hood to the instrument.

16.18   AUTOMATION WITH SINGLE CAVITY MODE

The system uses 24 positions with two racks of 12, one on each side of the cavity. The vessels are loaded into the cavity, and then a device clamps down on the vessel. Methods are selected and assigned to each vessel position by either drop and drag or highlight and assign. The system then runs the appropriate digestion program, including heating and cooling. Any residual gas is vented before it returns it to the original position.

In addition, the system can be interrupted at any time to digest a rush sample. For safety purposes, the system will stop after processing the current sample so the operator can place new sample into any position and tell the system to prep that sample next.

16.19   SAMPLING PROCEDURES FOR MERCURY

It's also worth pointing out that elements-specific sample-preparation techniques might also be necessary. For example, when preparing samples for the determination of mercury, care must be taken not to lose the analyte because of its volatility. This is especially relevant when carrying out microwave digestion. Some of the steps taken to minimize losses of mercury include the use of hydrochloric acid in the dissolution step to produce an excess of chloride ions, or the addition of gold (typically one ppm or less) to stabilize the mercury in solution.[11]

Under the right chemistry conditions, mercury can also be determined by the cold vapor (CV) generation technique. This technique is used in commercially available mercury analyzers, where the mercury in solution is reduced to its atomic state and the elemental mercury vapor is detected using either atomic absorption or atomic fluorescence. These instruments are extremely sensitive and are capable of carrying out both the chemistry and detection steps online, in an automated manner. In conventional mercury analyzers, the samples must either be liquids or brought into solution using a dissolution step, which limits their applicability for the analysis of solid or powdered pharmaceutical materials. However, there is a variation of this method, which can handle solids directly. In this approach, the sample is first heated up to 900 °C in a combustion furnace to volatilize the sample, then swept into a catalyst to release the mercury vapor, and concentrated onto the surfaces

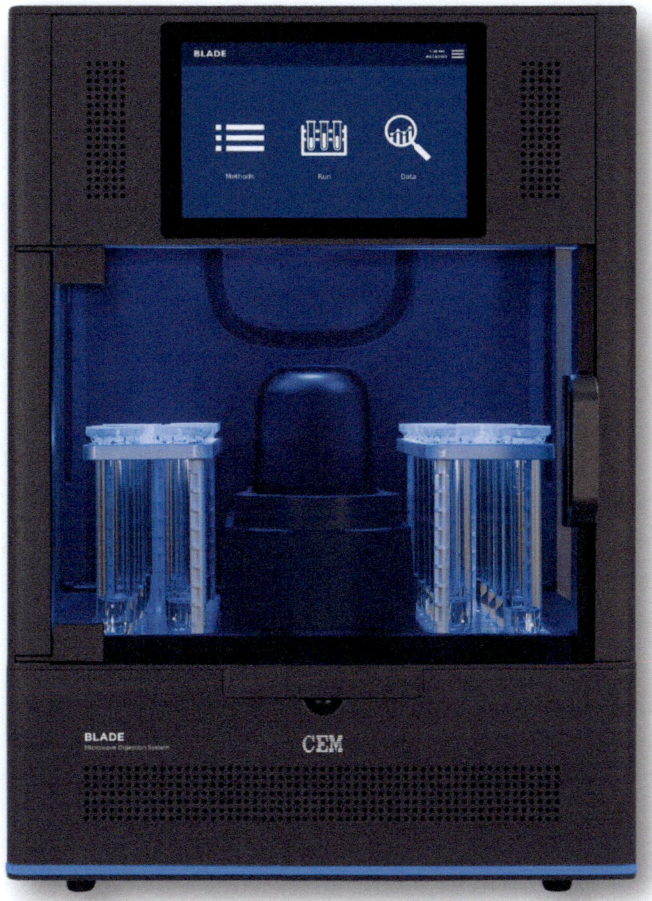

FIGURE 16.5 Commercially available single cavity mode MW technology.

(Courtesy of CEM Corp)

of a gold amalgamation trap, where it is eventually heated and swept into an atomic absorption for detection and quantitation.[12]

16.20 REAGENT BLANKS

The increasing demand for lower and lower detection limits is a major consideration for any laboratory performing trace metals analysis, where maximum contaminant limits are extremely low (and probably going lower). The development of sophisticated instrumentation, such as ICP-OES and ICP-MS, provides the capability to achieve limits of detection previously unobtainable. However, the analytical instrument is just one component in achieving the lowest limits of detection using plasma spectrochemistry.

The other critical area that puts significant demands on a laboratory's overall detection capability is to ensure that the sample-preparation procedure does not contribute any additional sources of contamination. There are several factors to consider when looking to minimize contamination, and reducing reagent-blank levels when preparing samples for analysis by plasma spectrochemistry, including general lab cleanliness (reducing atmospheric dust), vessel cleanliness (acid washing), reagent choice (use of high-purity acids and chemicals), quality of materials and consumables (low in trace metals), not to mention the digestion procedure itself. All these factors make a contribution to achieving the lowest possible detection capability when carrying out ultra-trace elemental determinations by ICP-OES and ICP-MS. However, this is not something many lab personnel necessarily think about, particularly if they do not have experience carrying out ultra trace analysis. Unfortunately the industry is so new that typical instrument operators only have a couple of years' experience at most, and many of them just a few months of running very sophisticated instrumentation. In addition, they may have been used to running less sensitive techniques like ICP-OES and AA, where the cleanliness requirements are not so strict. For that reason, it's always going to be challenging for operators who are not used to working in the ultra trace environment to fully understand all the analytical issues. And if that wasn't enough of a challenge, they are also probably working in labs that in most cases were not designed for ultra trace elemental analysis. The reader would be well advised to read Chapter 17, which deals with ways to reduce contamination when carrying out trace element analysis.

16.21   FINAL THOUGHTS

Modern plasma spectrochemical techniques will generate trace element data of the highest quality. However, with any solution technique, the data will only be as good as the sample presented to the instrument. For that reason, it is very important to ensure that the portion of the sample being analyzed is reflective of the bulk sample, especially for botanical-type samples, which are not homogeneous in nature.

This chapter has shown that robust closed-vessel microwave digestion techniques are critically important to the overall analytical procedure. There is no question that traditional rotor-based technology demonstrates good recovery for many elements. Highly reactive samples, such as oils and organic samples, can be digested even in large sample amounts, ensuring reliable analysis. In addition, by using a high-pressure, rotor-based system with a multi-position carousel, a high level of reproducibility can be achieved, even for volatile elements such as As and Hg.

However, for labs that might be handling a more diverse and complex range of samples such as herbal products, cannabis, pharmaceuticals or environmental-type samples, a rotor-based system might be somewhat restrictive, because similar samples have to be batched together to ensure they are being digested under the optimum conditions. Therefore, if there is a need to digest many differing sample matrices simultaneously in the same run, an SRC system could be the best solution, or, if your workload allows it, perhaps the single cavity mode technology might be the best approach.

Finally, it's important to emphasize that the sample-preparation and sample-digestion steps are littered with additional sources of contamination and errors, which are exaggerated with certain types of samples because they are natural absorbers of elemental contaminants from the growing, manufacturing and sample-processing steps.

It is not the intent to favor one microwave digestion approach over another. There are many companies that offer microwave digestion equipment, to offer advice to help users make the right decision based on their sample diversity and workload. For more detailed application material, please refer to the contact information for microwave digestion companies in the chapter on "Useful Contacts", where you will find a multitude of application notes for many different types of sample matrices.

FURTHER READING

1. B. Kratochvil and J. K. Taylor, *Analytical Chemistry*, **53**(8), 925–938A, 1981.
2. The National Institute for Occupational Safety and Health (NIOSH), *Safe Use of Hydrogen Fluoride and Hydrofluoric Acid*, www.cdc.gov/niosh/ershdb/emergencyresponsecard_29750030.html.

3. Microwave Digestion of Pharmaceutical Samples Followed by ICP-MS Analysis for USP Chapters <232> and <233>, *CEM Corp. Application Note*, https://cem.sharefile.com/download.aspx?id=sc6d12fed0084ab4b#.
4. S. Hussein and T. Michel, The Application of Single-Reaction-Chamber Microwave Digestion to the Preparation of Pharmaceutical Samples in Accordance with USP <232> and <233>, *Spectroscopy Magazine*, Special Issue, **27**(10), 2012, www.spectroscopyonline.com/application-single-reaction-chamber-microwave-digestion-preparation-pharmaceutical-samples-accordanc?id=&sk=&date=&pageID=3.
5. Microwave Digestion and Trace Metals Analysis of Cannabis & Hemp Products, *CEM Application Note*, August 2019, http://cem.com/media/contenttype/media/literature/ApNote_MARS6_Cannabis_Foods_ap0159_2.pdf.
6. R. Boyle and E. Farrell, Selecting Microwave Digestion Technology for Measuring Heavy Metals in Cannabis Products, *Cannabis Science and Technology*, 1(3), September/October 2018, www.cannabissciencetech.com/metals/selecting-microwave-digestion-technology-measuring-heavy-metals-cannabis-products.
7. R. Lockerman et al., Sample Preparation and Trace Elemental Analysis of Cannabis and Cannabis Products, *CEM Corp, Poster #ThP11, Plasma Winter Conference*, Tucson, AZ, 2020.
8. *Ethos UP Product Note/Brochure*, Milestone Inc., https://milestonesci.com/ethos-up-microwave-digestion-system/.
9. *Single Reaction Chamber: US Patent Number 5,270,010: Milestone Inc.*, Shelton, CT.
10. *The BLADE Microwave Digestion System*, CEM Corp, https://cem.com/en/future-speed-of-the-blade-microwave-digestion-system.
11. *Mercury Preservation Techniques: US Environmental Protection Agency (EPA)*, www.epa.gov/esd/factsheets/mpt.pdf.
12. R. J. Thomas, The Impact of Illegal Artisanal Gold Mining on the Peruvian Amazon: Benefits of Taking a Direct Mercury Analyzer into the Rain Forest to Monitor Mercury Contamination, *AP Column, Spectroscopy Magazine*, March, 2019, www.spectroscopyonline.com/impact-illegal-artisanal-gold-mining-peruvian-amazon-benefits-taking-direct-mercury-analyzer-rain-fo.

17 A Practical Guide to Reducing Errors and Contamination Using Plasma Spectrochemistry

There are many factors that influence the ability to get the correct result with any trace element technique, particularly in today's high-throughput cannabis testing laboratories. Unfortunately with plasma-spectrochemistry techniques, the problem is magnified even more because of their extremely high sensitivity. So, in order to ensure that the data reported is an accurate reflection of the sample in its natural state, the analyst must be aware of all the potential sources of contamination that could negatively impact the quality of the measurements. This chapter addresses this topic and suggests ways to minimize sources of errors in the generated data.

Analytical laboratories face more challenges and regulations than ever before as accreditation bodies issue an increasing number of guidelines and regulatory agencies increase the number of elements that need to be reported, while the levels of detection required decrease. A lot of times, it's the effort and money invested in deciphering the data and determining its validity and accuracy. This is probably more challenging in the cannabis testing industry, where there are few regulated methods and where labs are being set up almost overnight and equipped with very sophisticated instrumentation, sometimes with very little understanding of the ultra trace analytical environment they are being used in.

The methodology to reduce laboratory error and reduce contamination is composed of three primary steps: first, understanding data validity criteria, the application of appropriate statistics and the understanding of the use and terminology of analytical processes. The second step is correct selection and use of standards appropriate for the type and range of analysis being performed. The third step is the understanding of sample-preparation and sources of contamination that can cause error to be introduced into analyses.

17.1 UNDERSTANDING DATA ACCURACY AND PRECISION

All analytical laboratories pursue "good" data and "true" values. The reality is that true values are never absolute. True values are obtained by perfect and error-free measurements that do not exist in reality. Instead, the expected, specified or theoretical value becomes the accepted true value. Analysts then compare the observed or measured values against that accepted true value to determine accuracy or "trueness" of the data set.

Often, accuracy and precision are used in the same context when discussing data quality. Accuracy and precision are very different assessments of data and the acquisition process. Accuracy is the measurement of individual or groups of data points in relationship to the "true" value. In essence, accuracy is how close your data gets to the target and is often expressed as either a form of numerical or percent difference of the observed result and the target or "true" value.

Precision, on the other hand, is the measurement of a data set for how well the data points relate to each other. It is the measure of how clustered the data points fall within the target range and is often expressed as a calculation of standard deviation of the data set in some form. Precision is an important tool for the evaluation of instrumentation and methodologies by determining how data is produced after varied replications. The differences between accuracy and precision are exemplified in Figure 17.1.

Repeatability and reproducibility measure the quality of the data, method or instrumentation by examining the precision under the same (minimal difference) or different (maximal difference) test conditions. Repeatability (or test–retest reliability) is the measurement of variation arising when all the measurement conditions are kept constant, such as the same location, the same procedure, the same operator or the same instrument under the same conditions in repetition over a short period of time. Several standards organizations such as ASTM define the parameters of repeatability, intermediate precision and reproducibility in order to create and publish test methods[1,2] (Table 17.1).

Reproducibility is the measurement of variation arising in the same measurement process occurring across different conditions such as location, operator or instruments, and over long periods of time.

Another way of looking at accuracy and precision is in terms of measurement of different types of error.

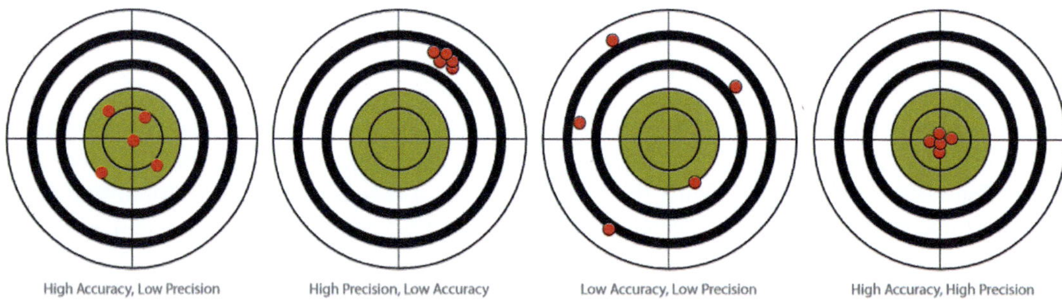

FIGURE 17.1   Representations of accuracy and precision.

TABLE 17.1

Conditions for Precision

	Repeatability	**Intermediate precision**	**Reproducibility**
Laboratory	Same	Same	Different
Operator	Same	Different	Different
Apparatus	Same	Same in type or actual apparatus	Different
Time between replicates	Less than a day	Multiple days	Not specified

17.2   ESTIMATING ERROR

The most common misconception regarding analytical data revolves around the concepts of error, mistakes and uncertainty. In general, an error is a deviation or difference between the estimated or measured value and the true, specified or theoretically correct value. If accuracy is the measurement of the difference between a result and a "true" value, then error is the actual difference or the cause of the difference. The estimation of error can be calculated in two ways, either as an absolute or relative error. Absolute errors are expressed in the same units as the data set and relative errors are expressed as ratios such as percent, fractions, etc.

TABLE 17.2

Accuracy and Precision for Absolute and Relative Errors

	Absolute	Relative
Accuracy	$E_{abs} = X_o - X_t$	$E_{rel} = X_o - X_t/X_t$
Precision	σ of data set or value taken from a curve	RSD or CV of data set

Absolute accuracy error is the true value subtracted from an observed value and is expressed in the same units as the data. For example, if a stated expected true value of an analysis is 5 ppm but the resulting value is 6 ppm, then the absolute error for that data point is 1 ppm. Relative accuracy error is the true value subtracted from the observed value and the result is divided by the true value. Errors in precision data are most commonly calculated as some variation of the standard deviation of the data set. An absolute precision error calculation is based on either the standard deviation of a data set or values taken from a plotted curve. A relative precision error is most commonly expressed as relative standard deviation (RSD) or the coefficient of variance (CV or %RSD) of the data set (Table 17.2).

17.3 TYPES OF ERRORS

There are many types of errors associated with scientific and statistical analyses. The most common error in regards to data is **observational or measurement errors**, which are the difference between a measured valve and its true value. Most measured values contain an inherent aspect of variability as part of the measurement process, which can be classified as either systematic error or random error.

Systematic (or determinate) errors are introduced inaccuracies from the measurement process or analytical system. There are some basic sources for systematic error in data. These sources are operator or analyst, apparatus and laboratory environment and method or procedure. Systematic errors can often be reduced or eliminated. Operator or analyst errors can occur due to inattentiveness, lack of training or misinformation. Operator or analyst errors are most often called mistakes. Apparatus or laboratory-environment errors can occur with improper maintenance, substandard laboratory environment and materials (i.e., improper volumetric, improper calibration, poor environmental temperature and humidity controls, etc.). Method or procedure errors can occur with poor method validation or lack of periodic updates as equipment or materials change. In cases where systematic errors lead to results in a data set that trend higher or lower than the "true" value, the difference is considered to be a bias. A positive bias creates a trend where results are higher than the expected value, while a negative bias displays results lower than the expected value. Determinate errors and bias can often be either corrected or adjusted for in the instrumentation or procedure.

Random (or indeterminate) errors lead to measured values that are inconsistent with repeated measurements. Random or indeterminate errors arise from random fluctuations and variances in the measured quantities and occur even in tightly controlled analysis systems or conditions. It is not possible to eliminate all sources of random error from a method or system. Random errors can, however, be minimized by experimental or method design. For instance, while it is impossible to keep an absolute temperature in a laboratory at all times, it is possible to limit the range of temperature changes. In instrumentation, small changes to the electrical systems from fluctuations in current, voltage and resistance cause small continuing variations, which can be seen as instrumental noise. The measurement of these random errors is often determined by the examination of the precision of the generated data set. Precision is a measure of statistical variability in the description of random errors. Precision analyzes the data set for the relationship and distance between each of the data points, independent of the "true" or estimated value of the data to identify and quantify the variability of the data.

Accuracy is the description of systematic errors and is a measure of statistical bias that causes a difference between a result and the "true" value (trueness). A second definition, recognized by ISO, defines accuracy as a combination of random and systematic errors, which then requires high accuracy to also have high precision and high "trueness". An ideal measurement method, procedure, experiment or instrument is both accurate and precise, with measurements that are all close to and clustered around the target or "true" value. The accuracy and precision of a measurement value is a process validated by the repeated measurements of a traceable reference standard or reference material.

17.4   STANDARDS AND REFERENCE MATERIALS

A standard in a laboratory setting often referred to as an actual chemical or physical material that is a known or characterized material used to confirm identity, concentration, purity or quality, called a metrological standard. A metrological standard is the fundamental example or reference for a unit of measure. Simply stated, a standard is the "known" to which an "unknown" can be measured. Metrological standards fall into different hierarchical levels.

1. **Primary standard:** The definitive example of its measurement unit to which all other standards are compared and whose property value is accepted without reference to other standards of the same property or quantity.[3,4] Primary standards of measure, such as weight, are created and maintained by metrological agencies and bureaus around the world (NIST, etc.).
2. **Secondary standards:** Close representations of primary standards that are measured against primary standards. Many chemical-standards companies create chemical standards against a primary weight set in order to create secondary standards traceable to that primary standard.
3. **Working standards:** Created against or with secondary standards to calibrate equipment.

There are also many standards designated as reference materials, reference standards or certified reference materials, which are materials manufactured or characterized for a set of properties and are traceable to a primary or secondary. If the material is a certified reference material, then it must be accompanied by a certificate that includes information on the material's stability, homogeneity, traceability and uncertainty.[4,5]

Stability is when a chemical substance is nonreactive in its environment during normal use. A stable material or standard will retain its chemical properties within the designated "shelf life" or within its expiration date if it is maintained under the expected and outlined stability conditions. A material is considered to be unstable if it can decompose, volatilize (burn or explode) or oxidize (corrode) under normal stated conditions.

Homogeneity is the state of being of uniform composition or character. Reference materials can have two types of homogeneity: within-unit homogeneity or between-unit (or lot) homogeneity. Within-unit homogeneity means there is no precipitation or stratification of the material that cannot be rectified by following instructions for use. Some reference materials can settle out of solution but are still considered homogeneous if they can be redissolved into the solution by following the instructions for use (i.e. sonicate, heat, shake, etc.). Between-unit or lot homogeneity is the homogeneity found between separate packaging units.

Traceability is the ability to trace a product or service from the point of origin, from the manufacturing or service process, through to final analysis, delivery and receipt. Reference-material producers must ensure that the material can be traced back to a primary standard.

Uncertainty is the estimate attached to a certified value that characterizes the range of values where the "true value" lies within a stated confidence level. Uncertainty can encompass random effects such as changes in temperature, humidity, drift accounted for by corrections and variability in performance of an instrument or analyst. Uncertainty also includes the contributions from within-unit and between-unit homogeneity, changes due to storage and transportation conditions and any uncertainties arising from the manufacture or testing of the reference material.

17.5 USING STANDARDS AND REFERENCE MATERIALS

Certified standards or certified reference materials (CRMs) are materials produced by standards providers that have one or more certified values with uncertainty established using validated methods, and are accompanied by a certificate. The uncertainty characterizes the range of the dispersion of values that occurs through the determinate variation of all the components, which are part of the process for creating the standard.

CRMs have a number of uses, including validation of methods, standardization or calibration of instrument or materials and for use in quality control and assurance procedures. A calibration procedure establishes the relationship between a concentration of an element and the instrumental or procedural response to that element.

A calibration curve is the plotting of multiple points within a dynamic range to establish the element response within a system during the collection of data points. One element of the correct interpretation of data from instrumental systems is the effect of a sample matrix upon an instrumental analytical response. The matrix effect can be responsible for either element suppression or enhancement. In analysis where the matrix can influence the response of an element, it is common to match the matrix of analytical standards or reference materials to the matrix of the target sample to compensate for matrix effects.

Different approaches to using calibration standards may need to be employed to compensate for the possible variability within a procedure or analytical system.

- Internal standards are compounds that are either similar in character or analogs of the target elements that have a similar analytical response. Internal standards are *added to the sample prior to analysis*. This type of standard allows the variation of instrument response to be compensated for by the use of a relative response ratio established between the internal standard and the target element.
  - Examples: Isotopic forms of elements, elements similar to the target, which are not in the sample.
  - Standard addition or spiking standard—an internal standard added to overcome matrix responses, instrument responses and the elemental responses that are indistinguishable from each other, as the element concentration nears the lower limit of detection or quantitation. A target standard can then be added in known concentration to compensate for the matrix or instrument effects to bring the signal of the target element into a quantitative range.
- External standards are multiple calibration points (customarily three or more points) that contain standards or known concentrations of the target elements and matrix components and exist *outside of the test samples*. Depending on the type of analytical techniques, linear calibration curves can be generated between response and concentration, which can be calculated for the degree of linearity or the correlation coefficient (r). An R-value approaching 1 reflects a higher degree of linearity. Most analysts accept values of > 0.999 or better as acceptable correlation.

17.6 CALIBRATION CURVES

Calibration curves are often affected by the limitations of the instrumentation. Data can become biased by calibration points, the instrument's limits of detection, quantitation and linearity and by the response of the system versus its baseline (signal-to-noise).

- Limit of detection (LOD): lower limit of a method or system, the target of which can be detected as different from a blank with a high confidence level (usually over three standard deviations from the blank response).
- Limit of quantitation (LOQ): lower limit of a method or system, the target of which can be reasonably calculated where two distinct values between the target and blank can be observed (usually over ten standard deviations from the blank response) (Figure 17.2.).
- Signal to noise (s/n): response of an element measured on an instrument as a ratio of that response to the baseline variation (noise) of the system. Limits of detection are often recognized as target responses that have three times the response of baseline noise or s/n >/= 3. Limits of quantitation are recognized as target responses that have ten times the response of baseline noise, or s/n >/= 10.

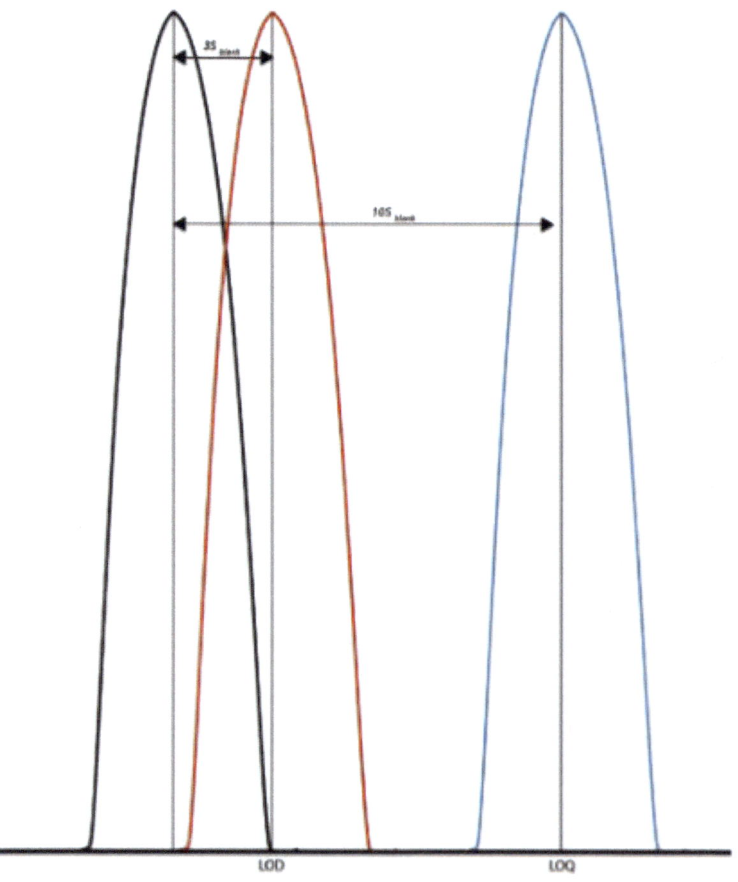

FIGURE 17.2   Limits of detection and quantitation.

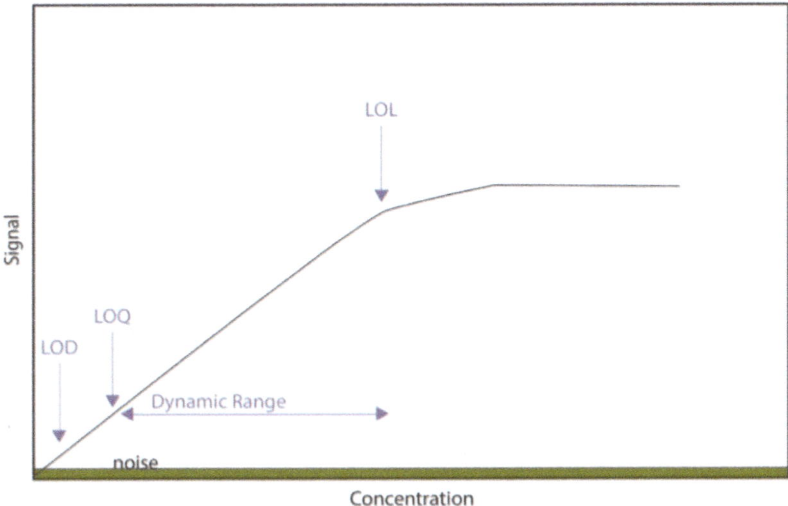

FIGURE 17.3 Calibration-curve limits and range.

- Limits of linearity (LOL): upper limits of a system or calibration curve where the linearity of the calibration curve starts to be skewed, creating a loss of linearity (Figure 17.3). This loss of linearity can be a sign that the instrumental detection source is approaching saturation.
- Dynamic range: an array of data values between the LOQ and the LOL, where the greatest potential for accurate measurements will occur.

The understanding of a system's dynamic range, the accurate bracketing of calibration curves within the range and around the target element concentration increases the accuracy of the measurements. If a calibration curve is created, that does not potentially bracket all the possible target data points; the calibration curve can be biased to artificially increase or decrease the results and create error. To create calibration curves and working standards, there must be an accurate process of converting units, calculating dilution targets and preparing dilutions.

17.7 DYNAMIC RANGE, CONCENTRATION AND ERROR

The first step in creating standards, working solutions and dilutions is to understand the dynamic range your analysis is targeting—is it ppb, ppm, percent?

If you are looking for a major component of the sample, then your standards have to be in percent levels or the samples must be diluted down to the correct concentrations. If target elements are trace elements or trace contaminants, then standards and calibration curves have to be diluted down to match the target within the dynamic range of the instrument technique.

Tables 17.3 and 17.4 show the basic conversions between different concentrations based on mass or volume.

17.8 LABORATORY SOURCES OF ERROR AND CONTAMINATION

Calibration curves are created by diluting standards into several target points along the dynamic range to cover the possible target results. Proper dilution of standards and samples is based on the understanding of basic dilution and volumetric procedures and dilution factors. Volumetric measurement is a commonly repeated daily activity in most analytical laboratories. Many processes

TABLE 17.3

Weight-to-Weight Concentrations

Name	Symbol	Equivalence			
Parts per thousand*	ppt*	g/kg	mg/g	μg/mg	ng/μg
Parts per million	ppm	mg/kg	μg/g	ng/mg	pg/μg
Parts per billion	ppb	μg/kg	ng/g	pg/mg	fg/μg
Parts per trillion	pptt	ng/kg	pg/g	fg/mg	ag/μg

TABLE 17.4

Weight-to-Volume Concentrations

Name	Symbol	Equivalence			
Parts per thousand*	ppt*	g/L	mg/mL	μg/μL	ng/nL
Parts per million	ppm	mg/L	μg/mL	ng/μL	pg/nL
Parts per billion	ppb	μg/L	ng/mL	pg/μL	fg/nL
Parts per trillion	pptt	ng/L	pg/mL	fg/μL	ag/nL

** Note:* Parts per thousand and parts per trillion both can use the ppt abbreviation; so use care to select the appropriate unit of measure.

in the laboratory, from sample preparation to standards calculation, depend on accurate and contamination-free volumetric measurements. Unfortunately, laboratory volumetric labware, syringes and pipettes are some of the most common sources of contamination, carryover and error in the laboratory.

The root of these errors is based on the four "I" errors of volumetric containers:

1. ***Improper use*** : measuring tool is not used correctly
2. ***Incorrect choice*** : measurement tool is not appropriate for the volume or type of measurement
3. ***Inadequate cleaning*** : carryover causes contamination
4. ***Infrequent calibration*** : measuring tool is not calibrated for use

These four "I's" can lead to error and contamination, which negate all intent of careful measurement processes.

Many errors can be avoided by understanding the markings displayed on the volumetric containers and choosing the proper tool for the job. There is a lot of information displayed on volumetric labware. Most labware, especially glassware, is designated as either Class A (analytical or quantitative) or Class B (general use) labware. If a critical measurement process is needed, then only Class A glassware should be used for measurement.

Other information that can be found on labware is the name of the manufacturer, country of origin, tolerance or uncertainty of the measurement of the labware and a series of descriptors as to how the glassware should be used. Labware can be marked with letters that designate the purpose of the container. If a volumetric is designed to contain liquid, it will be marked by either the letters TC or IN. Labware which is designated to deliver liquid will be marked by either the letters TD or EX. Sometimes there are additional designations, such as wait time or delivery time, inscribed on the labware. The delivery time refers to a period of time required for the meniscus to flow from the upper volume mark to the lower volume mark. The wait time refers to the time needed for the meniscus to come to rest after the residual liquid has finished flowing down from the wall of the pipette or vessel.

The second type of improper use and incorrect choice can be seen in the selection of pipettes and syringes for analytical measurements. Many syringe manufacturers recommend a minimum dispensing volume of approximately 10% of the total volume of the syringe or pipette. A study by SPEX CertiPrep showed that dispensing such a small percentage of the syringe's total volume created a large amount of error. The largest rates of error were seen in the smaller syringes of 10 and 25 μL. Dispensing 20% of the 10-μL syringe created a 23% error. Error only dropped down to below 5% as the volume dispensed approached 100%. In the larger syringes, measurements over 25% were able to see error in and around 1%. The larger syringes were able to get closer to the 10% manufacturer's dispensing minimum without a large amount of error, but the error did drop as the dispensed volume approached 100%.[6]

The third "I" of volumetric error is inadequate cleaning. Many volumetric containers can be subject to memory effects and carryover. In critical laboratory experiments, labware sometimes needs to be separated by purpose and use. Labware subject to high levels of organic compounds or persistent inorganic compounds can develop chemical interactions and memory effects. It is sometimes difficult to eliminate carryover from labware and syringes, even when using a manufacturer's stated instructions. For example, many syringes are cleaned by several repeated solvent rinses prior to use. A study of syringe carryover by SPEX CertiPrep showed that some syringes are subject to high levels of chemical carryover, despite repeated rinses.

The final source of error is infrequent calibration. Many laboratories have schedules of maintenance for equipment, such as balances and automatic pipettes, but often overlook calibration of reusable burettes, pipettes, syringes and labware. Under most normal use, labware often does not need frequent calibration, but there are some instances where a schedule of recalibration should be employed. Any glassware or labware in continuous use for years should be checked for calibration. Glass manufacturers suggest that any glassware used or cleaned at high temperatures, used for corrosive chemicals or autoclaved, should be recalibrated more frequently.

It is also suggested that, under normal conditions, soda-lime glass be checked or recalibrated every five years, and borosilicate glass after it has been in use for ten years. The error associated with the use of volumetric containers can be greatly reduced by choosing the correct volumetric for the task, using the tool properly and making sure the volumetric containers are properly cleaned and calibrated before use.

Inorganic analysts know that glassware is a source of contamination. Even clean glassware can contaminate samples with elements such as boron, silicon and sodium. If glassware, such as pipettes and beakers, are reused, the potential for contamination escalates. At SPEX CertiPrep, a study was conducted of residual contamination of our pipettes after being manually and automatically cleaned using a pipette washer.[6,9]

An aliquot of 5% nitric acid was drawn through a 5-mL pipette after the pipette was manually cleaned according to standard procedures. The aliquots were analyzed by ICP-MS. The results showed significant residual contamination still persisted in the pipettes, despite a thorough manual-cleaning procedure.

The experiment was repeated using a pipette washer especially made for use in parts-per-trillion analysis. The pipette washer repeated forced deionized water through the pipettes for a set time period. The pipettes were cleaned in the pipette washer, then the same aliquot of 5% nitric acid was drawn through the 5-mL pipettes. The aliquot was analyzed by ICP-MS. The automated washer reduced the contamination significantly as compared to the manual cleaning of the pipettes. The reduction of contamination by moving from manual cleaning to an automated cleaning process was clear. High levels of contamination of sodium and calcium (almost 20 ppb) dropped to < 0.01 ppb. Other common contaminants, including lead and iron, dropped from 5.4 and 1.6 ppb, respectively, to less than 0.01 ppb.

The reduction of contamination in labware can depend on the material of the labware and its use. Different materials contain many types of elemental and organic potential contamination, as seen in Table 17.5.[7] Trace inorganic analyses are best performed in polymer or high-purity quartz vessels, or fluorinated ethylene propylene (FEP), to minimize contact with borosilicate glass. Metals, such as Pb and Cr, are highly absorbed by glass but not by plastics. On the other hand, samples containing low levels of Hg (ppb levels) must be stored in glass or fluoropolymer, because Hg vapors diffuse through polyethylene bottles.

It is always important to know the best conditions for storage for a standard. The expiration dates on most standards reflect the manufacturer's confidence level of continued accuracy at proper storage conditions for the shelf life listed on the product. If a standard is not stored properly, it can affect the quality and accuracy of the standard. Elements within the packaging can slowly be leached from the packaging over time and change the value of the samples or standards. Some of the common packaging elements were examined in samples at the time of manufacturing and bottling, and again after one year, to determine if the packaging contributed to the contamination of the samples over time. After one year, the amounts of common elements, such as aluminum, calcium, iron, magnesium, sodium and silica, more than doubled in the solutions. Table 17.6 shows the changes of elemental concentrations in six common elements after a year's storage.

TABLE 17.5

Major Elemental Impurities Found in Laboratory Container Materials.[7]

Material	# Elements	Total ppm	Major impurities
Polystyrene-PS	8	4	Na, Ti, Al
Tetrafluoroethylene (TFE)	24	19	Ca, Pb, Fe, Cu
Low-density PE-LDPE	18	23	Ca, Cl, K, Ti, Zn
Polycarbonate-PC	10	85	Cl, Br, Al
Polymethyl Pentene-PMP	14	178	Ca, Mg, Zn
Fluorinated ethylene propylene (FEP)	25	241	K, Ca, Mg
Borosilicate glass	14	497	Si, B, Na
Polypropylene (PP)	21	519	Cl, Mg, Ca
High-density PE (HDPE)	22	654	Ca, Zn, Si

TABLE 17.6

Changes in Element Concentration after Storage (ppb).

Element	At manufacture	After one year
Al	1–10	10–40
Ca	1–10	20–100
Fe	0–5	5–50
Mg	0–10	10–40
Na	1–10	5–40
Si	5–15	15–5000

17.9 SOURCES OF LABORATORY CONTAMINATION AND ERROR

Contamination and error can occur at almost any point of the process, and then can be magnified as the method and analysis runs its course. In the past, issues of laboratory contamination were problematic, but now contaminants, even in trace amounts, can severely alter results. It is hard to imagine that such small amounts of contamination can dramatically change laboratory values.

Most questions about contamination come in the form of an inquiry about a particularly high result for some common contaminant. In most cases, the root of that contamination can be traced to a common source. The most common sources of standard and sample contamination, in addition to the previously discussed volumetric labware, are water, reagents and acids, laboratory environment, storage and personnel.

17.10 WATER QUALITY

Water is one of the most basic yet most essential laboratory components. There are many types, grades and intended uses for water. Water is most often used in two ways in the laboratory: as a cleaning solution and as a transfer solution for volumetric or gravimetric calibrations or dilutions. In both of these uses, the water must be clean in order to reduce contamination and the possibility of introducing error into the process. Poor-quality water can cause a host of problems, from creating deposits in labware or inadvertently increasing a target element or element concentration in solution.

The confusion starts when laboratories are unsure about which type of water they get from their water-filtration system. ASTM has guidelines that designate different grades of water. Table 17.7 shows the parameters for the four ASTM types of water.[8]

The actual type of water produced by a commercial laboratory water-filtration system can vary in pH, solutes and soluble silica. Critical analytical processes should always require a minimum quality of ASTM Type I water. All trace analysis standards, dilutions, dissolutions, extractions and digestions should be conducted with the highest-purity water. Analysts who use certified reference materials (CRMs) and perform quantitative analysis need to use quality water, in order not to contaminate their CRMs, standards and samples.

High-purity water is often achieved in several stages in multiple processes that remove physically and chemically potential contaminating substances. Municipal water supplies often test their own

TABLE 17.7

ASTM Designations for Reagent Laboratory Water.[8]

	ASTM Type			
Requirement	I	II	III	IV
Use	Critical laboratory applications and processes	General lab grade used for pH, buffers, feeding other water-polishing systems	Cleaning glassware, feeding water-polishing systems, water baths	Not good for lab use
Specific resistance (megohm/cm) (max)	18	1	4	0.2
pH	N/A	N/A	N/A	5.0 to 8.0
Sodium (max)	1 µg/L	5 µg/L	10 µg/L	50 µg/L
Total silica (max)	3 µg/L	3 µg/L	500 µg/L	high
Total organic carbon (max)	50 µg/L	50 µg/L	200 µg/L	N/A

water sources on an annual basis but that does not mean it is applicable for use in laboratory applications. Municipal water can become contaminated from its distribution point, especially when left static sitting in pipes, tubing and hoses. Water left stationary in a laboratory water system can be exposed to the leaching of elements and compounds from the pipes and hoses.

17.11   REAGENTS

An analytical laboratory often uses a large amount of various reagents, solvents and acids of varying quality and contamination levels. These chemical components of the laboratory can be a large economic investment for a laboratory, but also a large source of potential contamination. Just as in the case of water, there are different types or grades of chemicals, reagents, acids and solvents. Some designations are set forth by standards set by the US Pharmacopeia (USP) or the American Chemical Society (ACS). Other types or grades of material are designated by individual manufacturers based on intended use. Some grades are general laboratory grades with intended use for noncritical applications, while other grades usually high in purity and low in contaminants are designated for more critical analyses.

There are many persistent solvents that can be found in the lab which can cross-contaminate samples by their presence. Some persistent solvents include dichloromethane, which can cause chlorine contamination—and DMSO and carbon disulfide, which can add sulfur residues. There are solvents that react with air to form peroxides that can cause contamination and safety issues in the laboratory. Another lab reagent that can become both a potential danger and a potential contaminant in the lab are acids.

Acids are, by their nature, oxidizers, and many of the strongest acids are used in the processing of samples for inorganic analysis. Common acids in digestion and dissolution include perchloric acid, hydrofluoric acid, sulfuric acid, hydrochloric acid and nitric acid. Many of these acids are commercially available in several grades, from general laboratory or reagent grade to high-purity trace-metal grade. Acid grades often reflect the number of sub-boiling distillations the acid undergoes for purification before bottling. The more an acid is distilled, the higher the purity of the acid. These high-purity acids have the lowest amount of elemental contamination but can become very costly, at up to ten times the cost of the reagent grade of acids.

Often the question is asked if high-purity acids are necessary for sample preparation, if the laboratory is using a high-quality inductively coupled plasma mass spectrometry- (ICP-MS) grade CRM. Clean acids used in sample preparation, digestion and preservation can be very costly. But, the difference between the amounts of contamination in a low-purity acid and a high-purity acid can be dramatic. High-quality standards for use in parts-per-billion and parts-per-trillion analyses use the highest-purity acids available to reduce all possible contamination from the acid source.

An example of potential contamination is an aliquot of 5 mL of acid containing 100 ppb of Ni as a contaminant, used for diluting a sample to 100 mL; it can introduce 5 ppb of Ni into the sample.

To reduce contamination, it is recommended that high-purity acids be used to dilute and prepare standards and samples when possible. In addition to using pure acid, it is important that the chemist check the acid's certificate of analysis to identify the elemental contamination levels present in the acid. Some laboratories prefer to use blank subtractions to negate the background contamination, but blank subtraction for acids can only work in a range well over the instrumental level of detection. If blank subtraction causes an analytical result to fall below the instrument's level of detection, it should not be used.

17.12   LABORATORY ENVIRONMENT AND PERSONNEL

All laboratories believe they observe a level of laboratory cleanliness. Most chemists recognize there are inherent levels of contamination present in all laboratories. A common belief is that the small amounts of environmental and laboratory contamination cannot truly change the analytical results. To test the background level of contamination in a typical laboratory, samples of nitric acid were distilled in both a regular laboratory and in a cleanroom laboratory with special air-handling systems (HEPA filters). The nitric acid distilled in the regular laboratory had high amounts of aluminum, calcium, iron, sodium and magnesium contamination. Table 17.8 shows the acid distilled in the cleanroom displayed significantly lower amounts of most contaminants.[6]

Laboratory air also can contribute to the contamination of samples and standards. Common sources of air and particulate-matter contamination are from surfaces and building materials, such as ceiling tiles, paints, cement and drywall. Surface contaminants can be found in dust and rust on shelves, equipment and furniture. Dust contains many different earth elements such as sodium, calcium, magnesium, manganese, silicon, aluminum and titanium. Dust can also contain elements from human activities (Ni, Pb, Zn, Cu, As) and organic compounds like pesticides, persistent organic pollutants (POP) and phthalates. The dust and rust particles can contaminate open containers in the lab or enter containers by charge transfer from friction by the triboelectric effect. The triboelectric effect, or triboelectric charging, is when materials become charged after coming into contact with a second material creating friction. The most common example of this effect is seen when hair sticks to a plastic comb after a static charge is created. The polarity and the strength of the electrical charge are dependent upon the type of material and other physical characteristics. Many materials in the lab have strong positive or negative triboelectric charges, as seen in Figure 17.4.

TABLE 17.8

Elemental Impurities Found in Nitric Acid Distilled in Regular Labs Compared to a Clean Lab[6]

Element	Regular lab	Clean lab
Al	60	15
As	0.17	< 0.02
Ca	150	100
Cd	0.3	0.003
Cr	2.5	0.4
Cu	1.7	0.23
Fe	50	9
Mg	10	4
Mn	1.1	0.1
Mo	0.8	0.03
Pb	0.5	0.4
Sb	0.04	0.013
Zn	5.5	0.7

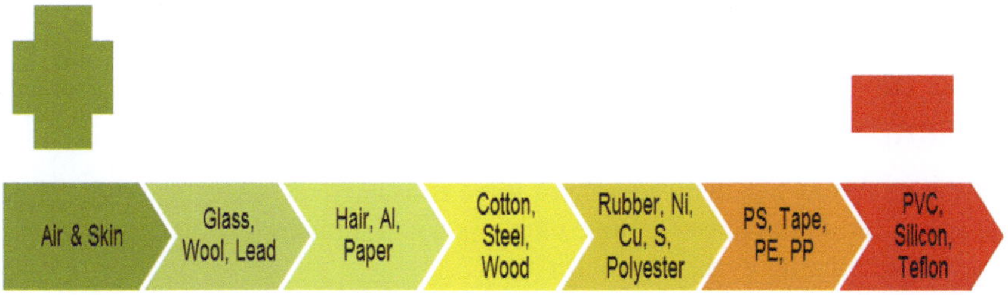

FIGURE 17.4 Triboelectric charge potential of common materials and particles in the laboratory.

TABLE 17.9

Common Elemental Contamination Sources

Element	Source
Aluminum	Lab glassware, cosmetics, dust, e-cigs
Bismuth	Cosmetics, lotions, medicines
Cadmium	Cigarettes, e-cigs, pigments, batteries
Cobalt	Surgical implants, dental prosthesis, jewelry
Copper	Algaecide used in swimming pools, e-cigs
Lead	Cosmetics, hair dye, jewelry, dust, cigarettes, e-cigs
Mercury	Mascara, dust
Selenium	Anti-dandruff shampoos
Zinc	Drugs, calamine lotion, cosmetics, powdered gloves

In the laboratory, materials like dust, air, skin and lead have extreme positive charges and can be attracted to the strong negative charge of Teflon™ or other plastic bottles when the bottle is opened, and creates a friction-inducing charge.

Laboratory personnel can add their own contamination from lab coats, makeup, perfume and jewelry. Aluminum contamination can come from lab glassware, cosmetics and jewelry. Many other common elements can be brought in as contamination from lotions, dyes and cosmetics. Even sweat and hair can cause elevated levels of sodium, calcium, potassium, lead, magnesium and many ions. If a laboratory is seeing unusually high levels of cadmium in the samples, it could be from cigarettes, pigments or batteries. If the levels of lead are out of range, contamination can be from paint, cosmetics and hair dyes. Table 17.9 shows potential sources of common elemental contamination from outside products.

Laboratory environment and personnel contamination can be reduced by limiting the use of personal-care products, jewelry and cosmetics, which could contain contamination and interfere with critical analyses. Lab coats can collect all types of contamination and should only be worn in the laboratory to avoid cross-contamination from other labs and the outside world. The laboratory surfaces should be kept clean. Deionized water can be used to wipe down work surfaces. Laboratory humidity can be kept above 50% to reduce static charge. An ethanol- or methanol-soaked laboratory wipe can be used to reduce static electricity as it evaporates.

Even with clean laboratory practices in place, erroneous results can often find their way into sample analysis. To eliminate some of these spurious results, replication of blanks and sample dilutions can be employed. The blank results should be averaged and the sample run values can either be minimally selected or averaged. The difference between the two values can then be plotted against a curve established against two more standards. A minimum of two standard points can be used if the chance of contamination is minimal, such as in the case of rare or uncommon elements. Additional standard points should be considered if the potential for contamination is high with common elements, such as aluminum, sodium and magnesium. Multiple aliquots of blanks and dilutions can also be employed to further minimize analytical uncertainty.

17.13 GENERAL PRINCIPLES AND PRACTICES

Laboratories should follow a general regime of three runs each of wash/rinse runs, blank runs and sample runs, as well as single runs of sample plus spike, and standard or spike runs without a sample to use as a control solution to evaluate recovery.

Analysts must realize that the cleanliness and accuracy of their procedures, equipment and dilutions affect the quality of the standards and samples. Many laboratories will dilute CRMs to use across an array of procedures and techniques. This in-house dilution of CRMs can be a savings to the laboratory, but in the final analysis can be a source of error and contamination.

CRM manufacturers design standards for particular instruments to obtain the highest level of accuracy and performance for that technique. They also use calibrated balances, glassware and instruments to ensure the most accurate standards are delivered to customers. Certifications, such as ISO 9001, ISO 17025 and ISO 17034, assure customers that procedures are being followed to ensure quality and accuracy in those standards. After those CRMs are in chemists' hands, it is then their responsibility to employ all possible practices to keep their analysis process free from contamination and error.

REFERENCES

1. ASTM E177–19, *Standard Practice for Use of the Terms Precision and Bias in ASTM Test Methods*, ASTM International, West Conshohocken, PA, 2019, www.astm.org.
2. ASTM E456–13A(2017)e3, *Standard Terminology Relating to Quality and Statistics*, ASTM International, West Conshohocken, PA, 2017, www.astm.org.
3. BIPM, IEC, IFCC, ILAC, IUPAC, IUPAP, ISO, and OIML, *The International Vocabulary of Metrology— Basic and General Concepts and Associated Terms (VIM)*, 3rd edition, JCGM, p. 200, 2012, www.bipm.org/vim.
4. ISO Guide 30, *Terms and Definitions used in Connection with Reference Materials*, https://www.iso.org/standard/46209.html
5. ISO Guide 17034, *General Requirements for the Competence of Reference Material Producers*, https://www.iso.org/standard/29357.html
6. SPEX CertiPrep Webinar, *Clean Laboratory Techniques*, www.spexcertiprep.com/webinar/clean-laboratory-techniques.
7. J. R. Moody and R. Lindstrom, Selection and Cleaning of Plastic Containers for Storage of Trace Element Samples, *Analytical Chemistry*, **49**, 2264, December 1977.
8. ASTM D1193–06(2018), *Standard Specification for Reagent Water*, ASTM International, West Conshohocken, PA, 2018, www.astm.org.
9. SPEX CertiPrep Application Note, *Understanding Measurement: A Guide to Error, Contamination and Carryover in Volumetric Labware, Syringes and Pipettes*, in publication.

18 Performance- and Productivity-Enhancement Techniques

Conventional sample-introduction systems, using a spray chamber and nebulizer, account for the majority of ICP-MS applications being carried out today. However, nonstandard sampling accessories such as laser ablation systems, flow injection analyzers, electrothermal vaporizers, cooled spray chambers, desolvation equipment, direct injection nebulizers, automated sample delivery systems, auto-dilutors and online chemistry procedures are considered critical to enhancing the practical capabilities of the technique. Initially regarded as novel sampling devices, they have since proved themselves to be invaluable for solving real-world application problems by enhancing the flexibility, performance and productivity of the technique. Chapter 18 describes the basic principles of these accessories and gives an overview of their practical capabilities. They include desolvation devices, autodilution equipment and automated sample-delivery accessories, including intelligent autosamplers and enhanced-productivity sampling tools, which can be very beneficial for testing labs that are looking to analyze a suite of samples in the most efficient and cost-effective manner.

It is recognized that standard ICP-MS instrumentation using a traditional sample-introduction system comprising a spray chamber and nebulizer has certain limitations, particularly when it comes to the analysis of complex samples. Some of these known limitations include the following:

- Inability to analyze solids directly.
- Contamination issues with samples requiring multiple sample-preparation steps.
- Liquid aerosol can impact ionization process.
- Total dissolved solids must be kept below 0.2%.
- If matrix has to be removed, it has to be done offline.
- Long washout times required for samples with a heavy matrix.
- Dilutions and addition of internal standards can be labor intensive and time consuming.
- Matrix components can generate severe spectral overlaps on many analytes.
- The analysis of slurries is very difficult.
- Matrix suppression can be quite severe with some samples.
- Spectral interferences generated by solvent-induced species can limit detection capability.
- Organic solvents can present unique problems.
- Sample throughput is limited by the sample-introduction process.
- Not suitable for the determination of elemental species or oxidation states.

Such were the demands of real-world users to overcome these kinds of problem areas, that instrument manufacturers developed different strategies based on the type of samples being analyzed. Some of these strategies involved parameter optimization or modification of instrument components, but it was clear that this approach alone was not going to solve every conceivable problem. For this reason, they turned their attention to the development of sampling accessories, which were optimized for a particular application problem or sample type. Over the past 10–15 years, this demand has led to the commercialization of specialized performance- and productivity-enhancement tools, not only manufactured by the instrument manufacturers themselves, but also by third-party vendors

DOI: 10.1201/9781003187639-18

specializing in these kinds of sampling techniques. The most common ones used today include the following:

- Laser ablation systems (LA)
- Flow injection analyzers (FIA)
- Electrothermal vaporizors (ETV)
- Chilled spray chambers and desolvation systems
- Direct injection nebulizers (DIN)
- Fast automated sampling procedures
- Auto dilution and calibration systems
- Automated sample identification and tracking

Let us take a closer look at some of these enhanced techniques to understand their basic principles and what benefits they bring to ICP-MS and ICP-OES. We can broadly categorize them as performance- and productivity-enhancing tools.

18.1 PERFORMANCE-ENHANCING TECHNIQUES LASER ABLATION

The limitation of ICP-MS to analyze solids, without dissolving the material, led to the development of laser ablation. The principle behind this approach is the use of a high-powered laser to ablate the surface of a solid and sweep the sample aerosol into the ICP mass spectrometer for analysis in the conventional way.[1]

Before I go on to describe some typical applications suited to laser ablation ICP-MS, let us first take a brief look at the history of analytical lasers and how they eventually became such a useful sampling tool. The use of lasers as vaporization devices was first investigated in the early 1960s. When light energy with an extremely high-power density interacts with a solid material, the photon-induced energy is converted into thermal energy, resulting in vaporization and removal of the material from the surface of the solid.[2] Some of the early researchers used ruby lasers to induce a plasma discharge on the surface of the sample and measure the emitted light with an atomic emission spectrometer.[3] Although this proved useful for certain applications, the technique suffered from low sensitivity, poor precision and severe matrix effects caused by non-reproducible excitation characteristics. Over the years, various improvements were made to this basic design with very little success,[4] because the sampling process and the ionization/excitation process (both under vacuum) were still intimately connected and interacted strongly with each other.

This limitation led to the development of laser ablation as a sampling device for atomic spectroscopy instrumentation, where the sampling step was completely separated from the excitation or ionization step. The major benefit is that each step can be independently controlled and optimized. These early devices used a high-energy laser to ablate the surface of a solid sample, and the resulting aerosol was swept into some kind of atomic spectrometer for analysis. Although initially used with atomic absorption[5,6] and plasma-based emission techniques,[7,8] it was not until the mid-1980s, when lasers were coupled with ICP-MS, that the analytical community sat up and took notice.[9] For the first time, researchers were coming up with evidence that virtually any type of solid could be vaporized, irrespective of electrical characteristics, surface topography, size or shape, and be transported into the ICP for analysis by atomic emission or mass spectrometry. This was an exciting breakthrough for ICP-MS, because it meant the technique could be used for the bulk sampling of solids, or, if required, for the analysis of small spots or micro-inclusions, in addition to being used for the analysis of solutions.

18.2 COMMERCIAL LASER ABLATION SYSTEMS FOR ICP-MS

The first laser ablation systems developed for ICP instrumentation were based on solid-state ruby lasers, operating at 694 nm. These were developed in the early 1980s, but did not prove to be successful for a number of reasons, including poor stability, low power density, low repetition rate and large beam diameter, which made them limited in their scope and flexibility as a sample-introduction device for trace element analysis. It was at least another five years before any commercial instrumentation became available. These early commercial laser ablation systems, which were specifically developed for ICP-MS, used the Nd:YAG (neodymium doped yttrium aluminum garnet) design, operated at the primary wavelength of 1064 nm—in the infrared.[10] They initially showed a great deal of promise because analysts were finally able to determine trace levels directly in the solid without sample dissolution. However, it soon became apparent that they did not meet the expectations of the analytical community, for many reasons, including complex ablation characteristics, poor precision, non-optimization for microanalysis and, because of poor laser coupling, they were unsuitable for many types of solids. By the early 1990s, most of the LAs purchased were viewed as novel and interesting, but not suited to solving real-world application problems.

These basic limitations in IR laser technology led researchers to investigate the benefits of shorter wavelengths. Systems were developed that were based on Nd:YAG technology at the 1,064-nm primary wavelength, but utilizing optical components to double (532 nm), quadruple (266 nm), and quintuple (213 nm) the frequency. Innovations in lasing materials and electronic design, together with better thermal characteristics, produced higher energy with higher pulse-to-pulse stability. These more advanced UV lasers showed significant improvements, particularly in the area of coupling efficiency, making them more suitable for a wider array of sample types. In addition, the use of higher-quality optics allowed for a more homogeneous laser beam profile, which provided the optimum energy density to couple with the sample matrix. This resulted in the ability to make spots much smaller and with more controlled ablations irrespective of sample material, which were critical for the analysis of surface defects, spots and micro-inclusions. Figure 18.1 shows the optical layout of a commercially available frequency-quintupled 213-nm Nd:YAG laser ablation system.

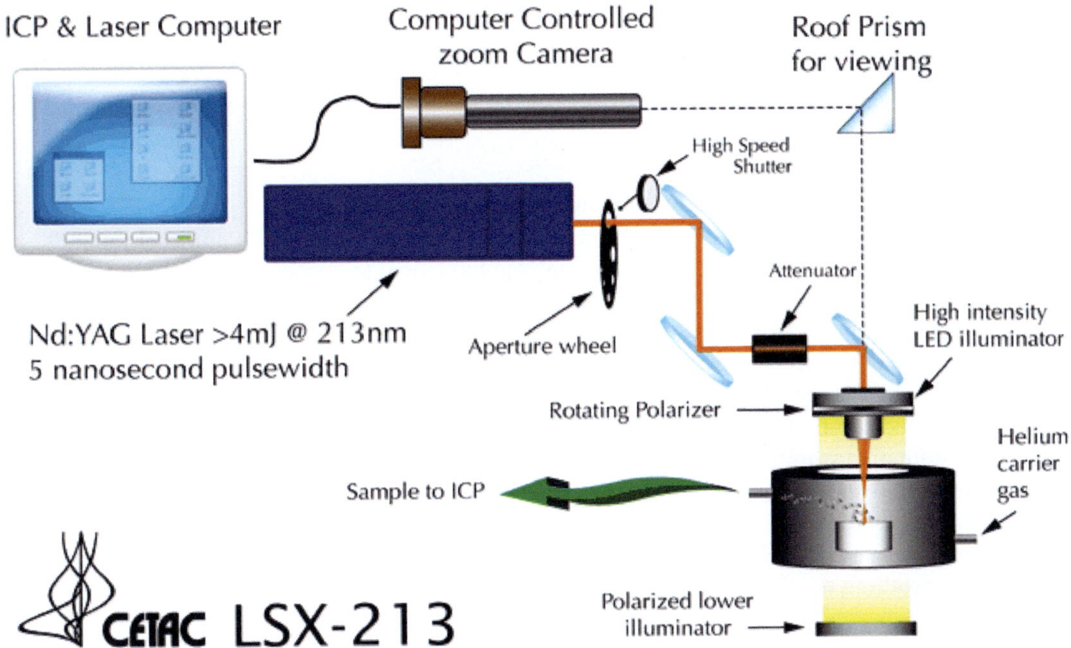

FIGURE 18.1   Schematic of a commercially available frequency-quintupled 213-nm Nd:YAG laser ablation system.

(Courtesy of Teledyne Cetac Technologies)

18.3   EXCIMER LASERS

The successful trend toward shorter wavelengths and the improvements in the quality of optical components also drove the development of UV gas-filled lasers, such as XeCl (308 nm), KrF (248 nm) and ArF (193 nm) excimer lasers. These showed great promise, especially the ones that operated at shorter wavelengths and were specifically designed for ICP-MS. Unfortunately, they necessitated a more sophisticated beam-delivery system, which tended to make them more expensive. In addition, the complex nature of the optics and the fact that gases had to be changed on a routine basis made them a little more difficult to use and maintain, and as a result required a more skilled operator to run them. However, their complexity was far outweighed by their better absorption capabilities for UV-transparent materials such as calcites, fluorites and silicates, smaller particle size, and higher flow of ablated material. There was also evidence to suggest that the shorter-wavelength excimer lasers exhibit better elemental fractionation characteristics (typically defined as the intensity of certain elements varying with time, relative to the dry aerosol volume) than the longer-wavelength Nd:YAG design, because they produce smaller particles that are easier to volatilize.

Even though excimer lasers are optically more complex than other designs, it's worth mentioning that today's instruments are far more rugged and robust compared to the earlier-designed systems, and as a result are being used for more and more routine applications. And because the higher-grade optical components used in the excimer technology are available at a more realistic cost, their price has come down quite significantly in the past few years.

18.4   BENEFITS OF LASER ABLATION FOR ICP-MS

Today there are a number of commercial laser ablation systems on the market designed specifically for ICP-MS, including 266-nm and 213-nm Nd:YAG and 193-nm ArF excimer lasers. They all have varying output energy, power density and beam profiles, and even though each one has different ablation characteristics, they all work extremely well, depending on the types of samples being analyzed and the data quality requirements. Laser ablation is now considered a very reliable sampling technique for ICP-MS, which is capable of producing data of the very highest quality directly on solid samples and powders. Some of the many benefits offered by this technique include the following:

- Direct analysis of solids without dissolution.
- Ability to analyze virtually any kind of solid material, including rocks, minerals, metals, ceramics, polymers, plastics, plant material and biological specimens.
- Ability to analyze a wide variety of powders by pelletizing with a binding agent.
- No requirement for sample to be electrically conductive.
- Sensitivity in the ppb to ppt range, directly in the solid.
- Labor-intensive sample-preparation steps are eliminated, especially for samples such as plastics and ceramics that are extremely difficult to get into solution.
- Contamination is minimized because there are no digestion/dilution steps.
- Reduced polyatomic spectral interferences compared to solution nebulization.
- Examination of small spots, inclusions, defects or micro-features on surface of sample.
- Elemental mapping across the surface of a mineral.
- Depth profiling to characterize thin films or coatings.

Let us now take a closer look at the strengths and weaknesses of the different laser designs with respect to application requirements.

18.5 OPTIMUM LASER DESIGN BASED ON THE APPLICATION REQUIREMENTS

The commercial success of laser ablation was initially driven by its ability to directly analyze solid materials such as rocks, minerals, ceramics, plastics and metals, without going through a sample-dissolution stage. Table 18.1 represents some typical multielement detection limits in NIST 612 glass generated with a 266-nm Nd:YAG design coupled to an ICP-MS system.[11] It can be seen that sub-ppb detection limits in the solid material are achievable for most of the elements. This kind of performance is typically obtained using larger spot sizes on the order of 100–1,000 μm in diameter, which is ideally suited to 266-nm laser technology.

However, the desire for ultratrace analysis of optically challenging materials, such as calcite, quartz, glass and fluorite, combined with the capability to characterize small spots and micro-inclusions, proved very challenging for the 266-nm design. The major reason is that the ablation process is not very controlled and precise, and as a result it is difficult to ablate a minute area without removing some of the surrounding material. In addition, erratic ablating of the sample initially generates larger particles (> 1 μm size), which are not efficiently ionized in the plasma and therefore contribute to poor precision.[12] Even though modifications helped improve ablation behavior, it was not totally successful, because of the basic limitation of the 266 nm to couple efficiently to UV-transparent materials. The drawbacks in 266-nm technology eventually led to the development of 213-nm lasers[13] because of the recognized superiority of shorter wavelengths to exhibit a higher degree of absorbance in transparent materials.[14]

Analytical chemists, particularly in the geochemical community, welcomed 213-nm UV lasers with great enthusiasm, because they now had a sampling tool that offered much better control of the ablation process, even for easily fractured minerals. This is demonstrated in Figures 18.2 and 18.3,

---

TABLE 18.1

Typical Detection Limits Achievable in NIST 612 SRM Glass Using a 266-nm Nd:YAG Laser Ablation System Coupled to an ICP Mass Spectrometer

Element	3σ DLs (ppb)	Element	3σ DLs (ppb)
B	3.0	Ce	0.05
Sc	3.4	Pr	0.05
Ti	9.1	Nd	0.5
V	0.4	Sm	0.1
Fe	13.6	Eu	0.1
Co	0.05	Gd	1.5
Ni	0.7	Dy	0.5
Ga	0.2	Ho	0.01
Rb	0.1	Er	0.2
Sr	0.07	Yb	0.4
Y	0.04	Lu	0.04
Zr	0.2	Hf	0.4
Nb	0.5	Ta	0.1
Cs	0.2	Th	0.02
Ba	0.04	U	0.02
La	0.05		

Source: Courtesy of Teledyne Cetac Technologies.

FIGURE 18.2   A 200-μm crater produced by the ablation of a NIST 612 glass SRM, using a 266-nm laser ablation system, showing excess ablated material around the edges of the crater.

(Courtesy of Teledyne Cetac Technologies)

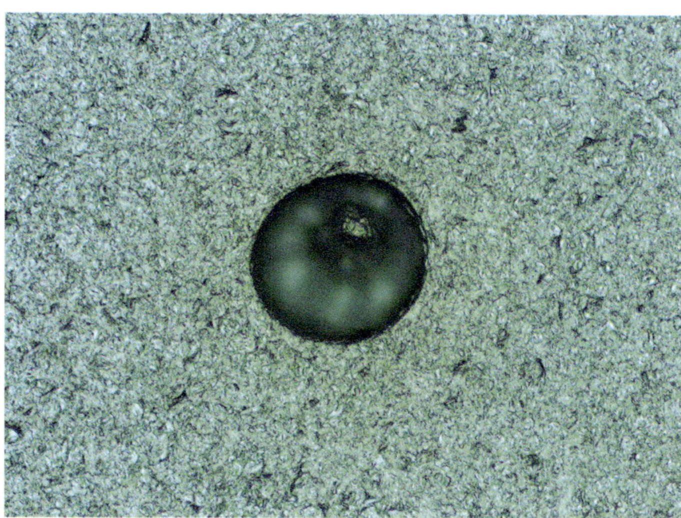

FIGURE 18.3   A 200-μm crater produced by the ablation of a NIST 612 glass SRM using a 213-nm laser system, showing a symmetrical, well-defined crater.

(Courtesy of Teledyne Cetac Technologies)

which show the ablation differences between 266 and 213 nm, respectively, for NIST 612 glass SRM. It can be seen that the 200-μm ablation crater produced with the 266-nm laser is irregular and shows redeposited ablated material around the edges of the crater, whereas the crater with the 213-nm system is very clean and symmetrical, with no ablated material around the edges. The absence of any redeposited material with the 213-nm laser means that a higher proportion of ablated material actually makes it to the plasma. Both craters are shown at 10× magnification.

This significant difference in crater geometry between the two systems is predominantly a result of the effective absorption of laser energy and the difference in the delivered power per unit area—also known as laser irradiance (or fluence per laser pulse width).[15] The result is a difference in depth penetration and the size/volume of particles reaching the plasma. With the 266-nm laser system, a high-volume burst of material is initially observed, whereas with the 213-nm laser, the signal gradually increases and levels off quickly, indicating a more consistent stream of small particles being delivered to the plasma and the mass spectrometer. Therefore, when analyzing this type of mineral with the 266-nm design, it is typical that the first 100 to 200 shots of the ablation process are filtered out to ensure that no data are taken during the initial burst of material. This can be somewhat problematic when analyzing small spots or inclusions, because of the limited amount of sample being ablated.

18.6 193-NM LASER TECHNOLOGY

The benefits of 213-nm lasers emphasize that matrix independence, high spatial resolution and the ability to couple with UV-transparent materials without fracturing (particularly for small spots or depth analysis studies) were very important for geochemical-type applications. These findings led researchers to study even shorter wavelengths, in particular, 193-nm ArF excimer technology. Besides their accepted superiority in coupling efficiency, a major advantage of the 193-nm design is that it utilizes a fundamental wavelength and therefore achieves much higher energy transfer, compared to an Nd:YAG solid-state system that utilizes crystals to quadruple or quintuple the frequency. Additionally, the less coherent nature of the excimer beam enables better optical homogenization, resulting in an even flatter beam profile. The overall benefit is that cleaner, flatter craters are produced down to approximately three to four μm in diameter, at energy densities up to 45 J/cm^2. This provides far better control of the ablation process, which is especially important for depth profiling and fluid-inclusion analysis. This is demonstrated in Figure 18.4, which shows a scanning electron

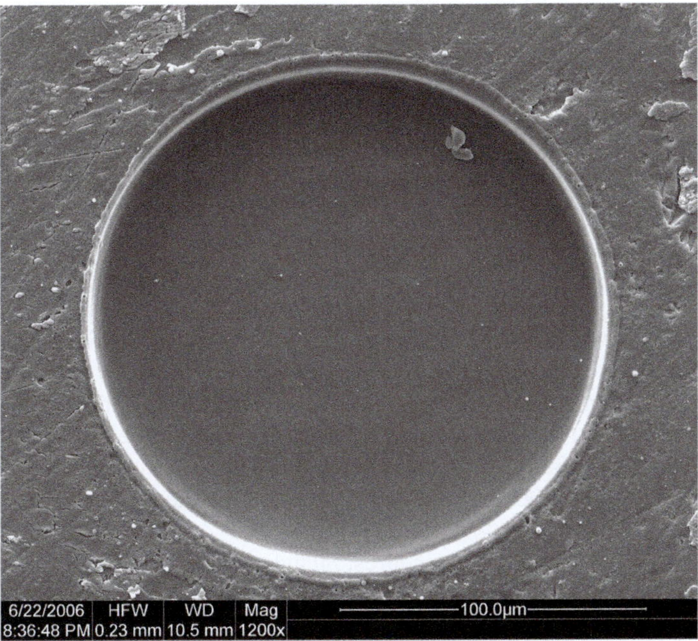

FIGURE 18.4 Scanning electron microscope (SEM) image (1200 magnification) of a 160-μm crater produced by the ablation (50 pulses) of NIST 1612 glass SRM using an optically homogenized flat beam, 193-nm ArF excimer laser system.

(Courtesy of Cetac Teledyne Technologies). (From SEM photo courtesy of Dr. Honglin Yuan, Northwest University, Xi'an, China.)

microscope (SEM) image (1200 magnification) of a NIST 1612 glass SRM ablated with a 193-nm ArF excimer laser using a highly homogenized, flattop optical-beam profile.[16] It can be seen that the 160-μm ablation crater produced by 50 laser pulses is extremely flat and smooth around the edges. The 213-nm Nd:YAG design just would not be capable of the kind of high precision required for depth analysis and small-spot/fluid-inclusion studies carried out by the geological community.

It is very important to emphasize that the geochemical community initially drove the design of excimer lasers, because they were interested in characterizing extremely small spots and inclusions on the surface of minerals and geological specimens, which was very difficult using either of the other approaches. As a result, the performance and the analytical capabilities were focused on their needs, which included the requirement for finely controlled, "homogenizer-flat" ablations with high sensitivity and split-second response. Also, fire-on-the-fly lasing synchronized to the stage's motion, combined with fast washout ablation cells, made high-precision depth profiling of spots, lines and areas possible, allowing for high spatial resolution elemental mapping. Additionally, the combination of ultrashort pulse length and the 193-nm wavelength produced very high coupling efficiency. This meant higher absorbance in a wide range of materials, which produced smaller particles on average than those produced by the 213-nm YAG design. This resulted in greater ionization within the plasma, leading to better sensitivity and less deposition at the ionization source. Today's commercial excimer lasers can ablate all materials, from opaque to highly transparent, including delicate powders, hard quartz and resilient carbonates with depth penetration in the tens of nanometers per shot. The beam energy profile is homogenized to ensure uniform ablations across the entire range of spot sizes and on a wide range of materials.

The benefits of laser ablation coupled with ICP-MS are now well documented by the large number of application references in the public domain, which describe the analysis of metals, ceramics, polymers, minerals, biological tissue, pharmaceutical tablets and many other sample types.[17–22] These references should be investigated further to better understand the optimum configuration, design and wavelength of laser ablation equipment for different types of sample matrices. It should also be emphasized that there are many overlapping areas when selecting the optimum laser system for the sample type. Roy and Neufeld published a very useful article that offers some guidelines on the importance of matching the laser hardware to the application.[23]

Note: a dedicated laser ablation, laser ionization time-of-flight mass spectrometer is now commercially available that does not use an ICP as an excitation source. For more information about this technique, please refer to Chapter 27 on "Emerging AS Techniques".

18.7   FLOW INJECTION ANALYSIS

Flow injection (FI) is a powerful front-end sampling accessory for ICP-MS that can be used for preparation, pretreatment and delivery of the sample. Originally described by Ruzicka and Hansen,[24] FI involves the introduction of a discrete sample aliquot into a flowing carrier stream. Using a series of automated pumps and valves, procedures can be carried out online to physically or chemically change the sample or analyte before introduction into the mass spectrometer for detection. There are many benefits of coupling FI procedures to ICP-MS, including the following:

- Automation of online sampling procedures, including dilution and additions of reagents.
- Minimum sample handling translates into a lower chance of sample contamination.
- Ability to introduce low sample/reagent volumes.
- Improved stability with harsh matrices.
- Extremely high sample throughput using multiple loops.

In its simplest form, FI-ICP-MS consists of a series of pumps and an injection valve preceding the sample-introduction system of the ICP mass spectrometer. A typical manifold used for microsampling is shown in Figure 18.5.

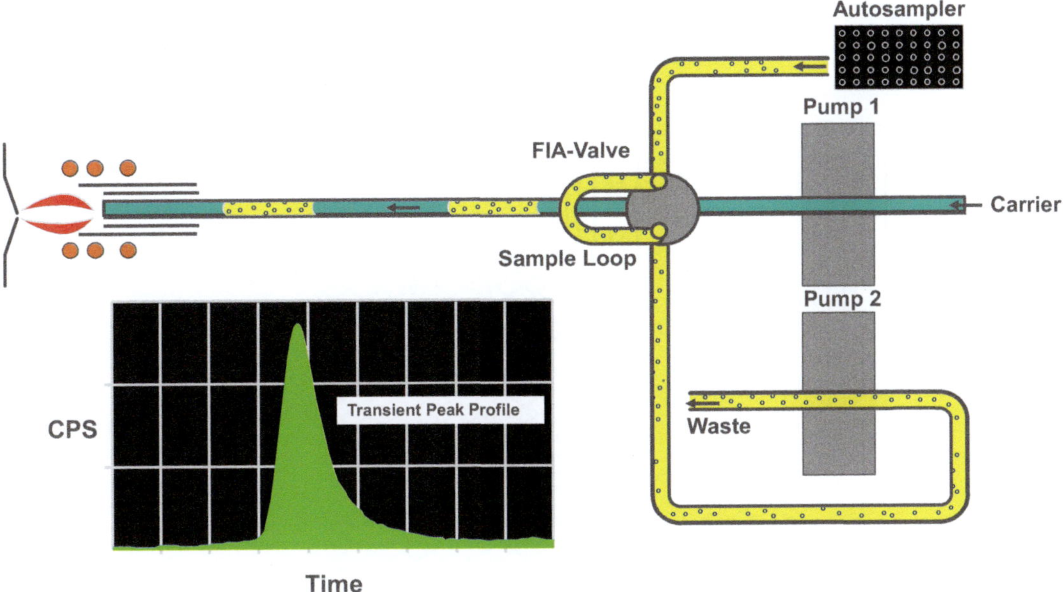

FIGURE 18.5   Schematic of a flow injection system used for the process of microsampling.

In the fill position, the valve is filled with the sample. In the inject position, the sample is swept from the valve and carried to the ICP by means of a carrier stream. The measurement is usually a transient profile of signal versus time, as shown by the signal profile in Figure 18.5.

The area of the signal profile measured is greater for larger injection volumes, but for volumes of 500 μL or greater, the signal peak height reaches a maximum equal to that observed using continuous solution aspiration. The length of a transient peak in flow injection is typically 20–60 s, depending on the size of the loop. This means that if multielement determinations are a requirement, all the data quality objectives for the analysis, including detection limits, precision, dynamic range and number of elements, etc., must be achieved in this time frame. Similar to laser ablation, if a sequential mass analyzer such as a quadrupole or single-collector magnetic-sector system is used, the electronic scanning, dwelling and settling times must be optimized in order to capture the maximum amount of multielement data in the duration of the transient event.[25] This can be seen in greater detail in Figure 18.6, which shows a 3D transient plot of intensity versus mass in the time domain for the determination of a group of elements.

Some of the many online procedures that are applicable to FI-ICP-MS include the following:

- Microsampling for improved stability with heavy matrices[26]
- Automatic dilution of samples/standards[27]
- Standards addition[28]
- Cold-vapor and hydride generation for enhanced detection capability for elements such as Hg, As, Sb, Bi, Te and Se[29]
- Matrix separation and analyte pre-concentration using ion exchange procedures[30]
- Elemental speciation[31]
- Maximize sample throughput

Flow injection coupled to ICP-MS has shown itself to be very diverse and flexible in meeting the demands presented by complex samples, as indicated in the foregoing references. However, one of

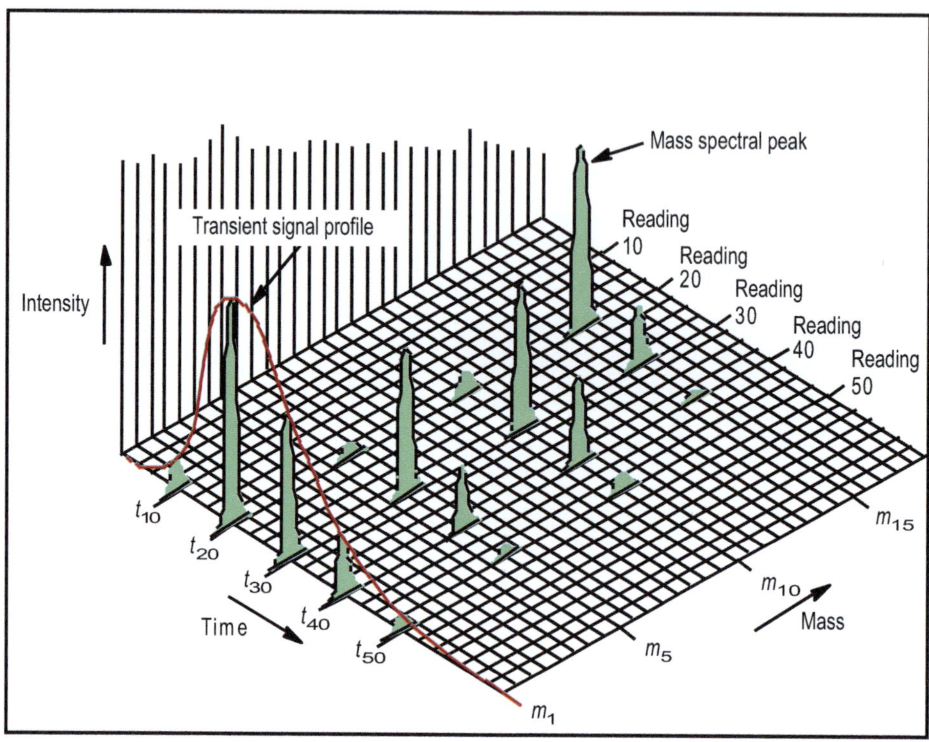

FIGURE 18.6   3D plot of intensity versus mass in the time domain for the determination of a group of elements in a transient peak.

(Courtesy of PerkinElmer Inc.)

the most interesting areas of research is in the direct analysis of seawater by flow-injection ICP-MS. Traditionally, seawater is very difficult to analyze by ICP-MS because of two major problems. First, the high NaCl content will block the sampler-cone orifice over time, unless a 10–20-fold dilution is made of the sample. This is not such a major problem with coastal waters, because the levels are high enough. However, if the sample is open-ocean seawater, this is not an option, because the trace metals are at a much lower level. The other difficulty associated with the analysis of seawater is that ions from the water, chloride matrix and the plasma gas can combine to generate polyatomic spectral interferences, which are a problem, particularly for the first-row transition metals.

Attempts have been made over the years to remove the NaCl matrix and pre-concentrate the analytes using various types of chromatography and ion exchange column technology. One such early approach was to use an HPLC system coupled to an ICP mass spectrometer utilizing a column packed with silica-immobilized 8-hydroxyquinoline.[32] This worked reasonably well, but was not considered a routine method, because silica-immobilized 8-hydroxyquinoline was not commercially available, and also spectral interferences produced by HCl and HNO_3 (used to elute the analytes) precluded determination of a number of the elements, such as Cu, As and V. More recently, chelating agents based on the iminodiacetate-acid functionality group have gained wider success, but are still not considered truly routine for a number of reasons, including the necessity for calibration using standard additions, the requirement of large volumes of buffer to wash the column after loading the sample and the need for conditioning between samples, because some ion exchange resins swell with changes in pH.[33–35]

However, a research group at the NRC in Canada has developed a very practical online approach, using a flow-injection sampling system coupled to an ICP mass spectrometer.[30] Using a special

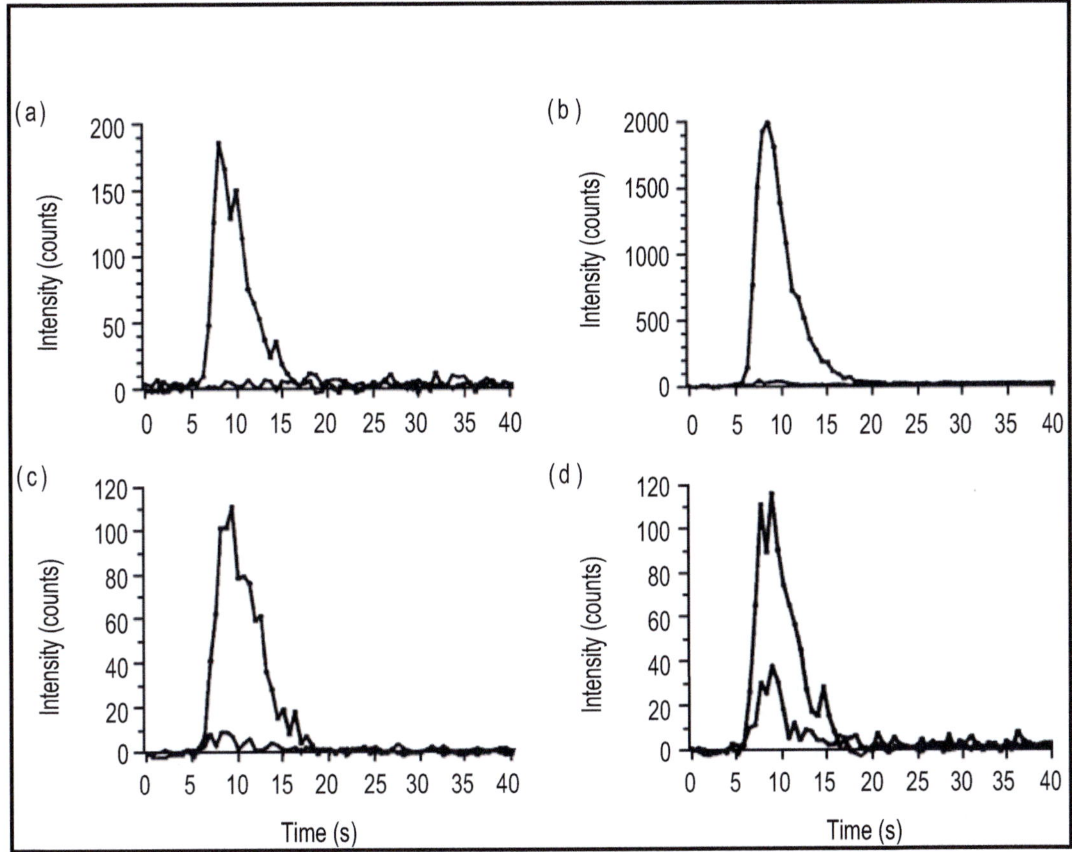

FIGURE 18.7 Analyte and blank spectral scans of (a) Co, (b) Cu, (c) Cd and (d) Pb in NASS-4 open-ocean seawater certified reference material, using flow injection coupled to ICP-MS.

(From S. N. Willie, Y. Iida and J. W. McLaren, *Atomic Spectroscopy*, **19**[3], 67, 1998)

formulation of a commercially available, iminodiacetate ion exchange resin (with a macroporous methacrylate backbone), trace elements can be separated from the high concentrations of matrix components in the seawater, with a pH 5.2 buffered solution. The trace metals are subsequently eluted into the plasma with 1M HNO_3, after the column has been washed out with deionized water. The column material has sufficient selectivity and capacity to allow accurate determinations at ppt levels using simple aqueous standards, even for elements such as V and Cu, which are notoriously difficult in a chloride matrix. This can be seen in Figure 18.7, which shows spectral scans for a selected group of elements in a certified reference material open-ocean seawater sample (NASS-4), and Table 18.2, which compares the results for this methodology with the certified values, together with the limits of detection (LOD). Using this online method, the turnaround time is less than 4 min per sample, which is considerably faster than other high-pressure chelation techniques reported in the literature.

18.8 ELECTROTHERMAL VAPORIZATION (ETV)

ETV for use with AA has proved to be a very sensitive technique for trace element analysis over the last three decades. However, the possibility of using the atomization/heating device for ETV sample introduction into both ICP-OES and ICP-MS was identified in the late 1980s.[36] The ETV

TABLE 18.2

Analytical Results for NASS-4 Open-Ocean Seawater Certified Reference Material, Using Flow Injection ICP-MS Methodology

Isotope	LOD (ppt)	NASS-4 (ppb)	
		Determined	Certified
$^{51}V^+$	4.3	1.20 ± 0.04	Not certified
$^{63}Cu^+$	1.2	0.210 ± 0.008	0.228 ± 0.011
$^{60}Ni^+$	5	0.227 ± 0.027	0.228 ± 0.009
$^{66}Zn^+$	9	0.139 ± 0.017	0.115 ± 0.018
$^{55}Mn^+$	Not reported	0.338 ± 0.023	0.380 ± 0.023
$^{59}Co^+$	0.5	0.0086 ± 0.0011	0.009 ± 0.001
$^{208}Pb^+$	1.2	0.0090 ± 0.0014	0.013 ± 0.005
$^{114}Cd^+$	0.7	0.0149 ± 0.0014	0.016 ± 0.003

Source: From S. N. Willie, Y. Iida and J. W. McLaren, *Atomic Spectroscopy*, **19**(3) 67, 1998.

sampling process relies on the basic principle that a carbon furnace or metal filament can be used to thermally separate the analytes from the matrix components and then sweep them into the ICP mass spectrometer for analysis. This is achieved by injecting a small amount of the sample (usually 20–50 μL via an autosampler) into a graphite tube or onto a metal filament. After the sample is introduced, drying, charring and vaporization are achieved by slowly heating the graphite tube/metal filament. The sample material is vaporized into a flowing stream of carrier gas, which passes through the furnace or over the filament during the heating cycle. The analyte vapor re-condenses in the carrier gas and is then swept into the plasma for ionization.

One of the attractive characteristics of ETV for ICP-MS in particular is that the vaporization and ionization steps are carried out separately, which allows for the optimization of each process. This is particularly true when a heated graphite tube is used as the vaporization device, because the analyst typically has more control of the heating process and as a result can modify the sample by means of a very precise thermal program before it is introduced to the ICP for excitation and/ or ionization. By boiling off and sweeping the solvent and volatile matrix components out of the graphite tube, spectral interferences arising from the sample matrix can be reduced or eliminated. The ETV sampling process consists of six discrete stages: sample introduction, drying, charring (matrix removal), vaporization, condensation and transport. Once the sample has been introduced, the graphite tube is slowly heated to drive off the solvent. Opposed gas flows, entering from each end of the graphite tube, and then purges the sample cell by forcing the evolving vapors out of the dosing hole. As the temperature increases, volatile matrix components are vented during the charring steps. Just prior to vaporization, the gas flows within the sample cell are changed. The central channel (nebulizer) gas then enters from one end of the furnace, passes through the tube and exits out of the other end. The sample-dosing hole is then automatically closed, usually by means of a graphite tip, to ensure no analyte vapors escape. After this gas-flow pattern has been established, the temperature of the graphite tube is ramped up very quickly, vaporizing the residual components of the sample. The vaporized analytes either re-condense in the rapidly moving gas stream or remain in the vapor phase. These particulates and vapors are then transported to the ICP in the carrier gas, where they are ionized by the ICP for analysis in the mass spectrometer.

Another benefit of decoupling the sampling and ionization processes is the opportunity for chemical modification of the sample. The graphite furnace itself can serve as a high-temperature reaction vessel, where the chemical nature of compounds within it can be altered. In a manner similar to that used in atomic absorption, chemical modifiers can change the volatility of species

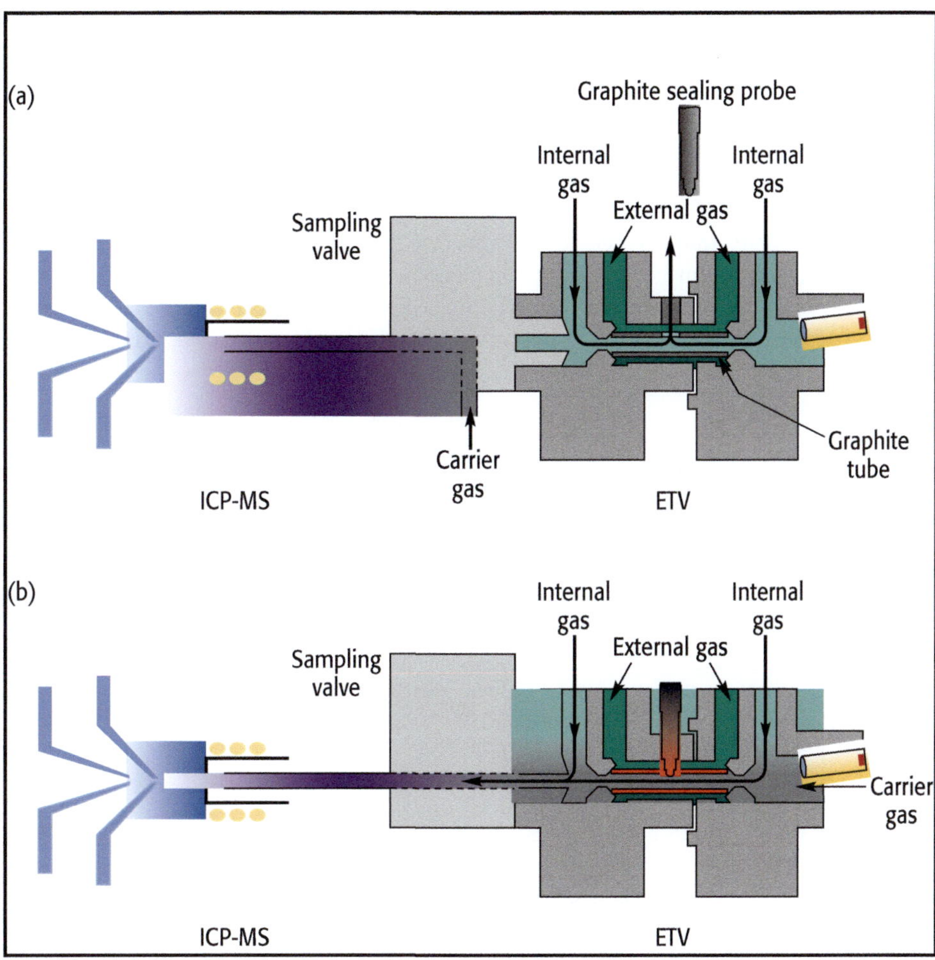

FIGURE 18.8   A graphite furnace ETV sampling device for ICP-MS, showing the two distinct steps of sample pretreatment (a) and vaporization (b) into the plasma.

(Courtesy of PerkinElmer Inc.)

to enhance matrix removal and increase elemental sensitivity.[37] An alternative gas, such as oxygen, may also be introduced into the sample cell to aid in the charring of the carbon in organic matrices such as biological or petrochemical samples. Here, the organically bound carbon reacts with the oxygen gas to produce CO_2, which is then vented from the system. A typical ETV sampling device, showing the two major steps of sample pretreatment (drying and ashing) and vaporization into the plasma, is seen schematically in Figure 18.8.

Over the past 20 years, ETV sampling for ICP-MS has mainly been used for the analysis of complex matrices including geological materials,[38] biological fluids[39] and seawater,[40] which have proved difficult or impossible by conventional nebulization. By removal of the matrix components, the potential for severe spectral and matrix-induced interferences is dramatically reduced. Even though ETV-ICP-MS was initially applied to the analysis of very small sample volumes, the advent of low-flow nebulizers has limited its use for this type of work.

An example of the benefit of ETV sampling is in the analysis of samples containing high concentrations of mineral acids such as HCl, HNO_3 and H_2SO_4. Besides physically suppressing analyte signals, these acids generate massive polyatomic spectral overlaps, which interfere with many analytes, including As, V, Fe, K, Si, Zn and Ti. By carefully removing the matrix components with the ETV device, the determination of these elements becomes relatively straightforward. This is illustrated in Figure 18.9, which shows a spectral display in the time domain for 50 pg spikes of a selected group of elements in concentrated hydrochloric acid (37% w/w), using a graphite-furnace-based ETV-ICP-MS.[42] It can be seen in particular that good sensitivity is obtained for $^{51}V^+$, $^{56}Fe^+$ and $^{75}As^+$, which would have been virtually impossible by direct aspiration because of spectral overlaps from $^{39}ArH^+$, $^{35}Cl^{16}O^+$, $^{40}Ar^{16}O^+$ and $^{40}Ar^{35}Cl^+$, respectively. The removal of the chloride and water from the matrix translates into ppt detection limits directly in 37% HCl, as shown in Table 18.3.

It can also be seen in Figure 18.9 that the elements are vaporized off the graphite tube in order of their boiling points. In other words, magnesium, which is the most volatile, is driven off first, whereas V and Mo, which are the most refractory, come off last. However, even though they emerge

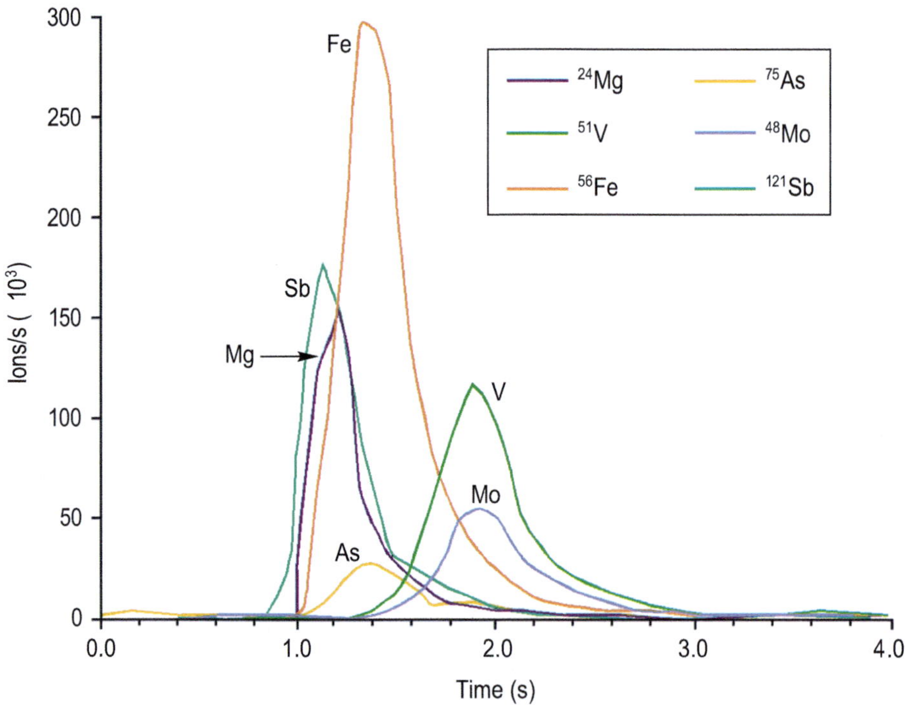

FIGURE 18.9   A temporal display of 50 pg of Mg, Sb, As, Fe, V and Mo in 37% hydrochloric acid by ETV-ICP-MS.

(From S. A. Beres, E. R. Denoyer, R. Thomas, P. Bruckner, *Spectroscopy*, **9**(1), 20–26, 1994)

TABLE 18.3

Detection Limits for V, Fe and As in 37% Hydrochloric Acid by ETV-ICP-MS

Element	DL (ppt)
$^{51}V^+$	50
$^{56}Fe^+$	20
$^{75}As^+$	40

Source: S. A. Beres, E. R. Denoyer, R. Thomas, P. Bruckner; *Spectroscopy*, **9**(1), 20–26, 1994.

at different times, the complete transient event lasts less than 3 s. This physical time limitation, imposed by the duration of the transient signal, makes it imperative that all isotopes of interest be measured under the highest signal-to-noise conditions throughout the entire event.

The rapid nature of the transient has also limited the usefulness of ETV sampling for routine multielement analysis, because realistically only a small number of elements can be quantified with good accuracy and precision in less than 3 s. In addition, the development of low-flow nebulizers, desolvation devices, automated online chemistry, cool-plasma technology and collision/reaction cells and interfaces has meant that multielement analysis can now be carried out on difficult matrices without the need for ETV sample introduction. This has limited its use with ICP-MS to more research-oriented applications.

However, more recently, ETV has been coupled with ICP-OES for the determination of heavy metals in cannabis. The major benefit is that no sample digestion is required, and the cannabis sample can be simply weighed directly into a sampling boat and automatically inserted into a furnace, where it is thermally pretreated before the analyte particles are swept into the plasma for excitation and detection by optical emission. The benefit of ICP-OES detection over ICP-MS is that larger sample weights can be taken, generating a longer temporal transient signal resulting in an optimized measurement protocol. In addition, with ICP-OES, the analyte particles are not going directly into the spectrometer (as with ions going into an ICP-MS), but instead are generating excited atoms and photons in the plasma, which are then being detected and identified by the optical system. Its main drawback is that ICP optical emission is being used for the measurement, which means the detection capability will be compromised compared to ICP-MS. For more information about this novel approach, it's worth checking out this webinar on the benefits of ETV-ICP-OES for measuring heavy metals directly in cannabis-related samples.[41]

18.9 CHILLED SPRAY CHAMBERS AND DESOLVATION DEVICES

Chilled/cooled spray chambers and desolvation devices are becoming more and more common in ICP-MS, primarily to cut down the amount of liquid entering the plasma in order to reduce the severity of the solvent-induced spectral interferences such as oxides, hydrides, hydroxides and argon/solvent-based polyatomic interferences. They are very useful for aqueous-type samples, but probably more important for volatile organic solvents, because there is a strong possibility that the sample aerosol would extinguish the plasma unless modifications are made to the sampling procedure. The most common chilled spray chambers and desolvation systems being used today include the following:

- Water-cooled spray chambers
- Peltier-cooled spray chambers
- Ultrasonic nebulizers (USN)
- Ultrasonic nebulizers (USN) coupled with membrane desolvation
- Specialized microflow nebulizers coupled with desolvation techniques

Let us take a closer look at these devices.

18.10 WATER-COOLED AND PELTIER-COOLED SPRAY CHAMBERS

Water-cooled spray chambers have been used in ICP-MS for many years and are standard on a number of today's commercial instrumentation to reduce the amount of water or solvent entering the plasma. However, the trend today is to cool the sample using a thermoelectric device called a Peltier cooler. Thermoelectric cooling (or heating) uses the principle of generating a hot or cold environment by creating a temperature gradient between two different materials. It uses electrical energy via a solid-state heat pump to transfer heat from a material on one side of the device to a different material on the other side, thus producing a temperature gradient across the device (similar to a household air-conditioning system). Peltier cooling devices, which are typically air-cooled (but water cooling is an option), can be used with any kind of spray chamber and nebulizer, but commercial products for use with ICP-MS are normally equipped with a cyclonic spray chamber and a low-flow pneumatic nebulizer.

The main purpose of cooling the sample aerosol is to reduce the amount of water or solvent entering the plasma by lowering the temperature of the spray chamber. This can be a few degrees below ambient or as low −20°C, depending on the type of samples being analyzed. This can have a three-fold effect: first, it helps to minimize solvent-based spectral interferences, such as oxides and hydroxides formed in the plasma, and second, because very little plasma energy is needed to vaporize the solvent, it allows more energy to be available to excite and ionize the analyte ions. There is also evidence to suggest that cooling the spray chamber will help minimize signal drift due to external environmental temperature changes in the laboratory.

Cooling the spray chamber to as low as −20°C by either Peltier cooling or a recirculating system using ethylene glycol as the coolant is particularly useful when it comes to analyzing some volatile organic samples. It has the effect of reducing the amount of organic solvent entering the interface, and when combined with the addition of a small amount of oxygen into the nebulizer gas flow, it is beneficial in reducing the buildup of carbon deposits on the sampler-cone orifice and also minimizing the problematic carbon-based spectral interferences.[43]

Some systems also have the ability to increase the temperature of the spray chamber above ambient. Studies have shown that the sensitivity for many analytes is enhanced by a factor of 2–3-fold by running the spray chamber as high as 60°C, a feature that is particularly important for samples with limited volume, or viscous matrices such as engine or edible oils.

18.11 ULTRASONIC NEBULIZERS

Ultrasonic nebulization was first developed in the late 1980s for use with ICP optical emission.[44] Its major benefit was that it offered an approximately 10× improvement in detection limits, because of its more efficient aerosol generation. However, this was not such an obvious benefit for ICP-MS, because more matrix entered the system compared to a conventional nebulizer, increasing the potential for signal drift, matrix suppression and spectral interferences. This was not such a major problem for simple aqueous-type samples, but was problematic for real-world matrices. The elements that showed the most improvement were the ones that benefited from lower solvent-based spectral interferences. Unfortunately, many of the other elements exhibited higher background levels, and as a result showed no significant improvement in detection limit. In addition, the increased amount of matrix entering the mass spectrometer usually necessitated the need for larger dilutions of the sample, which again negated the benefit of using a USN with ICP-MS for samples with a heavier matrix. This limitation led to the development of an ultrasonic nebulizer fitted with an additional membrane desolvator. This design virtually removed all the solvent from the sample, which dramatically improved detection limits for a large number of the problematic elements and also lowered metal oxide levels by at least an order of magnitude.[45]

The principle of aerosol generation using an ultrasonic nebulizer is based on a sample being pumped onto a quartz plate of a piezoelectric transducer. Electrical energy of 1–2 MHz frequency is coupled to the transducer, which causes it to vibrate at high frequency. These vibrations disperse the sample into a fine-droplet aerosol, which is carried in a stream of argon. With a conventional ultrasonic nebulizer, the aerosol is passed through a heating tube and a cooling chamber, where most of the sample solvent is removed as a condensate before it enters the plasma. If a membrane desolvation system is fitted to the ultrasonic nebulizer, it is positioned after the cooling unit. The sample aerosol enters the membrane desolvator, where the remaining solvent vapor passes through the walls of a tubular micro-porous membrane. A flow of argon gas removes the volatile vapor from the exterior of the membrane, while the analyte aerosol remains inside the tube and is carried into

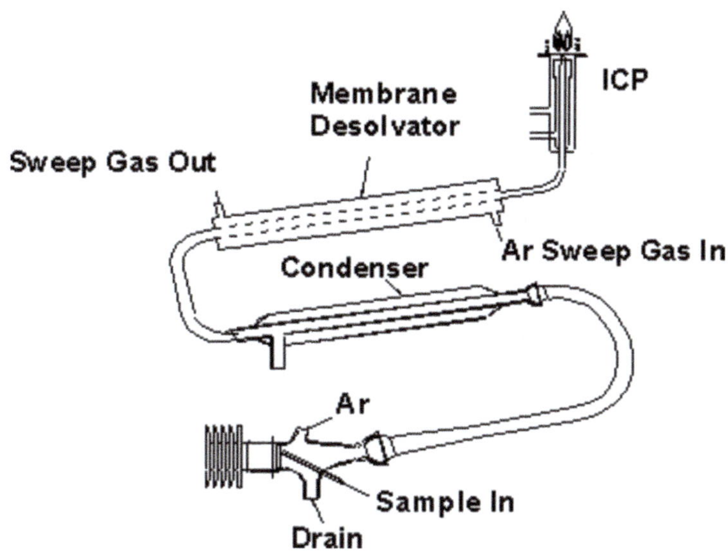

FIGURE 18.10 Schematic of an ultrasonic nebulizer fitted with a membrane desolvation system.

(Courtesy of Teledyne Cetac Technologies)

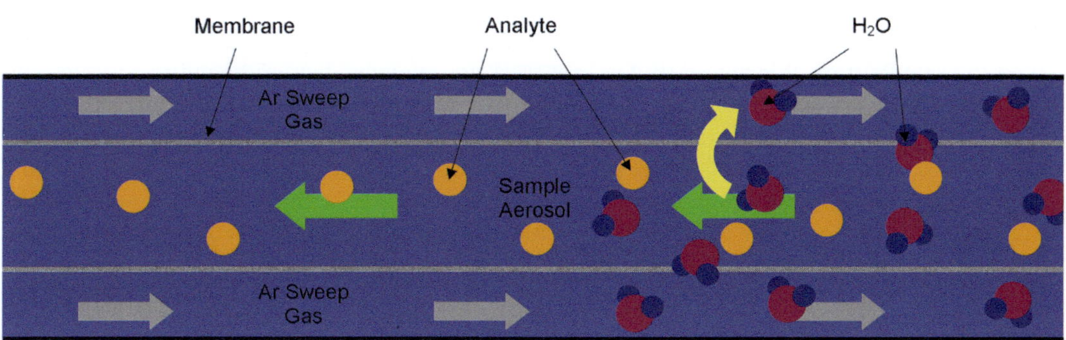

FIGURE 18.11 Principles of membrane desolvation showing the water molecules passing through a microporous membrane and being swept away by the argon gas, while the analyte is transported through the tube to the plasma.

(Courtesy of Elemental Scientific Inc.)

the plasma for ionization. Membrane desolvation systems also have the capability to add a secondary gas such as nitrogen, which has shown to be very beneficial in changing the ionization conditions to reduce levels of oxides in the plasma. The combination of membrane desolvation with an ultrasonic nebulizer can be seen more clearly in Figure 18.10, and Figure 18.11 shows the principles of membrane desolvation with water vapor as the solvent.

For ICP-MS, the system is best operated with both desolvation stages working, although for less demanding ICP-OES analysis, the membrane desolvation stage can be bypassed if required. The power of the system when coupled to an ICP mass spectrometer can be seen in Table 18.4, which compares the sensitivity (counts per second) and signal-to-background ratio of a membrane desolvation USN with a conventional cross-flow nebulizer for two classic solvent-based polyatomic interferences, $^{12}C^{16}O_2^+$ on $^{44}Ca^+$ and $^{40}Ar^{16}O^+$ on $^{56}Fe^+$, using a quadrupole ICP-MS system.

TABLE 18.4

Comparison of Sensitivity and Signal/Background Ratios for Two Analyte Masses—44Ca+, 56Fe+—Using a Conventional Cross-Flow Nebulizer and a USN Fitted with a Membrane Desolvation System

Analytical mass	Cross-flow nebulizer (cps)	Signal/BG	USN with membrane desolvation (cps)	Signal/BG
25 ppb $^{44}Ca^+$ (BG subtracted)	2,300	2,300/7,640	20,800	20,800/1730
$^{12}C^{16}O_2^+$ (BG)	7,640	**= 0.30**	1,730	**= 12.0**
10 ppb $^{56}Fe^+$ (BG subtracted)	95,400	95,400/868,00	262,000	262,000/8,200
$^{40}Ar^{16}O^+$ (BG)	868,000	**= 0.11**	8,200	**= 32.0**

Note: Signal/BG is calculated as the background subtracted signal divided by the background.

Source: Courtesy of Teledyne Cetac Technologies.

It can be seen that for the two analyte masses, the signal-to-background ratio is significantly better with the membrane-desolvated ultrasonic nebulizer than with the cross-flow design, which is a direct result of the reduction of the solvent-related spectral background levels. This approach is even more beneficial for the analysis of organic solvents because when they are analyzed by conventional nebulization, modifications have to be made to the sampling process, such as the addition of oxygen to the nebulizer gas flow, use of a low-flow nebulize and probably external cooling of the spray chamber. With a membrane desolvation USN system, volatile solvents such as isopropanol can be directly aspirated into the plasma with relative ease. However, it should be mentioned that depending on the sample type, this approach does not work for analytes that are bound to organic molecules. For example, the high volatility of certain mercury and boron organometallic species means that they could pass through the micro-porous PTFE membrane and never make it into the ICP-MS. In addition, samples with high dissolved solids, especially ones that are biological in nature, could possibly result in clogging the micro-porous membrane unless substantial dilutions are made. For these reasons, caution must be used when using a membrane desolvation system for the analysis of certain types of sample matrices.

18.12 SPECIALIZED MICROFLOW NEBULIZERS WITH DESOLVATION TECHNIQUES

Microflow or low-flow nebulizers, which were described in greater detail in Chapter 6, are being used more and more for routine applications. The most common ones used in ICP-MS are based on the microconcentric design, which operate at sample flows of 20–500 μL/min. Besides being ideal for small sample volumes, the major benefit of microconcentric nebulizers is that they are more efficient and produce smaller droplets than a conventional nebulizer. In addition, many microflow nebulizers use chemically inert plastic capillaries, which make them well suited for the analysis of highly corrosive chemicals. This kind of flexibility has made low-flow nebulizers very popular, particularly in the semiconductor industry, where it is essential to analyze high-purity materials using a sample-introduction system that is free of contamination.[46]

Such is the added capability and widespread use of these nebulizers across all application areas that manufacturers are developing application-specific integrated systems that include the spray chamber and a choice of different desolvation techniques to reduce the amount of solvent aerosol entering the plasma. Depending on the types of samples being analyzed, some of these systems include a low-flow nebulizer, Peltier-cooled spray chambers, heated spray chambers, Peltier-cooled condensers and membrane desolvation technology. Some of the commercially available equipment includes the following:

- **Microflow nebulizer coupled with a Peltier-cooled spray chamber:** An example of this is the PC3 from Elemental Scientific Inc. (ESI). This system is offered with or without the nebulizer and utilizes a Peltier-cooled cyclonic spray chamber made from either quartz, borosilicate glass or a fluoropolymer. Options include the ability to reduce the temperature to −20°C for analyzing organics and a dual-spray chamber for improved stability.
- **Microflow nebulizer with heated spray chamber and Peltier-cooled condenser:** An example of this design is the Apex inlet system from ESI. This unit includes a microflow nebulizer, heated cyclonic spray chamber (up to 140°C), and a Peltier multipass condenser/cooler (down to −5°C). A number of different spray chamber and nebulizer options and materials are available, depending on the application requirements. Also, the system is available with Teflon or Nafion micro-porous membrane desolvation, depending on the types of samples being analyzed. Figure 18.12 shows a schematic of the Apex sample inlet system with the cross-flow nebulizer.

Microflow nebulizer coupled with membrane desolvation: An example of this is the Aridus system from Cetac Technologies. The aerosol from the nebulizer is either self-aspirated or pumped into a heated PFA spray chamber (up to 110°C) to maintain the sample in a vapor phase. The sample vapor then enters a heated PTFE membrane desolvation unit, where a counterflow of argon sweep gas is added to remove solvent vapors that permeate the micro-porous walls of the membrane. Nonvolatile sample components do not pass through the membrane walls, but are transported to the ICP-MS for analysis. Figure 18.13 shows a schematic of the Aridus II microflow nebulizer with membrane desolvation.

There is an extremely large selection of these specialized sample-introduction techniques, so it is critical that you talk to the vendors so they can suggest the best solution for your application

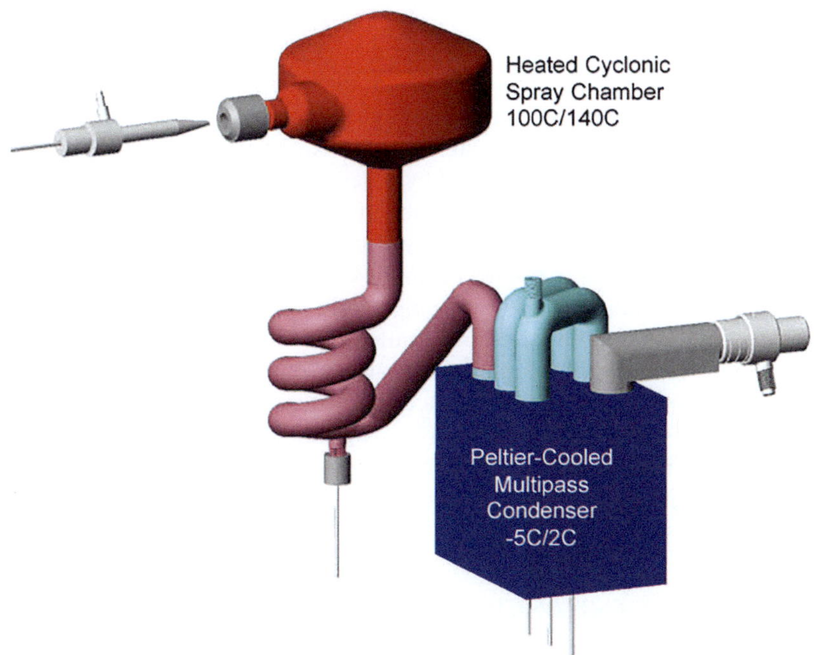

FIGURE 18.12 A schematic of the Apex sample inlet system.

(Courtesy of Elemental Scientific Inc.)

Aridus II Schematic

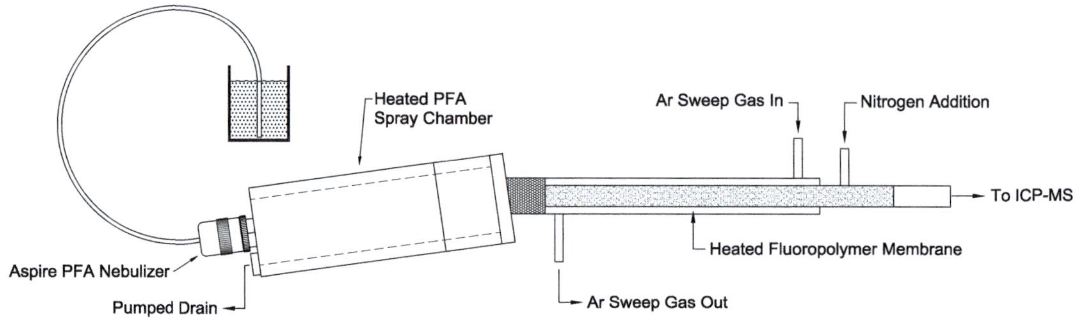

FIGURE 18.13   A schematic of the Aridus II microflow nebulizer with membrane desolvation. (Courtesy of Teledyne Cetac Technologies)

problem. They may not be required for the majority of your application work, but there is no question they can be very beneficial for analyzing certain types of sample matrices and for elements that might be prone to solvent-based spectral overlaps.

18.13   DIRECT INJECTION NEBULIZERS

Direct injection nebulization is based on the principle of injecting a liquid sample under high pressure directly into the base of the plasma torch.[47] The benefit of this approach is that no spray chamber is required, which means that an extremely small volume of sample can be introduced directly into the ICP-MS with virtually no carryover or memory effects from the previous sample. Because they are capable of injecting less than 5 µL of liquid, they have found a use in applications where sample volume is limited or where the material is highly toxic or expensive.

They were initially developed over 15 years ago and found some success in certain niche applications that could not be adequately addressed by other nebulization systems, such as introducing samples from a chromatography separation device into an ICP-MS or the determination of mercury by ICP-MS, which is prone to severe memory effects. Unfortunately, they were not considered particularly user-friendly, and as a result became less popular when other sample introduction devices were developed to handle microliter sample volumes. More recently, a refinement of the direct injection nebulizer has been developed, called the direct inject high-efficiency nebulizer (DIHEN), which appears to have overcome many of the limitations of the original design.[48] The advantage of the DIHEN is its ability to introduce microliter volumes into the plasma at extremely low sample flow rates (1–100 µL/min), with an aerosol droplet size similar to a concentric nebulizer fitted with a spray chamber. The added benefit is that it is almost 100% efficient and has extremely low memory characteristics. A schematic of a commercially available DIHEN system is shown in Figure 18.14.

18.14   PRODUCTIVITY-ENHANCING TECHNIQUES

With the increasing demand to analyze more and more samples, carry out automated dilutions and additions of internal standards, as well as performing online chemistry procedures, manufacturers of autosamplers and sample-introduction accessories are designing automated sampling systems to maximize sample throughput, minimize sample preparation times and increase productivity.[49,50,51]

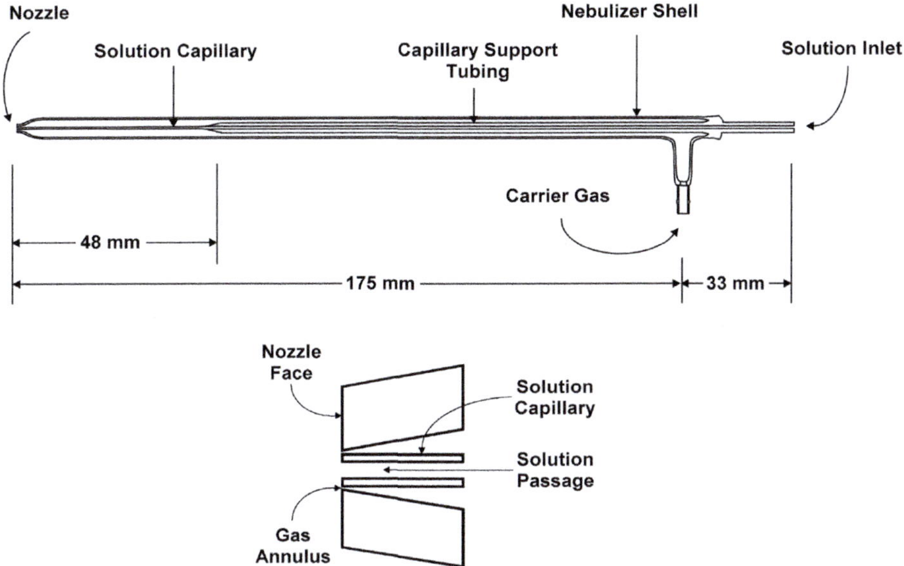

FIGURE 18.14　A schematic of a commercially available DIHEN system.
(Courtesy of Meinhard Glass Products)

Depending on the application requirements, this is being achieved in a number of different ways using a variety of components including multiport/switching valves, loops, vacuum/piston/syringe pumps, mixing chambers and ion-exchange/preconcentration columns. There are basically three approaches to enhancing productivity in ICP-MS, depending on the application requirements:

- Achieve faster analysis times by optimizing sample delivery to the instrument.
- Perform online dilutions, internal standard additions and calibrations to save manual operations.
- Carry out automated chemistry online to remove sample matrices and/or pre-concentrate the samples to reduce interferences and minimize labor-intensive, manual sample-preparation steps.

Let's take a more detailed look at each of these approaches.

18.15　FASTER ANALYSIS TIMES

This is basically a rapid sampling approach integrated into an intelligent autosampler, which significantly reduces analysis times by optimizing the sample delivery process to reduce the pre- and post-measurement time. There are a number of these systems on the market, which work slightly differently, but basically they all use piston/syringe/vacuum pumps and switching valves and loops to control the delivery of the sample and standards to and from the ICP-MS. Besides significantly faster analysis times, other benefits include improved precision/accuracy, reduced carry-over and longer lifetime of sample introduction consumables. Depending on the design of the system, some typical areas of optimization include:

Autosampler response is the time it takes for the instrument to send a signal to the autosampler to move the sample probe to the next sample. By moving the autosampler probe over to the next sample while the previous sample is being analyzed, a significant amount of time will be saved over the entire automated run.

Sample uptake is the time taken for a sample to be drawn into the autosampler probe and pass through the capillary and pump tubing into the nebulizer. By using a small vacuum pump to rapidly fill the sample loop, which is positioned in close proximity of the sample loop to the nebulizer, sample uptake time is minimized.

Signal stabilization is the time required to allow the plasma to stabilize after air has entered the line from the autosampler probe dipping in and out of the sample tubes (this can also be exaggerated if the pump speed is increased to help in sample delivery). However, if the pump delivering the sample to the plasma remains at a constant flow rate, and the injection valve ensures no air is introduced into the sample line, very little stabilization time is required.

Rinse-out is the time required to remove the previous sample from the sample tubing and sample-introduction system. So, if the probe is being rinsed during the sample analysis, minimal rinse time is needed.

Overhead time is the time spent by the ICP performing calculations and printing results, so if this time is used to ensure the previous sample has reached baseline, minimal rinse time is required for the next sample.

Another slightly different approach is to use a rapid-rinse accessory, based on a flow injection loop and a positive displacement piston pump coupled to an autosampler. With this system, the multiport valve switches between two positions. In the first position, a loop of capillary tubing is filled with the sample, while in the second position the sample is delivered to the nebulizer. This is seen in greater detail in Figure 18.15 (a), which shows the piston pump rapidly filling the sample loop. At the same time, rinse and internal standard solutions are delivered to the nebulizer, washing out the nebulizer and spray chamber, and ensuring that plasma stability is maintained. Figure 18.15 (b) shows the actual aspiration process where the valve switches position so that the rinse solution pushes the sample into the nebulizer. The internal standard is then mixed with the sample within the valve. At the same time, the autosampler probe and sample uptake tubing are rinsed by the piston pump.

There are a number of these enhanced-productivity sampling systems on the market, that all work in a slightly different way. However, they all have one thing in common, and that is they offer at least a two-fold reduction in analysis time compared to traditional autosamplers. Some of the other benefits that are realized with this time saving include:

- Improved precision because of no pulsing from peristaltic pump
- Better accuracy due to online dilution and addition of internal standards
- Constant flow of solutions to plasma reduces stabilization times
- Less sample volume used
- Lower argon consumption
- Reduced cost of consumables
- Less routine maintenance
- Much lower chemical waste

There is no question that all these benefits can make a significant improvement in the overall cost of analysis, especially in high-workload routine environmental laboratories, where high sample throughput is an absolute requirement.[52,53]

18.16   AUTOMATED IN-LINE AUTO-DILUTION AND AUTO-CALIBRATION

A new range of automated sampling accessories has recently been developed, which performs very precise and accurate online auto-dilutions and auto-calibration procedures using syringe/piston pumps.[54] Samples are rapidly and reproducibly loaded from each autosampler location into a sample loop. From there the sample is injected into a diluent liquid stream and transported to a tee located between the valve and nebulizer. The internal standard is added in the tee to obtain final dilution factors defined by the operator. At the heart of the system is a syringe pump, which delivers the

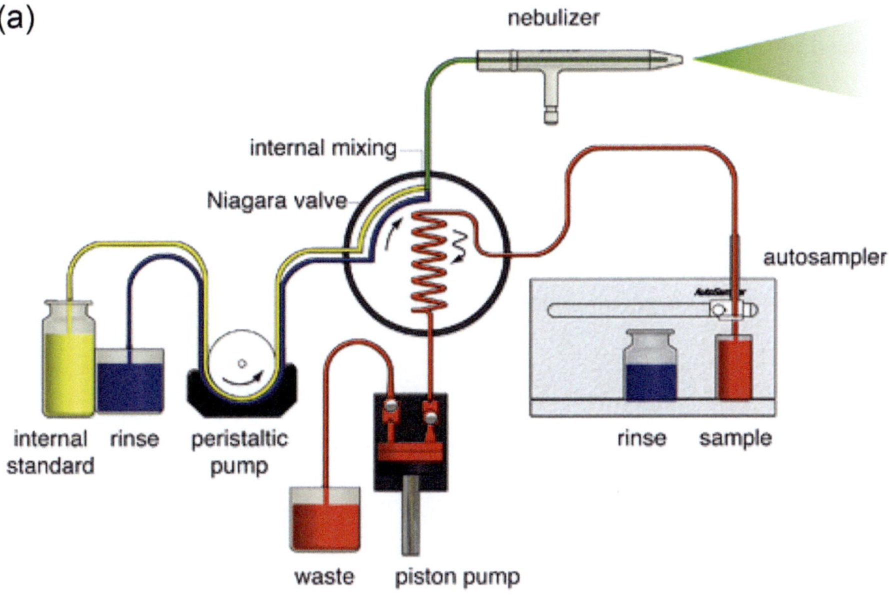

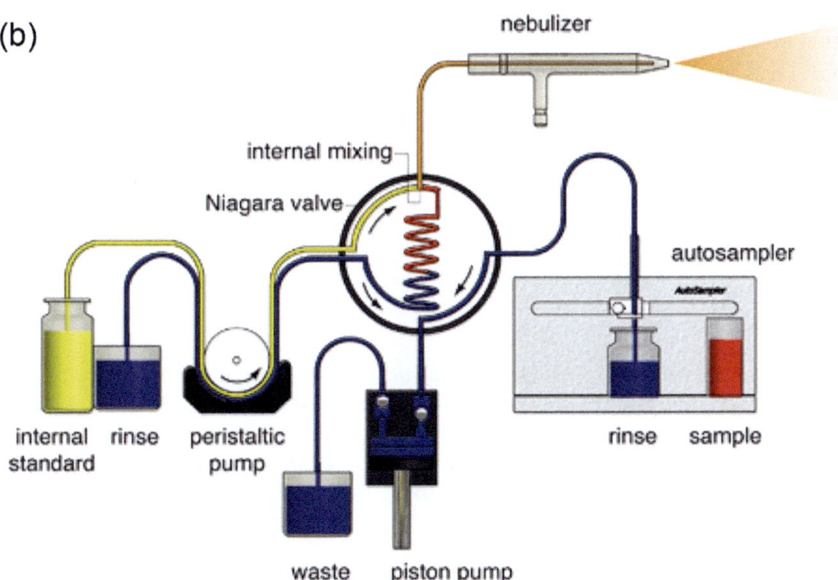

FIGURE 18.15 (A) The piston pump rapidly fills the sample loop, while at the same time, rinse and internal standard solutions are delivered to the nebulizer.

(Courtesy of Glass Expansion Inc)

FIGURE 18.15 (B) The valve switches position so that the rinse solution pushes the sample into the nebulizer, while the internal standard is mixed with the sample and the autosampler probe and sample uptake tubing are rinsed by the piston pump.

(Courtesy of Glass Expansion Inc)

sample over a wide range of flow rates, ensuring rapid and reliable in-line dilutions. The benefits of fully automated in-line auto-dilution and auto-calibration include:

- Real-time dilutions
- Dilution in valve head and tee
- No additional tube or reagents required
- Eliminates manual dilutions
- Rapid uptake and washout
- Lowers risk of contamination
- Sample analysis time constant, independent of dilution factor

18.17 AUTOMATED SAMPLE IDENTIFICATION AND TRACKING

A recent development in automation is in the area of advanced, automated sample identification and tracking systems that can accurately associate stored information with a sample throughout the sample collection/preparation/introduction process.[55] Through a series of four distinct stages, this technology uses barcodes to enter, store and reference data associated with a sample from initial collection to taring and final dilution.

For example, from the point of sample collection, the technology associates collection time and global positing system (GPS) location with sample container barcode. This then relates the information from its location and tracks it through the dilution process, inputting weight measurements before the sample-preparation process. The software then has the ability to automatically apply sample-digestion procedures according to sample-ID code. Finally, the system presents the samples for ICP-MS/ICP-OES analysis while confirming sample identity and providing legally defensible data for regulatory inspection.

This technology is ideally suited to any application area, such as the pharmaceutical or cannabis industry, where it is critical to keep track of the sample, as it moves from collection to preparation and dilution, etc., to finally being analyzed by the instrumentation of choice that is generating the data to make a decision about the maximum levels of elemental contaminants.

FURTHER READING

1. E. R. Denoyer, K. J. Fredeen, and J. W. Hager, *Analytical Chemistry*, **63**(8), 445–457A, 1991.
2. J. F. Ready, *Effects of High Power Laser Radiation*, Academic Press, New York, Chapters 3–4, 1972.
3. L. Moenke-Blankenburg, *Laser Microanalysis*, Wiley, New York, 1989.
4. E. R. Denoyer, R. Van Grieken, F. Adams, and D. F. S. Natusch, *Analytical Chemistry*, **54**, 26–30A, 1982.
5. J. W. Carr and G. Horlick, *Spectrochimica Acta*, **37B**, 1, 1982.
6. T. Kantor et al., *Talanta*, **23**, 585–590, 1979.
7. H. C. G. Human et al., *Analyst*, **106**, 265, 1976.
8. M. Thompson, J. E. Goulter, and F. Seiper, *Analyst*, **106**, 32–35, 1981.
9. A. L. Gray, *Analyst*, **110**, 551–555, 1985.
10. P. A. Arrowsmith and S. K. Hughes, *Applied Spectroscopy*, **42**, 1231–1239, 1988.
11. T. Howe, J. Shkolnik, and R. Thomas, *Spectroscopy*, **16**(2), 54–66, 2001.
12. D. Günther and B. Hattendorf, *Mineralogical Association of Canada—Short Course Series*, **29**, 83–91, 2001.
13. T. E. Jeffries, S. E. Jackson, and H. P. Longerich, *Journal of Analytical Atomic Spectrometry*, **13**, 935–940, 1998.
14. R. E. Russo, X. L. Mao, O. V. Borisov, and L. Haichen, *Journal of Analytical Atomic Spectrometry*, **15**, 1115–1120, 2000.
15. H. Liu, O. V. Borisov, X. Mao, S. Shuttleworth, and R. Russo, *Applied Spectroscopy*, **54**(10), 1435–1440, 2000.
16. *SEM Photo Courtesy of Dr. Honglin Yuan*, Northwest University, Xi'an, China.

17. S. E. Jackson, H. P. Longerich, G. R. Dunning, and B. J. Fryer, *Canadian Mineralogist*, **30**, 1049–1064, 1992.
18. D. Günther and C. A. Heinrich, *Journal of Analytical Atomic Spectrometry*, **14**, 1361–1366, 1999.
19. D. Günther, I. Horn, and B. Hattendorf, *Fresenius Journal of Analytical Chemistry*, **368**, 4–14, 2000.
20. R. E. Wolf, C. Thomas, and A. Bohlke, *Applied Surface Science*, **127–129**, 299–303, 1998.
21. J. Gonzalez, X. L. Mao, J. Roy, S. S. Mao, and R. E. Russo, *Journal of Analytical Atomic Spectrometry*, **17**, 1108–1113, 2002.
22. R. Lam and E. D. Salin, *Journal of Analytical Atomic Spectrometry*, **19**, 938–940, 2004.
23. J. Roy and L. Neufeld, *Spectroscopy*, **19**(1), 16–28, 2004.
24. J. Ruzicka and E. H. Hansen, *Analytic Chimica Acta*, **78**, 145–150, 1975.
25. R. Thomas, *Spectroscopy*, **17**(5), 54–66, 2002.
26. A. Stroh, U. Voellkopf, and E. Denoyer, *Journal of Analytical Atomic Spectrometry*, **7**, 1201–1207, 1992.
27. Y. Israel, A. Lasztity, and R. M. Barnes, *Analyst*, **114**, 1259–1264, 1989.
28. Y. Israel and R. M. Barnes, *Analyst*, **114**, 843–850, 1989.
29. M. J. Powell, D. W. Boomer, and R. J. McVicars, *Analytical Chemistry*, **58**, 2864–2871, 1986.
30. S. N. Willie, Y. Iida, and J. W. McLaren, *Atomic Spectroscopy*, **19**(3), 67–73, 1998.
31. R. Roehl and M. M. Alforque, *Atomic Spectroscopy*, **11**(6), 210, 1990.
32. J. W. McLaren, J. W. H. Lam, S. S. Berman, K. Akatsuka, and M. A. Azeredo, *Journal of Analytical Atomic Spectrometry*, **8**, 279–286, 1993.
33. L. Ebdon, A. Fisher, H. Handley, and P. Jones, *Journal of Analytical Atomic Spectrometry*, **8**, 979–981, 1993.
34. D. B. Taylor, H. M. Kingston, D. J. Nogay, D. Koller, and R. Hutton, *Journal of Analytical Atomic Spectrometry*, **11**, 187–191, 1996.
35. S. M. Nelms, G. M. Greenway, and D. Koller, *Journal of Analytical Atomic Spectrometry*, **11**, 907–912, 1996.
36. C. J. Park, J. C. Van Loon, P. Arrowsmith, and J. B. French, *Analytical Chemistry*, **59**, 2191–2196, 1987.
37. R. D. Ediger and S. A. Beres, *Spectrochimica Acta*, **47B**, 907–913, 1992.
38. C. J. Park and M. Hall, *Journal of Analytical Atomic Spectrometry*, **2**, 473–480, 1987.
39. C. J. Park and J. C. Van Loon, *Trace Elements in Medicine*, **7**, 103–107, 1990.
40. G. Chapple and J. P. Byrne, *Journal of Analytical Atomic Spectrometry*, **11**, 549–553, 1996.
41. Advances in XRF and ICP-OES Analysis for Plant Tissue and Soil Analysis Including Cannabis, *Spectro Instruments, Live Webcast*, February 26, 2020, www.spectro.com/landingpages-noindex/webinar-pages/icp-xrf-agronomy.
42. S. A. Beres, E. R. Denoyer, R. Thomas, and P. Bruckner, *Spectroscopy*, **9**(1), 20–26, 1994.
43. F. McElroy, A. Mennito, E. Debrah, and R. Thomas, *Spectroscopy*, **13**(2), 42–53, 1998.
44. K. W. Olson, W. J. Haas, Jr, and V. A. Fassel, *Analytical Chemistry*, **49**(4), 632–637, 1977.
45. J. Kunze, S. Koelling, M. Reich, and M. A. Wimmer, *Atomic Spectroscopy*, **19**, 5–13, 1998.
46. G. Settembre and E. Debrah, *Micro*, 67–71, June 1998.
47. D. R. Wiederin and R. S. Houk, *Applied Spectroscopy*, **45**(9), 1408–1411, 1991.
48. J. A. McLean, H. Zhang, and A. Montaser, *Analytical Chemistry*, **70**, 1012–1020, 1998.
49. *Characterization of a Customized Valve for Enhanced Productivity in ICP/ICP-MS*, Glass Expansion Application Note, www.geicp.com/site/images/application_notes/NiagaraApplicationsWhitePaper_Feb2010.pdf.
50. Automated Sample Introduction with the SC Fast, *ESI Application Note*, www.icpms.com/pdf/SC-FAST.pdf.
51. ASXpress Plus Rapid Sample Introduction System, *Cetac Technologies*, www.cetac.com/pdfs/Brochure_ASXpress-plus.pdf=www.cetac.com/pdfs/Brochure_ASXpress-plus.pdf.
52. *Improving Throughput of Environmental Samples by ICP-MS Following EPA Method 200.8*, ESI Application Note, www.elementalscientific.com/products/SC-FAST_enviro.html.
53. M. P. Field, M. LaVigne, K. R. Murphy, G. M. Ruiz, and R. M. Sherrell, *Journal of Analytical Atomic Spectrometry*, **22**, 1145, 2007.
54. *The Evolution of Automation: Prepfast Data Sheet*, Elemental Scientific Inc, www.icpms.com/products/prepfast.php.
55. *PlasmaTrax: Automated Sample Identification and Tracking System*, Application Note: Elemental Scientific Inc, Omaha, NE, www.icpms.com/pdfv1/15054-1%20PlasmaTRAX-2D%20Barcode%20Automation.pdf.

19 Coupling ICP-MS with Chromatographic Separation Techniques for Speciation Studies

The specialized sample-introduction techniques described in Chapter 18 were mainly developed as a result of a basic limitation of ICP-MS in carrying out elemental determinations on complex sample matrices or to enhance sample throughput. However, even though all of these sampling accessories significantly improved the flexibility, performance and productivity of the technique, they were still being used to measure the total metal content of the samples being analyzed. If the requirement is to learn more about the oxidation state or speciated form of the element, the trace metal analytical community had to look elsewhere for answers. Then, in the early 1990s, researchers started investigating the use of ICP-MS as a detector for chromatography systems, which triggered an explosion of interest in this exciting new hyphenated technique, especially for environmental and biomedical applications. In Chapter 19, we look at what drove this research and discuss the use of chromatographic separation techniques, including high-performance liquid chromatography (HPLC) with ICP-MS to carry out trace element speciation determinations. This chapter will be of particular interest, if there is a future requirement to measure different species or oxidation state of arsenic and/or mercury (or any other elemental contaminant) in a cannabis-related sample.

ICP-MS has gained popularity over the years, mainly because of its ability to rapidly quantitate ultratrace metal contamination levels. However, in its basic design, ICP-MS cannot reveal anything about the metal's oxidation state, alkylated form, how it is bound to a biomolecule or how it interacts at the cellular level. The desire to understand in what form or species an element exists led researchers to investigate the combination of chromatographic separation devices with ICP-MS. The ICP mass spectrometer becomes a very sensitive detector for trace element speciation studies when coupled to a chromatographic separation device based on HPLC, low-pressure liquid chromatography, ion chromatography (IC), gas chromatography (GC), size exclusion chromatography (SEC), capillary electrophoresis (CE), etc. In these hyphenated techniques, elemental species are separated based on their chromatographic retention, mobility or molecular size, and then eluted/passed into the ICP mass spectrometer for detection.[1] The intensities of the eluted peaks are then displayed for each isotopic mass of interest, in the time domain, as shown in Figure 19.1. The figure shows a typical time-resolved chromatogram for a selected group of masses.

There is no question that ICP-MS has allowed researchers in the environmental, biomedical, geochemical and nutritional fields to gain a much better insight into the impact of different elemental species on humans and their environment. Even though elemental speciation studies were being carried out using other atomic spectrometry (AS) detection techniques, it was the commercialization of ICP-MS in the early 1980s, with its extremely low detection capability, that saw a dramatic increase in the number of trace element speciation studies being carried out. Today, the majority of these studies are being driven by environmental regulations. In fact, the U.S. EPA has published a number of speciation methods involving chromatographic separation with ICP-MS, including Method 319.8 for the speciation of bromine compounds in drinking water and wastewater, and Method 6800 for the measurement of various metal species in potable and wastewaters by isotope

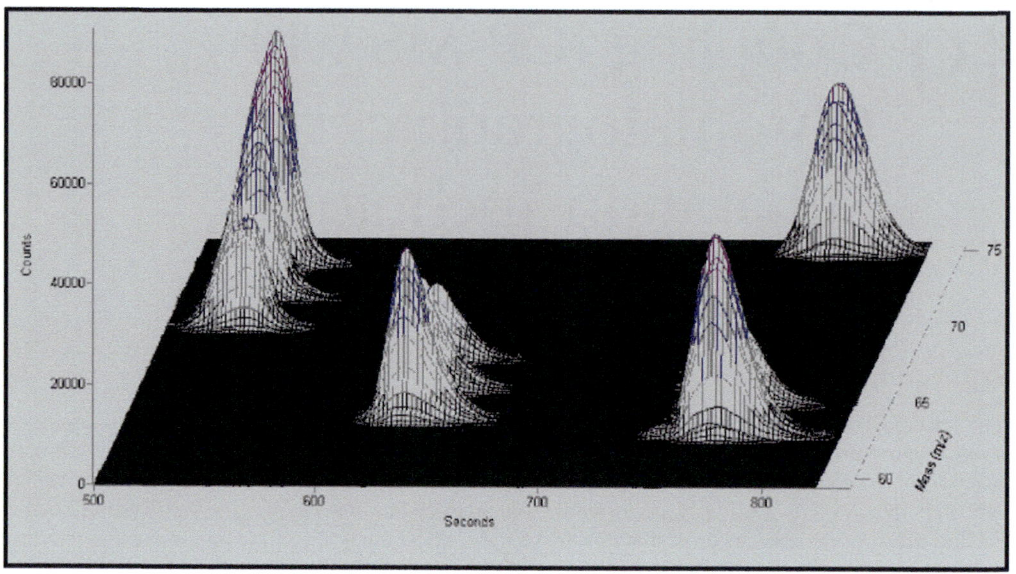

FIGURE 19.1 A typical time-resolved chromatogram generated using chromatography coupled with ICP-MS, showing a temporal display of intensity against mass.

(Courtesy of PerkinElmer Inc.)

dilution mass spectrometry. However, other important areas of interest include nutritional and metabolic studies, toxicity testing, bioavailability measurements, and now, with the new USP/ICH elemental impurity guidelines, the measurement of different arsenic and mercury species in pharmaceutical and nutraceutical materials. Speciation studies cross over many different application areas, but the majority of determinations being carried out can be classified into three major categories:

- **Measurement of different oxidation states:** For example, hexavalent chromium, Cr (VI), is a powerful oxidant and is extremely toxic, but in soil and water systems it reacts with organic matter to form trivalent chromium, Cr (III), which is the more common form of the element and an essential micronutrient for plants and animals.[2]
- **Measurement of alkylated forms:** Very often the natural form of an element can be toxic, although its alkylated form is relatively harmless, or vice versa. A good example of this is the element arsenic. Inorganic forms of the element such as As (III) and As (V) are toxic, whereas many of its alkylated forms such as monomethylarsonic acid (MMA) and dimethylarsonic acid (DMA) are relatively innocuous.[3]
- **Measurement of metallobiomolecules:** These molecules are formed by the interaction of trace metals with complex biological molecules. For example, in animal farming studies, the activity and mobility of an innocuous arsenic-based growth promoter are determined by studying its metabolic impact and excretion characteristics. So, measurement of the biochemical form of arsenic is crucial in order to know its growth potential.[4]

Table 19.1 represents a small cross section of both inorganic and organic species of interest classified under these three categories.

TABLE 19.1

Some Typical Inorganic and Organic Species That Have Been Studied by Researchers Using Chromatographic Separation Techniques

Oxidation states	Alkylated forms	Biomolecules
Se^{+4}	Methyl—Hg, Ge, Sn, Pb, As,	Organometallic complexes—
Se^{+6}	Sb, Se, Te, Zn, Cd, Cr	As, Se, Cd
As^{+3}	Ethyl—Pb, Hg	Metalloporphyrines
As^{+5}	Butyl, Phenyl,	Metalloproteins
Sn^{+2}	Cyclohexyl—Sn	Metallodrugs
Sn^{+4}		Metalloenzymes
Cr^{+3}		Metals at the cellular level
Cr^{+6}		
Fe^{+2}		
Fe^{+3}		

There are 400–500 speciation papers published every year, the majority of which are based on toxicologically significant elements such as As, Cr, Hg, Se and Sn.[5] The following is a small selection of some of the most recent research that can be found in the public domain:

- Determination of chromium (VI) in drinking water samples, using HPLC-ICP-MS[6]
- Determination of trivalent and hexavalent chromium in pharmaceutical and biological materials by IC-ICP-MS[7]
- The use of LC-ICP-MS in better understanding the role of inorganic and organic forms of selenium in biological processes[8]
- Identification of selenium compounds in contaminated estuarine waters using IC-ICP-MS[9]
- Determination of organoarsenic species in marine samples, using cation exchange HPLC-ICP-MS[10]
- Determination of biomolecular forms of arsenic in chicken manure by CE-ICP-MS[11]
- Analysis of tributyltin (TBT) in marine samples using HPLC-ICP-MS[12]
- Measurement of anticancer platinum compounds in human serum by HPLC-ICP-MS[13]
- Bioavailability of cadmium and lead in beverages and foodstuffs using SEC-ICP-MS[14]
- Investigation of sulfur speciation in petroleum products by capillary gas chromatography with ICP-MS detection[15]
- Analysis of methyl mercury in water and soils by HPLC-ICP-MS[16]
- Analysis of Arsenic and Mercury Species from Botanicals and Dietary Supplements using LC-ICP-MS [17]

19.1   HPLC COUPLED WITH ICP-MS

It can be seen from this brief snapshot of speciation publications that, by far, the most common chromatographic separation techniques being used with ICP-MS are the many different types of high-performance liquid chromatography, such as adsorption, ion exchange, size exclusion, gel permeation and normal- or reverse-phase chromatography. To get a better understanding of how the technique works, particularly when attempting to develop a routine method to simultaneously measure multiple species in the same analytical run, let us take a more detailed look at how the HPLC system is coupled to the ICP mass spectrometer. Figure 19.2 shows a typical setup of the hardware components.

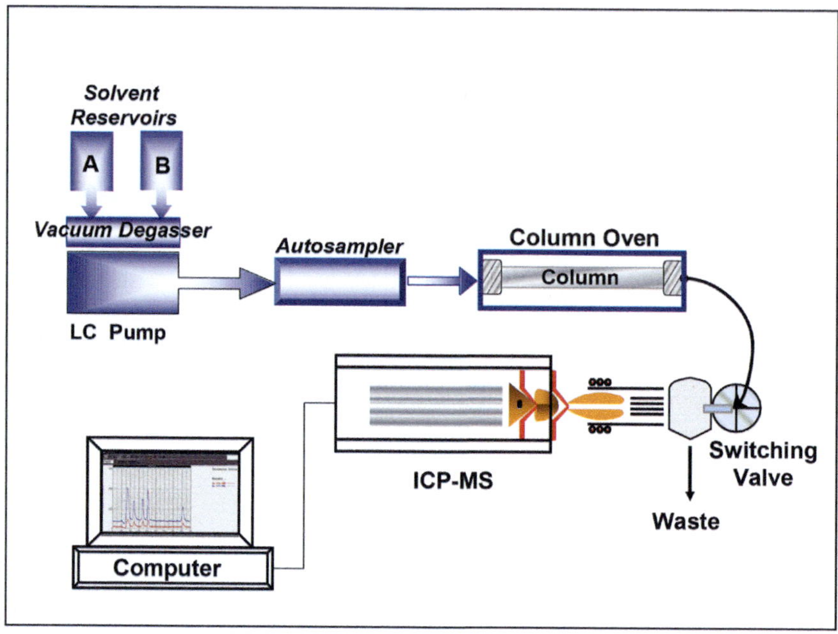

FIGURE 19.2 A typical configuration of an HPLC system interfaced with an ICP mass spectrometer. (Courtesy of PerkinElmer Inc.)

The coupling of the ICP-MS system to the liquid chromatograph hardware components is relatively straightforward, connecting a capillary tube from the end of the HPLC column, through a switching valve, to the sample-introduction system of the ICP mass spectrometer. However, matching the column flow with the uptake of the ICP-MS sample-introduction system is not a trivial task. Therefore, to develop a successful trace element speciation method, it is important to optimize not only the chromatographic separation, but also selection of the nebulization process to match the flow of the sample being eluted off the column, together with finding the best ICP-MS operating conditions for the analytes or species of interest. Let us take a closer look at this.

19.2 CHROMATOGRAPHIC SEPARATION REQUIREMENTS

Traditionally, the measurement of trace levels of elemental species by HPLC has been accomplished by separating the species using column separation technology and detecting them, one element at a time, as they elute. This approach works well for one element or species, but is extremely slow and time consuming for the determination of multiple species or elements, because the chromatographic separation process has to be optimized for each species being measured. Therefore, the limitation to multielement speciation is rarely the detection of the elements, but is usually the separation of the species. The inherent problem is that liquid chromatography works on the principle of equilibration between the species of interest, the mobile phase and the column material. Because the chemistries of elements and their species differ, it is difficult to find common conditions capable of separating species of more than one element simultaneously. For example, toxic elements of environmental interest, such as arsenic (As), chromium (Cr) and selenium (Se), have different reaction chemistries, thus requiring different chromatographic conditions to separate them. So, to achieve the analytical goal of fast, simultaneous measurement of different species of these elements in a single sample injection, the chromatographic separation has to be optimized.

Let us first examine the chromatography, focusing on two common forms of separation (ion exchange and reversed-phase ion-pairing chromatography) together with two commonly employed HPLC elution techniques (isocratic and gradient). Each process will be described, and the advantages and disadvantages of each will be discussed in the context of choosing a separation scheme to achieve our analytical goals. One final note before we discuss the separation process: it is extremely important, when preparing the sample, that the speciated form of the element not be changed or altered in any way. This is not such a serious problem with a simple matrix such as drinking water, which typically involves straightforward acidification. However, when more complex samples such as soils, biological samples or pharmaceutical matrices are being analyzed, it is quite common to use strong acids, oxidizing agents or high temperatures to get the samples into solution. So, maintaining the integrity of the valency/oxidation state or species should always be an extremely important decision when preparing a sample for speciation analysis.

19.3 ION-EXCHANGE CHROMATOGRAPHY (IEC)

In this technique, separation is based on the exchange of ions (anions or cations) between the mobile phase and the ionic sites on a stationary phase bound to a support material in the column. A charged species is covalently bound to the surface of the stationary phase. The mobile phase, typically a buffer solution, contains a large number of ions that have a charge opposite to that of the surface-bound ions. These mobile-phase ions, referred to as counter ions, establish equilibrium with the stationary phase. Sample ions passing through the column and having the same ionic charge as the counterion compete with the mobile-phase ions for sites on the column material, resulting in a disruption of the equilibrium and retention of that analyte. It is the differential competition of various analytes with the mobile-phase ions that ultimately produces species separation and chromatographic peaks, as detected by ICP-MS. Therefore, analyte retention is based on the affinity of different ions for the support material and on other solution parameters, including counterion type, ionic strength (buffer concentration) and pH. Varying these mobile-phase parameters changes the separation.

The principle of IEC is schematically represented in Figure 19.3 for an anion-exchange separation. In this figure, the anions are denoted by A⁻ and represent the analyte species, whereas the

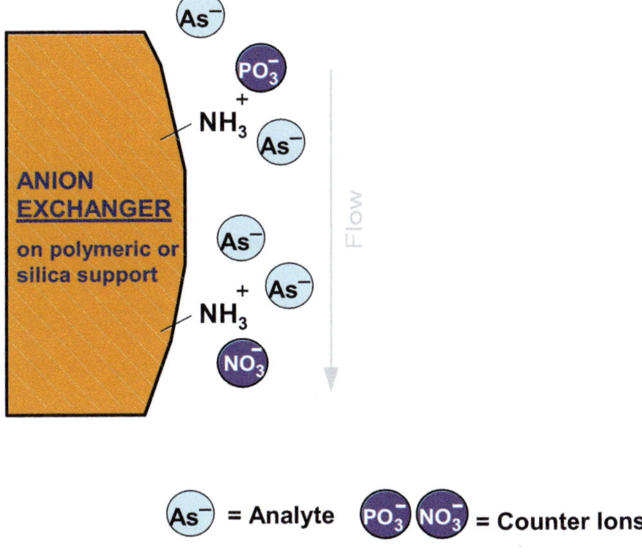

FIGURE 19.3 Principle of separation using anion-exchange chromatography.

counterions are symbolized by X^-. The main benefit of this approach is that it has a high tolerance to matrix components, and the main disadvantages are that these columns tend to be expensive and somewhat fragile.

19.4 REVERSED-PHASE ION-PAIR CHROMATOGRAPHY (RP-IPC)

Ion-pair chromatography typically uses a reversed-phase column in conjunction with a special type of chemical in the mobile phase called an *ion-pairing reagent*. "Reversed phase" essentially means that the column's stationary phase is nonpolar (less polar, more organic) than the mobile-phase solvents. For RP-IPC, the stationary phase is a carbon chain, most often consisting of 8 or 18 carbon atoms (C8, C18) bonded to a silica support. The mobile-phase solvents usually consist of water mixed with a water-miscible organic solvent, such as acetonitrile or methanol.

The ion-pairing reagent is a compound that has both an organic and an ionic end. To promote ionic interaction, the ionic end should have a charge opposite to that of the analytes of interest. For anionic IPC, a commonly used ion-pairing reagent is tetrabutylammonium hydroxide (TBAOH); for cationic IPC, hexanesulfonic acid (HSA) is often chosen. In principle, the ionized/ionizable analytes interact with the ionic end of the ion-pairing reagent, whereas the ion-pairing reagent's organic end interacts with the C18 stationary phase. Retention and separation selectivity are primarily affected by characteristics of the mobile phase, including pH, selection/concentration of ion-pairing reagent and ionic strength, as well as the use of additional mobile-phase modifiers. With this scheme, a reversed-phase column acts like an ion-exchange column.

The benefits of this type of separation are that the columns are generally less expensive and tend to be more rugged than ion-exchange columns. The main disadvantages, compared to IEC, are that the separation may not be as good, and there may be less tolerance to high sample matrix concentrations. A simplified representation of this approach is shown in Figure 19.4.

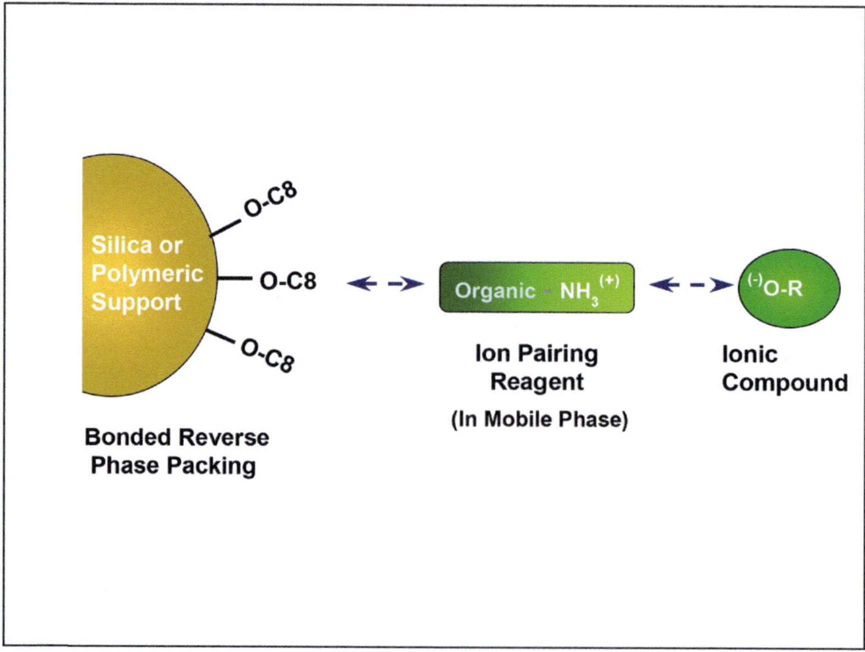

FIGURE 19.4 Principle of ion-pair chromatography (anionic mode).

19.5 COLUMN MATERIAL

As previously explained, several variables of the mobile phase can be modified to affect separations, both by IEC and RP-IPC. A very important consideration when choosing a column is the pH of the mobile phase used to achieve the separation. The pH is critical in the selection of the column for both IEC and RP-IPC, because the support materials may be pH-sensitive. One of the most common column materials is silica, because it is inexpensive, rugged and has been used for many years. The main disadvantage of silica columns is that they are useful only over a pH range of 2–8. Outside this range, the silica dissolves. For applications requiring pHs outside of this range, polymer columns are commonly used. Polymer columns are useful over the entire pH range. The downside to these columns is that they are typically more expensive and fragile. Columns consisting of other support materials are also available.

19.6 ISOCRATIC OR GRADIENT ELUTION

Another consideration is how best to elute the species from the column. This can be accomplished using either an isocratic or gradient elution scheme. Isocratic refers to using the same solvent throughout the analysis. Because of this, samples can be injected immediately after the preceding one has finished eluting. This type of elution can be performed on all HPLC pumping systems. The advantages of isocratic elutions are simplicity and higher sample throughput.

Gradient elution involves variation of the mobile-phase composition over time by a number of steps such as changing the organic content, altering the pH, changing concentration of the buffer or using a completely different buffer. This is usually accomplished by having two or more bottles of mobile phases connected to the HPLC pump. The pump is then programmed to vary the amount of each mobile-phase component. The mobile-phase components are mixed online before reaching the column, and therefore, a more complex pumping system is required to handle this task.

The main reason for using gradient elution is that the mobile-phase composition can be varied to fine-tune the separation. As a result, component separations are better, and chromatograms are generally shorter than with isocratic elutions. However, overall sample throughput is much lower than with isocratic separations because of the variation in mobile-phase composition. So, after an elution is complete, the mobile phase must be changed back to its original composition as at the start of the chromatogram. This means that the equilibrium between the mobile phase and the column must be reestablished before the next sample can be injected. If the equilibrium is not established, the peaks will elute at different times compared to the previous sample. The equilibration time varies with column, but it is usually between 5 and 30 min.

Therefore, if a fast, automated, routine method for the measurement of multispecies/elements is the desired analytical goal, it is often best to attempt an isocratic separation method first, because of the complexity of method development and the low sample throughput of gradient elution methods. In fact, a simultaneous method for the separation of As, Cr and Se species in drinking water samples was demonstrated by Neubauer and coworkers; they developed a method to determine inorganic forms of arsenic (As^{+3}, As^{+5}), chromium (Cr^{+3}, Cr^{+6}) and selenium (Se^{+4}, Se^{+6}, and $SeCN^-$) by reverse-phase ion-pairing chromatography with isocratic elution.[18] Details of the HPLC separation parameters/conditions they used are shown in Table 19.2.

It is important to emphasize that if these species were being measured on a single-element basis, the optimum chromatographic conditions would be different. However, the goal of this study was to determine all the species in a single multielement run so that the method could be applied to a routine, high-throughput environment. The 3.3-cm-long column was packed with a C8 hydrocarbon material (3-μm particle size). The mobile phase was a mixture of TBAOH, ammonium acetate (NH_4CH_2COOH), the potassium salt of EDTA and methanol. The pH was adjusted to 7.5 (prior to the addition of methanol) using 10% nitric acid and 10% ammonium hydroxide. The sample preparation consisted of dilution with the mobile phase and heating at 50–55°C to speed the formation of the Cr III–EDTA complex. For each analysis, 50 μL of sample was injected into the column, which resulted in a sample flow of 1.5 mL/min being eluted off the column into the ICP mass spectrometer.

TABLE 19.2

HPLC Separation Parameters/Conditions for Measuring Inorganic As, Cr and Se Species in Potable Waters

HPLC configuration	Quaternary pump, column oven and autosampler
Column	C8, reduced activity, 3.3 cm × 0.46 cm (3 μm packing)
Column temperature	35°C
Mobile phase	mM TBAOH + 0.15 mM NH_4CH_2COOH + 0.15 mM EDTA (K salt) + 5% MeOH
pH	7.5
pH adjustment	Dilute HNO_3, NH_4OH
Injection volume	50 mL
Sample flow rate	1.5 mL/min
Samples	Various potable waters
Sample preparation	Dilute with mobile phase (2–10 fold); heat at 50–55°C for 10 min

Source: K. R. Neubauer, P. A. Perrone, W. Reuter, R. Thomas, *Current Trends in Mass Spectroscopy*, May 2006.

19.7   SAMPLE-INTRODUCTION REQUIREMENTS

When coupling an HPLC system to an ICP mass spectrometer, it is very important to match the flow of sample being eluted off the column with the ICP-MS nebulization system. With today's choice of sample-introduction components, there are specialized nebulizers and spray chambers on the market that can handle sample flows from 20 μL/min up to 3000 μL/min. The most common type of nebulizer used for chromatography applications is the concentric design, because it is self-aspirating and it generates an aerosol with extremely small droplets, which tends to produce better signal stability compared to a cross-flow design. The choice of which type of concentric nebulizer to use should therefore be based on the sample flow coming off the column. If the sample flow is on the order of 1 mL/min, a higher-flow concentric nebulizer should be used, and if the flow is much lower, such as in nano- or microflow LC work, a specialized low-flow nebulizer should be used.

It is also very important that the dead volume of the sample-introduction process be kept to an absolute minimum to optimize peak integration of the separated species over the length of the transient signal. For this reason, the length of sample capillary from the end of the column to the nebulizer should be kept to a minimum; the internal volume of the nebulizer should be as small as possible; the connectors/fittings/valves should all have low dead volume; and a self-aspirating nebulizer should be used to avoid the need for peristaltic pump tubing. In addition, a spray chamber with a short aerosol path should be selected, which will not add additional dead volume to the method. However, depending on the total flow of the sample and the type of nebulizer, a spray chamber may not even be required. Cutting down on the sample-introduction dead volume and minimizing peak broadening by careful selection of column technology will ultimately dictate the number of species that can be separated in a given time—an important consideration when developing a routine method for high sample throughput. Figure 19.5 shows a high-efficiency concentric nebulizer (HEN) designed for HPLC-ICP-MS work, and Figure 19.6 shows the difference between the capillary of this nebulizer (on the right) and a standard concentric type A nebulizer (on the left).

Another important reason to match the nebulizer with the flow coming off the column is that concentric nebulizers are mainly self-aspirating. For this reason, the column flow must be high enough to ensure that the nebulizer can sustain a consistent and reproducible aerosol. On the other hand, if the column flow is too low, a makeup flow might need to be added to the column flow to meet the flow requirements of the nebulizer being used. This has the additional benefit of being able

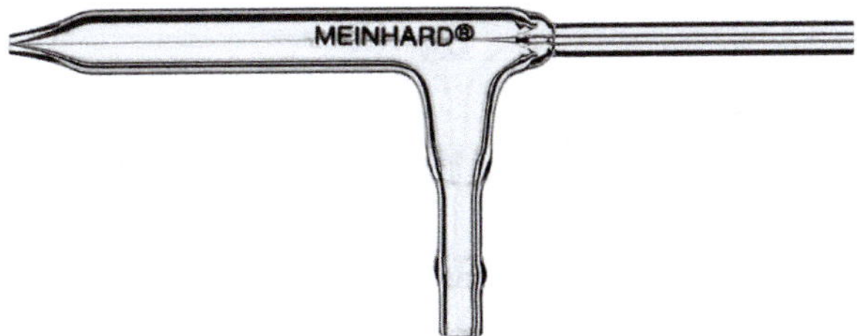

FIGURE 19.5　A high-efficiency concentric nebulizer (HEN) designed for HPLC-ICP-MS work.

(Courtesy of Meinhard Glass Products)

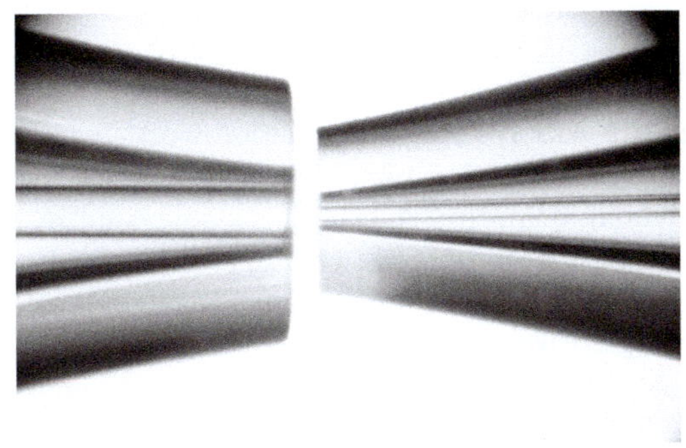

FIGURE 19.6　The difference between the capillary of a high-efficiency concentric nebulizer for HPLC work (on the right) and a standard concentric type A nebulizer (on the left).

(Courtesy of Meinhard Glass Products)

to add an internal standard after the column with another pump to correct for instrument drift or matrix effects in gradient-elution work.

19.8　OPTIMIZATION OF ICP-MS PARAMETERS

In the early days of trace element speciation studies using chromatography coupled with ICP-MS, researchers had no choice but to interface their own LC pumps, columns, autosamplers, etc., to the ICP mass spectrometer, because off-the-shelf systems were not commercially available. However, the analytical objectives of a research project are a little different from the requirements for routine analysis. With a research project, there are fewer time constraints to optimize the chromatography and detection parameters, whereas in a commercial environment, there are often financial penalties if the laboratory cannot be up, running real samples and generating revenue as quickly as possible. This demand, especially from commercial laboratories in the environmental and biomedical communities, for routine trace element speciation methods convinced the instrument manufacturers and vendors to develop fully integrated HPLC-ICP-MS systems.

The availability of these off-the-shelf systems rapidly drove the growth of this hyphenated technique, so much so that vendors who did not offer it with full application and hardware/software support were at a disadvantage. As this technique is maturing and is being used more as a routine analytical tool, it is becoming clear that the requirements of the ICP-MS system doing speciation analysis are different from those of an instrument carrying out trace element determinations using conventional nebulization. With that in mind, let us take a closer look at the typical requirements of an ICP mass spectrometer that is being utilized as a multielement detector for trace element speciation studies.

19.9 COMPATIBILITY WITH ORGANIC SOLVENTS

The requirements of the sample-introduction system, and in particular the nebulization process, have been described earlier in this chapter. However, some reverse-phase HPLC separations use gradient elution with mixtures of organic solvents such as methanol or acetonitrile. If this is the case, consideration must be given to the fact that some volatile organic solvents will extinguish the plasma.[19] Therefore, modifications to the sample introduction might need to be made, such as adding small amounts of oxygen to the nebulizer gas flow, or perhaps using a cooled spray chamber or a desolvation device to stop the buildup of carbon deposits on the sampler cone. Other approaches such as direct-injection nebulization have been used to introduce the sample eluent into the ICP-MS, but historically they have not gained widespread acceptance because of usability issues.

19.10 COLLISION/REACTION CELL OR INTERFACE CAPABILITY

Another requirement of the ICP-MS system for speciation work is the collision/reaction cell/interface capability. It is becoming clear that as more and more speciation methods are being developed, the ability to minimize polyatomic spectral interferences generated by the solvent, buffer, mobile phase, pH-adjusting acids/bases, the plasma gas, etc., is of crucial importance. Take, for example, the elements discussed earlier. As, Cr and Se are notoriously difficult elements for ICP-MS analysis because their major isotopes suffer from argon- and sample-based polyatomic interferences. Arsenic has only one isotope (m/z 75), which is difficult to quantify in chloride-containing samples because of the presence of $^{40}Ar^{35}Cl^+$. Low-level chromium analysis is difficult because of the presence of the $^{40}Ar^{12}C^+$ and $^{40}Ar^{13}C^+$ interferences, which overlap the two major isotopes of chromium at masses 52 and 53. These interferences are nearly always present, but are especially strong in samples with organic content. The argon dimers ($^{40}Ar^{40}Ar^+$, $^{40}Ar^{38}Ar^+$) at masses 80 and 78 interfere with the major isotopes of Se at mass 80 and 78, respectively, and bromine, which is usually present in natural waters, forms $^{79}BrH^+$ and $^{81}BrH^+$, which interferes with the Se masses at 80 and 82, respectively. The effects of these and other interferences have been reduced somewhat with conventional ICP-MS instrumentation, using alternate masses, interference correction equations, cool-plasma technology and desolvation techniques, but these approaches have not shown themselves to be particularly useful for these elements, especially at ultratrace levels.

For these reasons, it will be very beneficial if the ICP-MS instrumentation is fitted with a collision or reaction cell or interface and has the capability to minimize the formation of these undesired polyatomic interferences, using either collisional mechanisms with kinetic energy discrimination[20,21] or ion–molecule reaction kinetics with mass bandpass tuning,[22] or by introduction of a collision/reaction gas into the interface region.[23] The best approach will depend on the type of sample being analyzed, the number of species being determined and the detection-limit requirements, but there is no doubt that for multielement work, it is beneficial if one gas can be used for all the analyte species. This is exemplified by the research of Neubauer and coworkers described earlier, who used oxygen as the reaction gas to reduce the polyatomic spectral interferences described earlier to determine seven different species of As, Cr and Se in potable water. In fact, even though the oxygen was used to remove the argon-carbide and argon dimer interferences to quantify the chromium and selenium species, respectively, it was used in a different way to quantify the arsenic species. It was used to react with the arsenic ion to form the arsenic-oxygen molecular ion ($^{75}As^{16}O$) at mass 91, and move it away from the argon-chloride interference at mass 75. This novel approach has been reported many times in the literature[22,24] and has in fact been approved by the EPA in the recent ILM05.4 analytical procedure update for their Superfund, Contract Laboratory.

The instrumental conditions for the speciation analysis are shown in Table 19.3, and a plot of signal intensity versus time of the simultaneous separation of a 1 μg/L standard of As, Cr and Se is shown in Figure 19.7.

TABLE 19.3

ICP-MS Instrumental Conditions for the Speciation Analysis of As^{+3}, As^{+5}, Cr^{+3}, Cr^{+6}, Se^{+4}, Se^{+6} and SeCN in Potable Water Samples

Nebulizer	Quartz concentric
Spray chamber	Quartz cyclonic
RF power	1,500 W
Collision/reaction-cell technology	Dynamic-reaction cell
Reaction gas	O$_2$ = 0.7 mL/min
Analytical masses	AsO$^+$ (m/z 91), Se$^+$ (m/z 78), Cr$^+$ (m/z 52)
Analyte species	As^{+3}, As^{+5}, Cr^{+3}, Cr^{+6}, Se^{+4}, Se^{+6}, SeCN
Dwell time	330 ms (per analyte)
Analysis time	5.5 min

Source: K. R. Neubauer, P. A. Perrone, W. Reuter, R. Thomas, *Current Trends in Mass Spectroscopy*, May 2006.

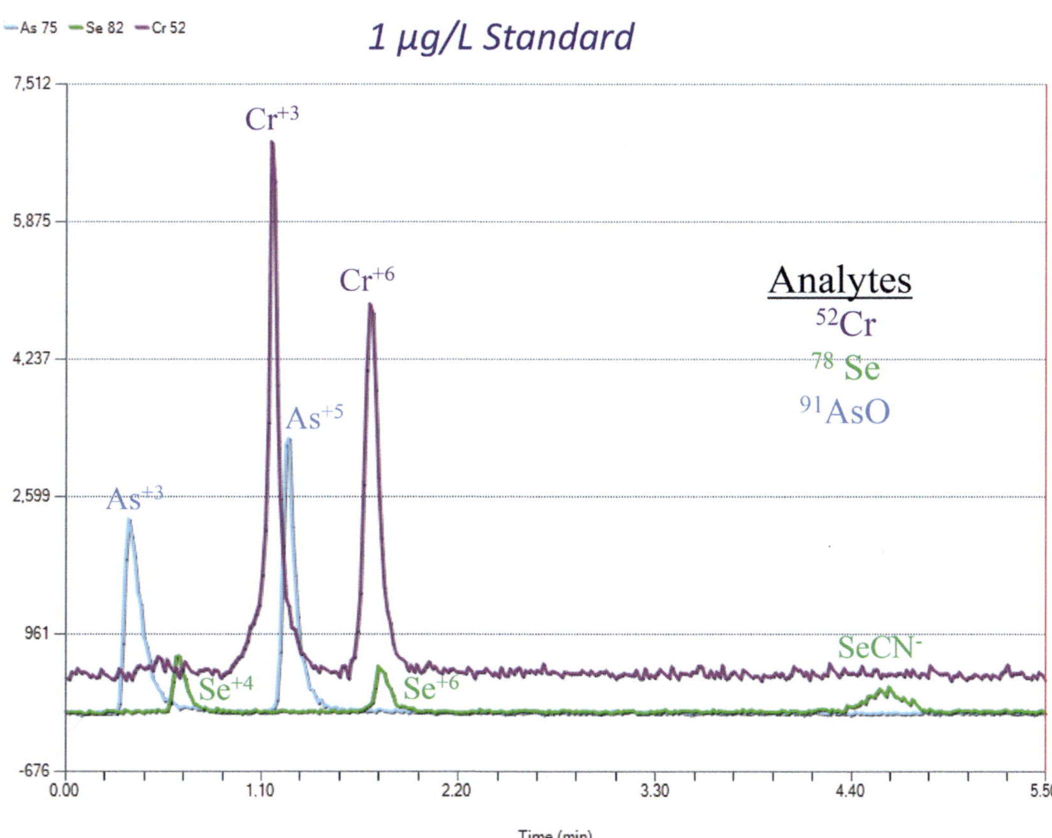

FIGURE 19.7   A plot of signal intensity versus time of the simultaneous separation of a 1 μg/L standard of As, Cr and Se species.

(From K. R. Neubauer, P. A. Perrone, W. Reuter, R. Thomas, *Current Trends in Mass Spectroscopy*, May 2006)

It should be emphasized that even though oxygen was used as the reaction gas in this study, many other collision/reaction strategies using inert gases such as helium and low-reactivity gases such as hydrogen have been successfully used for the determination of As, Cr and Se. The question is whether a single gas can be used to remove all the interferences to an acceptable level and allow quantitation at the trace level. If more than one gas is required, that is not such a major hardship, especially as all gas flows are under computer control and can be changed in a multielement run. However, if the method needs to be automated for a high-sample-throughput environment, it will be significantly faster if only one gas is used for interference removal, so that the collision/reaction-cell conditions need not be changed for each element.

19.11   OPTIMIZATION OF PEAK MEASUREMENT PROTOCOL

It can be seen from the chromatogram that the total separation time for the seven inorganic species is on the order of 2 min for the common oxidation states of the elements, and about 5 min if there is any selenocyanide in the sample. This means that all the peaks have to be integrated and quantified in a transient signal lasting 2–5 min. So, even though this is not considered a short transient such as an ETV or small-spot laser ablation work that typically lasts 2–5 s, it is not a continuous signal generated by a pneumatic nebulizer. For that reason, it is critical to optimize the measurement time in order to achieve the best multielement signal-to-noise ratio in the sampling time available. This is demonstrated in Figure 19.8, which shows the temporal separation of a group of elemental species in a chromatographic transient peak. The plot represents signal intensity against mass over the time period of the chromatogram. To get the best detection limits for this group of elements or species, it is very important to spend all the available time quantifying the peaks of interest.

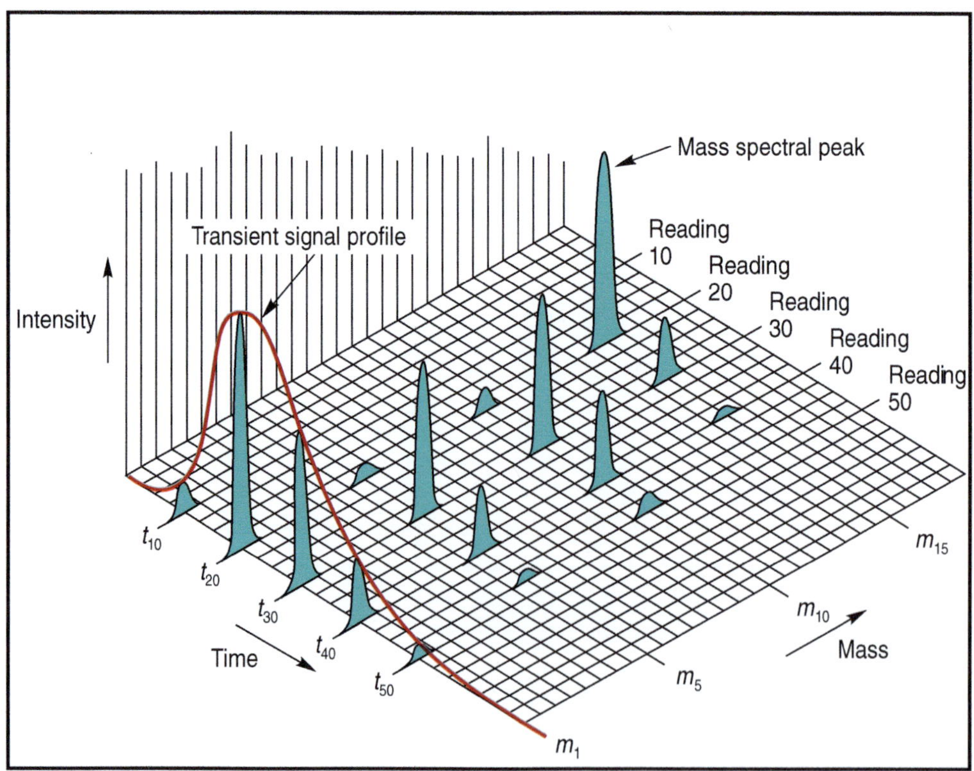

FIGURE 19.8   The temporal separation of a group of elemental species in a chromatographic transient peak.

For this reason, the quadrupole scanning/settling time and the time spent measuring the analyte peaks must be optimized to achieve the highest signal quality. This is described in greater detail in Chapter 15 on "Peak Measurement Protocol" and basically involves optimizing the number of sweeps, selecting the best dwell time and using short settling times to achieve the highest measurement duty cycle and maximize the peak signal-to-noise ratio over the duration of the transient event. If this is not done, there is a strong likelihood that the quality of the speciation data could be compromised. In addition, if the extended dynamic range is used to determine higher concentrations of elemental species, the scanning and settling time of the detector will also have an impact on the quality of the signal. For that reason, detectors that require two scans to characterize an unknown sample will use up valuable time in the quantitation process. This is somewhat of a disadvantage when doing multielement speciation on a chromatographic transient signal, especially if you have limited knowledge of the analyte concentration levels in your samples.

19.12   FULL SOFTWARE CONTROL AND INTEGRATION

In the early days of coupling HPLC components with ICP-MS, there were very few sophisticated communication protocols between the two devices. The sample was injected into the chromatographic separation system, and when the analyte species was close to being eluted off the column, the read cycle of the ICP mass spectrometer was initiated manually to capture the data using the instrument's time-resolved software. Processing and manipulation of the data was then carried out after the chromatogram had been captured, sometimes by a completely different software program. The HPLC and the ICP-MS were considered almost two distinctly different hardware and software devices, with very little communication between them. This was even more surprising considering that many of the ICP-MS vendors also offered LC equipment. As you can imagine, this was not the ideal scenario for a laboratory that wanted to carry out automated speciated analysis. As a result of this demand, manufacturers realized that unless they offered fully integrated hardware and software solutions for trace element speciation work, it was never going to be accepted as a routine analytical tool. Today, just about all the ICP mass spectrograph manufacturers offer fully integrated HPLC-ICP-MS systems. Some of the features available on today's instrumentation include the following:

- Full software and hardware control of both the chromatograph and the ICP-MS system from one computer
- Ability to check the status of the ICP-MS system when setting up the chromatography method or vice versa
- Computer control of the switching valve to allow the HPLC to be used in tandem with the ICP-MS normal sample-introduction system
- Real-time display of spectral data, including peak identification and quantitation, while the chromatogram is being generated
- Full data-handling capability, including manipulation of spectral peaks and calibration curves
- Comprehensive reporting options or exportable data to third-party programs

It is also important to emphasize that most vendors also offer integrated systems with turnkey application methods that are almost ready to run samples as soon as the instrument is installed. Although this is not available for all applications, it is becoming a standard offering for some of the more routine environmental applications. Manufacturers are also realizing that most analytical chemists who are experienced at trace element analysis may have very little expertise in chromatography. For that reason, they are providing full backup and customer support with HPLC application specialists as well as with the traditional ICP-MS product specialists.

19.13   FINAL THOUGHTS

Combining chromatography with ICP-MS has revolutionized trace metal speciation analysis. In particular, when HPLC or even low-pressure LC systems are coupled with the selectivity and sensitivity of ICP-MS, many elemental species at sub-ppb levels can now be determined in a single sample injection. When interference-reduction methods such as collision/reaction-cell or interface technology are available, it is possible to separate and detect different inorganic species of environmental significance in one automated run. By using a fully integrated, computer-controlled HPLC/ICP-MS system and optimizing the chromatographic separation, sample introduction and ICP-MS detection parameters, simultaneous quantitation can be carried out in a few minutes. There is no question that this kind of sample throughput will arouse the interest of the environmental, biomedical, nutritional and other application communities interested in trace element speciation studies, and help them realize that it is feasible to carry out this kind of analysis in a truly routine manner.

However, it should be emphasized that, historically, there has been limited demand for the determination of elemental species in cannabis-related products, pharmaceutical materials or dietary supplements. As a result, there is very little application material in the public domain describing the separation and quantitation of inorganic/organic forms of arsenic and mercury in these kinds of sample matrices. For that reason, if there is a need to determine these species, it is strongly advised that a person or a laboratory with expertise in speciated techniques initially develop the method, as many inter-laboratory round-robin studies have indicated that an optimized sample-preparation procedure and chromatographic separation technology are critical to get high-quality data.[25]

FURTHER READING

1. R. Lobinski, I. R. Pereiro, H. Chassaigne, A. Wasik, and J. Szpunar, *Journal of Analytical Atomic Spectrometry*, **13**, 860–867, 1998.
2. G. Cox and C. W. McLeod, *Mikrochimica Acta*, **109**, 161–164, 1992.
3. S. Branch, L. Ebdon, and P. O'Neill, *Journal of Analytical Atomic Spectrometry*, **9**, 33–37, 1994.
4. J. R. Dean, L. Ebdon, M. E. Foulkes, H. M. Crews, and R. C. Massey, *Journal of Analytical Atomic Spectrometry*, **9**, 615–618, 1994.
5. Z. A. Grosser and K. Neubauer, *Today's Chemist at Work*, May 2004.
6. Y. L. Chang and S. J. Jiang, *Journal of Analytical Atomic Spectrometry*, **9**, 858–865, 2001.
7. K. E. Lokits, D. D. Richardson, and J. A. Caruso, *Handbook of Hyphenated ICP-MS Applications*, Agilent Technologies, August 2007, www.chem.agilent.com/temp/rad5F0C7/00001681.PDF.
8. K. DeNicola, D. D. Richardson, and J. A. Caruso, *Spectroscopy*, **21**(2), 18–24, 2006.
9. D. Wallschlager and N. Bloom, *Journal of Analytical Atomic Spectrometry*, **16**, 1322–1326, 2001.
10. J. J. Sloth, E. H. Larsen, and K. Julshamn, *Journal of Analytical Atomic Spectrometry*, **18**, 452–459, 2003.
11. C. G. Rosal, G. Momplaisir, and E. Heithmar, *Electrophoresis*, **26**, 1606–1614, 2005.
12. L. Yang, Z. Mester, and R. E. Sturgeon, *Analytical Chemistry*, **74**, 2968–2975, 2002.
13. V. Vacchina, L. Torti, C. Allievi, and R. Lobinski, *Journal of Analytical Atomic Spectrometry*, **18**, 884–890, 2003.
14. S. Mounicou, J. Szpunar, R. Lobinski, D. Andrey, and C. J. Blake, *Journal of Analytical Atomic Spectrometry*, **17**, 880–886, 2002.
15. B. Bouyssiere, P. Leonhard, D. Pröfrock, F. Baco, C. Lopez Garcia, S. Wilbur, and A. Prange, *Journal of Analytical Atomic Spectrometry*, **5**, 700–702, 2004.
16. D. Chen, M. Jing, and S. Wang, *Handbook of Hyphenated ICP-MS Applications*, Agilent Technologies, August 2007, www.chem.agilent.com/temp/rad5F0C7/00001681.PDF.
17. B. Avula, Y. H. Wang, M. Wang, and A. Khan, Analysis of Arsenic and Mercury Species from Botanicals and Dietary Supplements using LC-ICP-MS, *Planta Medica*, **78**, 136–144, 2012.
18. K. R. Neubauer, P. A. Perrone, W. Reuter, and R. Thomas, *Current Trends in Mass Spectroscopy*, May 2006.

19. F. McElroy, A. Mennito, E. Debrah, and R. Thomas, *Spectroscopy*, **13**(2), 42–53, 1998.

20. M. Bueno, F. Pannier, M. Potin-Gautier, and J. Darrouzes, *Determination of Organic and Inorganic Selenium Using HPLC-ICP-MS*, Agilent Technologies Application Note—5989–7073EN, 2007, www.chem.agilent.com/temp/radBA32F/00001661.PDF.

21. S. McSheehy and M. Nash, *Determination of Selenomethionine in Nutritional Supplements Using HPLC Coupled to the XSeriesII ICP-MS with CCT*, Thermo Scientific Application Note—40745, 2005, www.thermo.com/eThermo/CMA/PDFs/Articles/articlesFile_26474.pdf.

22. J. Di Bussolo, W. Reuter, L. Davidowski, and K. Neubauer, *Speciation of Five Arsenic Compounds in Urine by HPLC-ICP-MS*, PerkinElmer LAS Application Note—D-6736, 2004, http://las.perkinelmer.com/content/applicationnotes/app_speciationfivearseniccompounds.pdf.

23. M. Leist and A. Toms, *Low Level Speciation of Chromium in Drinking Waters Using LC-ICP-MS*, Varian Inc. Application Note—29, www.varianinc.com/image/vimage/docs/products/spectr/icpms/atworks/icpms29.pdf.

24. D. S. Bollinger and A. J. Schleisman, *Atomic Spectroscopy*, **20**(2), 60–63, 1999.

25. M. L. Briscoe, T. M. Ugrai, J. Creswell, and A. T. Carter, An Interlaboratory Comparison Study for the Determination of Arsenic and Arsenic Species in Rice, Kelp, and Apple Juice, *Spectroscopy Magazine*, **30**(5), 2015.

20 Overview of the ICP-MS Application Landscape

Today, there are approximately 2,000 ICP-MS installations worldwide every year, performing a wide variety of applications, from routine, high-throughput multielement analysis to trace element speciation studies with high-performance liquid chromatography.[1] As more and more laboratories invest in the technique, the list of applications is getting significantly larger and more diverse. In previous editions of this book, I have given detailed information about the various ICP-MS application sectors being carried out by the user community. However, 40 years after the commercialization of the technique, it would be almost impossible to capture all of them. The most common ones include environmental monitoring, geochemical, metallurgical, petrochemical, petroleum, food, clinical, toxocology, semiconductor, industrial, energy, agricultural, nuclear, pharmaceutical and cannabis, but every year it seems that a new market has realized the potential of its capabilities. For that reason, I would like to give a summary of the ICP-MS market and application landscape, with the objective of highlighting current trends and where the technique might be heading in the future.

As a result of the widespread use and acceptability of ICP-MS, the cost of commercial instrumentation has dramatically fallen over the past 25 years. When the technique was first introduced, $250,000 was a fairly typical amount to spend, whereas today one can purchase a quadrupole or time-of-flight (TOF) system for under $150,000. Although it can cost a great deal more to invest in magnetic-sector technology or a "triple quadrupole" collision/reaction-cell instrument, most laboratories that are looking to invest in the technique should be able to justify the purchase of an instrument without price being a major concern. One of the benefits of this kind of price erosion is that, slowly but surely, the AA and ICP-OES user community is being attracted to ICP-MS, and as a result, the technique is being used in more and more diverse application areas. Figure 20.1 shows

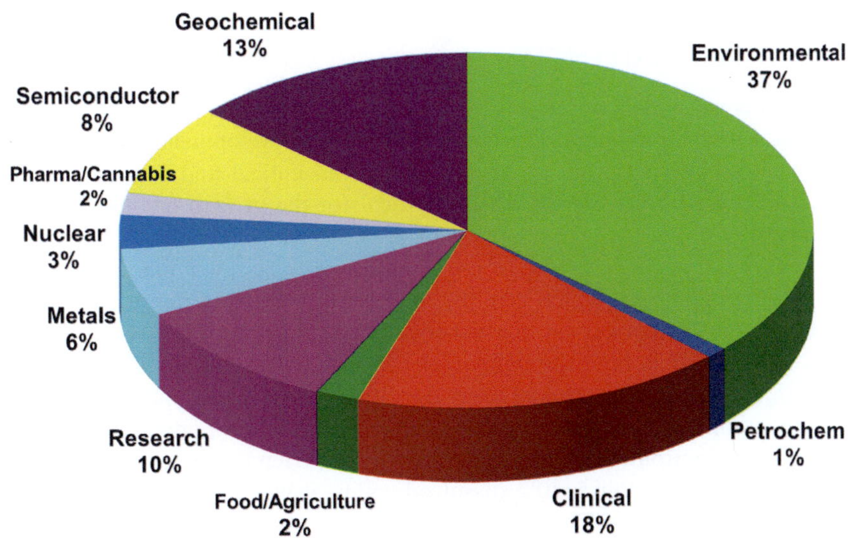

FIGURE 20.1   Worldwide ICP-MS application breakdown.

DOI: 10.1201/9781003187639-20

a percentage breakdown of the major market segments being addressed by ICP-MS on a worldwide basis.

Three points should be emphasized here. First, these data can be significantly different on a geographical or regional basis, because of factors such as a country's commitment (or lack of it) to environmental concerns or the size of a region's electronics or nuclear industry, for example. Second, many laboratories carry out more than one type of application, and as a result can be represented in more than one market segment. Finally, the research market segment has been listed as a separate category to show the instruments that are being used in an academic environment or for non-routine applications. However, many universities, federal organizations or corporate R&D groups might be using their instrumentation for research purposes in a particular application segment. For these reasons, these data should only be considered an approximation for comparison purposes.

20.1 APPLICATION CAPABILITY

Without question, extremely low detection capability, combined with increased sample throughput have been the main incentives for analytical chemists to invest in inductively coupled plasma mass spectrometry (ICP-MS). The technique is now considered a mature tool that is being used for the routine, high-throughput analysis of complex samples with very little or no sample preparation. In the early days, it was used by academic institutions and large corporate R&D laboratories. Nowadays, ICP-MS instrumentation is also being used by contract labs in the routine analysis of environmental, pharmaceutical and biomonitoring applications, among others. This has been made possible by greater automation on the sample-preparation side and seamless method-development tools, which have resulted in far less operator expertise required to run them. It is also worth pointing out that stricter regulations are requiring lower and lower detection limits (DLs), particularly in the semiconductor, biomonitoring, food and beverage market segments. In addition, for the past two to three years, the rapid growth in the legal-cannabis industry has meant cultivators and processors have to ensure the safety of consumer products by meeting strict state-based limits for contaminants like heavy metals. So the drive toward lower detection limits is just one factor, whereas the other is the analytical needs and demands of the customer. For example, the electronics industry is faced with the enormous challenge of "chasing zero", so manufacturers must measure the lowest level of elemental impurities in process chemicals and electronic devices to ensure the highest performance of their products.

20.2 ANALYTICAL CHALLENGES

As ICP-MS has become widely applicable in a variety of application areas, it has become more readily accessible to startup laboratories. Part of this is the development of turnkey methods for food, environmental, biomonitoring, geochemical, pharmaceutical and cannabis applications for inexperienced and novice users, which will be critical for the growth of this technique in these market segments.

Important considerations include ease of use, seamless software, low maintenance and the ability to handle different matrices, which give the analyst confidence in the results being generated and ensure they get the correct answer the first time and every time. The user community expects these qualities, particularly in high-throughput labs. They want to run the instrument for extended periods with minimum maintenance and short downtime while simultaneously running multiple samples, many containing high concentrations of mineral acids. It is important to emphasize that this kind of sampling environment is not really instrument friendly. When samples are being run that contain 5–10% nitric acid or 5–10% dissolved solids, it will take its toll on the instrument's sample-introduction components. Therefore, the instrument must be rugged enough to handle this kind of sample matrix, without contributing significantly to the routine maintenance of the system, while at the same time delivering high-quality data, day in and day out.

Therefore, ruggedness is critically important when designing an ICP-MS instrument. However, this is just one side of the analytical challenge. As mentioned previously, the semiconductor industry is the one application area where this kind of ruggedness is expected, but the lowest possible detection limits are also required. Just 15 years ago, 10 ppt was the guideline for elemental impurities in semiconductor process chemicals; today, the industry is requiring 1–2 ppt. This puts unique demands on the design of any instrument used for this kind of analysis.

Consequently, with the push for lower and lower detection capabilities, it is becoming a major challenge not only to develop instrumentation that has high analyte sensitivity and extremely low background noise, but also to have an ultraclean lab and sample-preparation environment that sets the stage for the detection of elements at such low levels. Both factors are absolutely critical. The trend towards ease of use has meant that operators have less time to be involved with the analytical procedure, and therefore are often not versed in the nuances of working in the ultra-trace environment and won't necessarily recognize issues that require user intervention, should they arise. For that reason, instruments have to offer maintenance reminders that mitigate potential problems before they actually happen.

20.3 MAJOR TRENDS

The increasing demand for state-of-the-art ICP-MS equipment that offers practical solutions has been mainly driven by regulatory requirements. Whatever the industry, there's a huge need for the detection of lower and lower impurities of inorganic pollutants like heavy metals. For example, many regulations require the testing of soil for toxic metals before any crops can be grown in the soil. In addition, industrial effluent must be regulated to know what is being discharged into waste streams. This is particularly the case with nanomaterials, which are finding their way into the environment. Experts have not extensively studied the long-term impact of inorganic nanoparticles on plants and animals, so going forward, industry clearly must have a much better understanding of this issue.

Another example is the mining of minerals such as titanium dioxide, which is used as a whitening agent and silicon dioxide, used for its drying and anticaking properties in the manufacture of pharmaceutical excipients, fillers and tablet binders. There are new regulations that require manufacturers to measure 24 elemental impurities at the microgram (μg) level in their products. If they do not comply with these permitted-daily-exposure (PDE) limits, they could be liable for significant fines or even the closure of their manufacturing plant. So, there are clearly many roles in which metals play a critical role in the world we live in, and, as a result, the quantification and regulation of these metals and metalloids is required. Hence the importance of using very sensitive techniques like ICP-MS as well as performance- and productivity-enhancing accessories to help us carry out those extremely low measurements.

There are two major trends to address those needs. One is at the high end, where the semiconductor and the biomonitoring markets are chasing after the lowest detection limits, being able to analyze as many inorganic analytes as possible. This is where you see significant research being carried out to develop offerings with the lowest detection limits, maximum interference reduction and optimum flexibility—in other words, the highest-performance instruments.

A separate trend involves the routine analysis of food, environmental, pharmaceutical and cannabis-related samples, for which ICP-MS sensitivity is already sufficient to meet their requirements. For these labs, the trend is toward improved robustness, less routine maintenance, more seamless integration with automated sampling solutions and ease of use, which is clearly a very different application segment from ultralow background equivalent concentrations and detection capability.

20.4 WHAT IS DRIVING ICP-MS DEVELOPMENT?

Since many academic groups are not developing hardware intended for commercialization these days, the trend in these institutions seems to be focused more on application development. This is in contrast to the 1970s and 1980s, when, for example, the fundamentals of ICP-MS research were carried out predominantly in academia. However, more recently, we have seen a shift toward ICP-MS development being carried out mostly by manufacturers, and much of the cutting-edge application work being carried out or supported by academia. A good example of this is the expansion of ICP-MS into the field of nanomaterials, and specifically detecting nanoparticles in environmental and biological samples over the past four to five years. Here there have been several excellent collaborations occurring between universities and manufacturers. In general, there is an increased interest from universities to expand into niche application areas that are not necessarily being addressed by vendors or other users, such as single-particle and single-cell analysis to monitor the metal content within single cells to better understand the mechanism of how metals such as platinum play a role in cancer-treatment research.

Another example is in the pharmaceutical industry, where USP Chapters <232>, <2232> and <233> and ICH Q3D Guidelines have opened up new markets for ICP-MS through recent regulations for elemental-impurity testing in drug products and dietary supplements. And an extension of this application is the testing of cannabis and cannabinoid products for heavy metal contaminants, such as lead (Pb), cadmium (Cd), arsenic (As) and mercury (Hg), using regulations initiated by the pharmaceutical industry. However, unlike the manufacture of drugs, there is no real tracking of the sources of heavy metals in cannabis, so it is not uncommon to see product recalls because of elevated levels of elemental contaminants. As a result of this, it has become critical to monitor the cannabis plants and products for a wider panel of elemental contaminants that are worthy of consideration using ICP-MS. Again, this is clear indication that some application sectors are being stimulated by the demands of the industry and supported and driven by the manufacturers, as opposed to coming from university-directed research.

20.5 FUTURE DIRECTION

Instrument vendors are firmly committed to the ICP-MS marketplace, with sales growing at approximately 6% per year compared to about 3% for ICP-OES. If we look at the main market segments, such as semiconductor, biomonitoring, environmental, geological, industrial, food, pharmaceutical and cannabis, it is clear that demand for ICP-MS is still growing rapidly. This growth is achieved mainly by collaborating with key market-driven customers to ensure that they are addressing their future needs, and using this feedback to help them design innovative products, whether related to application development, hardware improvements or software enhancements.

The aim of this chapter has therefore been to give the reader a snapshot of the ICP-MS application landscape, and a better understanding of why it is still the fastest-growing trace element technique available today. If there is one common theme that runs through many of these applications, it is the unparalleled detection limits the technique has to offer, for both the total and speciated form of the element. When this is combined with its rapid multielement characteristics, isotopic measurement capability, freedom from interferences and ease of use, it is clear that ICP-MS will continue to dominate the application landscape.

FURTHER READING

1. Global ICP-MS System Market Research Report 2022, *Industry Research, SKU ID: QYR-21299553*, July 12, 2022, www.industryresearch.biz/global-icp-ms-system-market-21299553.

21 Fundamental Principles and Applications of Atomic Absorption and Atomic Fluorescence

Atomic absorption spectrometry (AAS) is the most mature of all the atomic spectroscopic techniques, having been developed and commercialized in the late 1950s. There are literally hundreds of thousands of instruments being used for many diverse and varied applications across the globe. Atomic fluorescence spectrometry (AFS) is a complementary technique with a recent development of commercially available systems on the market for the determination of Pb, Cd, As and Hg. AAS is composed of predominantly single-element techniques, which limits its use for multielement analysis. However, AFS can offer multielement capability, by surrounding the atomization source with a number of element-specific photon sources.

When a small suite of elements is required, or the sample workload is not so high, AAS can be a very cost-effective technique to use. Conventional atomic absorption typically uses an air (or nitrous oxide) acetylene flame as the atomization source, but by using a graphite furnace in place of a flame, lower detection limits can be achieved, in some cases as well as ICP-MS. Additionally, dedicated AFS systems using vapor-generation atomization can be used for measuring elements that readily form volatile hydrides/cold vapor (Pb, As, Cd, Hg Sb, Se, Sn, Te). This chapter will present the fundamental principles of both atomic absorption and atomic fluorescence, and summarize their strengths and weaknesses with a particular emphasis on their use for the determination of the classic heavy metals, Pb, Cd, As and Hg.

In the world of atomic spectroscopy instrumentation, there are four techniques used quantitatively: atomic absorption, atomic emission, mass spectrometry and atomic fluorescence. Unlike the plasma techniques, atomic absorption and fluorescence require element specific light sources for the energy required to cause atoms of the selected elements in the light path to go through excitation.

The spectrometers for atomic absorption can support a variety of atomizers: flames, electrically heated furnaces and vapor generators. In terms of the instruments, flame atomic absorption is the oldest of the techniques. With a team led by Sir Alan Walsh, the current form of the flame atomic absorption instrument was developed in the 1950s. This work revolutionized the quantitative determination of more than 65 elements, and led to the variations of the instruments in use today.

Unlike plasmas, in most cases, the temperature of the atomizer does not supply sufficient energy to cause the resulting atoms to go through excitation. Instead, element-specific lamps are required to provide this extra energy. The lamp emits light energy at the wavelengths specific to the element, and that energy is absorbed if there are atoms of that element in the light path. This absorption causes the atoms to go through excitation. The excited state is not a stable state for the atoms so they also go through decay and light energy is emitted. The process that occurs in the flame is shown in Figure 21.1.

The amount of light absorbed is defined by comparing the light intensity before the atomizer, I_o, to the amount of light intensity after the atomizer, I_f. The relationship is shown in Figure 21.2. As the number of atoms increases due to increased concentration in the sample, the amount of light absorbed increases. It is this difference in the light intensity absorbed from the lamp that is actually being measured with each reading. The amount of light absorbed by solutions of known

DOI: 10.1201/9781003187639-21

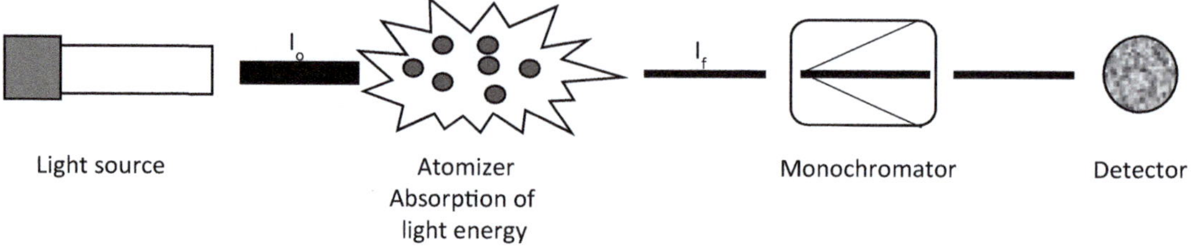

FIGURE 21.1   Basic components of atomic absorption.

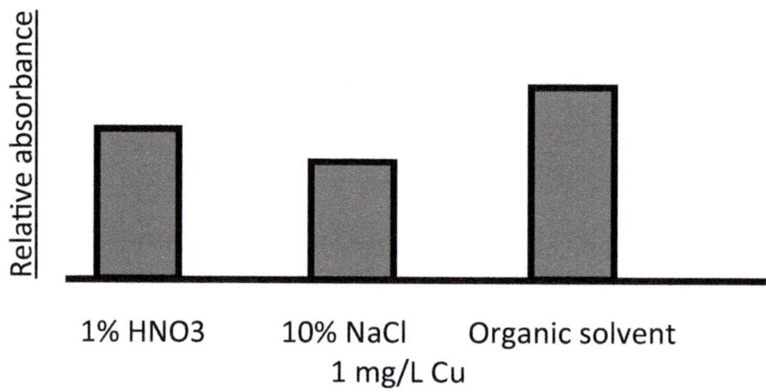

FIGURE 21.2   Inaccuracy produced by transport interferences in flame AAS.

concentrations can be compared to the amount of light absorbed by unknown solutions and hence the concentration of the analyte in real samples can be determined.

Transmittance is defined as a ratio of the lamp intensity after the atomizer, I_f, to the lamp intensity before the atomizer, I_o, or I_f/I_o, and

$$\%T = 100 * I_f/I_o.$$

Therefore, the percent absorption is

$$\% A = 100 - \%T.$$

Absorbance is defined as the log (I_o/I_f) and is a convenient form of measurement. This term follows a linear relationship with concentration according to Beer's law:

$$\text{Absorbance} = (\text{concentration, c}) * (\text{an absorption coefficient, a}) * (\text{path length, b}).$$

The absorption coefficient is dependent on the element, the wavelength chosen, the instrument configuration and operational parameters. The path length is the length of the atomizer.

While Beer's law defines a linear relationship between concentration and absorbance, when the absorbance becomes too high, the relationship becomes nonlinear. The point at which this nonlinearity occurs is specific to the element and the wavelength at which the measurement is made. As long as the absorbance increases with the increase in concentration, a nonlinear curve fit can be used for calibration.

21.1   FLAME AAS

As with plasma techniques, flame AAS (FLAAS), requires the sample in liquid form and utilizes nebulizers and spray chambers. The nebulizer mixes the liquid sample with air to form an aerosol mist that is sprayed into the spray chamber. A small part of the aerosol is dried as it is transported into the flame via a burner head aligned in the light path. The heat of the flame is used primarily to break molecular bonds. The most commonly used flames are air-acetylene and nitrous oxide-acetylene. A few elements (such as Cs, K and Na) will excite merely by the heat of the flame and not require a lamp. In these cases, the flame atomic emission signal can be read as the excited atoms return to ground state. But most elements are determined in the atomic absorption mode.

21.2   ADVANTAGES OF FLAAS

There are many advantages to using flame atomic absorption. The cost of the instrumentation and consumables required are much less than the other atomic spectroscopy techniques. Consumables consist primarily of the lamps, which have to be replaced periodically, and the gas supplies. Nebulizers, spray chambers and burner heads are also considered consumables, but with care they can be used for many years. Flame AAS is also a technique that is easy to use and does not require a great deal of operator skill to produce accurate results. There is minimal maintenance of the hardware required.

While there are some multielement atomic absorption instruments on the market, the number of elements that can be determined in one method is generally limited. Most AA spectrometers can determine only one element at a time. While this is considered by many to be a disadvantage, analysis by flame AAS is very quick. Literally, per element, results for each sample can be obtained in a matter of a few seconds. On most modern instruments, progressing from one element to the next can be automated with the use of autosamplers. Therefore, a great deal of data for multiple elements can be obtained in a relatively short time, with minimal analyst interaction. It is reasonable to expect 125–150 results per hour.

For the laboratory that needs to determine high concentrations, FLAAS is very versatile. By choosing a less sensitive wavelength or by merely turning the angle of the burner head (thus, reducing the path length) the analyst can easily calibrate to and determine high concentrations.

On the lower concentration end, there are many elements that can produce acceptable IDL values (10 μg/L or less): Ag, Be, Ca, Cr, Cu, K, Mg, Na, to name a few. Actual reporting limits will vary with sample matrix, as well as the instrumental conditions.

As with all the atomic spectroscopy techniques, there are interferences associated with flame AAS. These are well understood and documented so that the technique has become "cookbook" in nature. It is up to the analyst to pay attention to the information provided and follow the appropriate guidelines.

21.3   FLAAS INTERFERENCES AND THEIR CONTROL

In atomic absorption, interferences are divided into two main categories: non-spectral and spectral. Non-spectral interferences have to do with how the atom population is achieved, and include transport (or physical), chemical and ionization. Spectral interferences are those that contribute to the element signal producing an incorrect result and are primarily background absorption and emission interferences.

TRANSPORT (PHYSICAL) INTERFERENCES occur when the sample matrix is sufficiently different from that of the standards used for calibration, causing different amount of sample per time being transported to the flame. This can result in different absorbance readings for the same concentration of an analyte in sample versus the calibration solutions, as shown in Figure 21.3. This is the same affect we see with the plasma techniques, but are typically less severe in FLAAS.

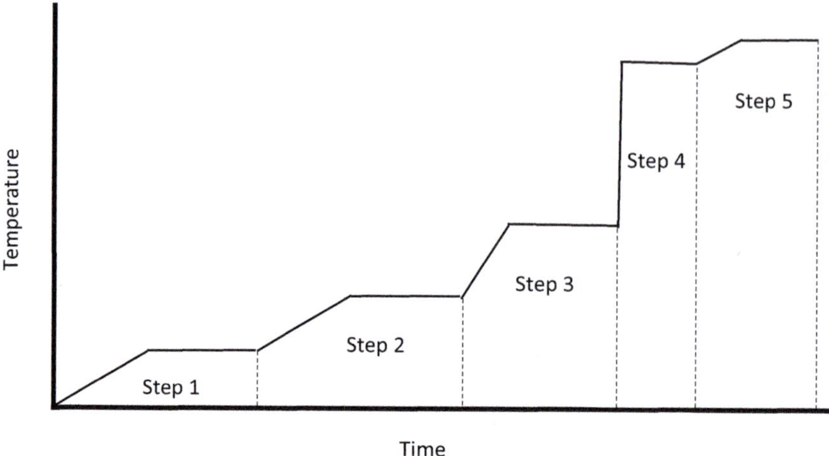

FIGURE 21.3   Typical furnace program cycle. Steps 1 and 2 dry the sample, step 3 is pyrolysis, step 4 is atomization and the atomic absorption signal is read and step 5 is the cleanout step.

There are two ways of alleviating transport issues when using FLAAS. If possible, match the matrix of the calibration solutions to those of the samples. For example, if the sample is an organic solvent (such as xylenes), then prepare the standards from an organometallic stock in the same solvent.

Matrix matching is not always possible or reasonable. The other option is to do a method of standard additions, where the calibration standards are prepared in the presence of the sample matrix. The preparation for this is more time consuming and does require that all the absorbances are in the linear range. Transport interferences are sample specific and will affect all the elements determined for that matrix.

CHEMICAL INTERFERENCES occur when the heat of the flame is not sufficient to break the analyte molecular bonds to create the free atom population. The most commonly used flame, air-acetylene, reaches temperatures in the range of 2,100 °C to 2,400 °C. Unfortunately, not all chemical bonds are easily broken at these temperatures, and low results are obtained.

One example is the determination of Ca, which forms strong molecular bonds with a number of sample constituents. One of these strong bonds is $Ca_3(PO_4)_2$. These bonds are not easily broken by the air-acetylene flame, so the atom population is not formed. However, a bond that is easily broken in the air-acetylene flame is $CaCl_2$. So for this interference, the $Ca_3(PO_4)_2$ can be converted to $CaCl_2$ by adding an excess of a *releasing reagent* ($LaCl_3$). This is added to all solutions (calibration blanks and standards, samples, QC solutions, etc.) to be analyzed to convert the form of the Ca to the form where the molecular bonds are broken to create the free atoms:

$$Ca_3(PO_4)_2 + 2LaCl_3 \rightarrow 3CaCl_2 + 2LaPO_4$$

There are many commonly determined elements that form these strong molecular bonds: Al, Ba, Ca, Ge, Mg, Mo, Si, Sn, V, Zr. In most cases, addition of the releasing reagent does not change the chemistry of the element sufficiently. So another option is to use a hotter flame, nitrous oxide-acetylene. This flame reaches higher temperatures in the range of 2,600 °C to 2,800 °C, and is sufficient to break the stronger bonds.

The use of the hotter flame can lead to another interference once the molecular bonds are broken. With the heat of the nitrous oxide-acetylene flame, there can be sufficient energy to cause some of the free atoms to go through ionization, reducing the atom population and leading to incorrect results. The solution for this type of interference is to add an *ionization buffer*, an excess of an element that is easily ionized, to all solutions to be analyzed. The elements used for the buffer generally will also ionize in the air-acetylene flame and include Cs, K, La, Li and Na.

Chemical and ionization interferences are element specific and well documented for flame AAS. Even the new user should have no trouble compensating for them by using the information provided by the instrument manufacturer.

SPECTRAL INTERFERENCES: Due to the specificity of the bandwidth setting for each element at its wavelengths, spectral interferences in terms of spectral overlaps are generally not a problem in atomic absorption. The most common type of spectral interference in atomic absorption is ***background absorption***, where something other than the analyte prevents the light from the lamp getting to the detector. Most often in flame AAS, this is undissociated molecules and tiny particulates that can occur in the flame when analyzing samples with high dissolved solids. This is not a major problem in flame AAS unless 1) the sample matrix has more than 1% TDS, 2) the element wavelength is below 250 nm and 3) the concentration of the element determined is less than 1 mg/L. An example of the necessity of using background correction is the determination of Zn, at 214.9 nm, in seawater samples. Naturally occurring Zn in seawater is typically < 50 μg/L. The absorbance due to background may be as much as 30 μg/L. Without background correction, the matrix would contribute to a large portion of the reading.

When the background readings do interfere with obtaining accurate analyte readings, most instruments have background-correction systems. The most common of these for FLAAS is continuum-source background correction. This correction technique utilizes a secondary lamp, usually deuterium (D2), installed in the instrument. The D2 lamp emits continuously over a broad wavelength range (180–350 nm) rather than at discreet wavelengths, as do the element lamps. The instrument then monitors intensities from both the D2 lamp and the element lamp. Absorption of the D2 lamp is essentially background only. Absorption of the element lamp is background and analyte. The instrument electronics then perform a correction as follows:

Element lamp absorbance – D2 lamp absorbance = Corrected Analyte Absorbance
(background + analyte) – (background only)

Background readings, if present, are generally small in flame AAS, and continuum source typically works well. Two other types of correction, Zeeman and Smith-Hieftje, can also be used but are more commonly used for graphite-furnace analysis.

Another potential spectral interference comes from the brightness of the flame when determining elements at visible wavelengths, particularly when the sample matrix causes an increase in the brightness as well. An example where this can be a problem is the determination of barium in a high-sodium sample. In order to break molecular bonds, barium requires a nitrous oxide-acetylene flame, which is already bright. The most sensitive wavelength for barium for flame AAS is at 553 nm, the visible region. And high Na adds additional brightness to the flame. A combination of these three factors can "blind" the detector, making the analysis difficult. Diluting the sample, narrowing the bandwidth or choosing a different wavelength if possible can help with this particular problem.

21.4 DISADVANTAGES OF FLAAS

One potential disadvantage of flame AAS is the volume of sample required for the determination of the elements of interest. While much depends on a number of different factors (shown shortly), uptake volumes are usually 2–7 mL of sample per minute.

1. Number of replicate readings desired for a mean value
2. Read-time settings
3. Type of nebulizer and spray chamber
4. Whether an autosampler is used

So for the lab that is sample limited, this could pose a problem when determining several elements.

Another disadvantage is the flame itself. While instruments today have many safety interlocks, it is still an open flame using flammable gases and should never be left completely unattended when the flame is on. There should always be an analyst in the area that can "keep an eye" on the instrument during analysis, even when using an autosampler.

The greatest limitation of flame AAS for many of the elements of interest is poor sensitivity. As reporting levels are being forced to go lower, flame AAS cannot achieve the levels required for many elements. A number of them that require reporting levels lower than it is possible to achieve with FLAAS include, but are not limited to: As, Hg, Pb, Sb and Se. To achieve lower levels of quantitation, the flame portion of the instrument can be replaced with alternate ways of creating and measuring the atomic signal. Two of these ways are to use graphite furnace or vapor generation.

21.5 GRAPHITE-FURNACE AAS

When lower levels of quantitation are required, one option is to replace the flame with an electrically heated furnace, an alternate technique developed in the 1960s. Because of the nature of the furnace materials, this is commonly known as graphite-furnace AAS (GFAAS), sometimes known as electro thermal atomization (ETA). Some instruments sold today are manufactured for furnace analysis only and some for flame analysis only. Other instruments have capabilities for both, though only one at a time can be used.

When using GFAAS, sample is pipetted into a graphite tube through an opening, the dosing hole of the tube, and then heated through a series of programmable steps to: dry the sample solvent, eliminate some of the sample matrix, without losing any of the analyte (the pyrolysis step), break the molecular bonds so that the atomic absorption signal can be read (the atomization step) and clean out any remaining sample residue. A typical furnace cycle is shown in Figure 21.4. There are many designs of furnace systems and the associated graphite components. The steps can vary not only with design, but with the element determined and the sample matrix analyzed.

It is important that the sample be dried completely with no boiling or spattering. While in some cases there may be only one dry step, a two-step drying program can insure complete drying of the sample. The times and temperatures used are dependent on the sample matrix and the furnace design. The furnace is ramped up to these temperatures and then held to insure dryness. Argon flows into the ends of the graphite tube to expedite the elimination of the moisture out the dosing hole of the tube.

The pyrolysis step has also been called ash, char and pretreatment. During pyrolysis, the goal is to eliminate as much of the sample matrix as possible before the atomization step, without losing any of the element. Again there is a ramping up to the desired temperature. Times and temperatures are dependent on the sample matrix and the element determined. Argon continues to flow into the tube ends to eliminate the excess matrix by pushing it out the dosing hole.

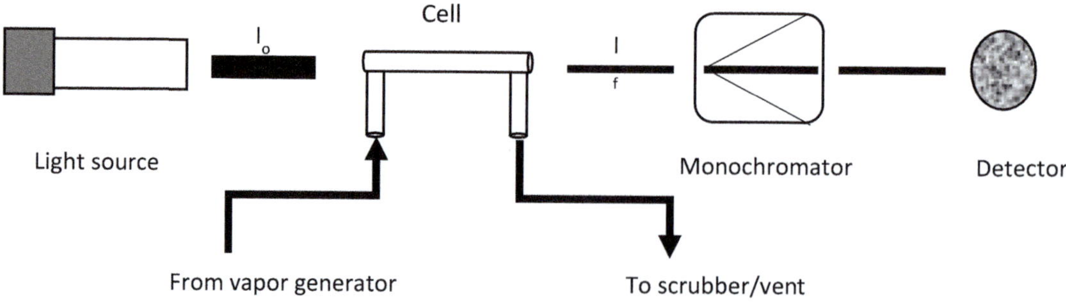

FIGURE 21.4 Basic configuration of vapor-generation atomic absorption.

The atomization temperature must be sufficient to break the molecular bonds to form the free atom population. To minimize interference effects, it is important that the element volatilize into a thermally stable environment. So in this step, to heat the tube as quickly as possible, there is no ramping time to the set temperature. Typically, the gas flow is stopped in the furnace during atomization to give the longest residence time, and the highest signal, possible.

The last step, cleanout, is to eliminate any remaining matrix (and in some cases analyte) from the graphite tube before injection of the next sample. Again, argon flows in the tube during this step.

In addition to lamps, consumables now include the graphite components, particularly the graphite tube (cuvette) which degrades with repeated cycles and must be changed on a routine basis to insure consistent results. High temperatures, particularly held for extended times, high acid concentrations and high dissolved solids in the samples are contributing factors for how fast the graphite tube degrades. Depending on these parameters, the tube may last for only a few hundred to over 1,000 cycles. The contact cylinders housing the graphite tube (and also made of graphite) will also have to be changed periodically, but much less frequently than the tube itself.

It is imperative when running a furnace analysis that each step of the analysis from injection to injection be as reproducible as possible. This includes sample deposition. Autosamplers in graphite furnace are not just a convenience, but a necessity for good reproducibility of results.

21.6 ADVANTAGES OF GFAAS

Because all of the analyte in the volume pipetted into the tube is atomized and the atoms are retained in the tube for an extended time, sensitivity and reportable levels are very good with graphite furnace. Detection limits are typically 100 to 1,000 times lower than with flame AAS.

For applications where the sample size is limited, another advantage is the small sample volumes required for analysis. Typical sample volumes range from 1 to 100 μL per injection. When less sensitivity is required, less sample volume can be used. When more sensitivity is required, more volume can be used.

With GFAAS, no flammable gases are used, but rather argon. So this is an analysis that can be left completely unattended. To use their instruments to full advantage, many labs with instruments that have both flame and furnace capabilities choose to run their flame analysis during the day and their furnace analysis overnight.

21.7 GFAAS INTERFERENCES AND THEIR CONTROL

For very simple matrices, method development for graphite-furnace analysis can be relatively simple. But as the matrix becomes more complicated with higher dissolved solids in the samples, method development can be time consuming. This is because, unlike flame AAS, interference effects in graphite furnace can vary with each sample matrix and the element determined.

Interferences again are separated into non-spectral, having to do with how the atom population is formed, and spectral, where something other than the analyte prevents the light from the lamp getting to the detector. The latter, unlike in FLAAS, can be very severe in furnace analysis.

The non-spectral interferences, which affect how the atom population is achieved, are typically controlled using a set of conditions that allows the accurate analysis of even very complicated sample matrices. The first of these conditions is to use a good-quality graphite tube. The tubes are generally coated with pyrolytic graphite to prevent interaction of the element with the graphite surface. Poor quality of graphite leads to contamination issues, poor reproducibility and short tube life.

It is necessary to delay the volatilization (breaking molecular bonds) of the analyte until the environment in the tube has become thermally stable. One way to accomplish this is to pipette the sample onto a secondary surface, a platform, inside the tube. The platform surface heats more slowly than the tube wall, allowing the temperature inside the tube to stabilize before volatilization of the analyte. Another style tube is thicker in the middle than at the ends. The thicker portion will again heat more slowly.

Two modes of signal collection can be used: peak height and peak area. While peak height certainly is indicative of the concentration of atoms in the vapor phase during the atomization step, it is also very dependent on the rate at which the molecular bonds are broken. The element may be bound in different molecular states as compared to the calibration solution, causing the molecular bonds to break more quickly or more slowly than the standards. Therefore, it is better to use peak area, which is rate independent, for furnace determinations.

Many elements are volatile, their molecular bonds broken at relatively low temperatures. Arsenic, selenium, cadmium and lead are just a few of these. In order to use a pyrolysis temperature high enough to eliminate troublesome matrix components during this step, it becomes necessary to add a "modifier" to stabilize these elements so that no losses of the element occur before the atomization step.[1] Modifiers have been used since the early days of graphite furnaces, and most elements will require them, except when analyzing the simplest of solutions. Most modifiers and the amount of modifier used are dependent on the element determined, but some may also depend on the sample matrix analyzed. The most common analyte-stabilizing modifiers currently used are Pd, and Pd mixed with $Mg(NO_3)_2$, as these are effective for a variety of elements typically determined by GFAAS.[2] Some typical pyrolysis temperatures with and without a modifier are shown in Table 21.1.[3]

There are also modifiers that can change the nature of the sample matrix, generally to make it more volatile so that it can be eliminated at a lower temperature. One example is the use of NH_4NO_3 as a modifier to react with samples with a high salt concentration. NaCl has a melting point of 1,400 °C, so it is more thermally stable than many elements and difficult to eliminate without analyte loss. When NH_4NO_3 is added in excess to all solutions analyzed, it will react with the NaCl to produce NH_4Cl and $NaNO_3$, both of which have melting points below 400 °C. The matrix now is less stable than the elements determined and can be eliminated without analyte loss. Note that it may still be necessary to stabilize the analyte, so two modifiers may be required: one to stabilize the analyte and the other to make the matrix more volatile.

Spectral interferences fall into two categories: emission intensity from the tube wall and background absorption. Emission intensity is minimized by making sure the furnace is aligned properly in the light path. As the analyte signal is increased using graphite furnace, so the background signal also increases. Therefore, background correction is always used with furnace analysis. While continuum-source correction can be used as it is with flame AAS, there are limitations to its effectiveness for GFAAS. With continuum source, there is not full wavelength coverage using D2 lamps.

TABLE 21.1

Recommended Pyrolysis Temperatures with and Without a Modifier for Some Elements Note: These temperatures are for standards and the addition of a real sample matrix can change the recommendation. These temperatures also vary with instrument design and are shown for comparisons only.[4,5]

Element	Pyrolysis °C no modifier used	Pyrolysis °C Pd/Mg(NO$_3$)$_2$ modifier
Arsenic	200	1,200
Cadmium	400	500
Chromium	1,200	1,500
Copper	900	1,200
Lead	450	1,200
Selenium	200	1,100

Continuum source also does not correct accurately for cases of structured background, and it cannot correct for very high background signals (greater than absorbances of about 1.7).

As samples analyzed by graphite furnace have increased in complexity, instrument manufacturers offer more efficient background correction systems to be able to handle samples that produce complicated background signals. The Smith-Hieftje type of correction is based on the principle that as the lamp is pulsed to very high currents, there is a loss of analyte sensitivity due to self-absorption. Readings are taken during atomization as the lamp is pulsed between normal and high currents and the following correction is made:

$$\text{Normal Current} - \text{High Current} = \text{Corrected Analyte Signal}$$
$$(\text{Analyte} + \text{Background}) - (\text{Background only})$$

This type of correction does not require any additional specialized equipment, as only the element lamp is involved. Of course, pulsing the lamp to high currents does affect lamp life. Another limitation of this type of correction is that not all elements go through self-absorption as efficiently as others, and there can be a significant loss of sensitivity. As measurements are made slightly off the elements' exact wavelength, it also does not correct well for cases of structured background within the bandpass.

The third type of background correction, Zeeman, is based on the principle that when atomic spectra are placed in a magnetic field, they become split into three or more π and σ components, while molecular structures remain unchanged. To date, this is the most effective form of background correction for atomic absorption instruments, as it can correct for even structured background within the bandpass. The atomizer is placed in a strong magnetic field and quickly pulsed on and off. When the magnet is off, light is absorbed by both the element and background. When the magnet is on, there is only background absorption. Correction is made as follows:

$$\text{Magnet Off} - \text{Magnet On} = \text{Corrected Analyte Signal}$$
$$(\text{Analyte} + \text{Background}) - (\text{Background only})$$

This type of correction requires the addition of a strong magnet and the electronics to control it, adding additional cost to the instrumentation. There is some loss of sensitivity due to elemental splitting patterns. In most cases this is minor, but in a few cases it is significant, as in the case of Cu, where the loss is more than 40%.

21.8 ADVANTAGES OF GFAAS

Because all of the analyte in the volume pipetted into the tube is atomized and the atoms are retained in the tube for an extended time, sensitivity and reportable levels are very good with graphite furnace. Detection limits are typically 100 to 1,000 times lower than with flame AAS.

For applications where the sample size is limited, another advantage is the small sample volumes required for analysis. Typical sample volumes range from 1 to 100 μL per injection. When less sensitivity is required, less sample volume can be used. When more sensitivity is required, more volume can be used.

With GFAAS, no flammable gases are used, but rather argon. So this is an analysis that can be left completely unattended. To use their instruments to full advantage, many labs with instruments that have both flame and furnace capabilities choose to run their flame analysis during the day and their furnace analysis overnight.

Another advantage of GFAAS is the ability to determine solid samples directly, without dissolution.[6] While a weighed portion of the solid can be inserted into the furnace in a variety of ways, the easiest way is to mix the solid with a liquid portion and pipette directly in the form of a "slurry".

The advantage of this method of introducing the solid into the furnace is that calibration is achieved with simple aqueous standards.

21.9   DISADVANTAGES OF GFAAS

A disadvantage of GFAAS can be the time required for method development. While manufacturers give a set of conditions to be used for each element, unless the samples analyzed have very low dissolved solids, these are just starting conditions. The times and temperatures used for drying the samples and the pyrolysis step generally have to be optimized for the matrix. The modifier strength, or even which modifier to use, may have to be changed. Less frequent changes for the atomization step are required, but again this is always dependent on the sample matrix.

Probably the greatest disadvantage of the graphite-furnace technique is the time it takes to do an analysis once the method has been developed. In contrast to flame, this is a very slow technique. The cycle time from reading to reading can range from one to five minutes, dependent primarily on the sample matrix. Most of this time is spent in the dry and pyrolysis steps. The greater the volume pipetted and the higher the dissolved solids, the longer these steps become. If the analyst has only one or two elements to determine with few samples to be analyzed, perhaps this is not an issue. But when eight to ten elements (or more) have to be determined, or there are many samples to be analyzed, this is very time consuming. Many laboratories that have this situation may have several furnace systems running at the same time.

21.10   VAPOR-GENERATION AAS

In addition to using graphite furnace to improve detection limits, for some elements there is also vapor generation. This falls into two categories, cold-vapor and hydride generation. In both cases, the element of interest is reacted with a strong reducing reagent and then stripped into a vapor phase (usually air or argon). The vapor is swept through a cell (glass or quartz) in the light path and the atomic absorption measurement is made using peak height or peak area mode as shown in Figure 21.5.

Cold vapor is specific for one element: mercury. It is the only element that can exist as free ground-state atoms at room temperature. The reducing reagent, either stannous chloride or sodium borohydride, is reacted with the sample. The gas is then bubbled through this solution to release the free mercury atoms into the gas phase, which is then swept into the cell and the atomic absorption signal is read.

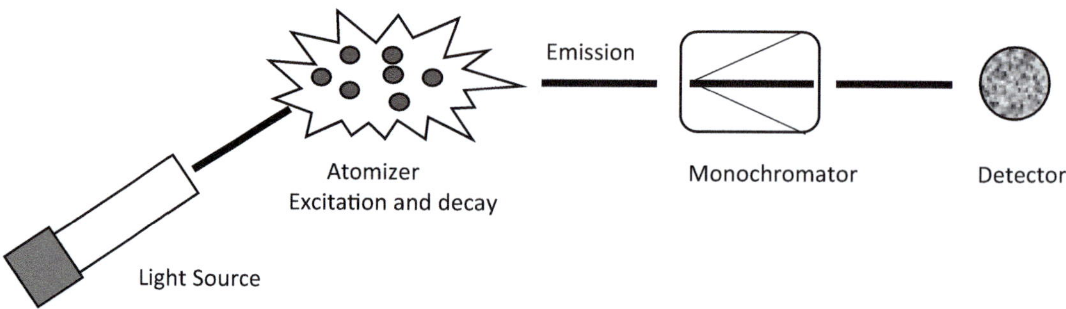

FIGURE 21.5   Basic components of atomic fluorescence.

21.11 ADVANTAGES OF COLD-VAPOR AAS

As flame AAS and even graphite furnace are not very sensitive for mercury, this technique allows the analyst to achieve reportable levels of approximately 0.02 μg/L. This can be improved further by sweeping the mercury-rich vapor over gold or gold-platinum wire or gauze. The Hg bonds to the gold, thereby trapping and concentrating the mercury. The gold is then heated to release a pre-concentrated amount of mercury. While early systems were very manual, this type of analysis can now be automated.

21.12 DISADVANTAGES OF COLD-VAPOR AAS

The obvious limitation of this is that it is only applicable to the determination of mercury since it is the only element to release free atoms at room temperature. It also is necessary to have the vapor-generation system, an additional expense.

21.13 HYDRIDE-GENERATION AAS

This technique is similar to cold vapor in that the element of interest is reacted with a reductant, typically sodium borohydride, then stripped into the gas phase. The difference is that the element is not released as free atoms but rather bound in a hydride bond. Therefore, the quartz cell through which the vapor will travel must be heated to break these bonds. Heating of the cell can be achieved with the flame portion of the instrument or the cell can be inserted into an electrically heated device.

21.14 ADVANTAGES OF HYDRIDE-GENERATION AAS

The elements that can be determined using this technique are: As, Bi, Ge, Pb, Sb, Se, Sn and Te. These elements can also be determined using graphite furnace with similar detection limits. However, samples with high dissolved solids are often challenging to eliminate the interferences they produce without losing some of the analyte during the furnace pyrolysis step. Vapor generation separates the matrix from the analyte to produce a cleaner analysis. Table 21.2 lists a comparison of potential flame, furnace and vapor generation IDL values. These values are shown for comparison only and are dependent on instrument and instrumental parameters.

TABLE 21.2

Instrument Detection Limit Comparisons (μg/L)

Element	FLAAS IDL µg/L	GFAAS IDL µg/L	Vapor Generation IDL μ g/L
Arsenic	150	0.05	0.03
Bismuth	50	0.05	0.03
Mercury	300	0.6	0.009
Antimony	45	0.05	0.15
Selenium	100	0.05	0.03
Tellurium	30	0.1	0.03

Data from PerkinElmer "Atomic Spectroscopy: A Guide to Selecting the Appropriate Technique and System"[7].

21.15 DISADVANTAGES OF HYDRIDE-GENERATION AAS

Again, the most obvious limitation is the limited number of elements that can be determined using hydride generation. In addition, results for each element are very dependent on the valence state of the element. Therefore, proper preparation of the samples is critical. Unfortunately, the same preparation does not work for all the hydride elements, so different preps may be needed for the different elements determined.

21.16 HYPHENATED TECHNIQUES

Atomic absorption can also be used by coupling more than one technique. To achieve lower limits of quantitation using flame AAS, samples can be first run through a metal chelating column to pre-concentrate elements of interest as the bulk of the matrix passes through. The pre-concentrated elements can then be stripped from the column with a mobile phase and introduced for flame analysis. The degree of improved limits of quantitation is dependent on the total volume passed through the column and the capacity of the column resin.

Graphite furnace can be coupled with vapor generation to improve reportable concentrations. This can be particularly advantageous when the sample matrix is very high in dissolved solids and difficult to eliminate in the pyrolysis step, resulting in overwhelming background signals during the atomization step. In this combined technique, the vapor that has stripped the analyte from the matrix is collected, "trapped", in the graphite tube and can be pre-concentrated in this fashion. Once the desired volume of vapor has been collected, the furnace is raised to an appropriate atomization temperature to drive it off the tube surface and breaking molecular bonds and the signal is read.

21.17 ATOMIC FLUORESCENCE

Another technique in atomic spectroscopy is atomic fluorescence. It shares properties with atomic absorption in that it has a light source, atomizer, monochromator and detector. The element specific energy from the light source is absorbed by the ground-state atoms, causing them to be excited. Rather than determining the amount of light being absorbed by the element in the atomizer, as the excited atoms return to ground state, they emit photons of light and this emission, or fluorescence, resulting from the decay is measured. As shown in Figure 21.6, the element-specific light source is at an angle so as not to interfere with the fluorescence process. To carry out multielement analysis using AFS, many lamps are positioned around the atomization/excitation source.

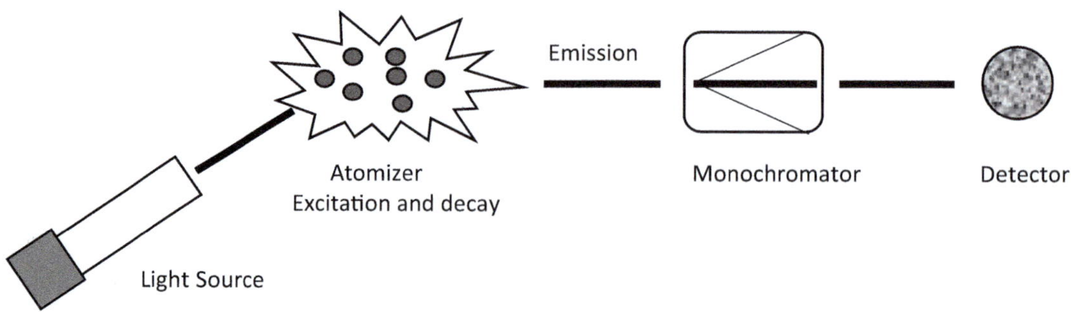

FIGURE 21.6 Basic components of atomic fluorescence.

TABLE 21.3

IDL Values Using Vapor Generation with Atomic Fluorescence, $\mu g/L$[8,9]

Elements	IDL (μ g/L)
As, Bi, Pb, Sb, Se, Sn, Te	0.01
Cd, Hg	0.001
Ge	0.05
Zn	1

21.18   ADVANTAGES AND DISADVANTAGES OF AFS

Background signals in AFS are very low, resulting in improved sensitivity as compared to absorption. However, the fluorescence measurement can be quenched by many molecular forms, so separation of the matrix from the analyte may be desirable. Commercially available vapor-generation AFS systems have been used successfully for the analysis of a variety of sample types, including water, soil, plants, pharmaceuticals, food, biologicals, petrochemicals and cannabis.[8,9] However, it's limited to the determination of those elements which work well using vapor generation such as As, Bi, Cd, Ge, Hg, Pb, Sb, Se, Sn, Te and Zn. Table 21.3 reflects typical average instrument detection limit (IDL) values that have been obtained using commercially available AFS systems. Actual performance will be dependent on lab conditions, sample type, reagents and instrument setup.

21.19   FINAL THOUGHTS

Atomic absorption and atomic fluorescence are both viable options for measuring heavy metals in any sample matrix, particularly for labs that don't have a high sample workload. It's unlikely that flame atomic absorption has the sensitivity for most state-based limits, but graphite-furnace and hydride-generation AA would offer the required detection capability and could be very attractive options. Furthermore, a dedicated AFS combined with a vapor-generation system could also be a viable option for labs with a limited budget. However, it should be emphasized that if the suite of heavy metal requirements increases due to workload or stricter regulations, these techniques would be limited in their elemental range.

FURTHER READING

1. R. Ediger, Atomic Absorption Analysis with the Graphite Furnace Using Matrix Modification, *Atomic Absorption Newsletter*, **14**, 127–145, 1975.
2. L. M. Voth-Beach and D. E. Shrader, Investigations of a Reduced Palladium Chemical Modifier for Graphite Furnace Atomic Absorption Spectrometry, *JAAS*, **2**, 45–50, 1987.
3. G. Schlemmer and B. Welz, Palladium and Magnesium Nitrates, a More Universal Modifier for Graphite Furnace Atomic Absorption Spectrometry, *Spectrochimica Acta Part B*, **41**(11), 1157–1165, 1986.
4. B. Welz, G. Schlemmer, and J. R. Mudakavi, Palladium Nitrate-Magnesium Nitrate Modifier for Graphite Furnace Atomic Absorption Spectrometry. Part 1. Determination of Arsenic, Antimony, Selenium and Thallium in Airborne Particulate Matter, *JAAS*, **3**, 93–97, 1988.
5. B. Welz, G. Schlemmer, and J. R. Mudakavi, Palladium Nitrate-Magnesium Nitrate Modifier for Graphite Furnace Atomic Absorption Spectrometry. Part 2. Determination of Arsenic, Cadmium, Copper, Manganese, Lead, Antimony, Selenium and Thallium in Water, *JAAS*, **3**, 695–701, 1988.
6. N. J. Miller-Ihli, Solids Analysis By GFAAS, *Analytical Chemistry*, **64**(20), 964A–968A, 1992.
7. *Guide to Techniques and Applications*, PerkinElmer Inc, www.perkinelmer.com/PDFs/Downloads/BRO_WorldLeaderAAICPMSICPMS.pdf.
8. *Lumina 3500 Atomic Fluorescence Spectrometer*, Aurora Illuminating Systems Elemental Analysis Instruments, 2020, www.aurorabiomed.com/wpcontent/uploads/2019/07/LUMINA_3500.pdf.
9. *PF7 Heavy Metals Analyzer*, Persee Analytics Inc., http://perseena.com/index/prolist/pid/33/c_id/43.html.

22 Fundamental Principles, Method Development and Operational Requirements of ICP-Optical Emission Spectroscopy

Since its introduction over 40 years ago, inductively coupled plasma optical emission spectroscopy (ICP-OES) has significantly changed the capabilities of elemental analysis. This technique combined the energy of an argon-based plasma with an optical spectrometer and detection system capable of measuring low-level emission signals, which allowed laboratories to perform rapid, automated, multielement analyses at trace concentrations.[1] This was approximately ten years before the introduction of the first commercial inductively coupled plasma-mass spectrometer, so ICP-OES became the workhorse instruments in many laboratories required to perform elemental analysis at trace level concentrations. Chapter 22, gives a detailed description of the fundamental principles together with method development optimization procedures and operational requirements of this technique.

The advantage in using an atmospheric pressure ICP source for making optical emission measurements was first published in 1964,[2] and the sensitivity, speed of analysis, ease of use and tolerance to high levels of dissolved solids are advantages that laboratories continue to rely on more than half a century later.[3,4] The success of the technique itself can be measured by the fact that thousands of ICP-OES instruments have been installed between 1983 and 2021, which have resulted in approximately 80,000 publications, (results courtesy of Google Scholar search). That published literature features elemental determinations in a variety of sample matrices in industries including environmental, nuclear, mining and geochemistry, materials testing, semiconductor, industrial, petrochemical, clinical and toxicological, food safety and pharmaceutical.

22.1 BASIC DEFINITIONS

A full glossary exists at the end of this book for purposes of defining terms used throughout the text; however, several terms are defined here to ensure clarity while reading this chapter. Several optical emission techniques exist, based on atmospheric discharges, which include inductively coupled plasmas (ICP), direct coupled plasmas (DCP), microwave-induced plasmas (MIP), DC arcs and AC sparks. Each discharge is generated via a different mechanism and has its own inherent advantages and disadvantages; however, a comparative discussion of these techniques is outside the scope of this text. The remainder of this chapter will focus solely on ICP-OES.

It should be noted that ICP-AES and ICP-OES are terms that are sometimes used interchangeably; however, the former term can be a source of error and confusion. The term ICP-AES refers to "atomic emission spectroscopy", which nominally excludes emission contributions from other species such as ions and molecules. The latter term refers to "optical emission spectroscopy" and is more commonly used, as it includes emission from multiple contributors. Only the term ICP-OES will be used in this text.

DOI: 10.1201/9781003187639-22

22.2   PRINCIPLES OF EMISSION

For most ICP-OES applications, a sample is delivered to the instrument's plasma in the form of an aerosol. As the aerosol travels from the base of the plasma to its tail, it travels through a variety of heated zones where it gets desolvated (unless delivered as a dry aerosol), vaporized, atomized and ionized. Further time spent in the plasma allows the atoms and ions to absorb additional energy, which excites an outer electron and produces excited state species. Relaxation back to a ground-state atom produces energy in the form of a photon. This production of photons from excited atoms and ions forms the basis for atomic emission measurements. There are many species in a sample that may absorb energy from the plasma and produce emission spectra. These species include atoms, ions and molecules. For the purpose of this section, the contribution from molecular emission will be excluded. All references to emission will include the contribution from atoms and ions only.

22.3   ATOMIC AND IONIC EMISSION

Elemental analysis by ICP-OES relies on the emission from excited atoms and ions within a sample. Argon plasmas contain ~15.8 eV of energy, which is sufficient to remove one or two electrons from the outer orbital of most atoms. This results in the presence of both atoms and ions in the plasma, all of which are in their ground (lowest level) energy state. Excitation, and subsequent emission, occurs when a species' absorbed energy from the plasma is released in the form of wavelength-specific photons.

A simplified schematic of atomic absorption and emission is illustrated in Figure 22.1. The horizontal lines represent energy levels in an atom. The lowest horizontal line and the four remaining horizontal lines represent the ground state and excited states, respectively. If included in the schematic, additional horizontal lines to represent ionic ground and excited states would be illustrated above the atomic excited states. The vertical arrows represent an energy transition for an electron, following the absorption or emission of a photon. The length of each vertical arrow correlates to the amount of energy involved in the transition.

As the schematic indicates, absorbed energy can shift electrons to different excited states, both atomic and ionic. Relaxation of these excited electrons produces energy in the form of photons.

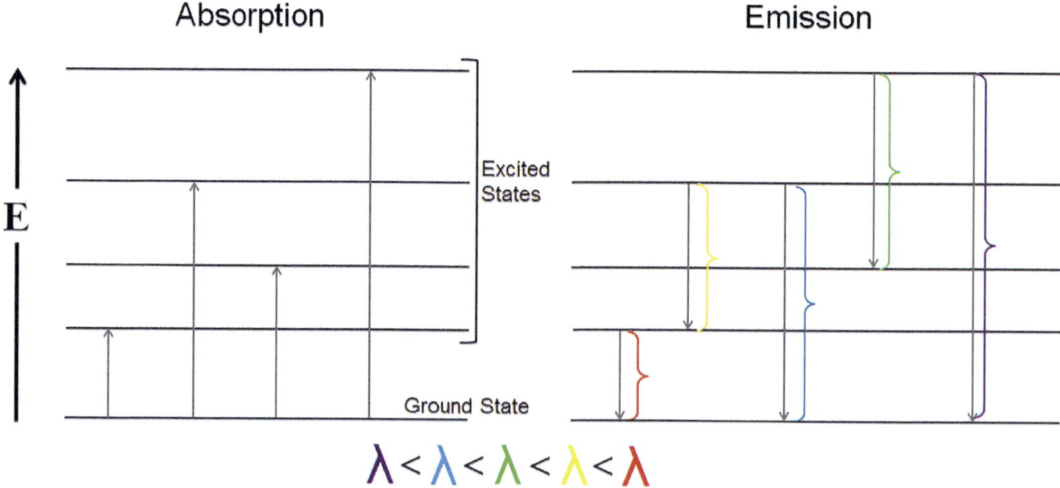

FIGURE 22.1   Diagram depicting energy transitions involved in an atom's absorption and emission of energy.

Photons vary in energy and can be correlated to their associated emission wavelength using Einstein's equation,[5] which relates the energy of light and its frequency according to:

$$E = hv$$

where E represents the energy of light, h represents Planck's constant and v represents the frequency of light. In optical emission spectroscopy, it is more practical to speak in terms of wavelength, so the term c/λ can be substituted to yield:

$$E = hc/\lambda$$

where E represents energy in Joules, h represents Planck's constant in units of Joule seconds, c represents the speed of light in meters per second and λ represents the wavelength in units of meters. From this equation, it becomes clear that each emitted photon is wavelength specific and represents the inverse relationship between energy and wavelength. These emission wavelengths represent the energy levels that are characteristic to each element, thus making optical emission spectroscopy a useful technique for identifying and quantifying elements in unknown samples.

22.4 INSTRUMENTATION

Commercially available spectrometers collect emission at wavelengths from 165 to 1,100 nm. Early elemental analysis did not include the collection of optical emission in the vacuum ultraviolet (VUV) region. Wavelengths below 190 nm are absorbed by oxygen, water vapor and other components in the ambient atmosphere, and early instrument designs did not sufficiently purge the optical path between the torch and the entrance slit of the spectrometer. The use of a completely purged optical path was first proposed in the early 1970s, where the use of high-purity nitrogen made it feasible to collect optical emission in the VUV wavelength region.[6,7]

22.5 SAMPLE INTRODUCTION

The "Achilles heel" of most plasma spectrochemical techniques lies with sample introduction.[8] This fundamental drawback was noted in the literature in the 1980s, however, it is an issue that holds true even decades later. Improving sample-introduction systems is a research topic that has been investigated for many years, the outcome of which has produced many nebulizer and spray chamber designs, as well as introduction systems that incorporate flow-injection and internal-standardization techniques. Research in this area is ongoing as there are still a number of areas in which sample introduction falls short. Introduction-system designs have made little progress in the way of improving sample throughput and reducing carryover from memory-prone elements. A number of systems allow for online addition of internal standards, however, many of those systems suffer from inconsistent mixing between the internal standard and carrier solutions. Flow-injection techniques have been incorporated into a number of introduction systems, however, the use of a relatively small, defined sample volume at the sample flow rates typically used with ICP-MS and ICP-AES results in short-lived transient signals that do not allow for the measurement of more than a few analytes at a time.

As with all analytical atomic-spectroscopic techniques, sample introduction is a critical step that strongly influences, and often dictates, analytical figures of merit such as sensitivity, precision and stability.[9] Improving the efficiency of sample introduction is an ongoing research topic, however, it has proven to be a nontrivial task, as the processes involved within both the nebulizer and spray chamber are numerous and complicated.

Alternative methods of sample introduction, including flow injection, flow injection with hydride generation and laser ablation, have been employed with the intent to improve plasma spectrochemical measurements. Flow injection is a technique that provides a continuous flow of solution to the nebulizer, which allows sample uptake and stabilization in the plasma to take place more rapidly, thereby increasing sample throughput. Furthermore, injection of small sample volumes into a relatively clean carrier stream reduces the amount of salt introduced into the instrument, which can reduce matrix effects and improve limits of detection.[10] A drawback to this setup is that the introduced sample is of finite volume, and the measured signal becomes transient. This is an undesirable situation when measuring a large suite of elements with a sequential instrument, as many of the elements will be measured off the peak center, which degrades the signal-to-noise ratio and, therefore, the sensitivity of the measurement.[11,12] In an effort to retain both steady-state signal analysis and rapid wash-in and washout, a relatively large-volume sample loop should be employed for analysis.

Hydride generation provides a way to remove interfering matrix components to allow only the analyte(s) of interest to be presented to the analytical instrument. Hydride-generation methods are well established[13–16] and rely on the generation of volatile gaseous hydrides, following a reaction with a reducing agent at an appropriate pH. Hydride generation can be performed online or offline prior to analysis and it is a technique that can be used for pre-concentration as well as matrix removal for the determination of hydride-active elements.[17,18]

Laser ablation, when combined with ICP-OES or ICP-MS, is a sampling technique that allows elemental impurities to be quantified in their native solid sample. In this technique, a laser is focused on a prepared sample to remove (ablate) surface material in the form of fine particles. The particles are swept into the instrument using an argon or helium carrier gas, where they are atomized/ionized and excited, similar to the processes that would occur with an aerosol from a liquid sample.[19]

The popularity of this technique stems from its ability to provide a sample aerosol from materials that are difficult to dissolve, such as metals, refractory materials, glasses and insoluble alloys. Additional advantages include simplified sample-preparation procedures with a reduced risk for contamination or loss, a reduced sample-size requirement and the ability to determine the spatial distribution of the quantified elements.[20,21] A significant challenge with laser ablation is in obtaining calibration standards that are matrix-matched to the unknown samples and contain elements at concentrations that bracket those of the analytes of interest.[22] The development of synthetic-liquid standards is relatively straightforward; however, this task is nontrivial for solid-based techniques.

The driving forces behind the design of sample-introduction systems include improved sensitivity and detection limits, improved precision across the working mass range and fewer interferences from matrix effects. Much of the design efforts have centered around improvements in nebulizers and spray chambers, however, significant progress has diminished in the recent years.[23]

22.6 AEROSOL GENERATION

Most ICP-OES applications require the delivery of a sample to the instrument's plasma in the form of an aerosol. The components in the sample-introduction system, which typically include a pump, nebulizer, spray chamber and torch, must work in a complementary fashion to convert a bulk sample solution into a body of micrometer-sized droplets. The pump (peristaltic or syringe) consistently delivers solution to the nebulizer, which uses a high-velocity gas stream to break the solution into small droplets. The spray chamber then filters the droplets by diameter, allowing only the smallest droplets to be transferred to the torch, where they are desolvated, vaporized and atomized in the plasma. A proper aerosol should consist of as many small (< 10-μm diameter) droplets as is possible, to maximize the efficiency of desolvation once the sample reaches the plasma.

The generation of a proper aerosol is a significant factor in determining the quality of the resulting emission signals and data.[24] Sample aerosols may vary in their density of droplets; however, a proper aerosol should maintain a consistent shape and density during data collection to promote analytical measurements with optimal precision. Two example aerosols are pictured in Figure 22.2 for illustrative purposes. They are generated from nebulizers with two different structural designs, which produces differences in their shape and droplet density.

FIGURE 22.2   Examples of proper sample aerosols, generated by a concentric nebulizer and a PFA micro-flow concentric nebulizer.

(top with kind permission from Meinhard Glass Products, bottom with kind permission from Elemental Scientific Inc.)

22.7   NEBULIZERS

The main role of the nebulizer is to generate an aerosol of small droplets from the interaction between a sample stream and a high-velocity gas stream.[25] Ideally, the nebulizer will transport an aerosol with a narrow drop size distribution into the spray chamber to allow the spray chamber to more efficiently filter out the large aerosol droplets.[26] Since the droplets must be desolvated, vaporized and atomized during their short residence time in the plasma, the spray chamber allows passage of only small droplets (typical diameter $< 10\ \mu m$).[27,28]

In an effort to overcome the limits associated with sample introduction, much effort has been put into the design of nebulizers and spray chambers. Typical introduction systems consist of a nebulizer–spray chamber arrangement in which the nebulizer is pneumatic.[29] The relative simplicity and low cost of this setup make it the preferred choice for ICP sample introduction, however, its associated drawbacks include low analyte transport efficiency (~1–2%), high sample consumption (1–2 ml/min^{-1}) and relatively high retention of some elements.[30] An image of a typical pneumatic nebulizer is illustrated in Figure 22.3.

Microconcentric nebulizers (MCN)[31–36] were designed to improve the gas–liquid interaction and reduce the size distribution of droplets formed in the aerosol.[37,38] These nebulizers, which operate

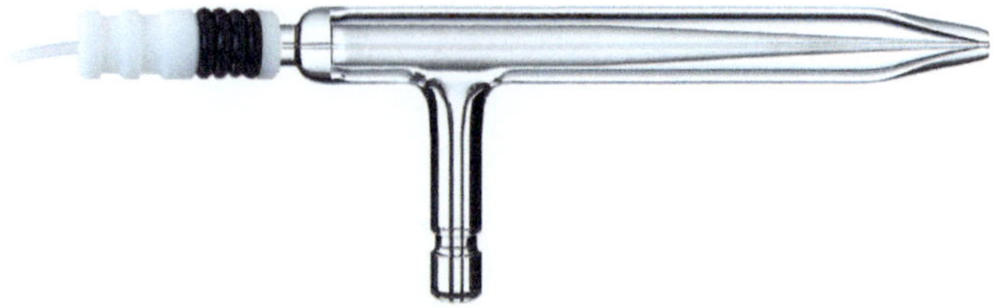

FIGURE 22.3   Image of a typical pneumatic nebulizer.

(With kind permission from Glass Expansion)

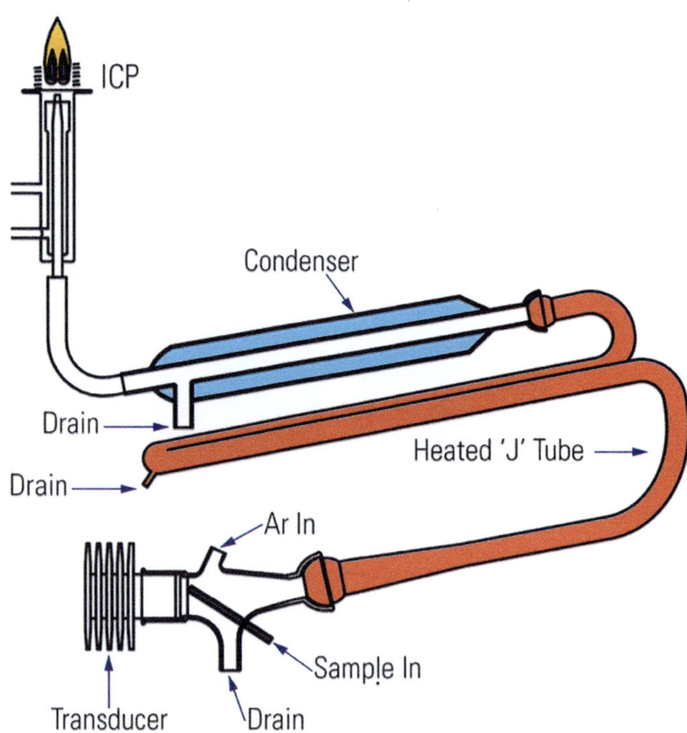

FIGURE 22.4   Schematic to illustrate the basic operation of an ultrasonic nebulizer.

(Used with kind permission from Teledyne Cetac Technologies)

at lower sample flow rates compared to those typically used with pneumatic nebulizers, include: high-efficiency nebulizers (HEN),[39–42] oscillating capillary nebulizers (OCN)[43–45] and sonic spray nebulizers (SSN).[46] These nebulizers have a relatively low dead volume and operate at normal nebulizer gas pressures, which improves analyte transport efficiency regardless of whether organic or aqueous solvents are used.[47,48]

Ultrasonic nebulizers (USN)[49–53] were developed in an effort to improve efficiency of aerosol generation. These nebulizers are highly efficient at producing a large volume of small droplets with a narrow size distribution.[54] The aerosol volume is sufficiently large that most USNs are used with desolvation systems to reduce the introduction of water vapor in the case of aqueous sample analyses,[55,56] and to reduce solvent loading and carbon deposition in the analysis of samples containing organic solvents or high concentrations of dissolved salts.[57,58] A schematic of an ultrasonic nebulizer is shown in Figure 22.4.

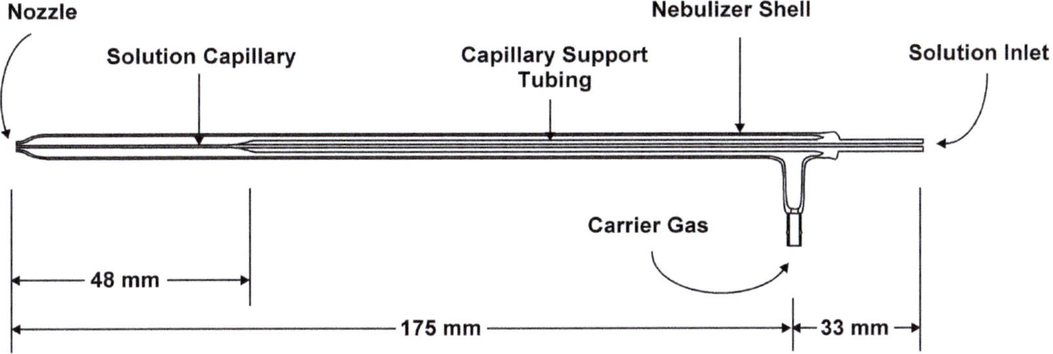

FIGURE 22.5 Schematic DIHEN nebulizer.

(Used with kind permission from Meinhard)

Nebulizers have also been designed in which no spray chamber is present, and samples are injected directly into the plasma. These direct-injection nebulizers (DIN)[59–63] and direct-injection high-efficiency nebulizers (DIHEN)[64–67] improve the sample transport efficiency to 100%, even at relatively low flow rates. Direct injection also reduces the dead volume associated with spray chambers, which increases the response time of the measurement and reduces memory effects.[68–71] A significant drawback of direct-injection nebulizers is their vulnerability toward samples containing high concentrations of dissolved salts or volatile solvents, which cause plasma instability and tip clogging.[72] The large-bore direct-injection high-efficiency nebulizer (LB-DIHEN)[73] was designed to reduce its susceptibility to blockage, however, the larger inside diameter of the nebulizer capillary produces a relatively large droplet-size distribution, which degrades both the precision and detection limits.[74] An example of a direct-injection nebulizer, specifically, a quartz DIHEN, is pictured in Figure 22.5.

22.8 SPRAY CHAMBERS

Much effort has been put into spray-chamber design, as spray chambers are responsible for the loss of > 90% of the aerosol produced by the nebulizer.[75] Improvements have focused on directing the flow of aerosol from the nebulizer to maximize the efficiency with which larger droplets are filtered out.[76] Reverse-flow, commonly called Scott-type double-pass,[77,78] spray chambers are the popular choice for relatively simple, low-cost introduction. The double-pass design is particularly useful for samples containing high concentrations of dissolved salts, as the transport efficiency is relatively low compared to other spray-chamber designs.[79] An example double-pass spray chamber is illustrated in Figure 22.6.

Cyclonic designs have been employed to improve transport efficiency, precision and detection limits.[80,81] Popular cyclonic spray-chamber designs include a flow spoiler, or dimple,[82] to disrupt the flow of aerosol within the spray chamber. Computer modeling of the fluid dynamics within cyclonic spray chambers suggests that the presence of three spoilers creates a "virtual cyclone",[83] which reduces interaction between the aerosol and the walls of the spray chamber, thereby improving transport efficiency and reducing memory effects. A commonly employed cyclonic spray chamber is illustrated in Figure 22.7. Spray chambers, without (left) and with (right) a flow spoiler, are illustrated for comparison.

Single-pass or cylindrical-type spray chambers have been designed for use in low-flow introduction.[84–86] These spray chambers provide high efficiency and reduced memory effects,[87] however, this design is unsuitable for conventional ICP analysis and is typically used when electrophoretic or

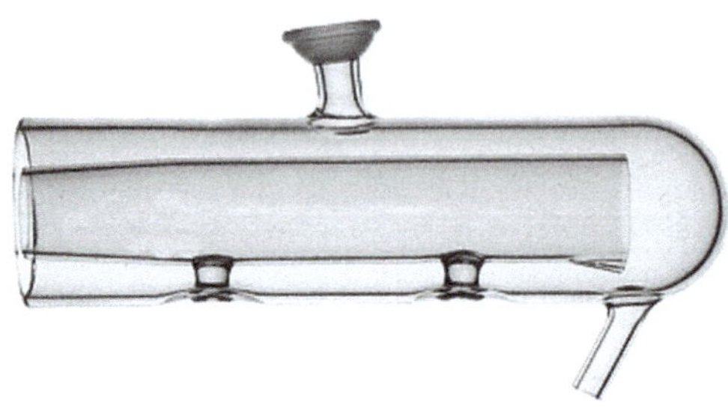

FIGURE 22.6   Image and schematic of a typical Scott-type double-pass spray chamber.

(With permission from Precision Glassblowing)

FIGURE 22.7   Cyclonic spray chamber without (top) and with (bottom) a knockout tube.

(With kind permission from Glass Expansion)

chromatographic separations are employed in conjunction with ICP detection.[88,89] Spray chambers have also been blamed for the retention of elements such as B and Hg.[90–93] As discussed previously, one approach to solving this problem has been in the removal of the spray chamber. Other approaches have involved the use of a thermostated spray chamber. Water-cooled spray chambers have been used to reduce aerosol desolvation and deposition along the walls of the spray chamber, thereby reducing memory effects and reducing oxide formation in the plasma.[94–96] This logic has been used in the application of heated spray chambers for desolvating and, therefore, improving the efficiency of aerosol generation in aqueous samples.[97–99]

22.9 TORCHES

The torch is the final stop in the sample's journey from sample cup to plasma, and often has a major influence on the analytical performance of the instrument in terms of sensitivity, detection limits and plasma robustness. Torch design for inductively coupled plasmas has been investigated for more than 50 years. The original ICP torch was based on Reed's design[100–102] and introduced the coolant and auxiliary gas flows in a tangential direction. This method of gas introduction was thought to produce fluid dynamics with maximum stability and ion density.[103]

Reed's torch design was modified by Greenfield and colleagues,[104,105] who constructed a torch containing three concentric tubes to generate and sustain an annular-shaped plasma. The outer and intermediate tubes were made out of silica and the center tube was constructed out of borosilicate glass. High-purity gas was introduced tangentially, similar to the Reed torch; however, gas was not introduced into the outer tube until after the plasma had formed. Only the intermediate tube was purged during ignition.

Fassell made minor modifications, and the resulting torch design[106] is used by most commercial-instrument manufacturers, even today. The Fassell torch design evolved to a design similar to that shown in Figure 22.8 The torch consists of three concentric quartz tubes: the outer tube, the intermediate tube and the center tube (or injector). The geometry of these tubes and their positioning with respect to each other and to the load coil, along with the gas-flow settings, directly affect the formation, stability and sustainability of the plasma.[107] This makes torch design critical to data quality in elemental analysis.

The inner tube, or injector, provides an inlet for introducing the sample aerosol into the plasma. The center tube provides a structured path for the auxiliary gas, which is used to shift the base of the plasma further away from the injector. The outer tube, or coolant, houses the largest volume of gas, which is used to generate and sustain the plasma. This gas also provides sufficient cooling to prevent the torch from melting or contributing background emission signals.

Since the introduction of the Fassell torch, minor changes have been made to torch design. Injector tubes have been constructed with both a tapered and a parallel end, and they are available with a variety of inside diameters. Small-diameter injectors are typically used for applications involving solvents that have a relatively high vapor pressure and are likely to overload and extinguish the plasma. Larger-diameter injectors are typically used for the analysis of relatively clean samples to increase the aerosol volume delivered to the plasma, which increases the resulting emission signals.

The intermediate tube has been produced with both a parallel and a tulip-shaped design. Research has been done to investigate the effect of the intermediate tube shape on the shape and stability of the plasma, however, both designs are still employed.[108] A parallel intermediate tube allows the auxiliary gas flow to remain constant along the length of the torch to reduce turbulence at the base of the plasma and to allow the auxiliary gas to reach a true laminar flow. This reduction in turbulence produces a plasma with a relatively flat base, and aids in the penetration of the nebulizer gas and, therefore, the sample, into the plasma.

Torches are available as a single piece or in a demountable design. The single-piece torch (refer to Figure 22.8 for an example) is generally easier for operator use and ensures that the injector will be fully centered in the intermediate tube; however, the individual parts of the torch cannot be replaced, and the injector diameter cannot be changed. Demountable torches must be manually assembled and aligned to ensure the injector is centered in the torch body. This style of torch is

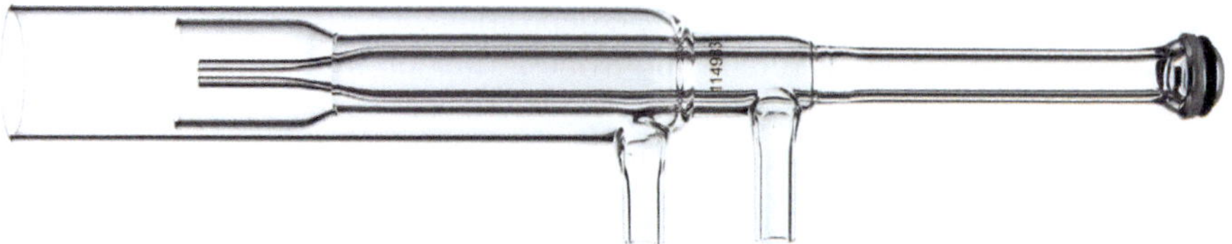

FIGURE 22.8 Schematic for a standard, commercially available ICP-OES torch.

(With kind permission from Glass Expansion)

FIGURE 22.9 Demountable ceramic-torch parts.

(With kind permission from Glass Expansion)

advantageous and more cost effective for applications that benefit from the use of multiple injector diameters, or that are likely to clog the torch injector.

Ceramic torches have been designed for use in measuring organic solvents, extremely high levels of dissolved solids and other samples that devitrify a standard quartz torch.[109] Commercial ceramic torches are available in fully demountable designs, which allow quartz and ceramic pieces to be combined to meet the needs of each specific application. An example of commercially available ceramic-torch components is illustrated in Figure 22.9.

22.10 SPECTROMETERS

The optical spectrometer is the heart of the modern ICP instrument and is comprised of the fore optics and the polychromator, with the detector attached. The purpose of the optical system is to separate the ICP source emission into element-specific wavelengths and to focus the resolved light onto the detector with high efficiency and with minimal stray-light contributions.[110] This should be accomplished with a minimum amount of absorption, scattering and optical aberration to maximize the light throughput to the detector.

22.11   FORE-OPTICS

The transfer of light from the plasma into the spectrometer is a critical first step in obtaining results with maximum sensitivity and signal-to-background ratios. Of equal importance is the transfer of emission with minimal contributions from molecular emission and stray-light sources. The gap between the plasma and the spectrometer's entrance optics can be a challenging environment to control. In addition to protecting the entrance optics from the plasma's intense heat, the plasma–spectrometer interface must effectively replace the ambient air to prevent it from absorbing emission wavelengths below ~200 nm. For a horizontally mounted torch, the interface must also remove the relatively cool portion of the plasma ("tail") to minimize interferences from self-absorbed analytes and atoms/ions that recombine and produce molecular emission.[111]

The purpose of the fore-optics is to focus emitted light from the plasma onto the entrance slit as efficiently as possible and to provide a light path suitable for transmitting in either radial or axial plasma views, or both, for systems that have that capability. In dual-view systems, changing plasma views should be a fast, automated process to avoid undesirable increases in sample-analysis time and subsequent degradation of productivity and cost efficiency.

Properly designed fore-optics should exhibit a number of characteristics to ensure optimal instrument performance. The light from the plasma should be transferred into the instrument with a high level of efficiency to maximize the sensitivity of the instrument. This requires a minimum number of optics, as each light-reflecting surface will absorb or scatter a small amount of light. Each additional optic reduces the transfer efficiency, which reduces the instrument's achievable detection limits.[112]

The fore-optics should have optimum focus for better sensitivity and detection limits. Photons must be collected from the analytical zone of the plasma—the area of the plasma where the maximum amount of analyte emission and the minimum amount of emission from the plasma exists. Once collected, the light must be properly focused onto the entrance slit of the spectrometer to maximize the amount of light that travels through the spectrometer.

There should be a minimal number of moving parts for greater stability. A small amount of movement is required to optimize the viewing height of a radially configured plasma, and to switch between axial and radial light collection in a dual-view instrument. The movement needs to be rapid and highly reproducible to minimize the degradation on sample throughput and precision.

The fore-optics should be robust and provide thermal protection from external heat sources, such as the plasma and the atmosphere in the torch box. Excessive heat transfer to the optics will produce instrument drift, resulting in QC failures and require frequent recalibration.

22.12   OPTICAL DESIGNS

The earliest ICP spectrometers are based on the Paschen–Rünge design.[113] In this configuration, all of the optics (diffraction grating, entrance slit, exit slit) are permanently attached, and the exit slits are mounted along a portion of a Rowland circle.[114] In its original design, this type of spectrometer was referred to as a direct reader and consisted of a polychromator, with photomultiplier tube (PMT) detectors positioned on the focal plane behind a series of exit slits. One exit slit and PMT was used for each analytical wavelength being measured.[115] This optical layout, which is illustrated in Figure 22.10, allowed several emission wavelengths to be measured simultaneously and in a matter of a few seconds.

Despite the ability to acquire emission signals with remarkable speed, early spectrometers with this optical configuration had relatively poor spectral resolution. Furthermore, most spectrometers utilized photomultiplier tube (PMT) detectors at that time, which were responsive over narrow wavelength ranges. This restriction required the end user's desired elements to be selected prior to

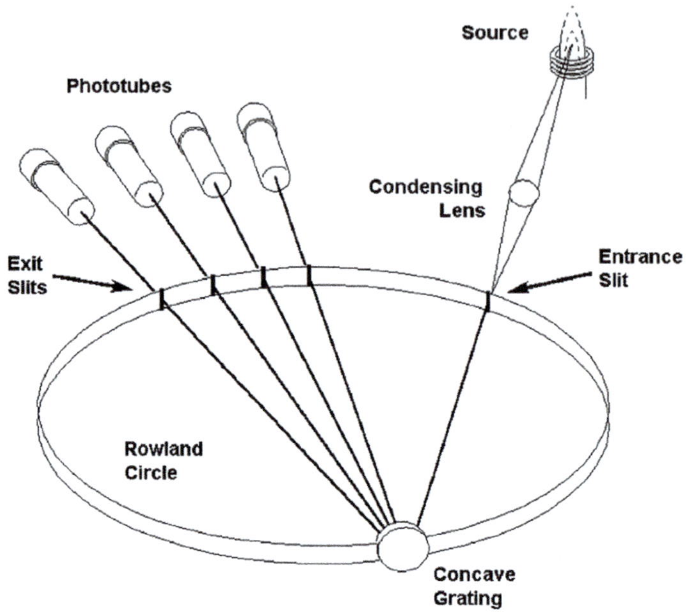

FIGURE 22.10   Schematic of a Paschen–Rünge optical design.[116]

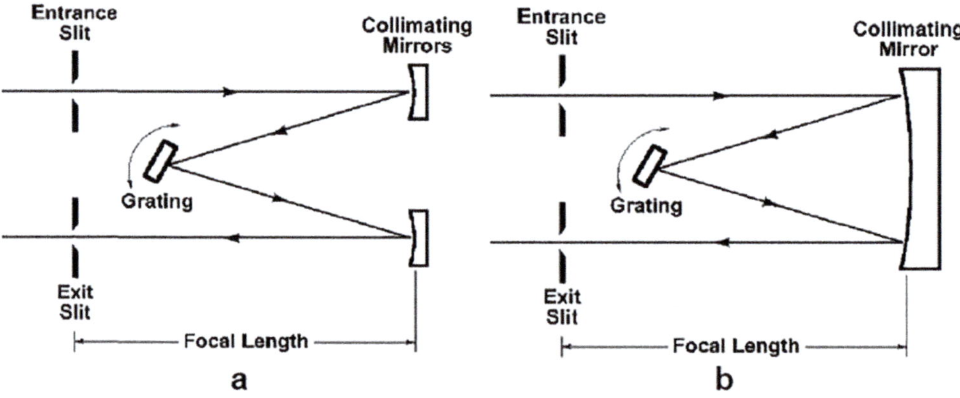

FIGURE 22.11   Schematic of Czerny–Turner (left) and Ebert (right) optical designs.[115]

the instrument being built and limited the instrument to a maximum of 20–30 emission lines due to the space required for having an exit slit and PMT for each wavelength.[116] This optical design poses additional challenges with internal standardization and interference corrections. The wavelengths being used for internal standard measurements and for interference calculations must have been included in the instrument at the time of manufacture.

The interference challenges and wavelength restrictions associated with the Paschen–Rünge spectrometer led to the development of Czerny–Turner and Ebert designs. The fixed, concave grating layout of the Paschen–Rünge spectrometer is replaced with a scanning, dispersive optic in the Czerny–Turner and Ebert designs.[117–120] In this arrangement, a dispersive optic, such as a plane grating, diffracts light onto either one or two collimating mirrors (Ebert and Czerny–Turner, respectively). The focused light is then directed onto a single exit slit. This layout is represented in Figure 22.11.

The wavelength flexibility of these monochromator-based systems is an advantage over polychromators; however, their measurement versatility is offset by the relatively long sample-analysis times. The use of a scanning optic dictates that wavelengths be accessed and measured sequentially, which means a single sample could require several minutes to analyze.

An additional challenge with this spectrometer design was in the spectral aberrations resulting from the use of spherical collimating mirrors.[121] Image distortions, such as astigmatism, spherical aberration, coma aberration or elongated slit images, are inherent to the traditional design.[122] Light from the grating strikes the mirror at an angle that prevents the rays from meeting at a common focal point. To prevent these image distortions from creating errors in the spectral results, compensating optics are often used. Example schematics of typical Ebert and Czerny–Turner spectrometers are illustrated in Figure 22.11.

The early 1980s saw the development of an echelle-based optical spectrometer.[123] This design utilizes a high-dispersion echelle grating in addition to a second dispersive optic (such as a grating or a prism) to separate incident polychromatic light into a two-dimensional spectrum, known as an echellogram. Unlike conventional diffraction gratings, echelle gratings are blazed with a relatively low groove density and relatively high angles of incidence to disperse light into its constituent wavelengths at higher spectral orders. While this dispersive behavior produces exceptional resolution, the resulting spectral orders overlap and are of no use without a second, cross-dispersive optic.

In 1983, the first commercially available, benchtop echelle optical system was released, significantly improving the optical dispersion and resolution capabilities of optical emission spectrometers.[124] Due to the high resolution of these systems over relatively short focal lengths, echellograms could be enlarged to fill the relatively large surface area of a photomultiplier tube (PMT) detector, or reduced to fill the small active surface area of a solid-state array detector. Further advances in instrument design resulted in spectrometers that could access more than one wavelength at the same time, allowing simultaneous and sequential measurements, or a combination of both, using a single optical platform.[125] Examples of simultaneous and sequential echelle-based optical spectrometers are illustrated in Figure 22.12.

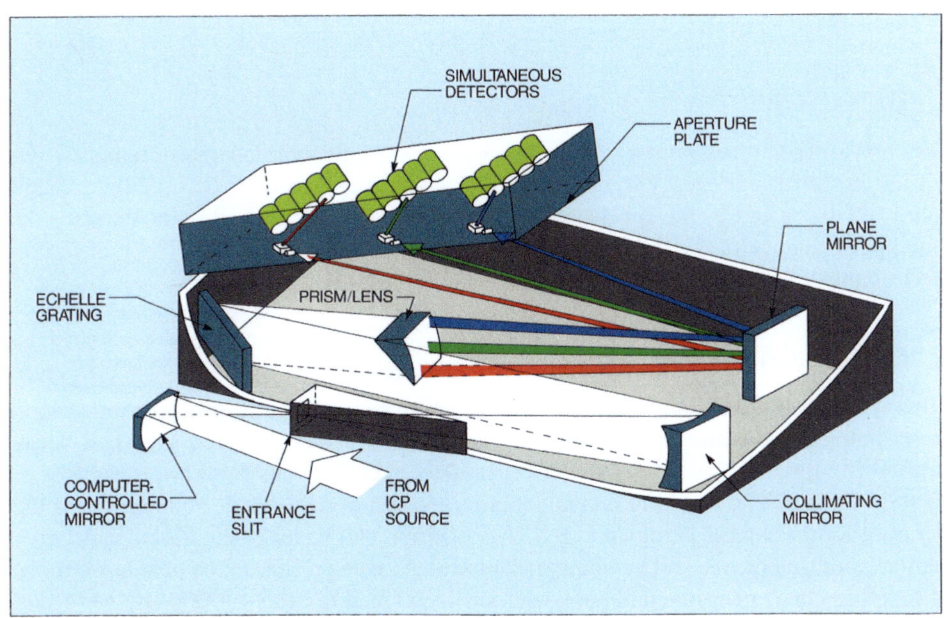

FIGURE 22.12 Schematic of a simultaneous (top) and sequential (bottom) echelle optical system. (With kind permission from Teledyne Leeman Labs)

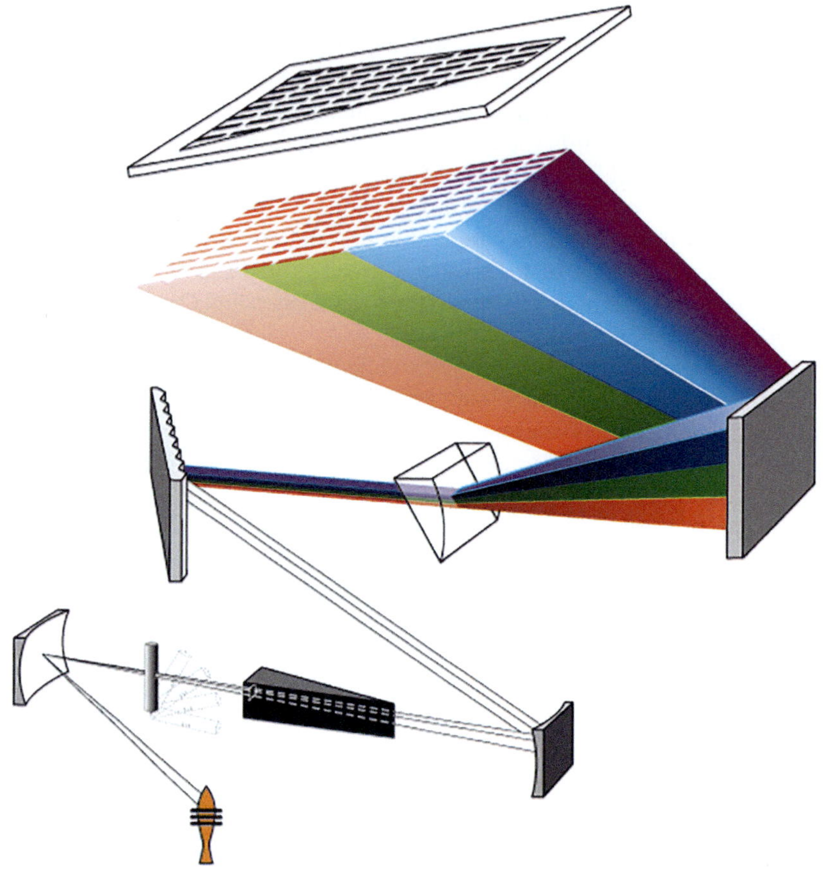

FIGURE 22.12   (Continued)

22.13   DETECTORS

Detectors are a crucial component to the success of optical emission measurements. Once the plasma transfers sufficient energy to the sample to generate photons, the emitted light is collected, collimated and diffracted by the spectrometer, where it is then transferred to the detector. The efficiency and performance of the detector then determine the quality and sensitivity of the resulting measured signals.

22.14   HISTORICAL PERSPECTIVE

The earliest published research involving optical emission measurements utilized photographic-plate detectors. Photographic-plate detectors operate via the photoelectric effect, so-called because the light emitted from the sample is of sufficient energy to eject electrons from the surface of a photo-reactive solid surface. The resulting image contains a series of dark bands and represents the emission spectrum for the sample being analyzed. The position and width of the dark bands can be used to determine which elements, and at what concentrations, were present in the original sample.[126]

Photographic-plate spectrometers have been utilized since the early 1900s and were successful in analyzing complex spectra, such as steel. Spectrographs were designed with a variety of optical systems to achieve various levels of resolution and to accommodate photographic plates of various sizes.[127]

Photographic plates offered the first permanent record of emission spectra from a variety of excitation sources including arc, spark and ICP. The photographic emulsion coated on its surface reacts over a wide energy range, allowing a single plate to record emission from most (or all) analytes of interest in a single sample run. A drawback to the detector technology, however, was its nonlinear response and its relatively low quantum efficiency (1–3%).[128]

The early 1930s saw the development of photomultiplier tube (PMT) detectors, and by the late 1970s, photomultiplier tubes were being used fairly routinely in atomic emission spectrometers.[129–131] By the early 1980s, technological advances were made to produce photodiode-array (PDA) detectors, which possessed multichannel measurement capabilities. These photodiodes were responsive in both the ultraviolet and visible wavelength ranges and had a wider dynamic range compared to their photomultiplier predecessors. This technology made a fundamental deviation in the way in which it measured spectral emission. Unlike its photomultiplier tube predecessor, which produces a measureable current from incoming radiation, the PDA is able to generate and store that photo-generated charge.[132,133]

During the late 80s and early 90s, charge-transfer devices (CTDs) such as charge-coupled devices (CCDs) and charge-injection devices (CIDs) began to replace PMTs and PDAs in analytical spectroscopy. These devices respond more uniformly across relatively wide wavelength ranges, and they have the ability to perform simultaneous background measurements. Today, CTD detectors are used almost exclusively in optical spectrometers to quantify spectral emission from a variety of analytes over a wide range of wavelengths.[134–137]

22.15 PHOTOMULTIPLIER TUBES

Over the years, photomultiplier tubes (PMTs) have been the most commonly used detectors for ICP-OES instruments. These detectors typically consist of a sealed vessel, with a series of emission dynodes placed between an anode and a large-surface area cathode. The dynodes are mounted in series and kept at voltage potentials that get progressively more positive than that of the cathode. A photon that strikes the cathode with a sufficient amount of energy dislodges an electron. The ejected electrons accelerate toward the first dynode in the series, which causes a cascade of electrons to be ejected and accelerated toward the next dynode. The process repeats until the multitude of electrons reaches the anode, which generates a measurable electrical pulse.[138] An example of a photomultiplier tube detector is illustrated in Figure 22.13.

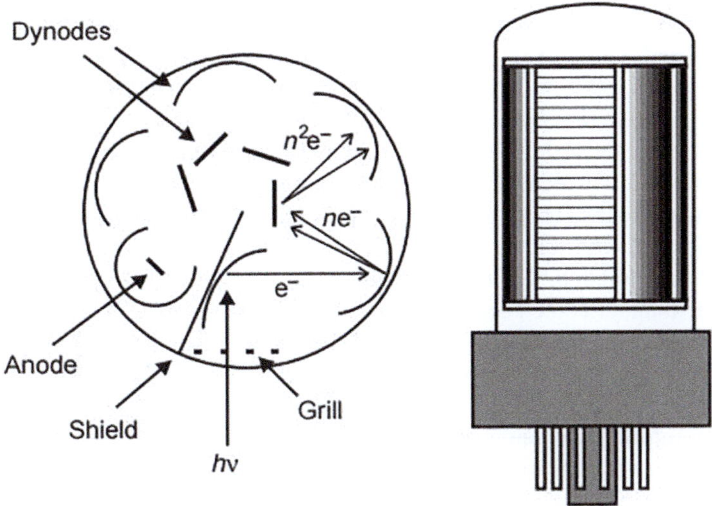

FIGURE 22.13 Schematic and basic function of a photomultiplier tube.[138]

This type of detector has long been known for its long linear working range and its near-noiseless gain. The detector's design gives it the ability to generate more than a million electrons from a single photon, which results in a signal gain of 10^6–10^8. The gain is achieved with almost no measureable dark current or background noise, making it well suited for small signals, such as those from trace-level concentrations of analytes.

These detectors lack simultaneous multielement collection capability; however, their single-channel design allows the signal-to-noise ratio to be optimized for each analyte that is being measured. If simultaneous emission detection is desired, multiple PMT detectors can be positioned behind multiple exit slits and used in parallel. The metal-oxide construction of the cathode makes the PMT responsive over a relatively short wavelength range, however, a combination of materials can be used to widen the response range.

22.16 PHOTODIODE ARRAYS

Photodiode arrays (PDAs) were available for use in making spectrochemical measurements by 1980.[139] Similar to photomultiplier tubes, these devices generate an electrical current. They do not have the same gain and signal response as photomultiplier tubes; however, they offer the advantage of being able to store photo-generated charge, which allows them to integrate emission signals.[140]

Photodiode arrays do not have the sensitivity and signal gain achievable with photomultiplier tubes; however, their responses are linear over a relatively wide range and they can be operated in a high-speed mode that allows for parallel readout (similar to the random access readout feature of charge-injection-device detectors).[141] These devices also require cooling to reduce contributions from dark current and fixed-pattern noises; however, signal-to-noise ratios could be improved if measurements could be made with longer integration times and fewer signal averages.[142]

22.17 CHARGE-TRANSFER DEVICES

Charge-transfer-device (CTDs) detectors are solid-state devices, consisting of an array of photoactive elements, which convert incident radiation to an electrical charge, which is stored and measured. These photoactive elements, also known as pixels, can be configured as a linear or two-dimensional array, and can be manufactured in a variety of sizes, to meet the requirements of the instrument.[143] Among other benefits, the structure of these array-based devices provides them with the ability to measure more than one wavelength simultaneously, which allows for true simultaneous background correction and real-time internal standardization.[144] In some applications, the efficiency of the background subtraction is sufficient to remove the contribution from background emission that is significantly larger than that for the analyte(s) being measured.

As the name implies, these devices employ a charge-transfer process to read out the accumulated charge. Charge transfer is executed via one of two methods: (1) inter-cell charge transfer, which shifts charge from where the charge accumulated to a single output amplifier, or (2) intra-cell charge transfer, which measures the voltage change induced by shifting charge within the detector element where it was first accumulated.[145] Charge-transfer devices generate dark current when in use; however, operating them under cooled conditions reduces the dark current contribution to a level that is almost immeasurable, even when emission signals are integrated for long periods of time.[146]

F. L. J. Sangster and K. Teer invented the first charge-transfer device at Philips Research Laboratories almost 50 years ago. Unbeknownst to them, it would become a device that had a significant impact on future developments and advancements in atomic spectroscopy.[147] These devices were originally referred to as bucket brigade devices (BBDs) to analogize the method of storing and transferring signals to a line of people passing buckets of water to fill a storage container or to put out a small fire. The potential for using these devices in imaging sensors was quickly realized, and they became a memory-storage device with significant use.[148]

In less than a year, Willard Boyle and George Smith, working at Bell Laboratories at the time, improved the BBD's method for processing signals, and introduced the charge-coupled device (CCD).[149] The BBD, which used transistors to transfer charge between a series of capacitors, was improved with the CCD, which transferred charge between capacitive bins (also known as "wells") across the surface of a metal-oxide semiconductor (MOS). This invention would earn them a Nobel Prize for Physics and would find mainstream use in imaging devices, such as telescopes, digital cameras and video cameras.[150,151]

By 1973, Hubert Burke and Gerald Michon, working at General Electric, had introduced the charge-injection-device (CID) detector. For the next two decades, charge-injection devices were used as imagers in machine-vision applications where digital-image-processing capabilities were required. By 1990, CID devices were adapted for use in astronomy, astrophysics and microscopy, where detector requirements included high sensitivity and spectral accuracy, large dynamic range and low dark current levels, even during integration periods lasting several hours.[152] The most significant difference between charge-coupled devices and charge-injection devices is their readout mechanism, which leads to unique benefits and challenges in atomic spectroscopy applications. Each type of device will be discussed separately in the following sections.

22.18   CHARGE-COUPLED DEVICES

Charge-coupled devices are compact charge-transfer devices that have found use as solid-state imagers in a variety of applications. In its original design, the CCD was constructed for the storage of digital information for use in devices such as digital cameras and video recorders. Once in use for routine consumer applications, it became clear that the CCD possessed the ability to convert digital charges to analog signals, broadening its utility to include industrial and scientific applications.[153,154] Figure 22.14 illustrates the structure of a typical CCD device.

The mechanism for transferring charge through a typical CCD detector is illustrated in Figure 22.15. When the device is read out, charge that has accumulated across the entire device must be transferred

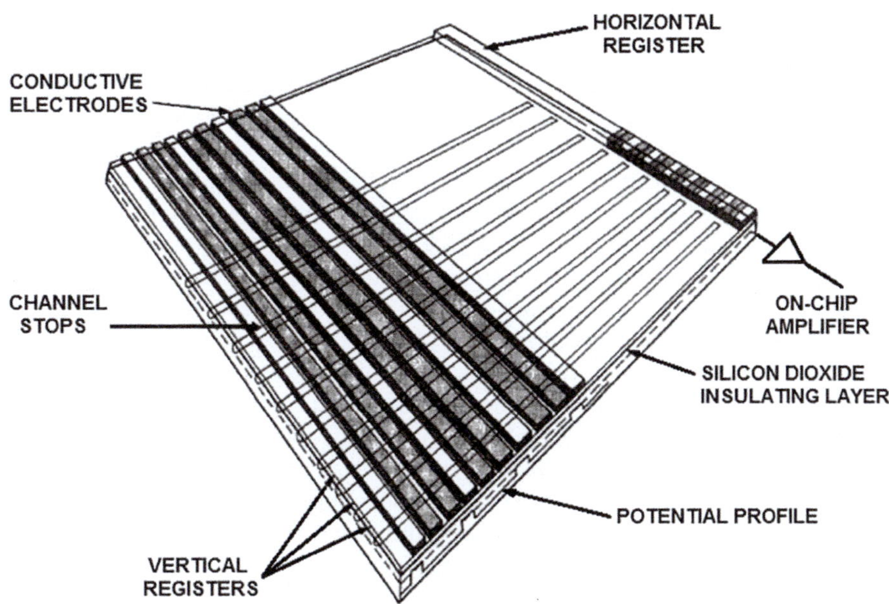

FIGURE 22.14   Basic schematic of a charge-coupled device.[155]

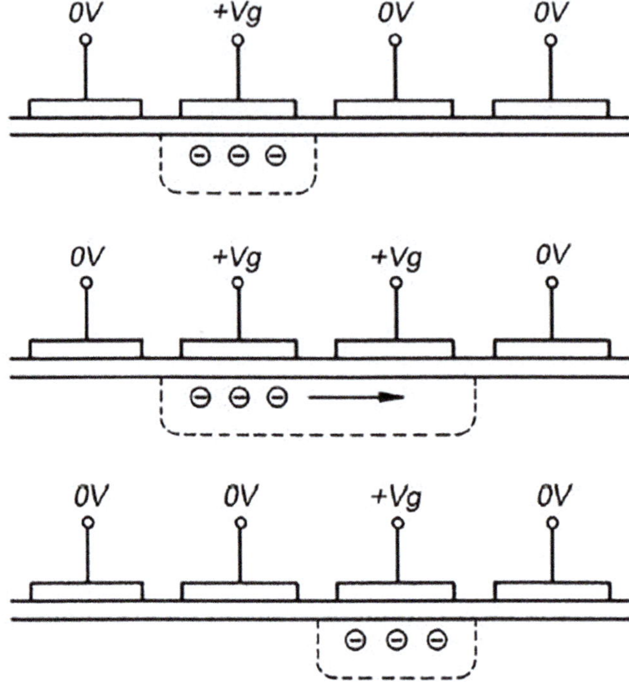

FIGURE 22.15 Charge-transfer mechanism in a typical charge-coupled device.[156]

and measured, even if only a small portion of the array needs to be read out. To accomplish this, a small bias voltage is applied to one of the pixels, which creates an area of depleted charge in the substrate directly below. A greater bias voltage is applied to the adjacent pixel, which creates a similar depletion area and promotes the transfer of charge. The process repeats itself until the accumulated charge is transferred through the entire device, until the charge is shifted to the output amplifier for analog-to-digital conversion and measurement with a digital processor. This readout process destroys the stored charge and prevents it from being measured again in the future.[156]

The structure of a charge-coupled device, combined with its readout mechanism, allows it to read its stored charge with a high level of uniformity, which translates to high-quality spectral images. However, it is this structure and readout that make the CCD susceptible to a phenomenon known as blooming. Blooming occurs when one of the device's detector elements (pixels) accumulates more charge than it can store and spills the excess charge into nearby pixels. This occurrence can significantly impact measurements for emission signals that are being measured simultaneously with intense emission from nearby wavelengths.[157]

To reduce the risk of blooming, a variety of modifications have been implemented. Some CCD devices have been assembled to include structures that are designed to remove and drain excess charge from pixels before they saturate and spill that charge into nearby pixels. These structures, often referred to as charge drains or anti-blooming gates, surround and isolate active pixels to allow low-level and high-level emission signals to be accurately measured, even if it requires storage in nearby pixels on the same detector.[158] These charge drains have been shown to be effective enough to prevent blooming, even when reading charge that's adjacent to pixels that accumulated charge 1,000 times their saturation level.[159]

Another method employed to reduce the likelihood of charge spillage was to use a series of separate linear CCD arrays, working as a cohesive unit. This device, known as a segmented-charge-coupled-device

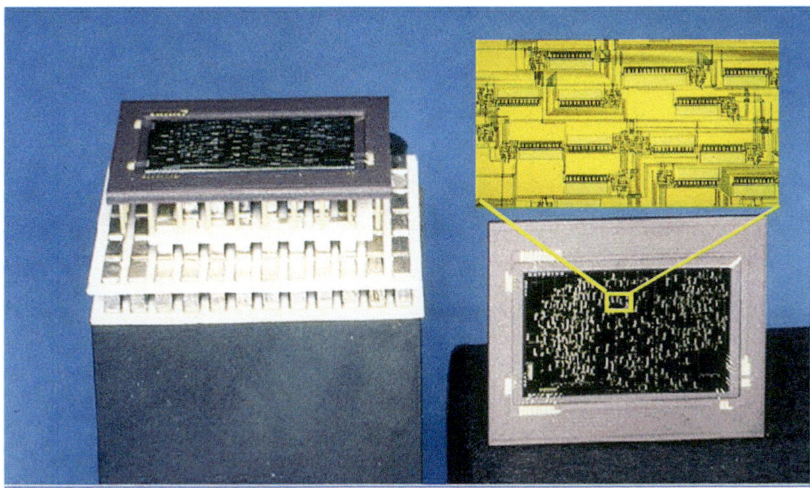

FIGURE 22.16   An SCD detector showing a photomicrograph close-up of the separate photosensitive linear arrays.[163]

(SCD) detector, is constructed such that a separate linear array is used for the measurement of one wavelength (up to three wavelengths if they occur in close proximity) for each of the elements on the periodic table. Using arrays that are physically separate from one another mimics the structure of a single CCD detector with built-in charge drains.[160,161] Size constraints limit the number of arrays that can be used; however, over 200 arrays can be combined for use in covering important analytical wavelengths across both the ultraviolet and visible wavelength regions.[162]

In addition to overcoming potential effects from blooming, the structure of an SCD allows it to measure a small number of wavelengths (or even a single wavelength) without reading the charge across the entire detector, in a method known as random-access integration. This feature increases the readout speed and dynamic range, compared to a single (non-segmented) CCD detector.[163] Figure 22.16 is an image of an SCD detector showing a photomicrograph close-up of the separate photosensitive linear arrays.[163]

22.19   CHARGE-INJECTION DEVICES

Charge-injection-device (CID) detectors fall into the category of complementary-metal-oxide-semiconductor (CMOS) technology. CIDs contain an array of detector elements that can store photo-generated charge. Each detector contains a light-sensitive area, as well as row and column electrodes that are addressable for purposes of storing and reading out charge. The electrodes are constructed from a conductive silicon material laid over an insulating layer, which provides a region for charge to be stored.[164] An example of a CID detector element is illustrated in Figure 22.17.

The mechanism for transferring charge through a typical CID detector is illustrated in Figure 22.18 (A–D). The four images represent a single pixel during the four-step charge-readout process. The pixel contains two photogates (labeled "Row" and "Column") that are used for storing and measuring photo-generated charge. When an analytical measurement commences, the detector starts collecting charge, based on the integration time set by the user (Step A). During this step, the incident radiation is converted to charge and collected in the column photogate. Once integration has begun, the detector measures the accumulated charge by shifting it between the row and column photogates (Steps B and C).

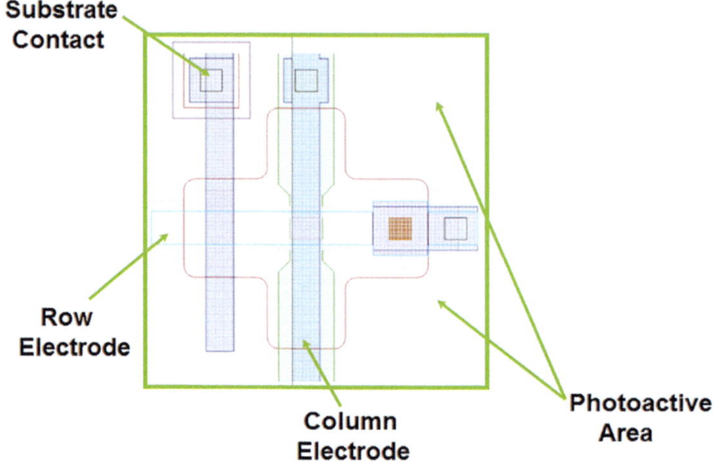

FIGURE 22.17   Schematic of an example detector element (pixel) within a CID detector. (With kind permission from Teledyne Leeman Labs)

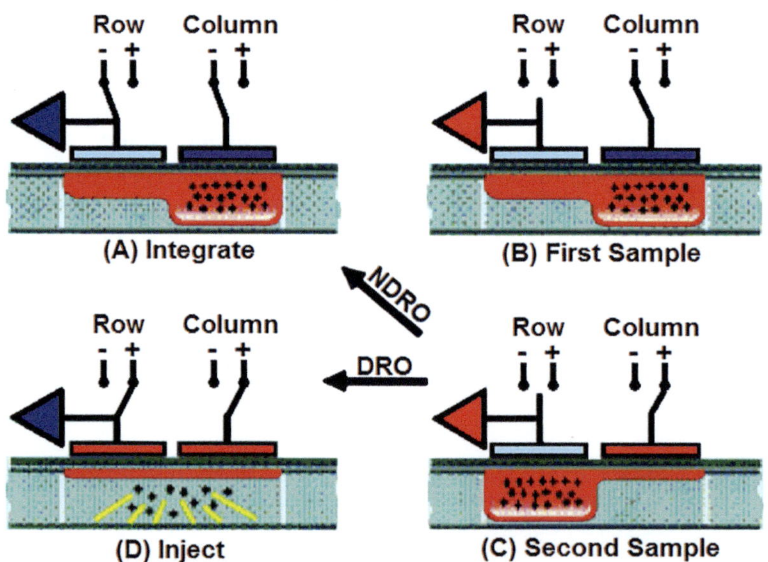

FIGURE 22.18   Schematic and operation of a CID pixel.[165]

If the pixel has accumulated enough charge to approach saturation, the charge can be cleared by injecting it into the substrate (Step D). This is a destructive readout (DRO) and resets the pixel to allow it to resume charge collection. If the pixel's charge is measured and is far from reaching its saturation point, the charge can be transferred back to the column photogate, where the pixel can continue collecting charge. This process is known as a nondestructive readout (NDRO), as the detector shifts the charge without destroying it.[165]

The structure of each pixel allows CID detectors to perform nondestructive readouts, which is the main functional difference between CID and CCD detectors. Measuring charge without destroying it allows each pixel to perform a nondestructive readout repeatedly during an analytical measurement, as needed. Since each pixel can sense and measure its own charge, the pixels within a CID can be randomly accessed for charge integration. These NDRO and random-access integration (RAI) features make these detectors well suited for a number of applications in spectroscopy.[166,167]

The presence of photogates on each detector element reduces the size of the photoactive area of each pixel, which decreases the total amount of charge that each pixel can accumulate before reaching saturation. This construction also decreases the uniformity of charge storage and readout across the array. Benefits to the structure of this detector are that it's inherently resistant to charge blooming, it allows for simultaneous wavelength measurements and background correction, plus simultaneous internal standardization for real-time drift correction.[168]

These devices can rapidly access all the addressable pixels within the array, which allows the detector to optimize the signal-to-noise ratio for each wavelength being measured. This optimization also increases the device's dynamic range to make it functional over several orders of magnitude. This feature is beneficial for applications in which both low- and high-intensity wavelengths must be measured simultaneously.[169] Figure 22.19 illustrates examples of typical CCD (top) and CID (bottom) devices.

FIGURE 22.19   CCD (top two) and CMOS (bottom two) images.

(With kind permission from Teledyne DALSA)

FIGURE 22.19   (Continued)

22.20   ANALYTICAL PERFORMANCE

The quality of any developed and optimized analytical method can be assessed by a number of different figures of merit, including accuracy, precision, sensitivity, speed, detection limits, working range, ease of use and reproducibility. Both the instrument operating conditions and method parameters can affect these performance characteristics, so it is important to be familiar with these factors, to understand their effects, on the analytical data being collected.

22.21   DEPENDENCE ON ENVIRONMENTAL OPERATING CONDITIONS

All commercially available ICP-OES instruments have cooling, venting, electrical and gas requirements for proper and safe operation. These instruments also prefer a certain set of room conditions with specified temperature, pressure, humidity and room vibration requirements. The environmental conditions will vary slightly between instruments; however, instruments should generally be operated under the following range of conditions.

The laboratory should be maintained at a temperature between 15 and 35 °C. Ideally, the temperature within the room will not change, particularly when the instrument is being used for analysis. If the room temperature cannot be maintained over the course of a day, the rate of temperature change should be minimal. Most instruments can tolerate temperature changes of one to three degrees per hour; however, significant temperature changes will produce wavelength drift.

The relative room humidity should be kept between 20–80% in a noncondensing and noncorrosive atmosphere. Most spectrometers remain constantly purged with a low flow of dry, high-purity gas; however, if the purge is disrupted or stopped and the gas inside the spectrometer exchanges with the air in the room, moisture in the air could deposit on the components inside the spectrometer and fog the optics.

Though ICP-OES instruments are designed with a certain level of robustness built in, the manufacturer's recommendations for environmental conditions should be followed for optimal instrument performance.

22.22 EXHAUST REQUIREMENTS

An exhaust system is used for removing heat and gases from the instrument for both safety and stability reasons. Heat generated by the system electronics must be removed to prevent them from overheating and malfunctioning while excess heat from the plasma must be withdrawn to prevent the temperature inside the torch box from rising and producing background emission from the torch.

The torch box area must be properly vented to remove gases generated in the plasma, which may contain ozone and other noxious substances. The positioning of the exhaust vent(s), the minimum and maximum draw required and the diameter of the vent tubing vary between instruments, so it is important to follow manufacturer recommendations.

22.23 ELECTRICAL REQUIREMENTS

Optical emission instruments require high voltage power to operate properly. If the instrument operates with a water-cooled load coil, a chiller or water recirculator may be used, which has a voltage requirement that is different from that for the instrument. A computer, which likely controls the instrument, may have its own set of electrical requirements. The power requirements, including voltage, frequency and phase, will vary slightly between instruments and the manufacturer requirements should be closely followed.

The performance and longevity of an ICP-OES can be affected by the quality of the power being delivered to the instrument. In severe cases, power fluctuations can cause severe damage to electronic components in the instrument. A power conditioner may be employed to protect the instrument from voltage sags, transient voltage issues and general regulation problems. An uninterruptible power supply (UPS) will allow the instrument to be properly shut down in the event of a power outage.

22.24 TEMPERATURE AND PRESSURE REQUIREMENTS

The spectrometer must be maintained at a constant temperature and pressure to prevent excessive wavelength drift during analysis. Spectrometers are typically heated to a temperature between 35 and 38 °C and the set point is maintained to a precision of approximately ± 0.1 °C. Maintaining the spectrometer at a temperature well above that in the laboratory reduces the effect of small temperature changes in the room on wavelength drift.

The spectrometer is typically purged with a relatively dry, high-purity gas to maintain a constant environment inside the spectrometer. A wide range of flow rates can be used for the purge gas, from 0.1–15 L/min; however, the pressure inside the spectrometer should be at equilibrium prior to starting a sample run to avoid wavelength drift due to pressure changes.

22.25 MAINTENANCE

Routine preventive maintenance is the one of the best ways to ensure proper instrument performance, as poor-quality data can often be traced back to the sample-introduction system. General good laboratory practices should be followed in terms of maintaining a clean laboratory, and any spills should be cleaned up immediately. Solutions should be properly stored and labeled, and standards and samples should be kept in covered vials or containers at the completion of a sample run to minimize spills, contamination or loss from volatilization. A clean-rinse solution should be run through the sample-introduction system for a suitable length of time at the end of each sample run to help reduce the frequency with which the sample-introduction system must be disassembled and thoroughly cleaned.

If a peristaltic pump is being used, all pump tubing should be regularly checked. Solution must move smoothly through the tubing, both into and out of the instrument, to minimize pulsations in the plasma. Tubing should be checked for distortions, flat spots and tears, and should be replaced if any are discovered. The platens on the pump should be properly tightened down on the pump tubing to allow solution to move through the tubing without any jerking movements.

The nebulizer should periodically be checked for cracks, blockages or leakages. The performance should be checked by pumping a clean solution through the nebulizer and visually inspecting the resulting aerosol to ensure it has the proper shape and lacks irregular pulsations. The nebulizer should be cleaned according to the manufacturer instructions and should be carefully stored when not in use to prevent breakages.

The spray chamber requires little to no maintenance; however, it should be visually checked for cracks on occasion. The spray chamber should be viewed during aerosol generation to ensure the inside surface is being properly wetted and large droplets are not beading up and disrupting the aerosol transfer. The drain tubing should fit tightly onto the spray chamber to allow waste solution to be properly removed.

The torch should be checked for leaks, cracks or other physical damage. The injector should be checked for deposits or blockages and thoroughly cleaned if any are present. If the torch has gas connections or o-rings, those should be inspected for damage or leakages. Torches will accumulate deposits during normal sample analysis and should be cleaned as necessary to remove them. Over time, small deposits will accumulate that do not get removed during cleaning. These deposits should not be of concern unless the instrument's performance is affected.

Most of the remaining parts of the instrument, including the RF generator, the spectrometer and the detector, require little or no maintenance other than an optical entrance window that requires occasional cleaning or air filters that need to be cleaned or replaced.

22.26 DEPENDENCE ON PLASMA OPERATING CONDITIONS

Optimum plasma operating conditions are critical in obtaining the highest-quality data. Plasma parameters will impact the stability and energy of the plasma and, therefore, the behavior of the elements, and are typically optimized to maximize the emission intensity for the analyte wavelengths in the method. Plasma conditions that are optimal for one type of sample are often different from those for another sample type, so the operating conditions should be carefully evaluated for each application.

Plasma parameters to consider for optimization are:

- RF power
- Nebulizer gas flow
- Auxiliary gas flow
- Coolant gas flow
- Pump settings
- Radial viewing height

It should be noted that, while important, plasma optimization is not as critical as the other method-development steps. ICP-OES methods typically are developed for the analysis of multiple elements. Each element prefers a slightly different set of plasma parameters to achieve its maximum intensity for the wavelength selected. Since ICP-OES instruments collect emission intensities from groups of elements simultaneously, it is not practical to select different plasma settings for each individual element. Therefore, the plasma settings for multielement analysis will be somewhat of a compromise for most of the elements in the method. Most modern ICP-OES instrumentation provides an automated optimization feature to rapidly and automatically select the best plasma parameters to maximize the emission intensity for as many elements as is possible in a multielement method. [170]

22.27 RF POWER

The quality of the RF generator and the RF power selected for use during analysis will determine the quality and precision of the resulting analytical measurements. While there may be more than one RF power setting that will produce high-quality data, this is an important plasma parameter because the RF field, in combination with the argon gas, is responsible for generating and maintaining the plasma source.

RF generators operate at a frequency of either 27.12 MHz or 40.68 MHz. These are the only two frequencies approved by the Federal Communications Commission (FCC) for RF generator operation, as other frequencies will interfere with established communication activities. There has been much debate over the relationship between an RF generator's operating frequency and its resulting performance;[171,172] however, there is no evidence to indicate that one frequency is beneficial over the other.

The stability of the plasma largely relies on the generator's ability to adjust to changing plasma conditions. Each time a new sample is introduced, the composition of the plasma rapidly changes. As the composition changes, the power required to adjust to the new plasma conditions changes and the RF generator must react accordingly. Matching the power being output by the generator to that required by the new plasma conditions is known as impedance matching and can be accomplished with two basic oscillator designs: (1) crystal-controlled (fixed-frequency) and (2) free-running (variable frequency).[173] There are benefits and detriments to both designs; however, the variable frequency design of free-running oscillators make them better able to generate a plasma from a cold start and better able to adapt to changes incurred by challenging sample matrices. A stable, successfully formed plasma is illustrated in Figure 22.20.

In addition to having a well-designed generator, an appropriate RF power must be selected for the application. Generally speaking, RF power is proportional to plasma temperature. Given the same

FIGURE 22.20 A close-up of a Thermo Jarrel Ash Atomscan 16 inductively coupled argon plasma's torch.

(With permission from Wblanchard)

set of gas flows and sample uptake parameters, a higher RF power produces a hotter, more energetic plasma. This is particularly advantageous for sample matrices that contain organic solvents or high levels of dissolved solids, which are notorious for causing plasma instabilities or blowouts. While a higher RF power produces a more robust plasma, this setting should still be carefully selected to avoid using a power that is higher than necessary. Increasing the RF power increases the amount of emission that gets produced from all components in the sample solution, which can increase the level of background emission, degrade the signal-to-noise ratio for the analytes of interest and degrade the background equivalent concentration (BEC).[174,175]

22.28 PLASMA GASES

While the plasma gases play an important role in generating and sustaining a stable plasma, the individual gas-flow settings can have an effect on the emission of some elements and on the overall data quality. The coolant gas serves two major purposes when the plasma is being operated: plasma generation and torch cooling. The coolant gas flows through the outer tube in a swirled pattern to help the plasma maintain its annular shape. A relatively high flow rate is used to sustain a stable plasma and to prevent the torch from overheating. Since the main function of the coolant gas is related to the operation of the plasma, the flow rate does not have a significant effect on the emission of most elements. Having said that, the flow rate should be chosen with logic and care. A setting that is too low will provide insufficient cooling of the torch. A slight overheating can cause an increase in background emission as blackbody radiation from the torch itself is emitted. A severely overheated torch will melt and will require replacement. A flow setting that is too high is wasteful of the coolant gas and may provide more turbulence in the plasma than is necessary. Flow rates for the coolant gas can range from 8–20 L/min; however, flow rates between 10 and 14 L/min are appropriate for most ICP-OES applications.

The auxiliary gas flows through the intermediate tube of the torch and supplements the coolant gas to shift the base of the plasma further away from the end of the inner tube (injector). Not all applications or torch configurations require the use of an auxiliary gas; however, it is useful during the analysis of sample matrices with high levels of dissolved solids, organic solvents or other matrix components that are known to cause a buildup of material inside the injector. The auxiliary gas adds turbulence to the plasma, which degrades the penetration of the nebulizer gas into the plasma. Therefore, ideal circumstances would dictate that minimal or no auxiliary gas is used. However, if the application requires an auxiliary gas, the flow rate should be set such that the plasma is far enough from the injector to avoid a blockage from forming, yet not so far that the plasma becomes unstable and at risk of being extinguished. Flow rates for the auxiliary gas, if being used, are typically between 0.1 and 1.5 L/min.

Of the three plasma gases, the nebulizer gas typically has the most significant effect on analyte emission and should be optimized carefully. The nebulizer gas flows through the center tube (injector) of the torch and carries the sample into the plasma. The gas must be able to penetrate the plasma and travel along its central channel to maximize the efficiency with which the sample is desolvated, vaporized, atomized, ionized and excited. Since the nebulizer gas carries the sample into the plasma, it stands to reason that increasing the nebulizer gas will transport more sample into the plasma, which will produce larger emission signals and more sensitive measurements. This is not the case, however. The nebulizer gas must be carefully selected based on the challenge of the application and analytes being measured. In addition to generating some turbulence, which affects plasma stability, the nebulizer gas transports the nebulized sample aerosol, which affects the load on the plasma. If the change in plasma load is too severe, the RF generator will not be able to adjust its power output to compensate, and severe plasma instability will result. In severe cases, the plasma load will be great enough to extinguish the plasma completely. Conversely, if the nebulizer gas flow is too low, the sample will not adequately penetrate the plasma, and emission signals will significantly degrade. Most instruments will allow the nebulizer gas flow to be set between 0.1 and 2 L/min; however, a setting of approximately

1 L/min (or the equivalent pressure if the nebulizer is not mass flow-controlled) is typically used for most applications. Elements from different groups on the periodic table will produce optimal emission with slightly different nebulizer-gas settings. Since most applications involve the analysis of multiple elements, a compromise must be made, and a nebulizer setting should be selected that will produce optimal sensitivity and detection limits for the overall method.

22.29 PUMP SETTINGS

Bulk solution is delivered to the instrument via either a peristaltic or a syringe pump. The speed and precision with which the solution is introduced to the nebulizer significantly affects the quality of the resulting aerosol. The pump speed, combined with the inside diameter of the pump tubing, dictate the rate of solution that is delivered to the nebulizer. While each nebulizer will have a recommended flow rate for sample introduction, an optimal pump speed should be selected, such that a consistent, stable aerosol is produced. If the pump speed is set such that the nebulizer aspirates solution faster than it is supplied via the pump, the nebulizer is being underfed ("starved"). Conversely, if the pump delivers solution faster than the nebulizer can aspirate it and produce an aerosol, the nebulizer is being flooded. Most nebulizers operate optimally when they are being starved for solution. Flooded nebulizers will produce an aerosol with a pulsing stream of large droplets that spill out the end. This "spitting" is from the excess solution that was pumped into the nebulizer but not converted to an aerosol. Some nebulizers, known as self-aspirating nebulizers, are designed to operate by drawing the solution through the pump tubing without assistance from a pump. Since these nebulizers do not require a pump for sample uptake, self-aspirating nebulizers are excluded from the text in this section.

The precision of the delivered solution will affect the precision of the aerosol. This is dictated by the style of the pump. A peristaltic pump contains a number of rollers that push solution through a flexible piece of tubing by alternatively compressing and relaxing the walls of the tubing. Each time the tubing is decompressed, a vacuum is created, and a miniscule amount of solution is drawn backward into the pump tubing. This causes a fluctuation in the volume of solution being delivered to the nebulizer, which produces an aerosol and resulting emission signals with a measurable pulsation.

A syringe pump can also be used for solution delivery to the nebulizer. A syringe pump is a type of infusion pump that delivers solution via small, rapid pulses. The pulsations are smaller and more frequent than those incurred with a peristaltic pump, which results in an aerosol and emission signal with smaller pump-related fluctuations. This concept is illustrated in Figure 22.21. A syringe pump

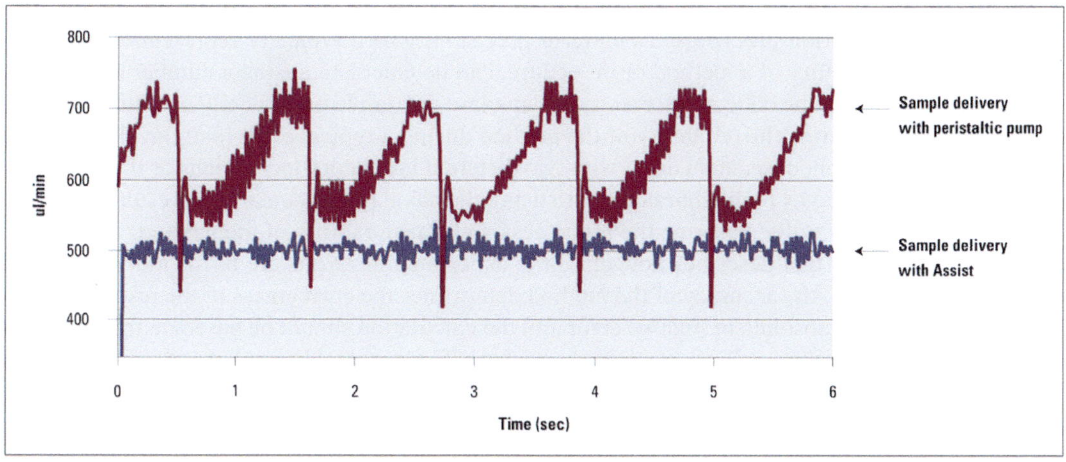

FIGURE 22.21   Signal fluctuation produced with a peristaltic pump versus a syringe pump.

(With kind permission from Glass Expansion)

will produce an aerosol with fewer fluctuations; however, pulsations from a peristaltic pump can be minimized with the use of a small-diameter pump and small pump rollers.

22.30   PLASMA VIEWING HEIGHT

Emission signals should be collected from a specific part of the plasma to maximize the signal to background ratio and sensitivity of each measurement. Often referred to as the "normal analytical zone", this section of the plasma consists of a temperature zone that is optimal for atomic and ionic emission. Axial measurements are made along the central channel of the plasma, which prevents emission from being collected from only one of the temperature zones. Radial measurements are made perpendicularly to the direction of the plasma, so the observation height, also known as the viewing height, should be carefully optimized.[176]

Viewing height affects the emission strength of different elements in different ways. Some would prefer a hotter plasma, which would render an optimal viewing height that is closer to the load coil. Other elements would prefer slightly cooler conditions, which would put the optimal viewing height further from the load coil and closer to the tail of the plasma. Unless the instrument is being used for single-element analysis, the plasma viewing height will be a compromise. This process is computer-controlled through the instrument's software; however, manual adjustments can sometimes be made to fully optimize the observation height of the measurement.

22.31   PRECISION AND ACCURACY

When discussing analytical performance, precision and accuracy are figures of merit that indicate the robustness and reliability of the instrument and of the method developed for analysis. The terms are often used together; however, it is important not to use them incorrectly or interchangeably. Precision is a term that describes the agreement between replicate results, which provides an indication of the reproducibility of the method. It is a figure of merit that can be obtained by repeating the measurement of a particular standard or sample and calculating the variation between the replicate results. To properly calculate precision, data for the same elements must be collected the exact same way each time.

Both short- and long-term precision can be calculated for a method. Short-term precision typically represents the variation between back-to-back measurements of a small number of standards and/or samples. This represents the reliability of the method for a particular application and, depending upon which solutions are included in the measurement, this can include the reliability of the sample-preparation procedure. Long-term precision, which typically represents the robustness and/or transferability of a method or procedure, can be calculated using a number of different approaches. Results can be collected for a chosen standard or sample periodically over the course of a sample run to determine the reliability of the method during a typical eight-hour day. Results can also be collected on nonconsecutive days, with two different laboratory technicians or in two different laboratories (as long as both laboratories are using the same instrumentation). Results collected under these conditions will determine the robustness and transferability of the method.

Accuracy is a term that describes how close the experimental results are to the known ("true") results. In other words, the accuracy of the method determines the correctness of the results. This is expressed as either an absolute or relative error and the calculation should be based on the measurement of a certified or other well-characterized standard. The acceptable level of inaccuracy should be determined during the method-development and optimization stage. All calculated results are going to possess a small amount of inaccuracy due to basic errors and uncertainties in preparing and analyzing samples. If the calculated accuracy does not meet the requirements of the method, the source of the error should be investigated. Errors can consist of two different types: random or systematic. Each type of error could be due to a number of sources and a combination of errors could be contributing to the overall issue, so troubleshooting inaccuracies should be approached with care.

Precision and accuracy should not be used interchangeably as they describe significantly different figures of merit. While it's desirable for a method to have excellent precision and accuracy, it is possible to have one without the other. Results that are precise but inaccurate could indicate that the instrument settings are properly chosen, but an issue exists with the calibration curve or there are interferences that have not been properly accounted for.

If a method has good accuracy, but poor precision, it is sometimes referred to as being "accurate in the mean". In this situation, the accuracy is a result of systematic imprecision and typically indicates an issue with the sample-introduction system (steering mirror is moving too slowly (dual-view methods only), sample uptake time is too short, sample flow rate is too high, platens on the peristaltic pump are not tightened properly).

22.32   DETECTION LIMITS

One of the most important, yet one of the more highly debated figures of merit for an analytical method is the limits of detection (LOD). For a given analyte, IUPAC defines the detection limit as the smallest change in the emission signal (x_L) that can be detected with statistical certainty.[177] In other words, this is the lowest signal that can be detected above the blank and is defined according to the following equation:

$$x_L = k\sigma_B$$

Where x_L is the net intensity, k represents a numerical multiplier that is selected according to the desired statistical confidence level and σ_B represents the standard deviation of replicate measurements of the blank solution. The net intensity can be converted to a detection-limit concentration using the following equation:[178]

$$LOD = (k\sigma_B)/S$$

where k represents the same multiplier as that in the previous equation, σ_B represents the standard deviation of the blank and S is the slope of the calibration curve for that analyte (i.e., the sensitivity). Kaiser[179] asserts that an appropriate value for k is 3, as that will represent a 95% confidence interval for the calculated detection limit for most applications, so 3 is the most commonly used value in this calculation.

The debate over the formula for calculating detection limits stems from its definition, which indicates that the concentrations must be calculated with statistical certainty.[180] Most accepted formulas for calculating detection limits are based on the IUPAC equation listed previously. The resulting concentrations are typically referred to as "instrumental detection limits" and allow detection limits to be compared between different instruments or between different conditions on the same instrument. While useful for comparison purposes, detection limits based on this formula are unrealistic and do not reflect those that would be obtained when measuring analyte signals in the presence of the sample matrix under investigation.

Suggested modifications have been made to reflect more conservative estimates of detection limits. The Environmental Protection Agency (EPA) suggests the calculation of a "method detection limit" based on seven replicate analyses of a blank fortified with the analytes of interest at two or three times the concentration of the calculated instrument detection limit.[181] This is a more conservative estimate of the detection limits and accounts for any signal degradation caused by the sample matrix. Currie[182] suggests calculating a "determination limit" by using the IUPAC definition of LOD with a multiplier of 10 instead of 3. Currie defines this as "a determination limit at which a given procedure will be sufficiently precise to yield a satisfactory quantitative estimate". Long and Winefordner[183] recommend multiplying the standard deviation of the blank by 6 to calculate a "limit of guarantee".

22.33 LIMIT OF QUANTITATION

The limits of quantitation (LOQ) are often calculated along with the limits of detection. The LOQ is a conservative estimate of the LOD and is sometimes used as the reporting limit for a method.[184] For each analyte, the LOQ represents the lowest concentration that can be measured and reported with a sufficient amount of precision and accuracy. The desirable level of precision and accuracy is somewhat arbitrarily defined, so there is no concrete formula for calculating the LOQ. However, the desired precision level is often around 10% (reported as a relative standard deviation), which means the LOQ is calculated using ten times the standard deviation of the blank.[185] By definition, the LOQ cannot be lower than the LOD and it should not be equal to the LOD.

22.34 BACKGROUND EQUIVALENT CONCENTRATION

A figure of merit that relates to the instrument detection limit is the background equivalent concentration (BEC). For a given analyte, the BEC is the concentration of that analyte, which produces a net signal (emission signal minus the contribution from the background) equal to the background signal at that wavelength. In other words, this is the analyte concentration that yields a signal-to-background ratio of 1.[186] There are a number of different formulas that exist for calculating BEC values; however, they can easily be determined using the calculated calibration curves. If the curves are plotted as measured emission intensity (x-axis) versus concentration (y-axis), the BEC for each analyte is represented by the y-intercept value. Since the BEC is based on the noise associated with the background signal (which is typically ~1% of the signal intensity of the background), a rough BEC calculation can be performed by multiplying the instrument detection limit by a factor of 30.

22.35 SENSITIVITY

Sensitivity is generally referred to as the ability to confidently measure small differences in analyte concentrations. The acceptable definition of sensitivity is outlined by the International Union of Pure and Applied Chemists (IUPAC) and refers to calibration sensitivity. Sensitivity can vary by element and is represented by the calculated calibration curve. Most calibrations that are used with ICP-OES applications are linear and can be represented using the following equation:

$$y = mx + b$$

where y represents the instrument response for a given analyte at a given concentration, m represents the calibration sensitivity (i.e., the slope of the calibration curve), x represents the analyte concentration and b is the measured signal for a blank solution. If the concentration of the signal is zero, the equation becomes $y = mx$ and the calibration sensitivity becomes independent of concentration.

Mandel and Stiehler[187] published a slightly different definition of sensitivity. Referred to as analytical sensitivity, this definition includes the precision of the measurement according to the following equation:

$$\gamma = m/s_S$$

where γ represents the analytical sensitivity, m represents the slope of the calibration curve and s_S represents the standard deviation of the measured signal. Since this term is based on the slope and standard deviation of the measurement, the sensitivity value is independent of the units that were used to measure the emission signals. A drawback to this definition is the inclusion of the signal standard deviation. Since s_S can vary with analyte concentration, analytical sensitivity is typically concentration dependent.

22.36 METHOD-DEVELOPMENT CONSIDERATIONS

As with many techniques, its success in the laboratory depends upon the quality of the methods that are developed for the application work being conducted. While ICP-OES instruments provide automated, intuitive operation with ppb-level detection limits and working ranges that extend over several orders of magnitude, developing a useful, well-optimized analytical method can be a manual, labor-intensive and time-consuming process.

Whether a method is being developed from scratch or an existing method is being further optimized, there are a number of parameters that should be taken into consideration to ensure the method is best suited for the intended application. These include:

- Analytical wavelengths
- Interferences
- Plasma parameters
- Data-acquisition parameters
- Validation of method

Let's take a look at each of these considerations, in turn, to understand how each factor plays a role in the quality of the final optimized method. It should be noted that the theory behind this approach is applicable to the development of an analytical method for any elemental analysis technique; however, some of the specific considerations will change. For example, if developing an analytical method for an inductively coupled plasma-mass spectrometry application, one would be choosing isotopes instead of emission wavelengths for measuring the analytes and internal standards.

22.37 ANALYTICAL-WAVELENGTH CONSIDERATIONS

When choosing appropriate wavelengths for the elements of interest, it's important to keep in mind the method's desired performance attributes. For example, if method development is taking place for quantifying trace elements in drinking water samples, the main performance attributes will likely be detection capability and accuracy. Therefore, the most sensitive wavelength should be chosen for each element and all potential interferences should be identified and carefully corrected. Alternatively, if a method is being developed to quantify elements in oil additives, a wider working range may be desired. In this case, wavelengths with lower sensitivity may need to be chosen to maximize the concentration range over which the elements can be calibrated.

A critical factor in determining the suitability of a wavelength is whether it suffers from interferences. Several types of interferences exist in ICP-OES, and their severity is dependent on the analyte wavelength, other elements present in the sample and the sample matrix itself.[188] Evaluating interferences is not a trivial task and must be carried out with a great deal of care and thought. Interferences that are erroneously identified or not compensated for correctly will result in poor-quality data. With the number of interference types and the variety of correction approaches possible, interference correction can quickly become an overwhelming task, particularly to an inexperienced operator. For this reason, when addressing interferences, the following wavelength selection procedure is recommended.

1. Choose several wavelengths for each element in the method
2. Collect analyte wavelength scans
3. Visually inspect the peaks for the data collected
4. Review data
5. Eliminate unsuitable wavelengths

Let's review this process in more detail. In the first step, it is recommended to choose two or three wavelengths for each element of interest. This includes both the analyte elements and any applicable internal-standard elements. Typically, the most sensitive wavelength for each element would be chosen, along with two wavelengths of slightly lower sensitivity. If a wide range of concentrations is expected to be encountered in the samples, a high-sensitivity wavelength might be chosen along with a low-sensitivity wavelength. In some cases, the sample matrix will produce an emission spectrum that is so complex, there may be only one or two suitable wavelengths (interference free and providing the desired working concentration range) to choose for the analysis.

Once a set of wavelengths has been chosen for evaluation, data should be collected to determine which wavelengths are best suited, based on the requirements of the application. It is recommended that three wavelength measurements be collected for each of the following solutions:

- A blank (calibration and method blanks, where applicable)
- A low-concentration calibration standard
- A high-concentration calibration standard
- A sample that represents each type of sample matrix to be analyzed

Data from the blank provides an emission profile of the matrix in the absence of analytes. Data from the calibration standards provides profiles for the elements of interest at low and high concentrations. Both sets of measurements are required for the data inspection outlined in steps 3 and 4 described earlier. The highest-concentration calibration standard should be used for this data collection to ensure that none of the selected wavelengths suffer from peak broadening or self-absorption.[189] Sample measurements produce emission profiles for the elements of interest in the presence of the sample matrix. One sample for each sample matrix type should be collected to ensure emission profiles are examined in every applicable sample matrix. For laboratories analyzing a wide range of sample matrices, collecting wavelength-measurement scans for each matrix type could require data collection for a large number of samples.

After wavelength data has been collected, they should be visually inspected to determine that the emission peaks have the proper shape and size. An ideal wavelength would produce the series of emission signals illustrated in Figure 22.22. In Figure 22.22, the emission from the blank

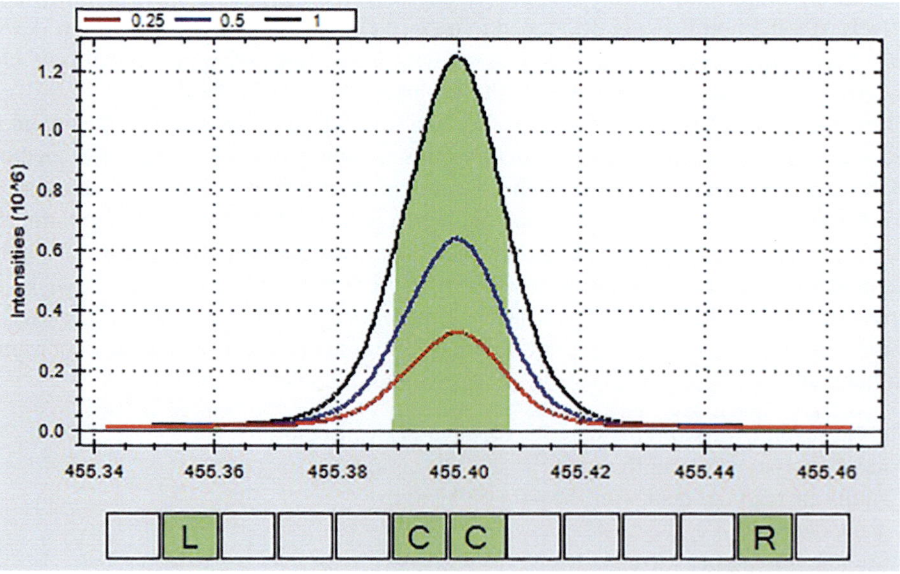

FIGURE 22.22   Examples of emission profiles for three calibration standards and a blank.

(With kind permission from Thermo Fisher Scientific)

produces a flat emission signal with a relatively low intensity. The emission signals for the calibration standards and representative samples should produce a symmetric, Gaussian-like shape with a single peak. Each peak should exhibit a proportional change in magnitude to match the difference in concentrations. For example, data from a standard containing elements at 1.0 ppm would be expected to have roughly twice the signal of that for a 0.5-ppm standard, while maintaining approximately the same shape.

Numerical data should be inspected to determine the intensity and precision at each wavelength. The intensity for each standard should increase at a rate that is proportional to the increase in its concentration. If a set of three wavelength measurements was collected for each solution, the measured intensities can be used to calculate the approximate precision for each measured standard. Calculated precision values should be within the acceptable limits for the application.

22.38 INTERFERENCES

Interferences are common with any plasma-based technique, particularly when trying to measure trace-level concentrations in a sample matrix, which contains high concentrations of elements that produce line-rich spectra, such as Fe, Ca and Si.[190] Sample matrices that are known to generate these types of interferences include geological, metallurgical and high-matrix environmental samples, such as soils or waste waters. The three common types of interferences in ICP-OES are: physical, chemical and spectral.

22.39 PHYSICAL INTERFERENCES

Physical interferences occur when the nebulization and/or transport efficiency of the standards differs from that of the samples. These differences in the physical characteristics of the matrices (density, viscosity, level of dissolved solids) can produce errors in the measured sample concentrations. Physical interferences are not wavelength specific and can be overcome by utilizing internal standards and preparing the calibration standards in a matrix that matches that of the samples.[191]

22.40 CHEMICAL INTERFERENCES

Chemical interferences occur when the standards behave differently from the samples as they enter the plasma. These types of interferences typically result from changes in temperature within the plasma and include easily-ionized-element (EIE) effects, molecular emission and plasma loading. Referring to Figure 22.23,[192] the plasma consists of several temperature zones, which translate to varying amounts of energy available for excitation of the ground-state atoms. The plasma is hottest and contains the greatest amount of energy at its base, which is where the sample is introduced. In this region, the plasma contains sufficient energy to atomize and ionize elements before they are excited. Ionization is particularly prevalent for elements in the first two groups of the periodic table, which all have relatively low first-ionization potentials. These are often referred to as "easily ionized elements" (EIE) and include elements like Na, K and Li.

When easily ionized elements are present at low concentrations, very few of the atoms become ionized, which means emission will take place from excited atoms. At high concentrations, a significant number of the atoms will become ionized, which shifts the emission wavelength such that it will predominantly occur from the excited ions. If left uncorrected, this interference will reduce the linear dynamic range of the calibration curve and can potentially produce highly inaccurate results.[193]

This interference can be corrected by adding an ionization suppressant (also known as an ionization buffer) to all the solutions prior to analysis. An ionization suppressant is a solution containing a high concentration of an easily ionized element (1,000 ppm Cs, for example). The solution will produce an extremely high concentration of cesium ions in the plasma, which, according to

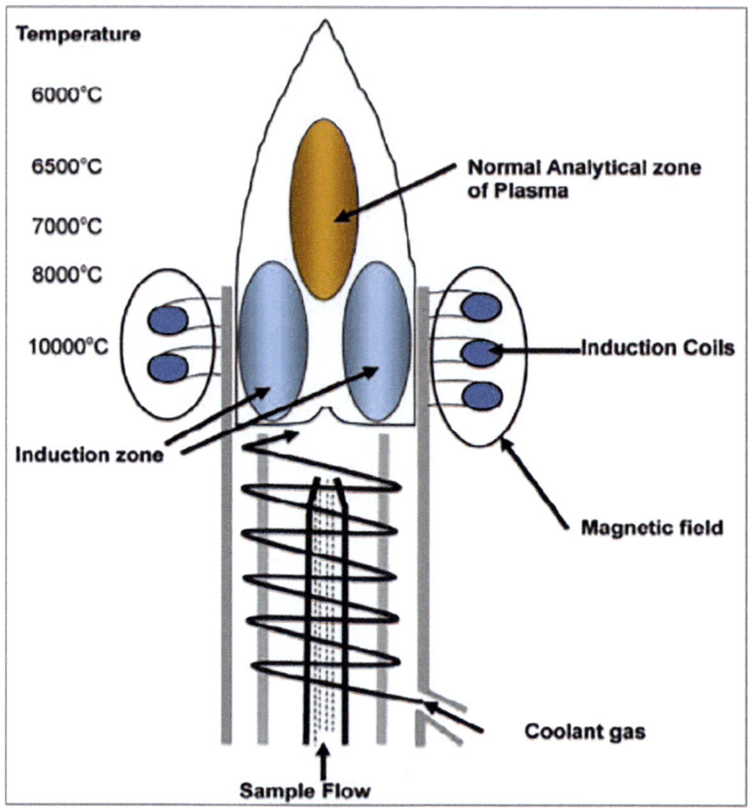

FIGURE 22.23   Different temperature zones within a plasma discharge.[192]

Le Châtelier's principle,[194] will reduce the concentration of ions and increase the concentration of atoms to provide a chemical-equilibrium balance in the plasma. Therefore, the element of interest will remain in its atomic state in the plasma.

If a dual-view instrument is being used, an additional step to take in correcting for this type of interference is to also measure emission from a radially configured plasma. EIE effects are much more significant for axial plasmas, as emission is being measured from species that are present in all temperature zones within the central channel of the plasma. If emission is measured perpendicularly to the direction of the plasma (radial view), emission will be measured from a single temperature zone.[195]

Converse to the formation of EIE interferences are the formation of molecular interferences. If we refer again to Figure 22.23, the tip (also known as the "tail") of the plasma is the coolest and least energetic part of the plasma.[196] In this region, atoms and ions that were formed in the hotter part of the plasma can combine to form molecules. These molecules can become excited and emit light, which will produce broadband, molecular emission spectra.

This type of interference is relatively straightforward to correct and requires no intervention from the analyst. Since the coolest part of the plasma is in its tail, removing that portion of the plasma prevents atoms and ions from traveling through a zone that's cool enough to allow them to combine and form molecular species. All modern ICP-OES instruments are set up to automatically and effectively remove the tail of the plasma. There are a variety of methods in which this is accomplished—a flow of inert gas that is counter to the direction of the plasma, a high-velocity shear gas or a cooled cone interface; however, all are effective in cutting off the tail of the plasma. An example of the way in which the tail is removed is illustrated in Figure 22.24.

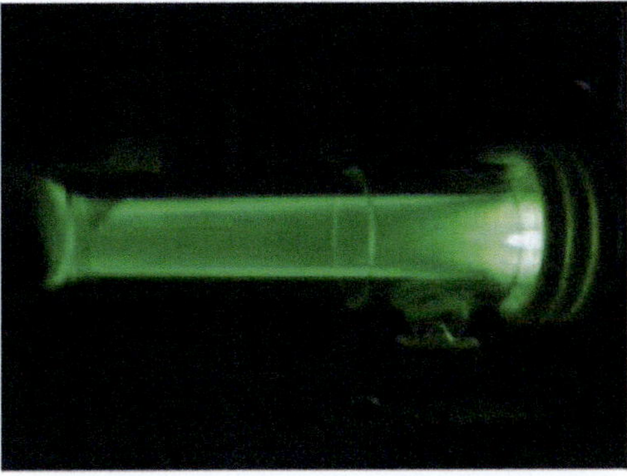

FIGURE 22.24   Image of a dual-view plasma to demonstrate removal of plasma tail.

(With kind permission from Thermo Fisher Scientific)

As with chemical interferences, another option for eliminating molecular interferences is to measure emission from a radially configured plasma. If emission is collected at 90 degrees to the plasma and the observation height is optimized, emission should be collected from the "normal analytical zone", which is well separated from the cooler tail of the plasma.

Plasma loading is an interference that is sometimes experienced when samples contain organic solvents or high levels of dissolved solids.[197] When a significant amount of material is introduced into the plasma, it cools the plasma and, in severe cases, can extinguish it completely. Cooling the plasma reduces the amount of available energy, which affects the emission intensity of the elements present. As a result, if samples contain concentrations of dissolved solids that are different from those in the standards, the emission intensity for the same analyte concentration will be different in the samples and standards, which could produce data inaccuracies.

This interference can usually be addressed by preparing standards in a matrix that matches that of the samples, utilizing proper internal standards, choosing sample-introduction components that were designed for the application and selecting sample-introduction conditions that minimize plasma loading.

22.41   SPECTRAL INTERFERENCES

The third and sometimes most challenging interferences are spectral. These interferences occur when emission from one or more species in the sample matrix overlaps with the emission for the analyte of interest. These interferences can result in a background shift, a partial peak overlap or a direct overlap. If the analyte is suffering from a direct spectral overlap, the interfering element must either be removed from all solutions prior to analysis, or an alternative wavelength must be selected. If the interfering element produces a simple background shift (either a flat, raised baseline or a sloping background), the analyte can be accurately quantified by the use of optimized background correction points.

If the analyte suffers from partial spectral overlap, a combination of background correction and mathematical correction may need to be employed. An example of partial spectral overlap is illustrated in Figure 22.25. The figure depicts data for a solution containing 1 ppm B in a matrix that contains 100 ppm Fe.

A shoulder is clearly visible on the right side of the main peak, indicating the presence of a spectral overlap on B at 249.773 nm. If the emission profile for the sample solution is overlaid with that for single-element solutions containing 1 ppm B and 100 ppm Fe, the image in Figure 22.26 is produced.

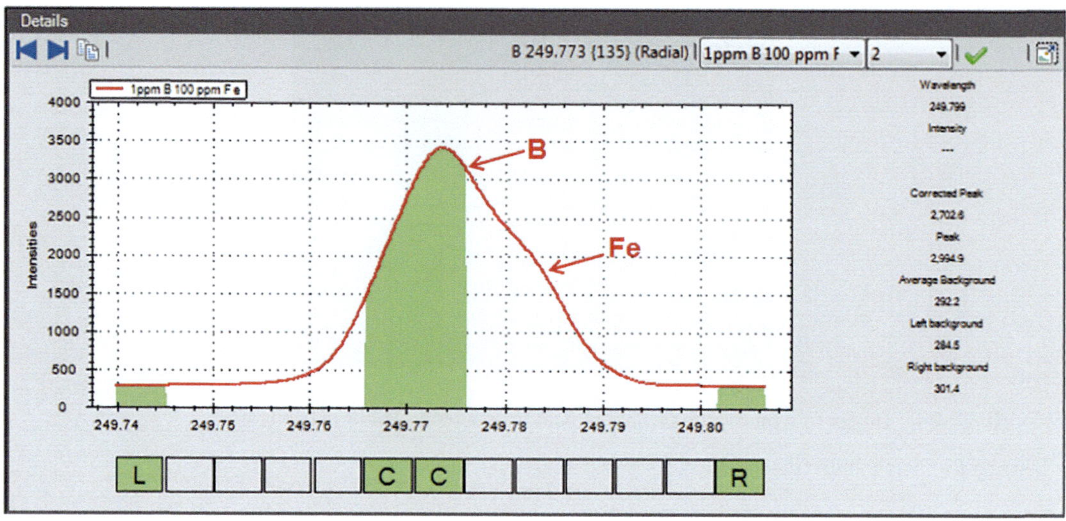

FIGURE 22.25 Emission profile of 1 ppm B in a matrix containing 100 ppm Fe.[198]

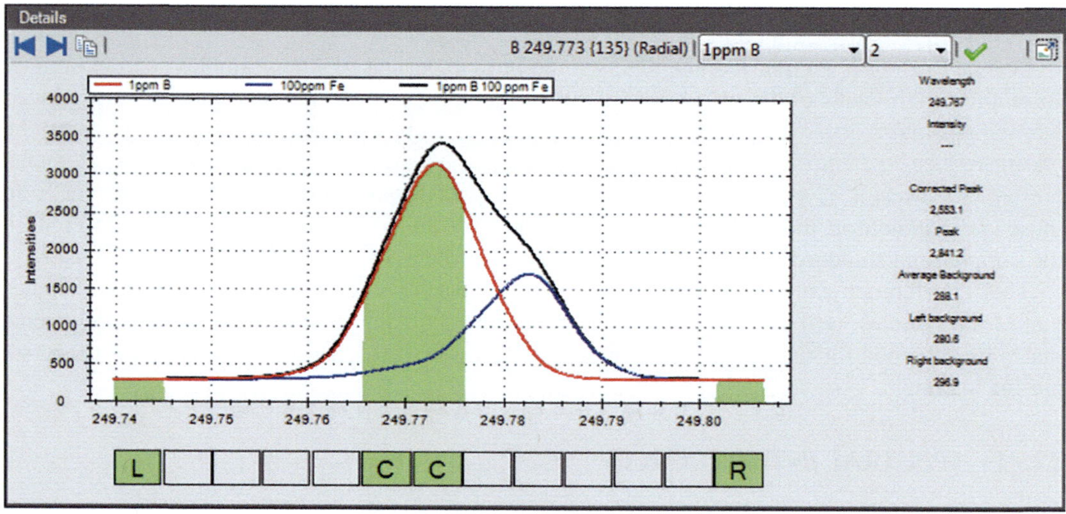

FIGURE 22.26 Emission profiles to demonstrate spectral overlap.[198]

In this figure, the red peak represents the emission profile for 1 ppm B at 249.773 nm, whereas the blue trace represents the emission peak for 100 ppm Fe at 249.782 nm. The black outline represents the combined emission from a sample containing 1 ppm B and 100 ppm Fe. The center green-shaded region represents the area under the peak, which will be included for data collection. This area includes a portion of the emission peak from Fe and, if left uncorrected, this analytical measurement might produce erroneous data.

The best way to correct for this interference is to select an alternative emission wavelength for B that is free from interferences. If an alternative wavelength cannot be used, an inter-element-correction (IEC) factor should be calculated to correct for the overlap of Fe on B. If IEC factors are to be used, they must be carefully calculated and applied to ensure spectral overlaps are being accurately corrected. Improperly calculated IEC factors can produce data that is more inaccurate than if no IEC factors were used.[199]

22.42 DATA ACQUISITION

Data-acquisition parameters play a crucial role in determining the quality of the resulting data. Even if the instrument is performing optimally, sensitive wavelengths have been selected for all analytes of interest and all interferences have been identified and removed, the analytical data will suffer significantly if the proper data-acquisition parameters are not chosen. For the purposes of this text, data-acquisition parameters will be restricted to the choice of axial or radial plasma viewing (if using a dual-view instrument), the integration time and the number of integrations. In general, an axial plasma is used for determining low, ppb-level concentrations, while a radial plasma view is used for higher (ppm-level) concentrations or to help compensate for chemical interferences.[200] The integration time for each plasma viewing configuration must be long enough to collect sufficient emission from all elements being measured. This would usually be dictated by the element with the weakest emission line or the element that is present at the lowest concentration. An axial integration time of 15 seconds and a radial integration time of 5 seconds is fairly typical to use. Finally, the number of integrations must be set. Keep in mind that if every sample is collected using a single integration, statistical analysis and precision data and limits of quantitation cannot be calculated and reported. Therefore two or more integrations must be collected for each sample in order to report this kind of information.

22.43 METHOD VALIDATION

The final step in the method-development process is to validate that the method will produce results that meet the figures of merit required for the application. Methodology that achieves good detection limits yet cannot obtain the required accuracy is unlikely to meet the data quality objectives of the analysis.

A detection limit study is always encouraged for any analytical method. Calculated detection limits and limits of quantitation allow reporting limits to be calculated, in order to determine the lowest concentrations that can be measured and reported for a suite of elements in a given sample matrix.[201]

Regardless of whether a detection-limit study is performed, the developed method needs to be validated prior to use with real samples. The best approach to validating an analytical method is to obtain a certified reference material (CRM) that is appropriate for the application. If an analytical method is valid for the application, data acquired for a suitable CRM should match the certified concentrations that accompany the CRM (within defined error limits).

If a suitable CRM is not available for a given application, a number of other check standards can be utilized. A calibration standard can be prepared and analyzed as a sample to see the concentration at which it was prepared. A calibration or QC (quality control) standard can be purchased from an external source and analyzed as a sample. Samples can be prepared and measured in duplicate to determine the precision of the method. Samples can be spiked with a known concentration of the elements of interest to determine whether the spiked elements can be measured and recovered at their expected concentrations. If interference corrections are being used, interference check standards can be prepared or purchased and analyzed to ensure that the correct results are obtained.

22.44 FINAL THOUGHTS

It should be emphasized that even though ICP-OES is a very useful analytical technique for measuring sub-ppm and ppb levels in different sample matrices, it might have limited use to determine ultralow-level elemental impurities in many biological and related materials. This is mainly due to the fact that the sample-digestion procedure and dilution factors involved can take the analyte levels in solution below the detection capability of the technique, unless an ICP-OES coupled with an ultrasonic nebulizer is being used. For this reason, ICP-MS is probably the safest choice to use. Chapter 24 will give this a little more clarity, by comparing the detection capability of ICP-MS, ICP-OES, AA and AF, particularly for the four classic heavy metals, Pb, Cd, As and Hg.

FURTHER READING

1. K. Ohls and B. Bogdain, *Journal of Analytical Atomic Spectrometry*, **31**, 22–31, 2016.
2. K. Ohls and B. Bogdain, *Journal of Analytical Atomic Spectrometry*, **31**, 22–31, 2016.
3. S. Greenfield, I. L. I. Jones, and C. T. Berry, *Analyst*, **89**, 713–720, 1964.
4. R. H. Wendt and V. A. Fassel, *Analytical Chemistry*, **37**(7), 920–922, 1965.
5. D. A. Skoog, F. J. Holler, and T. A. Nieman, *Principles of Instrumental Analysis*, Cengage Learning, Boston, MA, 2007.
6. G. F. Kirkbright, A. F. Ward, and T. S. West, *Analytica Chimica Acta*, **62**(2), 241–251, 1972.
7. G. F. Kirkbright, A. F. Ward, and T. S. West, *Analytica Chimica Acta*, **64**(3), 353–362, 1973.
8. R. F. Browner and A. W. Boorn, *Analytical Chemistry*, **56**, A786–A798, 1984.
9. A. Montaser, M. G. Minnich, J. A. McLean, H. Liu, J. A. Caruso, and C. W. McLeod, *Sample Introduction in ICP-MS*, Wiley-VCH, New York, 1998.
10. J. F. Tyson, *Analytica Chimica Acta*, **234**, 3–12, 1990.
11. E. R. Denoyer, *Atomic Spectroscopy*, **13**, 93–98, 1992.
12. E. R. Denoyer, *Atomic Spectroscopy*, **15**, 7–16, 1994.
13. G. G. Bortoleto and S. Cadore, *Talanta*, **67**, 169–174, 2005.
14. J. Muller, *Fresenius' Journal of Analytical Chemistry*, **363**, 572–576, 1999.
15. A. U. Shaikh and D. E. Tallman, *Analytical Chemistry*, **49**, 1093–1096, 1977.
16. M. A. Wahed, D. Chowdhury, B. Nermell, S. I. Khan, M. Ilias, M. Rahman, L. A. Persson, and M. Vahter, *Journal of Health, Population and Nutrition*, **24**, 36–41, 2006.
17. E. H. Evans, J. A. Day, C. Palmer, W. J. Price, C. M. M. Smith, and J. F. Tyson, *Journal of Analytical Atomic Spectrometry*, **22**, 663–696, 2007.
18. Z. Long, Y. Luo, C. Zheng, P. Deng, and X. Hou, *Applied Spectroscopy Reviews*, **47**, 382–413, 2012.
19. R. E. Russo, X. Mao, H. Liu, J. Gonzalez, and S. S. Mao, *Talanta*, **57**, 425–451, 2002.
20. J. R. Bacon, K. L. Linge, R. R. Parrish, and L. Van Vaeck, *Journal of Analytical Atomic Spectrometry*, **21**(8), 785–818, 2006.
21. O. T. Butler, J. M. Cook, C. F. Harrington, S. J. Hill, J. Rieuwerts, and D. L. Miles, *Journal of Analytical Atomic Spectrometry*, **21**(2), 217–243, 2006.
22. S. A. Wilson, W. I. Ridley, and A. E. Koenig, *Journal of Analytical Atomic Spectrometry*, **17**, 406–409, 2002.
23. E. H. Evans, J. A. Day, C. Palmer, W. J. Price, C. M. M. Smith, and J. F. Tyson, *Journal of Analytical Atomic Spectrometry*, **22**, 663–696, 2007.
24. J. Mora, S. Maestre, V. Hernandis, and J. L. Todoli, *Trends in Analytical Chemistry*, **22**(3), 123–132, 2003.
25. B. L. Sharp, *Journal of Analytical Atomic Spectrometry*, **3**, 613–652, 1988.
26. R. F. Browner and A. W. Boorn, *Analytical Chemistry*, **56**, A875–A888, 1984.
27. A. Montaser, M. G. Minnich, H. Liu, A. G. T. Gustavsson, and R. F. Browner, *Fundamental Aspects of Sample Introduction in ICP Spectrometry*, Wiley-VCH, New York, 1998.
28. G. Schaldach, L. Berger, I. Razilov, and H. Berndt, *Journal of Analytical Atomic Spectrometry*, **17**, 334–344, 2002.
29. J. W. Olesik and L. C. Bates, *Spectrochimica Acta, Part B*, **50**, 285–303, 1995.
30. L. Ebdon and M. R. Cave, *Analyst*, **107**, 172–178, 1982.
31. S. Augagneur, B. Medina, J. Szpunar, and R. Lobinski, *Journal of Analytical Atomic Spectrometry*, **11**, 713–721, 1996.
32. E. Debrah, S. A. Beres, T. J. Gluodenis, R. J. Thomas, and E. R. Denoyer, *Atomic Spectroscopy*, **16**, 197–202, 1995.
33. F. Vanhaecke, M. VanHolderbeke, L. Moens, and R. Dams, *Journal of Analytical Atomic Spectrometry*, **11**, 543–548, 1996.
34. J. W. Olesik and S. E. Hobbs, *Analytical Chemistry*, **66**, 3371–3378, 1994.
35. K. E. Lawrence, G. W. Rice, and V. A. Fassel, *Analytical Chemistry*, **56**, 289–292, 1984.
36. J. L. Todoli and J. M. Mermet, *Journal of Analytical Atomic Spectrometry*, **13**, 727–734, 1998.
37. A. Gustavsson, *Spectrochimica Acta Part B*, **39**, 743–746, 1984.
38. A. Gustavsson, *Spectrochimica Acta Part B*, **39**, 85–94, 1984.
39. H. Y. Liu and A. Montaser, *Analytical Chemistry*, **66**, 3233–3242, 1994.
40. H. Y. Liu, R. H. Clifford, S. P. Dolan, and A. Montaser, *Spectrochimica Acta Part B*, **51**, 27–40, 1996.

41. S. H. Nam, J. S. Lim, and A. Montaser, *Journal of Analytical Atomic Spectrometry*, **9**, 1357–1362, 1994.
42. H. Y. Liu, A. Montaser, S. P. Dolan, and R. S. Schwartz, *Journal of Analytical Atomic Spectrometry*, **11**, 307–311, 1996.
43. T. T. Hoang, S. W. May, and R. F. Browner, *Journal of Analytical Atomic Spectrometry*, **17**, 1575–1581, 2002.
44. P. W. Kirlew and J. A. Caruso, *Applied Spectroscopy*, **52**, 770–772, 1998.
45. L. Q. Wang, S. W. May, R. F. Browner, and S. H. Pollock, *Journal of Analytical Atomic Spectrometry*, **11**, 1137–1146, 1996.
46. M. Huang, H. Kojima, A. Hirabayashi, and H. Koizumi, *Analytical Sciences*, **15**, 265–268, 1999.
47. E. Debrah, S. A. Beres, T. J. Gluodenis, R. J. Thomas, and E. R. Denoyer, *Atomic Spectroscopy*, **16**, 197–202, 1995.
48. J. L. Todoli and V. Hernandis, *Journal of Analytical Atomic Spectrometry*, **14**, 1289–1295, 1999.
49. R. I. Botto and J. J. Zhu, *Journal of Analytical Atomic Spectrometry*, **9**, 905–912, 1994.
50. B. Budic, *Journal of Analytical Atomic Spectrometry*, **16**, 129–134, 2001.
51. P. Masson, A. Vives, D. Orignac, and T. Prunet, *Journal of Analytical Atomic Spectrometry*, **15**, 543–547, 2000.
52. M. A. Tarr, G. X. Zhu, and R. F. Browner, *Applied Spectroscopy*, **45**, 1424–1432, 1991.
53. R. J. Thomas and C. Anderau, *Atomic Spectroscopy*, **10**, 71–73, 1989.
54. Q. H. Jin, F. Liang, Y. F. Huan, Y. B. Cao, J. G. Zhou, H. Q. Zhang, and W. Yang, *Laboratory Robotics and Automation*, **12**, 76–80, 2000.
55. S. Yamasaki and A. Tsumura, *Water Science & Technology*, **25**, 205–212, 1992.
56. T. T. Nham, *American Laboratory*, **27**, 48L–48V, 1995.
57. I. B. Brenner, J. Zhu, and A. Zander, *Fresenius' Journal of Analytical Chemistry*, **355**, 774–777, 1996.
58. J. Kunze, S. Koelling, M. Reich, and M. A. Wimmer, *Atomic Spectroscopy*, **19**, 164–167, 1998.
59. S. C. K. Shum and R. S. Houk, *Analytical Chemistry*, **65**, 2972–2976, 1993.
60. S. C. K. Shum, R. Neddersen, and R. S. Houk, *Analyst*, **117**, 577–582, 1992.
61. S. C. K. Shum, H. M. Pang, and R. S. Houk, *Analytical Chemistry*, **64**, 2444–2450, 1992.
62. T. W. Avery, C. Chakrabarty, and J. J. Thompson, *Applied Spectroscopy*, **44**, 1690–1698, 1990.
63. D. R. Wiederin, F. G. Smith, and R. S. Houk, *Analytical Chemistry*, **63**, 219–225, 1991.
64. J. A. McLean, H. Zhang, and A. Montaser, *Analytical Chemistry*, **70**, 1012–1020, 1998.
65. M. G. Minnich and A. Montaser, *Applied Spectroscopy*, **54**, 1261–1269, 2000.
66. J. L. Todoli and J. M. Mermet, *Journal of Analytical Atomic Spectrometry*, **16**, 514–520, 2001.
67. E. Bjorn and W. Frech, *Journal of Analytical Atomic Spectrometry*, **16**, 4–11, 2001.
68. M. G. Minnich and A. Montaser, *Applied Spectroscopy*, **54**, 1261–1269, 2000.
69. A. C. S. Bellato, M. F. Gine, and A. A. Menegario, *Microchemical Journal*, **77**, 119–122, 2004.
70. M. J. Powell, E. S. K. Quan, D. W. Boomer, and D. R. Wiederin, *Analytical Chemistry*, **64**, 2253–2257, 1992.
71. S. E. O'Brien, J. A. McLean, B. W. Acon, B. J. Eshelman, W. F. Bauer, and A. Montaser, *Applied Spectroscopy*, **56**, 1006–1012, 2002.
72. J. A. McLean, M. G. Minnich, L. A. Iacone, H. Y. Liu, and A. Montaser, *Journal of Analytical Atomic Spectrometry*, **13**, 829–842, 1998.
73. B. W. Acon, J. A. McLean, and A. Montaser, *Analytical Chemistry*, **72**, 1885–1893, 2000.
74. C. S. Westphal, K. Kahen, W. E. Rutkowski, B. W. Acon, and A. Montaser, *Spectrochimica Acta, Part B*, **59**, 353–368, 2004.
75. G. Schaldach, L. Berger, I. Razilov, and H. Berndt, *Journal of Analytical Atomic Spectrometry*, **17**, 334–344, 2002.
76. B. L. Sharp, *Journal of Analytical Atomic Spectrometry*, **3**, 939–963, 1988.
77. C. Rivas, L. Ebdon, and S. J. Hill, *Journal of Analytical Atomic Spectrometry*, **11**, 1147–1150, 1996.
78. R. H. Scott, V. A. Fassel, R. N. Kniseley, and D. E. Nixon, *Analytical Chemistry*, **46**, 75–81, 1974.
79. D. R. Luffer and E. D. Salin, *Analytical Chemistry*, **58**, 654–656, 1986.
80. J. L. Todoli, S. Maestre, J. Mora, A. Canals, and V. Hernandis, *Fresenius' Journal of Analytical Chemistry*, **368**, 773–779, 2000.
81. X. H. Zhang, H. F. Li, and Y. F. Yang, *Talanta*, **42**, 1959–1963, 1995.
82. M. Wu and G. M. Hieftje, *Applied Spectroscopy*, **46**, 1912–1918, 1992.
83. G. Schaldach, H. Berndt, and B. L. Sharp, *Journal of Analytical Atomic Spectrometry*, **18**, 742–750, 2003.

84. H. Isoyama, T. Uchida, C. Iida, and G. Nakagawa, *Journal of Analytical Atomic Spectrometry*, **5**, 307–310, 1990.

85. H. Isoyama, T. Uchida, T. Niwa, C. Iida, and G. Nakagawa, *Journal of Analytical Atomic Spectrometry*, **4**, 351–355, 1989.

86. B. Bouyssiere, Y. N. Ordonez, C. P. Lienemann, D. Schaumloffel, and R. Lobinski, *Spectrochimica Acta, Part B*, **61**, 1063–1068, 2006.

87. B. Bouyssiere, Y. N. Ordonez, C. P. Lienemann, D. Schaumloffel, and R. Lobinski, *Spectrochimica Acta, Part B*, **61**, 1063–1068, 2006.

88. A. Prange and D. Schaumloffel, *Journal of Analytical Atomic Spectrometry*, **14**, 1329–1332, 1999.

89. D. Schaumloffel, J. R. Encinar, and R. Lobinski, *Analytical Chemistry*, **75**, 6837–6842, 2003.

90. A. Woller, H. Garraud, F. Martin, O. F. X. Donard, and P. Fodor, *Journal of Analytical Atomic Spectrometry*, **12**, 53–56, 1997.

91. Y. F. Li, C. Y. Chen, B. Li, J. Sun, J. X. Wang, Y. X. Gao, Y. L. Zhao, and Z. F. Chai, *Journal of Analytical Atomic Spectrometry*, **21**, 94–96, 2006.

92. A. Al-Ammar, R. K. Gupta, and R. M. Barnes, *Spectrochimica Acta, Part B*, **54**, 1077–1084, 1999.

93. A. Al-Ammar, R. K. Gupta, and R. M. Barnes, *Spectrochimica Acta, Part B*, **55**, 629–635, 2000.

94. R. L. Sutton, *Journal of Analytical Atomic Spectrometry*, **9**, 1079–1083, 1994.

95. H. Naka and H. Kurayasu, *Bunseki Kagaku*, **45**, 1139–1144, 1996.

96. P. Schramel, *Fresenius' Journal of Analytical Chemistry*, **320**, 233–236, 1985.

97. J. H. D. Hartley, S. J. Hill, and L. Ebdon, *Spectrochimica Acta, Part B*, **48**, 1421–1433, 1993.

98. W. Schron and U. Muller, *Fresenius' Journal of Analytical Chemistry*, **357**, 22–26, 1997.

99. A. R. Eastgate, R. C. Fry, and G. H. Gower, *Journal of Analytical Atomic Spectrometry*, **8**, 305–308, 1993.

100. T. B. Reed, *Journal of Applied Physics*, **32**, 821–824, 1961.

101. T. B. Reed, *Journal of Applied Physics*, **32**, 2534–2535, 1961.

102. T. B. Reed, *57th Annual Meeting American Institute of Chemical Engineers*, December 1964.

103. C. D. Allemand and R. M. Barnes, *Applied Spectroscopy*, **31**, 434–443, 1977.

104. S. Greenfield, I. L. Jones, and C. T. Berry, *Analyst*, **89**, 713, 1964.

105. S. Greenfield, *U.S. Patent No. 3*, 467, 471, September 16, 1969.

106. R. H. Scott, V. A. Fassel, R. N. Kniseley, and D. E. Nixon, *Analytical Chemistry*, **46**, 75, 1975.

107. R. Rezaaiyaan, G. M. Hieftje, H. Anderson, H. Kaiser, and B. Meddings, *Applied Spectroscopy*, **36**, 627–631, 1982.

108. P. W. J. M. Boumans, *Fresenius' Journal of Analytical Chemistry*, **299**, 337–361, 1979.

109. Thermo Scientific, *Radial Demountable Ceramic Torch for the Thermo Scientific iCAP 6000 Series ICP Spectrometer*, Thermo Fisher Product Technical Note: 43053, 2010, http://tools.thermofisher.com/content/sfs/brochures/D01563~.pdf

110. G. F. Larson, V. A. Fassel, R. K. Winge, and R. N. Kniseley, *Applied Spectroscopy*, **30**, 384–391, 1976.

111. F. V. Silva, L. C. Trevizan, C. S. Silva, A. R. A. Nogueira, and J. A. Nóbrega, *Spectrochimica Acta, Part B*, **57**, 1905–1913, 2002.

112. Thermo Scientific, *Thermo Scientific iCAP 7000 Plus Series ICP-OES: Innovative ICP-OES Optical Design*, Thermo Fisher Scientific Product Technical Note: 43333, 2016, https://tools.thermofisher.com/content/sfs/brochures/TN-43333-ICP-OES-Optical-Design-iCAP-7000-Plus-Series-TN43333-EN.pdf.

113. C. R. Runge and F. Paschen, *Abhandlungen der Koniglichen Akademie der Wissenschaften*, **1**, 1902.

114. G. L. Clark, *The Encyclopedia of Spectroscopy*, G. L. Clark (ed), Reinhold Publishing Corporation, New York, 1960.

115. J. F. James and R. S. Sternberg, *The Design of Optical Spectrometers*, Chapman and Hall Ltd, London, 1969.

116. C. B. Boss and K. J. Fredeen, *Concepts, Instrumentation, and Techniques in Inductively Coupled Plasma Optical Emission Spectrometry*, 2nd edition, Perkin-Elmer Corporation, Norwalk, CT, 1997.

117. K. Jankowski, A. Jackowska, A. P. Ramsza, and E. Reszke, *Journal of Analytical Atomic Spectrometry*, **23**, 1234–1238, 2008.

118. Y. Okamoto, *Analytical Sciences*, **7**, 283–288, 1991.

119. T. Maeda and K. Wagatsuma, *Microchemical Journal*, **76**, 53–60, 2004.

120. U. Engel, C. Prokisch, E. Voges, G. M. Hieftje, and J. A. C. Broekaert, *Journal of Analytical Atomic Spectrometry*, **13**, 955–961, 1998.

121. Q. Xue, *Applied Optics*, **50**, 1338–1344, 2011.

122. Q. Xue, S. Wang, and F. Lu, *Applied Optics*, **48**, 11–16, 2009.

123. J. A. C. Broekaert, Instrument Column, *Spectrochimica Acta*, **37B**, 359, 1982.

124. Plasma-Spec. *The Next Generation in Plasma Spectrometry*, Technical Bulletin of Leeman Labs, Inc., Lowell, MA.

125. X. Hou and B. T. Jones, *Encyclopedia of Analytical Chemistry*, R. A. Meyers (ed), John Wiley & Sons Ltd, Chichester, pp. 9468–9485, 2000.

126. K. Paech and M. V. Tracey, *Modern Methods of Plant Analysis*, **1**, Apringer-Verlog, Berlin, p. 177, 1956.

127. R. F. Jarrell, F. Brech, and M. J. Gustafson, *Journal of Chemical Education*, **77**, 592–598, 2000.

128. A. Scheeline, C. A. Bye, D. L. Miller, S. W. Rynders, and R. C. Owen, Jr., *Applied Spectroscopy*, **45**, 334–346, 1991.

129. A. T. Zander and P. N. Keliher, *Applied Spectroscopy*, **33**, 499–502, 1979.

130. C. Allemand, *ICP Information Newsletter*, **2**, 1, 1976.

131. C. Allemand, *ICP Information Newsletter*, **4**, 44, 1978.

132. R. B. Bilhorn, P. M. Epperson, J. V. Sweedler, and M. B. Denton, *Applied Spectroscopy*, **41**, 1125–1136, 1987.

133. Y. Talmi and R. W. Simpson, *Applied Optics*, **19**, 1401–1414, 1980.

134. F. M. Pennebaker, D. A. Jones, C. A. Gresham, R. H. Williams, R. E. Simon, M. F. Schappert, and M. B. Denton, *Journal of Analytical Atomic Spectrometry*, **13**, 821–827, 1998.

135. J. Marshall, A. Fisher, S. Chenery, and S. T. Sparkes, *Journal of Analytical Atomic Spectrometry*, **11**, 213R–238R, 1996.

136. Q. S. Hanley, C. W. Earle, F. M. Pennebaker, S. P. Madden, and M. B. Denton, *Analytical Chemistry*, **68**, 661A–667A, 1996.

137. J. V. Sweedler, K. L. Ratzlaff, and M. B. Denton, eds., *Charge Transfer Devices in Spectroscopy*, VCH Publishers, New York, 1994.

138. X. Hou and B. T. Jones, *Encyclopedia of Analytical Chemistry*, R. A. Meyers (ed), John Wiley & Sons Ltd, Chichester, pp. 9468–9485, 2000.

139. Y. Talmi and R. W. Simpson, *Applied Optics*, **19**, 1401–1414, 1980.

140. R. B. Bilhorn, P. M. Epperson, J. V. Sweedler, and M. B. Denton, *Applied Spectroscopy*, **41**, 1125–1136, 1987.

141. Y. Talmi, *Applied Spectroscopy*, **36**, 1–18, 1982.

142. E. D. Salin and G. Horlick, *Analytical Chemistry*, **52**, 1578–1582, 1980.

143. J. M. Harnly and R. E. Fields, *Applied Spectroscopy*, **51**, 334A–351A, 1997.

144. J. V. Sweedler, R. D. Jalkian, R. S. Pomeroy, and M. B. Denton, *Spectrochimica Acta, Part B*, **44B**, 683–692, 1989.

145. R. B. Bilhorn, J. V. Sweedler, P. M. Epperson, and M. B. Denton, *Applied Spectroscopy*, **41**, 1114–1125, 1987.

146. Q. Xue, S. Wang, and F. Lu, *Applied Optics*, **48**, 11–16, 2009.

147. F. L. J. Sangster and K. Teer, *IEEE Journal of Solid-state Circuits*, **SC-4**, 131–136, 1969.

148. A. J. P. Theuwissen, *Solid-State Imaging with Charge-Coupled Devices*, Kluwer Academic Publishers, New York, 2002.

149. W. Boyle and G. Smith, *Bell System Technical Journal*, **49**, 587–593, 1970.

150. W. Boyle and G. Smith, U.S. Patent 3792322, *Buried Channel Charge Coupled Devices*, February 12, 1974.

151. W. Boyle and G. Smith, U.S. Patent 3796927, *Three-Dimensional Charge Coupled Devices*, March 12, 1974.

152. P. M. Epperson, J. V. Sweedler, R. B. Bilhorn, G. R. Sims, and M. B. Denton, *Analytical Chemistry*, **60**, 327A–335A, 1988.

153. A. G. Milnes, Charge-Transfer Devices, in *Semiconductor Devices and Integrated Electronics*, Van Nostrand Reinhold Company Regional Offices, New York, pp. 590–642, 1980.

154. G. C. Holst and T. S. Lomheim, *CMOS/CCD Sensors and Camera Systems*, JCD Publishing, Oviedo, FL, 2007.

155. J. R. Janesick, *Scientific Charge-Coupled Devices*, SPIE—The International Society for Optical Engineering, Bellingham, WA, p. 24, 2001.

156. N. Waltham, CCD and CMOS Sensors, in *Observing Photons in Space: A Guide to Experimental Space Astronomy*, M. C. E. Huber, A. Pauluhn, J. L. Culhane, J. G. Timothy, K. Wilhelm, and A. Zehnder (eds), Springer Science & Business Media, New York, pp. 423–442, 2013.

157. J. V. Sweedler, R. D. Jalkian, R. S. Pomeroy, and M. B. Denton, *Spectrochimica Acta, Part B*, **44**, 683–692, 1989.

158. J. M. Mermet, A. Cosnier, Y. Danthez, C. Dubuisson, E. Fretel, O. Rogerieux, and S. Vélasquez, Tech Note: Design Criteria for ICP Spectrometry Using Advanced Optical and CCD Technology, *Spectroscopy*, **20**, 60–66, 2005.

159. J. Marshall, A. Fisher, S. Chenery, and S. T. Sparkes, *Journal of Analytical Atomic Spectrometry*, **11**, 213R–238R, 1996.

160. T. W. Barnhard, M. I. Crockett, J. C. Ivaldi, and P. L. Lundberg, *Analytical Chemistry*, **65**, 1225–1230, 1993.

161. T. W. Barnhard, M. I. Crockett, J. C. Ivaldi, P. L. Lundberg, D. A. Yates, P. A. Levine, and D. J. Sauer, *Analytical Chemistry*, **65**, 1231–1239, 1993.

162. I. B. Brenner and A. T. Zander, *Spectrochimica Acta Part B*, **55**, 1195–1240, 2000.

163. J. M. Harnly and R. E. Fields, *Applied Spectroscopy*, **51**, 334A–351A, 1997.

164. R. B. Bilhorn, J. V. Sweedler, P. M. Epperson, and M. B. Denton, *Applied Spectroscopy*, **41**, 1114–1125, 1987.

165. J. M. Harnly and R. E. Fields, *Applied Spectroscopy*, **51**, 334A–351A, 1997.

166. P. M. Epperson, J. V. Sweedler, R. B. Bilhorn, G. R. Sims, and M. B. Denton, *Analytical Chemistry*, **60**, 327A–335A, 1988.

167. G. R. Sims and M. B. Denton, *In Multichannel Image Detectors*, Y. Talmi (ed), ACS Symposium Series No. 236, Vol. 2, Chap. 5, American Chemical Society, Washington, DC, 1983.

168. S. Bhaskaran, T. Chapman, M. Pilon, and S. VanGorden, *SPIE Proceedings, Infrared Systems and Photoelectronic Technology III*, **7055**, 70550R, 2008.

169. R. B. Bilhorn and M. B. Denton, *Applied Spectroscopy*, **44**, 1538–1546, 1990.

170. Thermo Scientific, *Overcoming Interferences with the Thermo Scientific iCAP 7000 Plus Series ICP-OES*, Thermo Fisher Scientific Product Technical Note: 43332, 2016, https://tools.thermofisher.com/content/ sfs/brochures/TN-43332-ICP-OES-Overcoming-Interferences-iCAP-7000-Plus-Series-TN43332-EN. pdf.

171. K. E. Jarvis, P. Mason, T. Platzner, and J. G. Williams, *Journal of Analytical Atomic Spectrometry*, **13**, 689–696, 1998.

172. G. H. Vickers, D. A. Wilson, and G. M. Hieftje, *Journal of Analytical Atomic Spectrometry*, **4**, 749–754, 1989.

173. H. E. Taylor, *Inductively Coupled Plasma-Mass Spectrometry: Practices and Techniques*, Academic Press, San Diego, CA, 2001.

174. I. B. Brenner, A. Zander, M. Cole, and A. Wiseman, *Journal of Analytical Atomic Spectrometry*, **12**, 897–906, 1997.

175. I. B. Brenner, M. Zischka, B. Maichin, and G. Knapp, *Journal of Analytical Atomic Spectrometry*, **13**, 1257–1264, 1998.

176. L. C. Trevizan and J. A. Nobrega, *Journal of the Brazilian Chemical Society*, **18**, 1678–4790, 2007.

177. G. L. Long and J. D. Winefordner, *Analytical Chemistry*, **55**, 712A–724A, 1983.

178. J. Mermet and E. Poussel, *Applied Spectroscopy*, **49**, 12A–18A, 1995.

179. H. Kaiser, *Analytical Chemistry*, **42**, 53A, 1987.

180. V. Thomsen, D. Schatzlein, and David Mercuro, *Spectroscopy*, **18**, 112–114, 2003.

181. *EPA Method 2007*, Revision 4.4, 1994, www.epa.gov/sites/production/files/2015-08/documents/ method_200-7_rev_4-4_1994.pdf.

182. L. A. Currie, *Analytical Chemistry*, **40**, 586–593, 1968.

183. G. L. Long and J. D. Winefordner, *Analytical Chemistry*, **55**, 713A–724A, 1983.

184. J. M. Mermet, G. Granier, and P. Fichet, *Spectrochimica Acta Part B*, **76**, 221–225, 2012.

185. F. C. Garner and G. L. Robertson, *Chemometrics and Intelligent Laboratory Systems*, **3**, 53–59, 1988.

186. V. Thompsen, *Spectroscopy*, **27**, 3, 2012.

187. J. Mandel and R. D. Stiehler, *Journal of Research of the National Bureau of Standards*, **155**, A53, 1964.

188. G. F. Larson, V. A. Fassel, R. H. Scott, and R. N. Kniseley, *Analytical Chemistry*, **47**, 238–243, 1975.

189. Thermo Scientific, *High Performance Radio Frequency Generator Technology for the Thermo Scientific iCAP 7000 Plus Series ICP-OES*, Thermo Fisher Scientific Product Technical Note: 43334, 2016, https:// tools.thermofisher.com/content/sfs/brochures/TN-43334-ICP-OES-RF-Generator-iCAP-7000-Plus- Series-TN43334-EN.pdf.

190. H. G. C. Human and R. H. Scott, *Spectrochimica Acta Part B*, **31**, 459–473, 1976.
191. Thermo Scientific, *Overcoming Interferences with the Thermo Scientific iCAP 7000 Plus Series ICP-OES*, Thermo Fisher Scientific Product Technical Note: 43332, 2016, https://tools.thermofisher.com/content/sfs/brochures/TN-43332-ICP-OES-Overcoming-Interferences-iCAP-7000-Plus-Series-TN43332-EN.pdf.
192. R. F. Browner, *Fundamental Aspects of Aerosol Generation and Transport, in "Inductively Coupled Plasma Emission Spectrometry," Part II, "Applications and Fundamentals"*, Wiley-Interscience, New York, 1987.
193. M. W. Blades and G. Horlick, *Spectrochimica Acta Part B*, **36**, 881–900, 1981.
194. A. J. Miller, *Journal of Chemical Education*, **31**, 455, 1954.
195. M. H. Abdallah, R. Diemiaszonek, J. Jarosz, J. M. Mermet, J. Robin, and C. Trassy, *Analytica Chimica Acta*, **84**, 271–282, 1976.
196. I. B. Brenner and A. T. Zander, *Spectrochimica Acta Part B*, **55**, 1195–1240, 2000.
197. A. W. Boorn and R. F. Browner, *Analytical Chemistry*, **54**, 1402–1410, 1982.
198. Thermo Scientific, *Overcoming Interferences with the Thermo Scientific iCAP 7000 Plus Series ICP-OES*, Thermo Fisher Scientific Product Technical Note: 43332, 2016, https://tools.thermofisher.com/content/sfs/brochures/TN-43332-ICP-OES-Overcoming-Interferences-iCAP-7000-Plus-Series-TN43332-EN.pdf.
199. V. Thomsen, D. Mercuro, and D. Schatzlein, *Spectroscopy*, **21**(7), 2006, www.spectroscopyonline.com/interelement-corrections-spectrochemistry.
200. F. V. Silva, L. C. Trevizan, C. S. Silva, A. R. A. Nogueira, and J. A. Nóbrega, *Spectrochimica Acta Part B*, **57**, 1905–1913, 2002.
201. T. R. Dulski, Statistics and Specifications in "A Manual for the Chemical Analysis of Metals", *ASTM*, 200–201, 1996.

23 Other Complementary Atomic Spectroscopy Techniques

Even though atomic absorption (AA), inductively couple plasma optical emission (ICP-OES) and inductively couple plasma mass spectrometry (ICP-MS) are considered the most common atomic spectroscopic (AS) techniques, there are others which are worthy of discussion. In this chapter we will introduce the reader to four predominantly solid sampling techniques, X-ray fluorescence (XRF), X-ray Diffraction (XRD), laser induced breakdown spectrometry (LIBS) and laser ablation, laser ionization Time-of-Flight Mass Spectrometry (LALI-TOFMS); together with a novel atomic emission technique called microwave induced atomic emission spectrometry (MIP-AES), which is primarily used for liquids or digested solid samples. They all have their own strengths and weaknesses, but they are all complementary to ICP-MS so could be used if the lab's testing requirements change. Additionally, the four techniques that offer solid sampling capabilities could be used for on-site field testing such as the sorting of metal alloys or the elemental characterization of soils and rocks at remote locations.

23.1 X-RAY FLUORESCENCE

Over the past 30 years, X-ray fluorescence (XRF) has become the elemental technique of choice for the determination of low parts-per-million to percentage concentration levels in solid samples including metals, ores, rocks, soils, glasses, powders, plastics, ceramics, foodstuffs, pharmaceuticals and plant materials.[1]

The principles of XRF spectrometry are well documented in the literature.[2] A sample is irradiated with a beam of high-energy X-rays. As the excited electrons in the atom fall back to a ground state, they emit X-rays that are characteristic of those elements present in the sample, as shown in Figure 23.1.

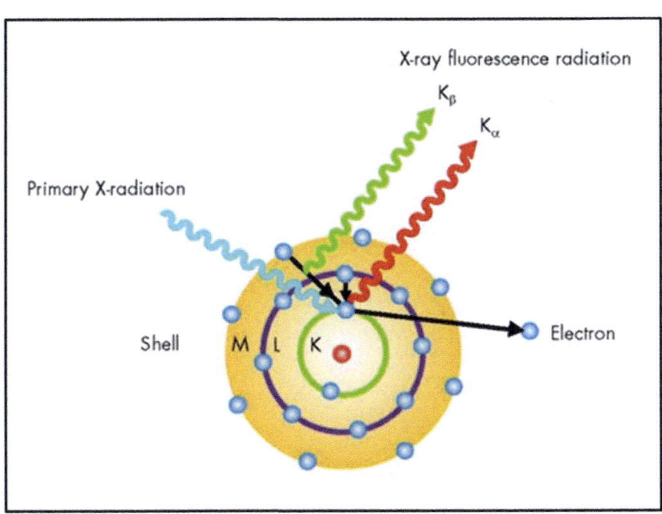

FIGURE 23.1 Principles of X-ray fluorescence (XRF).[3]

DOI: 10.1201/9781003187639-23

The individual X-ray wavelengths are then separated and measured via a system of crystals, optics, and detectors. Elemental concentrations in unknown samples are quantified by comparing the X-ray intensities against known calibration standards. The major benefit of XRF over other solid sampling techniques such as arc or spark emission is that it can analyze both conducting and non-conducting materials as well as inorganic and organic matrices with minimal sample preparation.[4,5]

23.2 XRF INSTRUMENTAL CONFIGURATION

The XRF technique is available in two separate configurations—energy-dispersive (EDXRF) and wavelength-dispersive (WDXRF). In the EDXRF approach, the intensity of the photon energy of the individual X-rays is detected and measured simultaneously using multi-channel data electronics.

Whereas in WDXRF spectrometry, the polychromatic beam emerging from a sample surface is dispersed into its monochromatic components or wavelengths with an analyzing crystal. A specific wavelength is then calculated from knowledge of the dispersion characteristics of the X-ray crystal. This design is shown in Figure 23.3.

It is generally recognized that WDXRF offers several advantages over EDXRF, including better spectral resolution, superior detection limits (particularly for the low mass elements) and the ability to determine major concentrations of elements with very high precision and accuracy. As a result, the higher performance and better capabilities of WDXRF means they are typically used for high end research-type applications[6].

On the other hand, EDXRF systems are smaller than WDXRF, they do not require any external utilities, such as chillers or gases, they typically use a smaller power supply and they have no moving parts. As a result of this compact footprint, EDXRF systems can be placed on a small bench, or using portable hand-held devices, can even be taken into the field to carry out remote site evaluations[7].

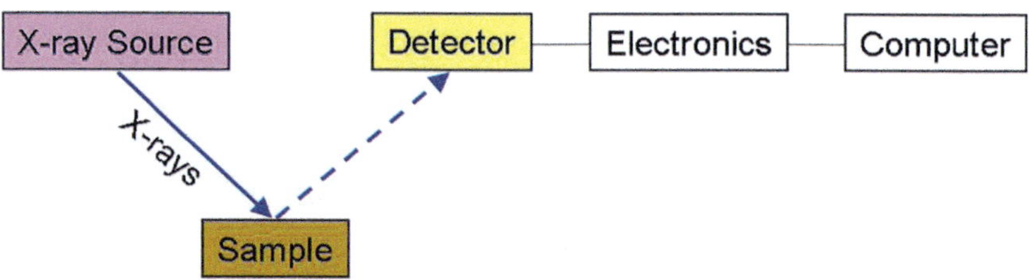

FIGURE 23.2   Simple schematic of energy-dispersive x-ray fluorescence (EDXRF) system.[3]

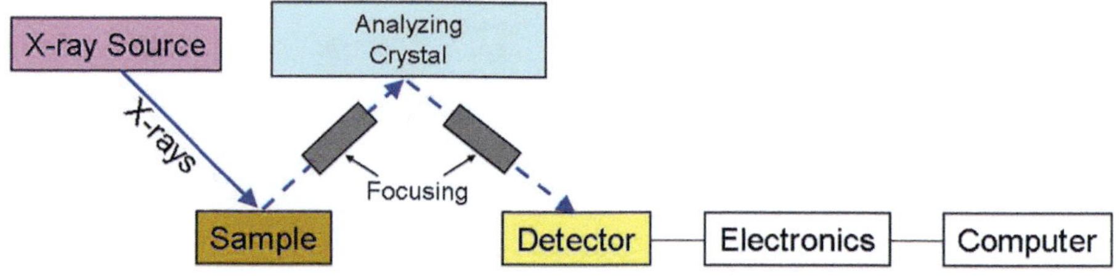

FIGURE 23.3   Simple schematic of wavelength-dispersive x-ray fluorescence (WDXRF).[8]

23.3 QUANTITATION BY XRF

XRF techniques do not require chemical pretreatment such as dissolution, and being a nondestructive analysis, can identify and determine a wide elemental range from low ppm up to percentage levels directly in solid, powdered or liquid samples. However, because the energy of the fluorescent X-rays corresponds directly to the atomic number of the element, the capability drops off at the low mass end, particularly for hand-held EDXRF systems.

Quantitation is conducted using external calibration standards containing varying concentrations of elements. XRF calibrations have the advantage of being suitable stable for long periods of time before requiring recalibration, in contrast to solution based AS techniques, which require recalibration procedures to be carried out at regular intervals. However, it should be emphasized that to ensure the highest accuracy in XRF, calibration standards should be of a similar matrix to the samples being measured.

EDXRF has been applied to the bulk analysis of a wide variety of organic-based samples such as pharmaceuticals botanicals, foods, plant materials and soil samples without the need for sample dissolution or chemical pretreatment required by ICP-OES or ICP-MS, which can be very labor intensive, and prone to human error and contamination of the sample. Because of these sample-digestion issues, the EDXRF technique is often used as a preliminary screening approach to determine if additional chemical analyses are required by one of the solution techniques. This is quite common in the pharmaceutical industry.[9] In fact, it could be an opportunity for portable units to be used as initial semi-quantitative tools for elemental contaminants in agricultural crops before the samples are sent to the independent testing lab for a more accurate assessment of the metal content by plasma spectrochemistry. In addition, EDXRF could also be used to characterize the soil for nutrients as well as levels of contaminant elements such as transition and heavy metals.

23.4 XRF DETECTION LIMITS

Detection capability of ED XRF is typically in the low-ppm (μg/g) range in the solid material. Furthermore, it can achieve lower detection limits if the measurement time is extended. Typical measurement times are in the range of 10–30 s, but a five- to ten-fold increase in integration time can show a two- to five-fold improvement. However, it should be emphasized that this represents detection capability directly in the solid material. For plasma spectrochemistry to achieve similar performance the solution detection limit must be in the order of 10 ppb (μg/L), if the EDXRF detection limit is 1ppm, assuming a sample weight of 1 g is digested and made up to 100 mL (100-fold dilution factor). Typical detection limits of some heavy metals in selected matrices are shown in Table 23.1.

TABLE 23.1

Typical Heavy Metal Detection Limits in ppm (µg/g) in Selected Matrices

	Pb	Cd	As	Hg	Cr	Ni
Plant materials[10]	1	N/M	N/M	N/M	2	1
Candy[11]	6	N/M	N/M	N/M	N/M	N/M
Soil[12]	15	8	7	13	30	40

Note: Measurement time 60s, N/M = not measured

23.5   SAMPLE PREPARATION FOR XRF

Solid, liquid and powder samples can be analyzed by XRF with the minimum of sample preparation. The only preparation required for XRF is reducing the sample to a size that fits in the sample cell or sample chamber. So basically, the larger the sample volume, the smaller the sampling error. Almost all solid samples can be analyzed directly by simply placing them in the sample chamber as is. Viscous liquid samples like oils are poured in a sample cell, with a supporting film at the bottom. Typical films are made from polypropylene of a few micrometers thickness.[13] The majority of pharmaceutical or botanical-type materials are organic matrices and have relatively low X-ray absorption, allowing for the relatively straight forward measurement of the elemental impurities. Powder samples are placed directly into the sample cell using the tapping method to remove voids in the sample. Coarse powders must be ground to a fine particle size, while non-homogenous samples should be ground by means of mortar and pestle or a grinding device, such as a ball mill. Pelletizing using a laboratory press and a binding agent is often a good way to keep powdered samples together.

23.6   X-RAY DIFFRACTION

A comparative technique to XRF is X-ray diffraction (XRD), which is an analytical technique that looks at the X-ray scattering from crystalline or polymorphic materials.[14] Each material produces a unique X-ray "fingerprint" of X-ray intensity versus scattering angle that is characteristic of its crystalline atomic structure. Qualitative analysis is possible by comparing the XRD pattern of an unknown material with a library of known patterns. Although its principles are different, XRD can be considered complementary to XRF. For example, XRF can tell you that a material is composed of iron and sulfur, while XRD can tell you that both iron sulfide (FeS) and elemental iron (Fe) are present. Furthermore, because XRD works with any crystalline solid, there is almost no limit to the types of materials that can be studied. Some of the materials that are typically characterized by XRD include chemicals, dusts, rocks, minerals, metals, cements, pigments and forensic samples.

One of the major applications of XRD is to identify and quantify specific compounds or phases in materials that are organic in nature, such as foodstuffs or pharmaceutical compounds.[15] These kinds of applications have evolved over the last few years mainly because of a new generation of solid-state detectors, higher-quality optics and the availability of sampling attachments. These technological improvements have led to both higher sensitivity, which is needed for the detection of minor and trace levels of crystalline compounds in pharmaceutical compounds, and higher resolution for improved spectral specificity in the presence of interfering species.

23.7   LASER INDUCED BREAKDOWN SPECTROSCOPY

Laser Induced Breakdown Spectroscopy (LIBS) has been commercially available since the early 2000s, but it's only been in the past five years where its application potential has been fully explored. Initially developed by a group of government scientists in the early 1980s,[16] it was first applied to the analysis of soils and hazardous waste sites, because of its remote sampling capabilities.[17] However, since it has been in the hands of the routine analytical community, it is now being used to solve real-world application problems.[18] Although the analytical capability, performance and elemental range of LIBS compares quite favorably with other AS techniques, it should be considered complementary (and a competitor) to both ICP-OES and XRF. The unique benefits of LIBS are in its ability to sample a very diverse range of matrices, both in a laboratory environment and at remote locations out in the field. Let's take a closer look at the fundamental principles and capabilities of LIBS in order to get a better understanding of where it fits into the atomic spectroscopy toolbox.

23.8   LIBS FUNDAMENTAL PRINCIPLES

LIBS is an atomic emission spectroscopic technique that utilizes a small plasma generated by a focused pulsed laser beam (typically from a Nd:YAG laser) as the emission source. The plasma formed is about ten times hotter than inductively coupled plasma and as a result will vaporize just about any material it comes in contact with. The energy from the plasma excites the vaporized sample, which results in a characteristic emission spectrum of the elements present in the sample, which is then optically dispersed and detected by optical components traditionally used in an emission spectrometer.[19] A typical optical layout of a LIBS system is shown in Figure 23.4, while Figure 23.5 exemplifies a generated emission spectrum.

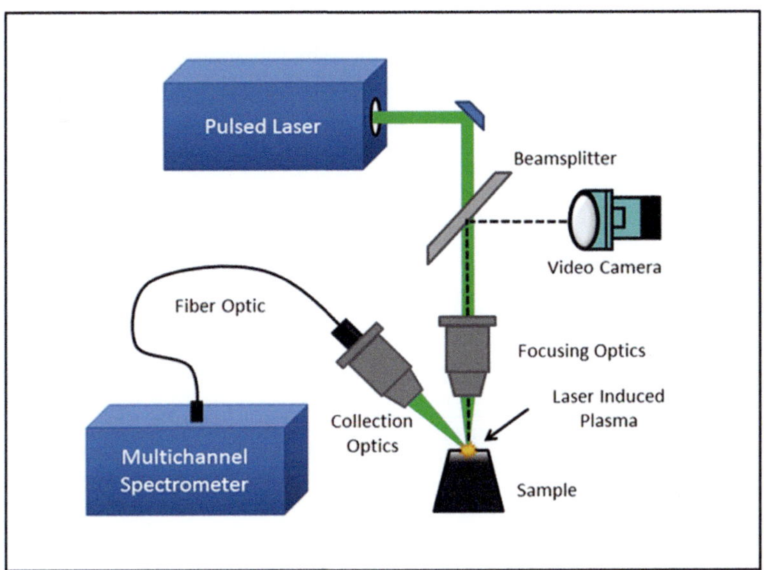

FIGURE 23.4 Optical configuration of a LIBS system.

(Courtesy of TSI Inc.)

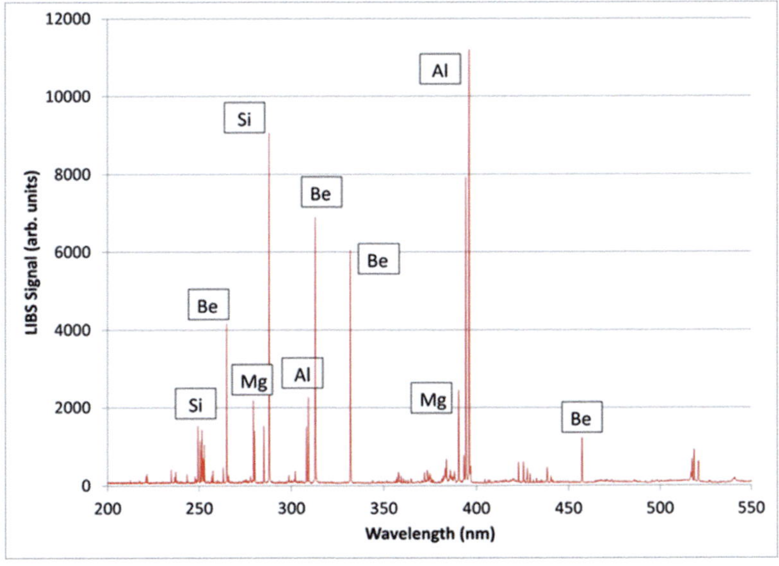

FIGURE 23.5 A typical LIBS emission spectrum.

(Courtesy of TSI Inc.)

23.9 LIBS CAPABILITIES

This capability has given LIBS some unique characteristics over other AS techniques including:

- Any type of solid, liquid or gaseous sample can be vaporized and excited, and is not limited to liquids like ICP-OES or conducting solids like arc/spark emission.
- Unlike XRF, the majority of the periodic table is available, even the low-mass elements like B and Li.
- The laser-induced plasma is readily generated in the open atmosphere negating the need for large argon flows or expensive consumables used in ICP instrumentation.
- The laser beam can be directed almost anywhere by the appropriate optical components or through a fiber optic, making it suitable for remote sampling applications such as hazardous waste sites or studies on other planets.
- The optical design makes it ideally suited to be put into a briefcase that can be carried around for field applications.
- Precise control of spot size and laser penetration, allows surface mapping and depth profiling studies.

23.10 LIBS APPLICATION AREAS

Some of the many applications that are being carried out today which have contributed to the growing popularity of LIBS include:

- The determination of major and minor elements in metallurgical samples
- Identification and classification of different materials using multivariate processing techniques
- Microanalysis of gemstones and precious minerals
- Trace element analysis of glass and quartz materials
- Minimally destructive analysis of forensic samples
- Uniformity of pharmaceutical ingredients
- Analysis of tablet coatings
- Analysis of botanicals and plant materials
- Direct analysis of coals
- Measurement of the metallic content of liquid baths or flowing process streams
- Remote field sampling of environmental, geological materials and soils

23.11 LIBS DETECTION CAPABILITY

Detection limits for LIBS are very similar to EDXF, with LIBS having an advantage of being able to measure lower levels of the lighter elements (below Mg in the periodic table). Additionally, similar to XRF, LIBS is predominantly a solid sampling technique. So, for a solution technique like ICP-MS to achieve similar performance (assuming a sample weight of 1 g is digested and made up to 100 mL), the detection limit must be in the order of 10 ppb (µg/L), if the LIBS detection limit is 1 ppm. Table 23.2 exemplifies some typical LIBS heavy metal limits of detection in ppm (µg/g) in selected matrices.

TABLE 23.2

LIBS Heavy Metal Detection Limits in Selected Matrices

	Pb	Cd	As	Cr	Cu
Fe-Mineral[20]	N/M	N/M	N/M	4	14
Slag[20]	N/M	10	N/M	7	N/M
Soil[21,22]	43	N/M	85	N/M	N/M
Fertilizer[23,24]	15	1	N/M	2	N/M
Rice[25]	44	3	N/M	N/M	N/M

23.12 LIBS ON MARS

There is no question that the most high-profile LIBS system is the ChemCam instrument onboard NASA's Curiosity Rover, which landed on the surface of Mars in the fall of 2012. Once the rock or mineral has been prepared for analysis by Curiosity's robotic arm, the ChemCam system can analyze the sample remotely from a distance of up to 20 feet away, by firing the laser pulses and directing the resulting emission from the generated plasma back to the on-board spectrometer via fiber optics for detection and quantitation.[26] This work will be extremely relevant to better understand the geological composition of Mars and ultimately how our solar system came into existence.[27]

23.13 MICROWAVE INDUCED PLASMA OPTICAL EMISSION SPECTROSCOPY

ICP-OES has become the dominant solution technique for carrying out the multielement analysis of samples that require high-ppb and low-ppm analyte levels. It was first developed for commercial use in the late 1970s as a response to a demand from the application community that flame atomic absorption was limited in its ability to determine a large suite of elements in a timely manner. Initially available as a simultaneous spectrometer and then in the more flexible sequential configuration, ICP-OES rapidly became the technique of choice for laboratories where high sample throughput, multielement analysis was the overriding requirement. There was still a place for flame atomic absorption (FAA) in laboratories where the sample workload was significantly lower, but there still appeared to be applications where either the analyte or sample throughput requirements were too high for FAA, or they were not high enough to justify the cost of an ICP-OES. Instrument vendors attempted to fill those gaps with either high cost, fully automated AAs or low-cost ICP-OES systems, but each approach tended to be a compromise.

However, this gap in the application toolbox has been filled with the development and commercialization of a microwave induced plasma optical emission spectrometer (MIP-OES). Microwave-generated plasmas have been used as gas chromatography detectors for decades and were explored as potential excitation sources for emission studies back in the 1970s, when the ICP was being investigated. However, it was found they had limited applicability, because they were not considered robust enough for introducing liquids. As a result of this limitation in MIP technology, the ICP became the dominant excitation source. To better understand this early limitation, let's take a closer look at how an MIP is generated.

A microwave-induced plasma basically consists of a quartz tube surrounded by a microwave waveguide or cavity. Microwaves produced from a magnetron fill the cavity and cause the electrons in the plasma support gas to oscillate. The oscillating electrons collide with other atoms in the flowing gas to create and maintain a high-temperature plasma. As in the inductively coupled plasmas, a high voltage spark is needed to create the initial electrons to create the plasma, which achieves temperature of approximately 5,000 K.

The limiting factor to their use was that with the low power and high frequency of the MIP, it was very difficult to maintain the stability of the plasma when aspirating liquid samples containing high levels of dissolved solids. Various attempts had been made over the years to couple desolvation techniques to the MIP, but only managed to achieve limited success.[28] However, the latest MIP technology appears to have overcome many of the limitations of the earlier designs. Let's take a closer look its fundamental principles.

23.14 BASIC PRINCIPLES OF THE MP-AES TECHNOLOGY

There is only one vendor of this technology (Agilent Technologies: 4210 MP). At the heart of this technology is a 2.5 GHz magnetron coupled into nitrogen plasma, which is created using compressed air and a nitrogen generator. By using the magnetic field rather than the electric field for excitation, extremely robust plasma is formed, which is capable of handling much higher dissolved solids than previous MIP designs. The microwave waveguide concentrates both the axial magnetic and radial electrical fields around the torch, which creates a conventional-looking plasma, allowing for a traditional inert concentric nebulizer and double pass spray chamber to be connected to the torch.

The optical system uses an axial viewing configuration, where the emission from the plasma is directed into a fast-scanning, 600-mm-focal-length, Czerny–Turner monochromator with a wavelength range 178–780 nm. The 2,400 lines/mm holographic grating, blazed at 250 nm for optimum UV performance, offers a resolution of 0.050 nm. The detection system is a back-lit CCD array detector, which is cooled to 0 °C using a thermoelectric Peltier device, and collects the analyte wavelengths and surrounding background spectra, allowing for simultaneous background correction.

23.15   BENEFITS OF MIP-AES

One of the biggest advantages of this particular design over ICP-OES is that it runs on compressed air, which means that no gaseous or liquid argon is required. This makes it very attractive for laboratories that are on a tight consumables' budget. However, the benefits over FAA are probably where it will receive the most attention.[29] These benefits include:

- Lower detection limits
- Wider linear dynamic range means larger concentration ranges can be determined
- Increased sample throughput
- Higher productivity, as it can be run unattended overnight
- No requirement to purchase or replace multiple hollow cathode lamps
- The ability to add other analytes to the suite of elements whenever the demand arises
- No safety concerns as neither acetylene nor nitrous oxide are required

The detection limit improvement over FAA is exemplified in Table 23.3.

23.16   TYPICAL APPLICATIONS OF MP-AES

There is no question, based on the application material in the public domain that the MP-AES system has been designed to fill the gap between FAA and ICP-AES, and particularly to address the many limitations of AA. Some of the most common applications being addressed by this technique

TABLE 23.3

Detection Limit Comparison (µg/L) of MP-AES with FAA

Element	FAA	MP-AES	Element	FAA	MP-AES
K	0.8	0.65	As	60	57
Ca	0.4	0.04	Cd	1.5	1.4
Mg	0.3	0.09	Cr	5	0.3
Na	0.3	0.12	Mn	1.0	1.05
Au	5	2.1	Pb	14	2.5
Pt	76	6.1	Sb	37	12
Pd	15	1.6	Se	500	77
Ag	1.7	1.2	Zn	1.6	3.1
Rh	4	0.5			

Note: 10 s integration time, except for As and Se, which were 30 s[30]

include the analysis of geological materials,[31] particularly if there is a requirement for remote sampling or when field studies are being carried out; the analysis of petrochemical samples using the addition of air to reduce carbon buildup on the torch; analysis of major, minor and trace elements in crops and plant materials;[32] non-regulated applications, such as the environmental monitoring of industrial waste streams; and the analysis of fruit juices[33] and agricultural samples appears to be some of the most common applications that can be readily addressed by using the MP-AES. For a full suite of application literature on this technique, please check out the following reference.[34]

Since the manuscript for this book was completed, a brand new microwave ICP-OES using an atmospheric plasma been developed which was introduced at the 2023 Pittsburgh Conference in Philadelphia. For more information on this technology, please refer to the following reference.

<u>Design and Performance Review of Innovative Microwave Inductively Coupled Atmospheric Plasma – Optical Emission Spectroscopy (MICAP™-OES 1000) Flexible Sample Matrix and Full Spectrum Acquisition,</u> E. Mohen J. Biesterve, D. Rehbein, C. Rivera, PittCon 2023, Philadelphia, PA, Mar. 2023, <u>https://radomcorp.com/wp-content/uploads/2023/03/PITTCON SAMPLE MATRIX poster 48x96 cmyk 09.pdf</u>

23.17 LASER ABLATION LASER IONIZATION TIME OF FLIGHT MASS SPECTROMETRY

Laser ablation, laser ionization coupled with TOF mass spectrometry (LALI-TOFMS) is a brand-new development that offers virtually the entire periodic table of the elements at mass-spec detection capability directly on solid samples. It has broad applicability for any testing lab that is looking to avoid sample-digestion procedures to measure a wide panel of trace elements in any type of solid sample. Currently, Exum Instruments is the only manufacturer of this technique with the Massbox™ laser ablation, laser ionization TOF mass spectrometer. So, let's take a closer look at this novel technology.

23.18 MASSBOX

The Massbox LALI-TOFMS is the first commercial instrument equipped with a Laser Ablation, Laser Ionization (LALI) source. The patented, dual-laser technology extracts and subsequently ionizes material in two discrete steps—first ablation and then ionization. Historically, LALI has been referred to by numerous acronyms in the literature, i.e., SALI[35,36], LDI[37,38], L²MS[39], LD-LPI-TOF-MS.[40] Most of the research regarding LALI was purely academic and performed between the late 1980s and early 1990s. At the time, cost and electronic limitations prohibited commercialization of LALI technology, and other techniques took center stage. However, recent advances in computing technology and miniaturized, high-powered, solid-state lasers make LALI much more commercially viable today.

23.19 BASIC PRINCIPLES LALI-TOF-MS

The first step of LALI uses a focused laser beam to ablate/desorb material from a solid sample surface (for more detailed information about laser ablation, please refer to Chapter 20). The Massbox utilizes an Nd:YAG desorption laser with an adjustable wavelength. Depending on the application, the desorption/ablation laser can be set to the fourth harmonic (266 nm) or fifth harmonic (213 nm) for ablation, or the fundamental 1064 nm wavelength for desorption. In a process exactly analogous to Laser Ionization Mass Spectrometry (LIMS) and Laser Induced Breakdown Spectroscopy (LIBS), the ablation/desorption laser generates an initial set of ion/electron pairs from a temporal plasma along with a neutral particle cloud that migrates normal to the sample surface.

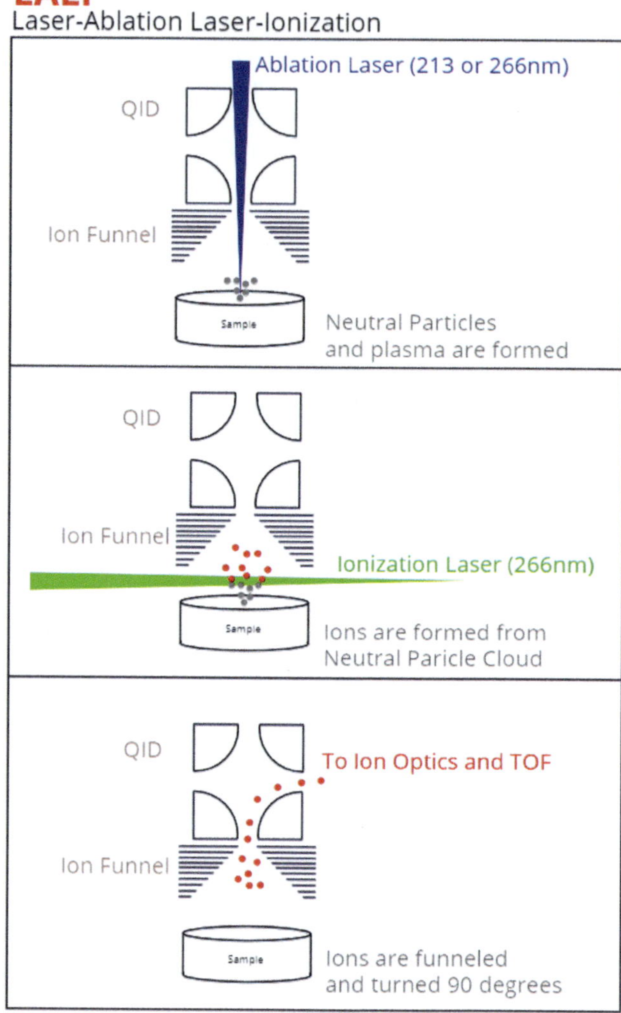

FIGURE 23.6 Laser Ablation Laser Ionization process flow.

(Courtesy of Exum Instruments)

A short delay (< 1 μs) allows plasma extinction and the dispersion of plasma-generated ions before a second laser is triggered for ionization (also Nd:YAG set to 266 nm). As shown in Figure 23.6, the ionization laser is aligned parallel to the sample surface and its beam is focused inside the neutral particle cloud. The focused beam from the ionization laser has an energy density $> 10^9$ W*cm^{-2}, which allows for ionization of neutral particles via multiphoton ionization (MPI). MPI differs from Resonant Enhanced Multiphoton Ionization (REMPI) in that with MPI the laser is not tuned to a specific elemental or molecular frequency for ionization.

By ionizing elements across a wide range of ionization energies, MPI serves as a highly efficient ion source and replaces the Ar plasma of ICP instruments. Once ionized, an ion funnel collects and focuses the ions in a low pressure (0.2–0.3 mbar) environment. After exiting the ion funnel, a Quadrupole Ion Deflector (QID) turns the ions 90 degrees and directs the ion beam through an Einzel Lens Stack and a quadrupole to further improve the beam shape. After the transfer quadrupole, the Massbox is equipped with a notch quadrupole filter. For applications that require high sensitivity, the notch filter increases dynamic range by selectively reducing the signal of up to four

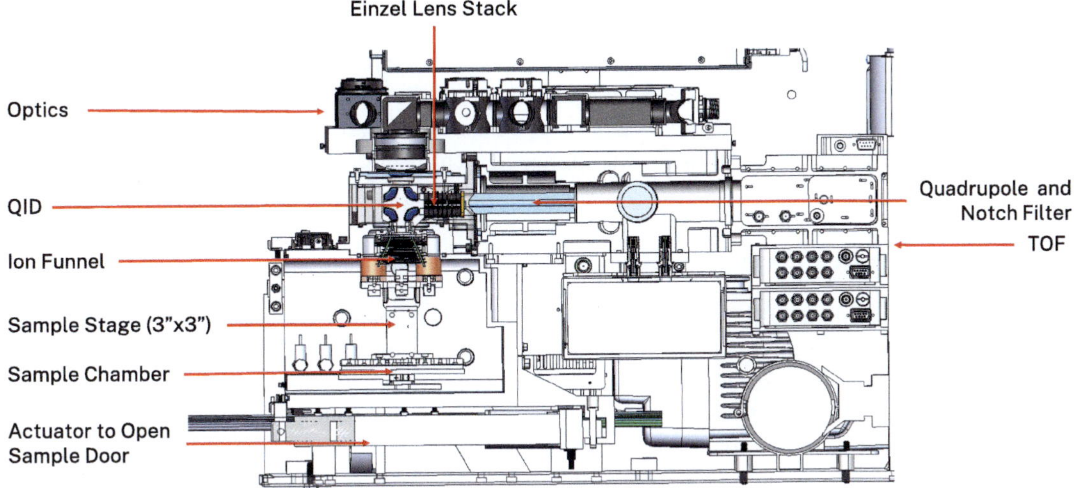

FIGURE 23.7   Simplified schematic diagram of Massbox instrument.

(Courtesy of Exum Instruments)

different ion masses (typically the most abundant matrix elements). Ions are then transferred to the reflectron time-of-flight (TOF) mass spectrometer which completes the mass analysis as exemplified in Figure 23.7 (for more information about TOF technology, please refer to Chapter 9).

23.20 MATRIX EFFECTS

Compared to LIBS and LIMS, which both use only the plasma generated by the laser, LALI has more efficient ionization and reduced matrix effects. LALI's ability to reduce/eliminate matrix effects makes quantification more straight-forward. In LIBS for example, matrix-matched standards are a critical requirement for reliable quantification. For applications that require quantification across a variety of matrices, this is a severe limitation of LIBS.

The Massbox's LALI source ionizes gas-phase particles within the neutral particle cloud instead of relying on the plasma generated during ablation. Compared to material ionized by the initial ablation plume, material in the neutral cloud is significantly less variable across different matrices. Ionization of neutral cloud material also results in stoichiometric accuracy, which enables quantification of a variety of sample matrices without the need for matrix-matched standards.

23.21 DIFFUSION AND TRANSPORT

Another major advantage of this design is that ionization occurs under vacuum in the sample chamber. The Massbox is a completely static system held at high-vacuum (~10^{-7} mbar) in the TOF and a pressure gradient to ~10^{-4} mbar in quadrupoles and the ion optics. The pressure in the sample chamber is maintained at ~0.2 mbar with an inert helium cooling gas system. The low pressure of the ion source results in a significant improvement in sensitivity because it greatly reduces losses associated with gas transport from atmospheric pressure to a vacuum system. Furthermore, without an inductively coupled plasma source, the Massbox does not require a carrier gas, which eliminates most polyatomic molecular isobaric interferences. Interference reduction improves the detection for a whole host of elements including Si, K, Ca, and Fe, which are traditionally problematic by ICP-MS. The removal of the plasma source also has the advantage that thermal emission of contaminant ions from the cones and/or injector is eliminated, which greatly improves the ability to determine many of the volatile elements including Na and Pb. Additionally, without a plasma source or any carrier gasses, LALI does not rely on components with dynamic fluctuations, which leads to long-term signal stability without regular instrument tuning.

23.22 INTERFERENCES

Figure 23.8 shows a LALI mass spectrum from a GSE-2G standard glass which is free of poly-atomic or multiple-charged interferences. As can be seen in the marked zoomed insets (27–30 amu, 80–100 amu, 130–50 amu, 17–180 amu), peak area ratios of the LALI mass spectra are isotopically correct across the entire periodic table, which makes identification easier and quantification more accurate.

23.23 TRANSMISSION EFFICIENCY

LALI provides considerable improvements in transmission efficiency of an ion beam compared to techniques that generate ions at atmospheric pressure. Ions generated by a plasma at atmospheric pressure are generally transferred in several stages before reaching the high-vacuum mass analyzer. Each stage transition of cones and/or lenses removes a significant portion of ions. For instance, LA-ICP-MS has a very high ionization efficiency for elements with a first ionization potential (FIP) less than 8 eV, but only ~1 in every 10^5–10^6 ions reach the detector (~0.01–0.001% transmission efficiency). The Massbox's LALI source is already under vacuum, so it does not suffer from transmission loss going from atmospheric pressure to vacuum. Removing the atmospheric/vacuum interface greatly improves transmission efficiency, allows for higher sensitivity, and further reduces matrix effects by removing plasma-ion spatial interaction effects.

23.24 INORGANIC AND ORGANIC ANALYSIS

LALI is capable of analyzing both organic and inorganic samples, which is difficult to carry out with other techniques that suffer from the problems previously discussed.[41] The ability to analyze organic compounds by LALI has been studied by many groups for applications including planetary missions and crude oil analysis.[39,42] The analysis of organics utilizes the infrared component of the Nd:YAG laser (1064 nm). The intense IR pulse flash heats the sample (10^8 K/s) to desorb intact material from the sample surface.[43] Following desorption, organic compounds are ionized via MPI using the secondary ionization laser as previously described. The ability to analyze both organics and inorganics in the same analytical run enables mapping (or bulk characterization) of both in the same sample almost simultaneously after a quick mode switch.

23.25 OPERATIONAL USE

Apart from its technical performance, this technology has been developed to be easily deployed across multiple industries and applications, which is a step-change in analytical capability over other lab-based approaches for trace element analysis. Because of its compact footprint (28" x 24" x 24"), it is field and lab portable, which can be deployed on a tabletop with no additional equipment required. In addition, its simplicity of design allows for a lower purchase price and operational costs compared to typical laser ablation ICP-MS systems.

23.26 USER INTERFACE

The instrument has a very straightforward user interface (UI). Figure 23.9 shows the sample chamber opening to allow the sample tray to be loaded and accommodating a 3" by 3" stage with nanometer (nm) precision.

A macro-camera then opens to take a high precision, spatially located picture. After the sample chamber door closes and begins to pump down to vacuum. Meanwhile, the high-resolution image from the macro-camera is loaded onto the touch screen interface. The macro-image is used to enable navigation around the sample(s), as shown in Figure 23.10. Moving back and forth on the

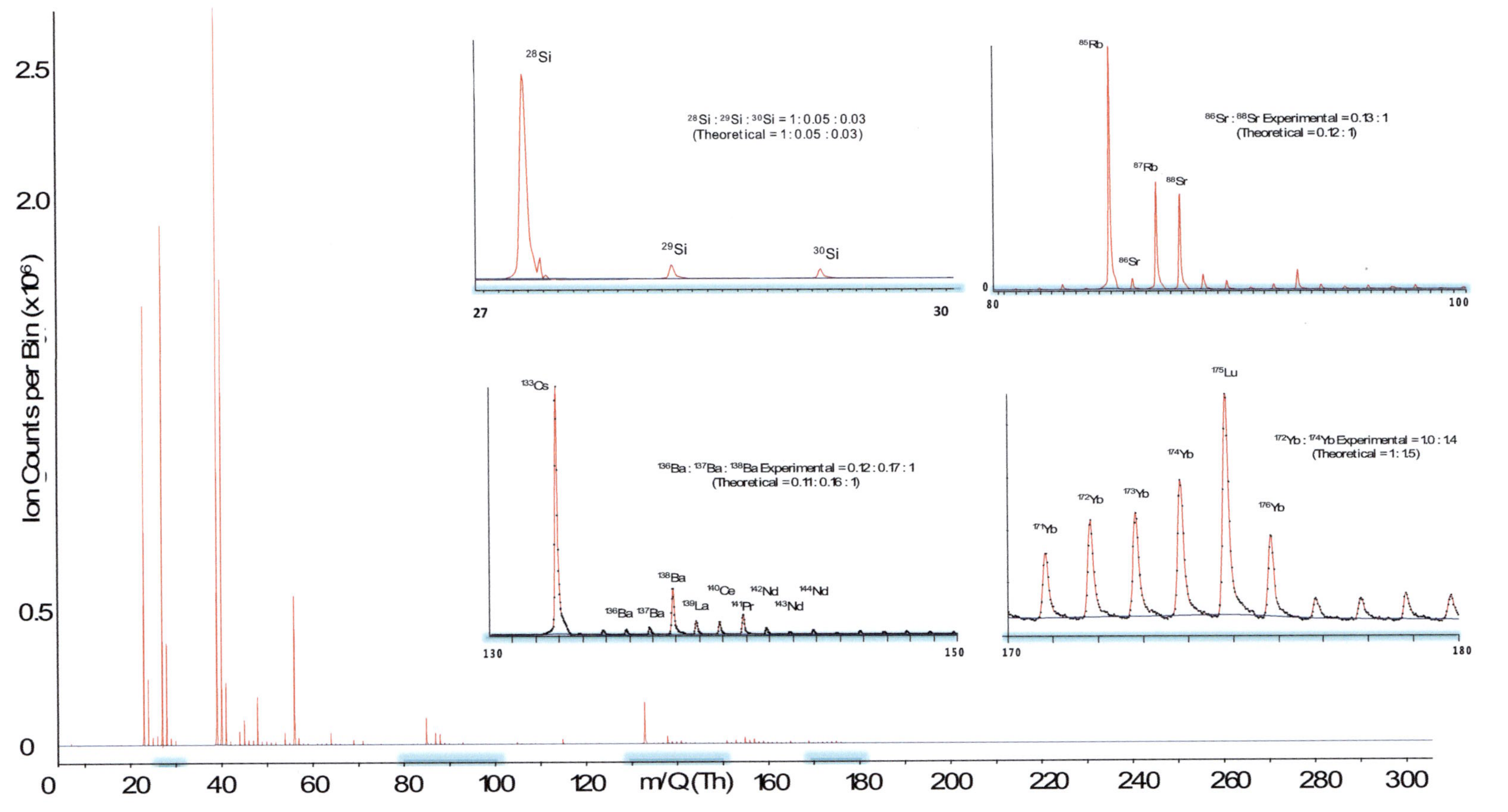

FIGURE 23.8 LALI mass spectrum and selected zoomed insets from GSE-2G standard glass.

(Courtesy of Exum Instruments)

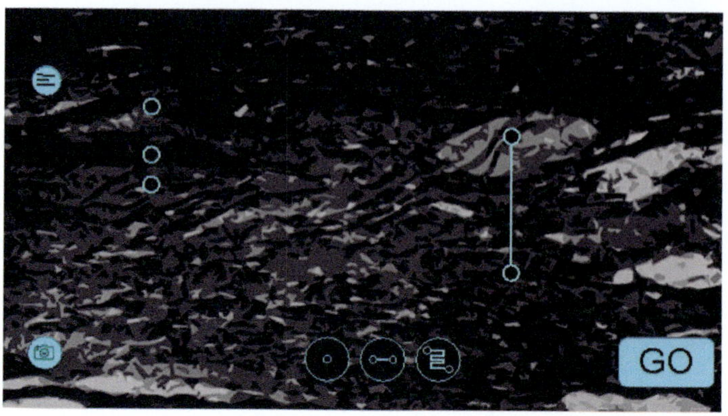

FIGURE 23.9 Schematic of sample loading, macro imaging and LALI source.

(Courtesy of Exum Instruments)

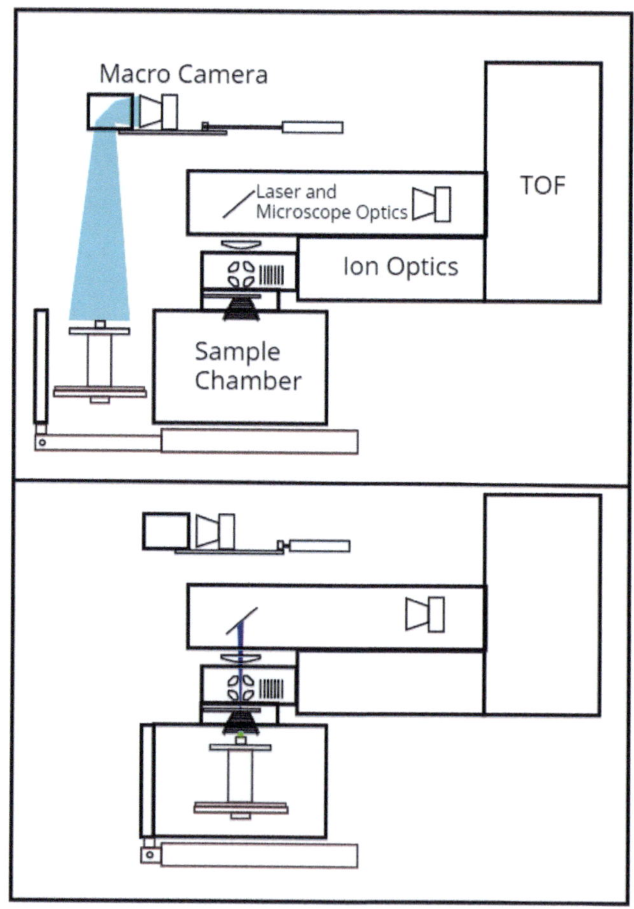

FIGURE 23.10 User interface used to identify where characterization occurs.

(Courtesy of Exum Instruments)

screen physically moves the sample stage within the chamber and aligns the desired area properly with the lasers. "Pinching" on the image zooms in and switches to a live microscopy image that allows precision when choosing an area to analyze. From the live view, spots, lines and/or rasters for maps are chosen. After selecting the type and number of sampling areas, analysis and data processing are automated, and all that remains is to interpret the results.

23.27 PERFORMANCE CAPABILITIES

The Massbox is the only LALI-TOF-MS instrument on the market, which can be used for both organic and inorganic analyses, which provides quantitative chemical mapping and/or bulk analysis without the need for complex quantification schemes or matrix matched standard materials. The instrument's simple design enables field portability, and its easy-to-use, intuitive interface makes it seamless to train novice users. These factors open up the technology to a wide range of potential applications, including the measurement of trace elements in high purity metals, geological samples, soils, biological specimens and much more. A recent study showed that it can achieve extremely low limits of quantitation directly in metallic powders and disks without having to digest the samples.[44]

The performance specifications of the instrument are shown in Table 23.4.

23.28 FINAL THOUGHTS

The atomic-spectroscopic tool kit seems to expand every few years with new, innovative technology and developments. It is a very active area of research for many vendors, which makes it more difficult for the user community to know which is the best technique to use to solve their specific application problems. Today's testing labs have many choices depending on their budget and

---

TABLE 23.4

Performance Specifications of the Massbox

Massbox specifications

Analysis type	Inorganic and organic
Sample type	Solid or pseudo solid
Elemental coverage	Li–U (3–238 amu)
Detection limits	10 ppb–1 ppm (Application dependent)
Mass range	1–20,000 mass units
Mass resolution	700–1100 or 6000–14000
Mass Accuracy	4 ppm
Notch filter channels	4
Mapping	80 mm × 80 mm × 30 mm 3-axis with nm repeatability and resolution
Laser spot size	Adjustable from 1 μm—250 μm
Repetition rate	1–50 Hz
Ablation laser power (266 nm)	10 mJ
Ablation laser power (213 nm)	3 mJ
Ionization laser power (266 nm)	10 mJ
Beam characteristics	Flattop Profile
Total analysis time (including pump down)	< 5 min
Sample holder	Customizable depending on application. Standard holder has nine sample slots
Quantification	Variable matrix materials resolved without complex standardization

sampling and detection requirements, include ICP-MS, ICP-OES, sampling accessories, graphite furnace AAS or vapor-generation AFS. However, this chapter has presented some complementary alternative approaches to consider, even though it might not be immediately obvious which technique would be best suited for a particular application. For that reason. I encourage you to learn more about them and if the opportunity arises, test them out and get some of your own samples run to see if it meets the application demands of your lab.

FURTHER READING

1. R. Jenkins, R. W. Gould, and D. Gedcke, *Quantitative X-ray Spectrometry*, Marcel Dekker, Inc., New York, 1995.
2. E. P. Bertin, *Introduction to X-Ray Spectrometric Analysis*, Plenum Press, New York, 1978.
3. Wikipedia Commons File, https://commons.wikimedia.org/wiki/File:Dmedxrfschematic.jpg.
4. H. W. Major and B. J. Price, *Plastics Engineering*, **46**(8), 37–39, 1990.
5. V. Thomsen and D. Schatzlein, *Spectroscopy*, **17**(7), 22–27, 2002.
6. R. Yellipedi and R. Thomas, New Developments in Wavelength-Dispersive X-Ray Fluorescence and X-Ray Diffraction for the Analysis of Foodstuffs and Pharmaceutical Materials, *Spectroscopy*, **21**(9), 2–6, September 2006.
7. The Applications of Hand-Held XRF Technology, *Thermo Fisher*, https://assets.thermofisher.com/TFS-Assets/CAD/Specification-Sheets/Niton%20XL5-Brochure.pdf.
8. Wikipedia Commons File, https://commons.wikimedia.org/wiki/File:Dmwdxrfschematic.jpg.
9. D. Davis and H. Furakawa, Using XRF as an Alternative Technique to Plasma Spectrochemistry for the New USP and ICH Directives on Elemental Impurities in Pharmaceutical Materials, *Spectroscopy*, **32**(7), July 2017.
10. H. L. Byers, L. J. McHenry, and T. J.Grund, XRF Techniques to Quantify Heavy Metals in Vegetables at Low Detection Limits, *Food Chemistry*, **1**, 30 March 2019, https://doi.org/10.1016/j.fochx.2018.100001.
11. A. M. Phipps, *Evaluation of X-ray Fluorescence for a Lead Detection Method for Candy*, UNLV, Dissertation, Environment of Public Health Commons, 2009, https://digitalscholarship.unlv.edu/thesesdissertations/963/.
12. R. L. Johnson, Real Time Demonstration of XRF Performance for the Paducha Gas Diffusion Plant, *Kentucky Research Consortium for Energy and Environment*, April 3, 2008, www.ukrcee.org/Challenges/Documents/Soil/RTD_XRF_Perf_Eval_Results_040708fin.pdf.
13. L. Oelofse, Measuring the Metal Content of Crude and Residual Oils by EDXRF: A Rapid Alternative Approach to ICP-OES, *Spectroscopy*, **33**(7), 1, July 2018.
14. *Basics of X-ray Diffraction*, www.thermo.com/eThermo/CMA/PDFs/Product/product-PDF_11602.pdf.
15. L. Perring, D. Andrey, M. Basic-Dvorzak, and J. Blanc, *Journal of Agricultural and Food Chemistry*, **53**, 4696–4700, 2005.
16. L. J. Radziemski, D. A. Cremers, and T. R. Loree, Detection of Beryllium by Laser-Induced Breakdown Spectroscopy, *Spectrochimica Acta B*, **38**(12), 349–355, 1983.
17. K. Y. Yamamoto, D. A. Cremers, M. J. Ferris, and L. E. Foster, Detection of Metals in the Environment Using a Portable Laser-Induced Breakdown Spectroscopy Instrument, *Applied Spectroscopy*, **50**(2), 222–233, 1996.
18. R. S. Harmon, R. E. Russo, and R. R. Hark, Applications of Laser-Induced Breakdown Spectroscopy for Geochemical and Environmental Analysis: A comprehensive Review, *Spectrochimica Acta, Part B*, **27**, 1–16, 2013.
19. D. Cremers and L. Radziemski, *Handbook of Laser Induced Breakdown Spectroscopy*, John Wiley & Sons, Ltd., Hoboken, NJ, 2006.
20. T. Hussain and M. A. Gondal, Laser Induced Breakdown Spectroscopy (LIBS) as a Rapid Tool for Materials Analysis, *Journal of Physics: Conference Series*, **439**, 012050, 2013, https://iopscience.iop.org/article/10.1088/1742-6596/439/1/012050/pdf.
21. C. Wu, et al., Quantitative Analysis of Pb in Soil Samples by Laser-induced Breakdown Spectroscopy with a Simplified Standard Addition Method, *Journal of Analytical Atomic Spectrometry*, **34**(7), 14, 2019, https://pubs.rsc.org/en/content/articlepdf/2019/ja/c9ja00059c.

22. J. H. Kwaka, et al., Quantitative Analysis of Arsenic in Mine Tailing Soils Using Double Pulse-Laser Induced Breakdown Spectroscopy, *Spectrochimica Acta Part B*, **64**, 1105–1110, 2009.

23. J. J. Mortvedt, Heavy Metal Contaminants in Inorganic and Organic Fertilizers, *Fertilizers and Environment*, **43**(1–3), 55–61, January 1995.

24. L. C. Nunez et al., Determination of Cd, Cr and Pb in Phosphate Fertilizers by Laser-Induced Breakdown Spectroscopy, *Spectrochimica Acta Part B: Atomic Spectroscopy*, **97**, 42–48, July 1, 2014.

25. P. Yang et al., High-Sensitivity Determination of Cadmium and Lead in Rice Using Laser-induced Breakdown Spectroscopy, *Food Chemistry*, **272**, 323–328, January 30, 2019. http://doi.org/10.1016/j.foodchem.2018.07.214, www.ncbi.nlm.nih.gov/pubmed/30309550.

26. NASA's Mars Science Laboratory website, http://mars.nasa.gov/msl/mission/instruments/spectrometers/chemcam/.

27. R. C. Weins and S. Maurice, The ChemCam Instrument Suite on the Mars Science Laboratory Rover Curiosity: Remote Sensing by Laser-Induced Plasmas, *Geochemical News*, June 2011, www.geochemsoc.org/publications/geochemicalnews/gn145jun11/chemcaminstrumentsuit

28. K. Jankowski, Evaluation of Analytical Performance of Low-power MIP-AES with Direct Solution Nebulization for Environmental Analysis, *Journal of Analytical Atomic Spectrometry*, **14**, 1419–1423, 1999.

29. Benefits of Transitioning from FAA to the 4200 MP-AES, *Agilent Technologies Application Note*, www.chem.agilent.com/Library/technicaloverviews/Entitled%20Partner/5991-3807EN.pdf.

30. S. Elliott, Overview of 4200 MPAES, *Agilent Webinar*, 2014, www.agilent.com/cs/library/eseminars/public/4200%20MP-AES%20for%20Mining%20Playlist.pdf.

31. *Determination of Major and Minor Elements in Geological Samples Using the 4200 MP-AES: Agilent Technologies Application Note*, www.chem.agilent.com/Library/applications/5991-3772EN.pdf.

32. Analysis of Major, *Minor and Trace Elements in Rice Flour Using the 4200 MP-AES: Agilent Technologies Application Note*, www.chem.agilent.com/Library/applications/5991-3777EN.pdf.

33. *Analysis of Major Elements in Fruit Juices Using the Agilent 4200 MP-AES with the Agilent 4107 Nitrogen Generator: Agilent Technologies Application Note*, https://hpst.cz/sites/default/files/oldfiles/5991-3613en-fruit-juice.pdf

34. *4210 MP-AES Application Library: Agilent Technologies*, www.agilent.com/en/products/mp-aes/mp-aes-systems/4210-mp-aes#literature.

35. C. H. Becker and K. T. Gillen, Can Nonresonant Multiphoton Ionization be Ultrasensitive? *Journal of the Optical Society of America*, **B2**, 1438, 1985.

36. C. H. Becker and K. T. Gillen, Surface Analysis by Nonresonant Multiphoton Ionization of Desorbed or Sputtered Species, *Analytical Chemistry*, **56**, 1671–1674, 1984.

37. G. R. Kinsel and D. H. Russell, Design and Calibration of an Electrostatic Energy Analyzer-time-of-flight Mass Spectrometer for Measurement of Laser-Desorbed Ion Kinetic Energies, *Journal of the American Society for Mass Specrometry*, **6**, 619–626, 1995.

38. S. A. Getty, et al., Compact Two-Step Laser Time-of-Flight Mass Spectrometer for in Situ Analyses of Aromatic Organics on Planetary Missions, *Rapid Communications in Mass Spectrometry*, **26**, 2786–2790, 2012.

39. P. Hurtado, F. Gamez, and B. Martinez-Haya, One- and Two-Step Ultraviolet and Infrared Laser Desorption Ionization Mass Spectrometry of Asphaltenes, *Journal of Energy & Fuels*, **24**, 6067–6073, 2010.

40. H. Sabbah, et al., Laser Desorption Single-Photon Ionization of Asphaltenes: Mass Range, Compound Sensitivity, and Matrix Effects, *Journal of Energy & Fuels*, **26**, 3521–3526, 2012.

41. B. Schueler and R. W. Odom, Nonresonant Multiphoton Ionization of the Neutrals Ablated in Laser Microprobe Mass Spectrometry Analysis of Gallium Arsenide, *Journal of Applied Physics*, **61**, 4652–4661, 1987.

42. J. H. Hahn, R. Zenobi, and R. N. Zare, Subfemtomole Quantitation of Molecular Adsorbates by Two-Step Laser Mass Spectrometry, *Journal of the American Chemical Society*, **109**, 2842–2843, 1987.

43. L. V. Vaeck and R. Gijbels, Laser Microprobe Mass Spectrometry: Potential and Limitations for Inorganic and Organic Micro-Analysis, *Fresenius Journal of Analytical Chemistry*, **337**, 743–754, 1990.

44. E. Williams, P. Willis, A. Koenig, and J. C. Putman, Using an Innovative Mass Spectrometer and Non-Matrix—Matched Reference Materials to Quantify Composition of Metal Disks and Powders, *Spectroscopy*, **37**(3), 9, 2022.

24 Performance Characteristics of Common Atomic Spectroscopy Techniques

Since the introduction of the first commercially available atomic absorption spectrophotometer (AAS) in the early 1960s, there has been an increasing demand for better, faster, easier-to-use, and more flexible trace element instrumentation. A conservative estimate showed that in 2022, the market for atomic spectroscopy (AS)-based instruments was well in excess of $1 billion in annual revenue, and that doesn't include after-market sales and service costs, which could increase this number by an additional $500 million. As a result of this growth, we have seen a rapid emergence of more sophisticated equipment and easier-to-use software. Moreover, with an increase in the number of manufacturers of AS instrumentation and its sampling accessories, together with the availability traditional solid-sampling techniques such as X-ray fluorescence (XRF) and newer techniques like microwave-induced optical emission spectroscopy (MIP-OES) and laser induced breakdown spectroscopy (LIBS), the choice of which technique to use is often unclear. It also becomes even more complicated when budgetary restrictions allow only one analytical technique to be purchased to solve a particular application problem. This chapter takes a look at the major AS techniques and compares their performance characteristics and in particular their detection limits, linear dynamic range and sample throughput capability.

In order to select the best technique for a particular analytical problem, it is important to understand exactly what the problem is and how it is going to be solved. For example, if the requirement is to monitor copper at percentage levels in a copper plating bath and it is only going to be done once per shift, flame atomic absorption would adequately fill this role. Alternatively, when selecting an instrument to measure 24 elemental impurities in pharmaceutical products, this application is clearly better suited for a multielement technique such as inductively coupled plasma optical emission spectrometry (ICP-OES). However, if the demand is for four heavy metal contaminants in cannabis, the decision is not so clear. ICP-MS could be the most suitable technique, but depending on the state's maximum action limits, the consumer product being tested, the sample workload or budgetary restrictions, other techniques such as graphite furnace AA or vapor-generation atomic fluorescence might be the best fit.

So when choosing a technique, it is important to understand not only the application problem, but also the strengths and weaknesses of the technology being applied to solve the problem. However, there are many overlapping areas between the major atomic spectroscopy techniques, so it is highly likely that for some applications, more than one technique would be suitable. For that reason, it is important to go through a carefully thought-out evaluation process when selecting a piece of equipment.[1,2]

We will not be discussing XRF, MIP-OES or LIBS techniques in this evaluation. They are all very useful techniques for many sample types and have been described at length in previous chapters. However, for the purpose of this comparison, we will be focusing on the most commonly used atomic spectroscopy techniques—flame atomic absorption (FAA), electrothermal atomization (ETA), hydride-generation atomic absorption (HGAA), hydride-generation atomic fluorescence (HGAF), ICP-OES and ICP-MS.

Let's first take a brief look at their fundamental principles. We have covered the techniques in greater detail in previous chapters, but it's useful to briefly compare the performance of these two plasma spectrochemical techniques (ICP-OES, ICP-MS) with flame atomic absorption (FAA),

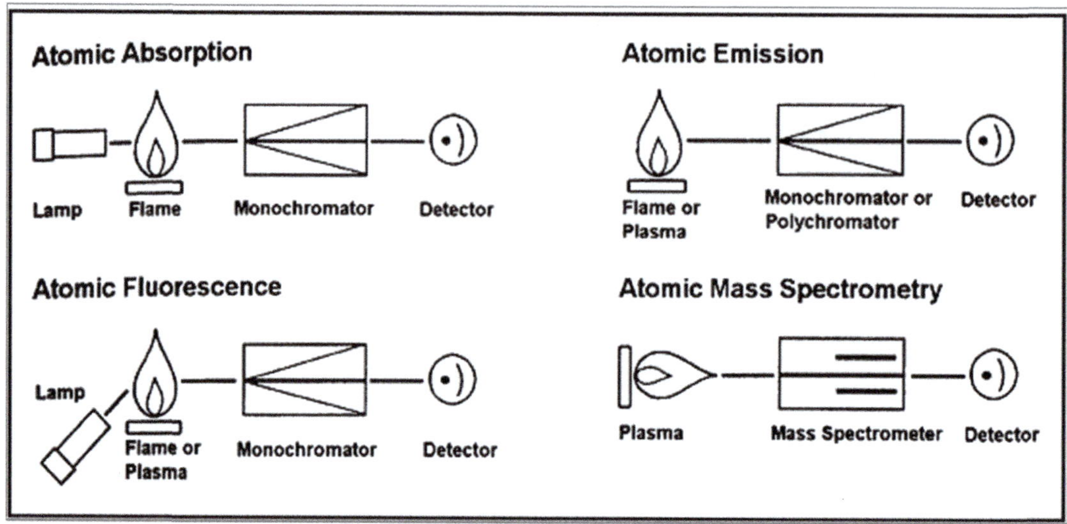

FIGURE 24.1   Simplified schematic of AA, AF, ICP-OES and ICP-MS.

electrothermal atomization (ETA) and hydride-generation atomic fluorescence (HGAF). Simplified schematics of each of these techniques are shown in Figure 24.1.

24.1 FLAME ATOMIC ABSORPTION

This is predominantly a single-element technique for the analysis of liquid samples that uses a flame to generate ground-state atoms. The sample is aspirated into the flame via a nebulizer and a spray chamber. The ground-state atoms of the sample absorb light of a particular wavelength, either from an element-specific, hollow cathode lamp or a continuum-source lamp. The amount of light absorbed is measured by a monochromator (optical system) and detected by a photomultiplier tube or solid-state detector, which converts the photons into an electrical signal. As in all AS techniques, this signal is used to determine the concentration of that element in the sample, by comparing it to calibration standards.

Flame AA typically uses about 2–5 mL/min of liquid sample and is capable handling in excess of 10% total dissolved solids, although for optimum performance it is best to keep the solids down below 2%. For the majority of elements, it's detection capability is in the order of 1–100 parts-per-billion (ppb) levels with an analytical range up to 10–1,000 parts per million (ppm), depending on the absorption wavelength used. However, it is not really suitable for the determination of the halogens, non-metals like carbon, sulfur and phosphorus and has very poor detection limits for the refractory, rare earth and transuranic elements. Sample throughput for 15 elements per sample is in the order of 10 samples an hour.

24.2 ELECTROTHERMAL ATOMIZATION

This is also mainly a single-element technique, although multielement instrumentation is now available. It works on the same principle as flame AA, except that the flame is replaced by a small, heated tungsten filament or graphite tube. The other major difference is that in ETA, a very small sample (typically, 50 µL) is injected onto the filament or into the tube, and not aspirated via a nebulizer and a spray chamber. Because the ground-state atoms are concentrated in a smaller area, more absorption takes place. The result is that ETA offers detection capability at the 0.01–1 parts-per-billion (ppb) level, with an analytical range up to 10–100 parts per billion (ppb).

The elemental coverage limitations of the technique are similar to flame AA technique. However, because a heated graphite tube is used for atomization in most commercial instruments, it cannot determine the refractory, rare earth and transuranic elements, because they tend to form stable carbides that cannot be readily atomized. One of the added benefits is that ETA can also analyze slurries and some solids due to the fact that no nebulization process is involved in introducing the sample. This technique is not ideally suited for multielement analysis, because it takes three to four minutes to determine one element per sample. As a result, sample throughput for four elements is in the order of three samples per hour.

24.3 HYDRIDE/VAPOR-GENERATION AA

Hydride generation (HG) is a very useful analytical technique to determine the hydride-forming metals, such as As, Bi, Sb, Se and Te, usually by AA (although detection by either ICP-OES or ICP-MS can also be used). In this technique, the analytes in the sample matrix are first reacted with a very strong reducing agent, such as sodium borohydride to release their volatile hydrides, which are then swept into a heated quartz tube for atomization. The tube is heated either by a flame or a small oven which creates the ground-state atoms of the element of interest and then measured by atomic absorption. When used with ICP-OES or ICP-MS, the volatile hydrides are passed directly in the plasma for excitation or ionization.

By choosing the optimum chemistry, mercury can also be reduced in solution in this way to generate elemental mercury. This is known as the cold-vapor (CV) technique. HG and CV AA can improve the detection for these elements over flame AA by up to three orders of magnitude, achieving detection capability in the order of 0.005–0.1 parts-per-billion (ppb) levels with an analytical range of 5–100 parts per billion (ppb), depending on the element of interest. It should also be pointed out that dedicated mercury analyzers (some using gold-amalgamation techniques), coupled with atomic absorption or atomic fluorescence, are capable of better detection limits. Because of the online chemistry involved, these techniques are very time-consuming and are normally used in conjunction with FAA, so they will most likely impact the overall sample throughput.

24.4 ATOMIC FLUORESCENCE

Hydride/vapor generation can also be used with atomic fluorescence, to improve the detection capability of the volatile elements/heavy metals. Atomic fluorescence shares properties with atomic absorption in that it has a light source, atomizer, monochromator and detector. The element-specific light energy is absorbed, taking the element through excitation. Rather than determining the amount of light being absorbed by the element in the atomizer, as the excited atoms return to ground state, they emit photons of light and this emission, or fluorescence, resulting from the decay is measured. The major difference is the element specific light source is at an angle so as not to interfere with the fluorescence process. To carry out multielement analysis using AFS, many lamps are positioned around the atomization/excitation source.

Background signals in AFS are very low, resulting in improved sensitivity and detection limit as compared to atomic absorption. However, the fluorescence measurement can be quenched by many molecular forms, so separation of the matrix from the analyte may be desirable. There are commercially available vapor-generation AFS systems on the market, which have been successfully used for the measurement of Pb, Cd, As and Hg in a wide variety of samples.[3,4]

24.5 RADIAL ICP-OES

Radially viewed ICP-OES is a multielement technique that uses a traditional radial (side-view) inductively coupled plasma to excite ground-state atoms to the point where they emit wavelength-specific photons of light that are characteristic of a particular element. The number of photons produced at an element-specific wavelength is measured by high-resolution optics and a photon-sensitive device, such as a photomultiplier tube or a solid-state detector. This emission signal is directly related to the concentration of that element in the sample. The analytical temperature of an ICP is about 6,000–7,000°K, compared to that of a flame or a graphite furnace, which is typically 2,500–3,500°K.

For the majority of elements, a radial ICP instrument can achieve detection capability in the order of 0.1–100 parts-per-billion (ppb) levels with an analytical range up to 10–1,000 parts per million (ppm), depending on the emission wavelength used. The technique can determine a similar number of elements to FAA but has the advantage of offering the capability of non-metals like sulfur and phosphorus, together much better performance for the refractory, rare earth and transuranic elements. The sample requirement for ICP-OES is approximately 1 mL/min, and is capable of aspirating samples containing up to 10% total dissolved solids, but for optimum performance, is usually kept below 2%. ICP-OES is a rapid multielement technique, so sample throughput for 15 elements per sample is in the order of 20 samples an hour.

24.6 AXIAL ICP-OES

The principle is exactly the same as radial ICP-OES, except that in the axial view, the plasma is viewed horizontally (end-on). The benefit is that more photons are seen by the detector and for this reason, detection limits can be as much as an order of magnitude lower, depending on the design of the instrument. The disadvantage is that the working range is also reduced by an order of magnitude. As a result, for the majority of elements, an axial ICP instrument can achieve detection capability in the order of 0.01–10 parts-per-billion (ppb) levels with an analytical range up to 1–100 parts per million (ppm), depending on the emission wavelength used. The other disadvantage of viewing axially is that more severe matrix interferences are observed, which means that the total-dissolved-solids content of the sample needs to be kept much lower. Sample flow requirements are the same as for radial ICP-OES.

It's also important to point out that most commercial ICP-OES instruments have both radial and axial capability built in. However, because of the different hardware configurations available, some instruments work by carrying out the analysis either using the radial or the axial view (sequentially), while others can use both the radial and axial view at exactly the same time (simultaneously), which for some applications may be advantageous. Strategic use of both radial and axial view, together with the optimum wavelength selection, can extend the analytical range by two to three orders of magnitude. Sample throughput will be approximately 20 samples per hour, the same as radial ICP-OES.

24.7 ICP-MS

The fundamental difference between ICP-OES and ICP-MS is that in ICP-MS, the plasma is not used to generate photons, but to generate positively charged ions. The ions produced are transported and separated by their atomic mass-to-charge ratio using a mass-filtering device such as a quadrupole or a magnetic sector. The generation of such large numbers of positively charged ions allows ICP-MS to achieve detection limits approximately three orders of magnitude lower than ICP-OES, even for the refractory, rare earth and transuranic elements. As a result, for the majority of elements, an ICP-MS instrument can achieve detection capability in the order of 0.0001–1 parts per billion (ppb) with an analytical range up to 0.1–100 parts per million (ppm), using pulse-counting measurement, but can be extended even further up to 100–100,000 ppm by using analog-counting techniques. However, it should be emphasized that if such large analyte concentrations are being measured, expectations should be realistic about also carrying out ultra-trace determinations of the same element in the same sample run.

The sample requirement for ICP-MS is approximately 1 mL/min, and is capable of aspirating samples containing up to 5% total dissolved solids for short periods with the use of specialized sampling accessories. However, because the sample is being aspirated into the mass spectrometer, for optimum performance, matrix components should ideally be kept below 0.2%. This is particularly relevant for laboratories that experience a high sample workload. Sample throughput will be approximately 15 samples per hour, for the determination of 24 elements in a sample.

24.8 COMPARISON HIGHLIGHTS

There is no question that the techniques like ICP-OES and ICP-MS are better suited for multielement analysis, especially in labs that are expected to be carrying out the analysis in a high-throughput, routine environment. If the best detection limits are required, ICP-MS offers the best choice followed by graphite-furnace AA (ETA). Axial ICP-OES offers very good detection limits for most elements, but generally not as good as ETA. Radial ICP-OES and flame AA show approximately the same detection-limits performance, except for the refractory, rare earth and the transuranic elements, for which performance is much better by ICP-OES. For mercury and those elements that form volatile hydrides, such as As, Bi, Sb, Se and Te, the cold-vapor or hydride-generation techniques offer exceptional detection limits. Table 24.1 gives a detailed comparison of detection

TABLE 24.1

A Comparison of Detection Limits between ICP-MS and the Other AS Techniques in μg/L (ppb)

Element	Flame AA	Hg/ Hydride	GFAA	ICP-OES	ICP-MS	Element	Flame AA	Hg/ Hydride	GFAA	ICP-OES	ICP-MS
Ag	1.5		0.005	0.6	0.002	Mo	45		0.03	0.5	0.001
Al	45		0.1	1	0.005	Na	0.3		0.005	0.5	0.0003
As	150	0.03	0.05	2	0.006	Nb	1500			1	0.0006
Au	9		0.15	1	0.0009	Nd	1500			2	0.0004
B	1000		20	1	0.003	Ni	6		0.07	0.5	0.0004
Ba	15		0.35	0.03	0.00002	Os				6	
Be	1.5		0.008	0.09	0.003	P	75000		130	4	0.1
Bi	30	0.03	0.05	1	0.0006	Pb	15		0.05	1	0.00004
Br					0.2	Pd	30		0.09	2	0.0005
C					0.8	Pr	7500			2	0.00009
Ca	1.5		0.01	0.02	0.05	Pt	60		2.0	1	0.002
Cd	0.8		0.008	1	0.003	Rb	3		0.03	5	0.0004
Ce				1.5	0.0002	Re	750			0.5	0.0003
CI					12	Rh	6			5	0.0002
Co	9		0.15	0.2	0.0002	Ru	100		1.0	1	0.0002
Cr	3		0.004	2	0.02	S				10	28
Cs	15				0.0003	Sb	45	0.15	0.05	2	0.0009
Cu	1.5		0.014	0.4	0.0002	Sc	30			0.1	0.004
Dy	50			0.5	0.0001	Se	100	0.03	0.05	4	0.0007
Er	60			0.5	0.0001	Si	90		1.0	10	0.03
Eu	30			0.2	0.00009	Sm	3000			2	0.0002
F					372	Sn	150		0.1	2	0.0005
Fe	5		0.06	0.1	0.0003	Sr	3			0.025	0.05	0.00002

(Continued)

TABLE 24.1

(Continued)

Element	Flame AA	Hg/ Hydride	GFAA	ICP-OES	ICP-MS	Element	Flame AA	Hg/ Hydride	GFAA	ICP-OES	ICP-MS
Ga	75			0.5	0.0002	Ta	1500			1	0.0005
Gd	1800			0.9	0.0008	Tb	900			2	0.00004
Ge	300			1	0.001	Te	30	0.03	0.1	2	0.0008
Hf	300			0.5	0.0008	Th				2	0.0004
Hg	300	0.009	0.6	1	0.016	Ti	75		0.35	0.4	0.003
Ho	60			0.4	0.00006	Tl	15		0.1	2	0.0002
I					0.002	Tm	15			0.6	0.00006
In	30			1	0.0007	U	15000			10	0.0001
Ir	900		3.0	1	0.001	V	60		0.1	0.5	0.0005
K	3		0.005	1	0.0002	W	1500			1	0.005
La	3000			0.4	0.001	Y	75			0.2	0.0002
Li	0.8		0.06	0.3	0.0001	Yb	8			0.1	0.0002
Lu	1000			0.1	0.00005	Zn	1.5		0.02	0.2	0.0003
Mg	0.15		0.004	0.04	0.0003	Zr	450			0.5	0.0003
Mn	1.5		0.005	0.1	0.00007						

Note: If there is no D/L data shown, it means the AS technique is not ideally suited to determine that element.

Courtesy of PerkinElmer Inc.

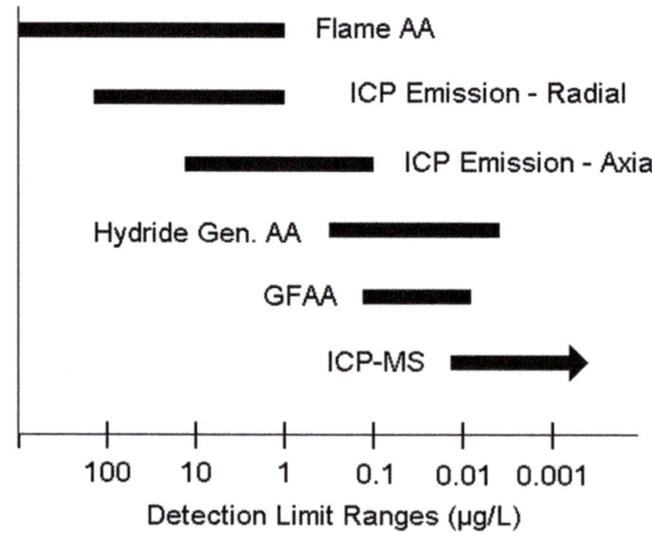

FIGURE 24.2 Typical AS detection-limit ranges.

limits of the four major AS techniques, while Figures 24.2 and 24.3 and Table 24.1 shows an overview of the detection capability, working analytical range and sample throughput of the major AS approaches, which are three of the metrics most commonly used to select a suitable technique. They should be considered an approximation and only used for guidance purposes.

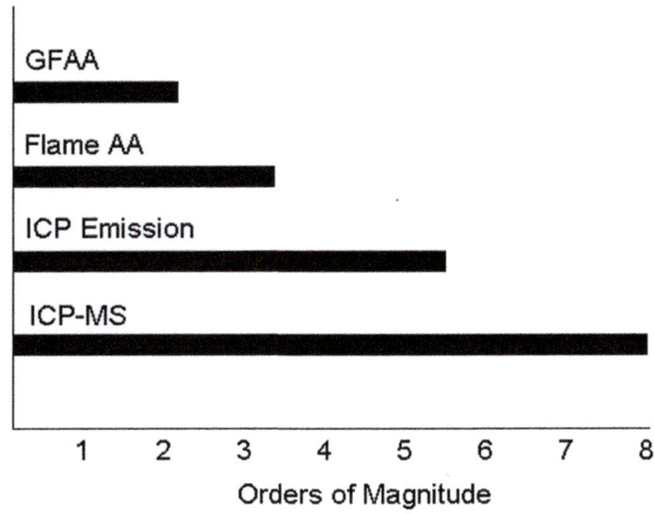

FIGURE 24.3   Typical AS analytical working ranges.

TABLE 24.2

Approximate Sample-Throughput Capability of AS Techniques

Technique	Elements at a time	Duplicate analysis (min)	Samples per hour (1 element)	Samples per hour (4 elements)	Samples per hour (15 elements)
FAA	1	0.3	150	30	10
ETA	1	5	12	3	1
AF	Up to 4	3	20	20	90
ICP-OES	Up to 70	3	20	20	20
ICP-MS	Up to 70	3	20	20	20

It is also worth mentioning that the detection limits of quadrupole-based ICP-MS, when used in conjunction with collision/reaction cell/interface and/or multiple mass separation/selection devices, is now capable of sub parts per trillion (ppt) detection limits, even for the elements such as Fe, K, Ca, Se, As, Cr, Mg, V and Mn, which traditionally suffer from plasma- and solvent-based polyatomic interference.

This is not meant to be a detailed description of each technique, but a basic understanding as to how they differ from each other. For a more detailed comparison of the fundamental principles and application strengths and weaknesses of ICP-MS and ICP-OES, please refer to the previous chapters.

FURTHER READING

1. R. J. Thomas, Choosing the Right Trace Element Technique: Do You Know What to Look For, *Today Chemist at Work*, October, 1999.
2. R. J. Thomas, Emerging Technology Trends in Atomic Spectroscopy Are Solving Real-World Application Problems, *Spectroscopy Magazine*, **29**(3), 42–51, 2014.
3. Lumina 3500 Atomic Fluorescence Spectrometer, *Aurora Illuminating Systems Elemental Analysis Instruments*, 2020, www.aurorabiomed.com/wpcontent/uploads/2019/07/LUMINA_3500.pdf.
4. *PF7 Heavy Metals Analyzer*, Persee Analytics Inc., http://perseena.com/index/prolist/pid/33/c_id/43.htm.

25 Running Costs of Common AS Techniques

Chapter 24 gave an overview of the performance differences of the major atomic-spectroscopic techniques, but selection of the optimum approach might also involve the cost of the instrumentation, particularly if funds are limited. Chapter 25 takes a look at the running costs of four of the most common AS techniques, particularly from the standpoint of the cost of consumables, gases and electricity.

For the purpose of this evaluation, let us make the assumption that the major operating costs associated with running AS instrumentation are the gases, electricity and consumable supplies. For comparison purposes, the exercise will be based on a typical laboratory running their instrument for two-and-a-half days (20 h) per week and 50 weeks a year (1,000 h per year). These data are based on the cost of gases, electricity and instrument consumables in the United States in 2021. They have been obtained from a number of publically available commercial sources, including suppliers of industrial and high-purity gases, independent utilities companies, a number ICP-MS instrument vendors and sample-introduction/consumable suppliers. It's also important to emphasize that these costs might vary slightly based on the different vendor instrument design/technology/being used.[1]

25.1 GASES

25.1.1 FAA

Most flame AA systems use acetylene (C_2H_2) as the combustion gas, and air or nitrous oxide (N_2O) as the oxidant. Air is usually generated by an air compressor, but the C_2H_2 and N_2O come in high-pressure cylinders. Normal atomic absorption-grade C_2H_2 cylinders contain 380 ft^3 (10,760 L) of gas. N_2O is purchased by weight and comes in cylinders containing 56 lbs of gas, which is equivalent to 490 ft^3 (13,830 L). A cylinder of C_2H_2 costs approximately \$200, whereas a cylinder of N_2O costs about \$70. These prices have remained fairly stable over the past few years. Normal C_2H_2 gas flows in FAA are typically 2 L/min when air is the oxidant and 5 L/min when N_2O is the oxidant. N_2O gas flows are on the order of 10 L/min.

25.1.2 Air—C_2H_2

Mixtures are used for the majority of elements, whereas an N_2O—C_2H_2 mixture has traditionally been used for the more refractory elements. So, for this costing exercise, we will assume that half the work is done using air—C_2H_2, and for the other half N_2O—C_2H_2 is being used. Therefore, a typical laboratory running the instrument for 1,000 h per year will consume 16 cylinders of C_2H_2, which is equivalent to \$3,200 per year, and 22 cylinders of N_2O costing \$1,500, making a total of \$4,700.

25.1.2.1 ETA

The only gas that the electrothermal atomization process uses on a routine basis is high-purity argon, which costs about \$100 for a 340 ft^3 (9630 L) cylinder. Typically, argon gas flows of up to 300 mL/min are required to keep an inert atmosphere in the graphite tube. At these flow rates, 540 h of use can be expected from one cylinder. Therefore, a typical laboratory running their instrument for 1,000 h per year would consume almost two cylinders costing \$200.

DOI: 10.1201/9781003187639-25

25.1.2.2   ICP-OES and ICP-MS

The consumption of gases in ICP-OES and ICP-MS is very similar. They both use a total of approximately 15–20 L/min (~1,000 L/h) of gaseous argon (inc. plasma, nebulizer, auxiliary and purge flows), which means a cylinder of argon (9,630 L) would last only about 10 h. For this reason, most users install a Dewar vessel containing a liquid supply of argon. Liquid argon tanks come in a variety of different sizes, but a typical Dewar system used for ICP-OES/ICP-MS holds about 240 L of liquid gas, which is equivalent to 6,300 ft^3 (178,000 L) of gaseous argon. (Note: The Dewar vessel can be bought outright, but are normally rented.) It costs about \$350 to fill a 240-L Dewar vessel with liquid argon. At a typical argon flow rate 17 L/min total gas flow, a full vessel would last for almost 175 h. Again, assuming a typical laboratory runs their instrument for 1,000 h per year, this translates to 6 fills at approximately \$350 each, which is equivalent to about \$2,100 per year. If cylinders were used, about 100 would be required, which would elevate the cost to almost \$10,000 per year.

Note: When liquid argon is stored in a Dewar vessel, there is a natural bleed-off to the atmosphere when the gas reaches a certain pressure. For this reason, a bank of argon cylinders is probably the best option for laboratories that do not use their instruments on a regular basis. Some of the newer ICP-OES instruments operate at approximately 60–70% argon consumption compared to older instruments. So this should be taken into consideration if this technology is being used.

Another added expense with ICP-MS is that if it is fitted with collision/reaction-cell technology, the cost of the collision or reaction gas will have to be added to the running costs of the instrument. Fortunately, for most applications, the gas flow is usually less than 5 mL/min, but for the collision/reaction-interface approach, typical gas flows are 100–150 mL/min. The most common collision/reaction gases used are hydrogen, helium and ammonia. The cost of high-purity helium is on the order of \$400 for a 300 ft^3 (8500 L) cylinder, whereas that of a cylinder of hydrogen or ammonia is approximately \$250. One cylinder of either gas should be enough to last 1,000 h at these kinds of flow rates. So, for this costing exercise, we will assume that the laboratory is running a collision/reaction cell/interface instrument, with an additional expense of \$650. If other collision/reaction gases are being used, these should be factored into the calculation.

It should also be pointed out that some collision/reaction cells require high-purity gases with extremely low impurity levels, because of the potential of the contaminants in the gas to create additional by-product ions. This can be achieved either by purchasing laboratory-grade gases and cleaning them up with a gas purification system (getter), or by purchasing ultra-high-purity gases directly from the gas supplier. If the latter option is chosen, you should be aware that ultra-high-purity helium (99.9999%) is approximately twice the price of laboratory-grade helium (99.99%), whereas ultra-high-purity hydrogen is approximately four times the cost of laboratory-grade hydrogen.

25.2   ELECTRICITY

Calculations for power consumption are based on the average cost of electricity, which is currently about \$0.10 per kilowatt hour (kW/h) in the United States. The cost will vary depending on the location and demand, but it represents a good approximation for this exercise. So the following formula has been used for calculating the cost of electricity usage for each technique:

Cost per kW hour (\$) * Power Consumption (kW) * Annual Usage (h)

25.2.1   FAA

The power in a flame AA system is basically used for the hollow cathode lamps and the onboard microprocessor that controls functions like burner head position, lamp selection, photo multiplier tube voltage, grating position, etc. A typical instrument requires less than 1kW of power. If it is used for 1,000 h per year, it will be drawing less than 1,000 kW total power, which is ~\$100 per year.

25.2.2 ETA

A graphite-furnace system uses considerably more power than a flame AA system because a separate power unit is used to heat the graphite tube. In routine operation, there is a slow ramp heating of the tube for ~3 min until it reaches an atomization temperature of about 2,700°C, requiring a maximum power of ~3 kW. This heating cycle combined with the power requirements for the rest of the instrument costs ~$300, for a system that is run 1,000 h per year.

25.2.3 ICP-OES AND ICP-MS

Both these techniques can be considered the same with regard to power requirements as the RF generators are of very similar design. Based on the voltage, magnitude of the electric current, and the number of lines used, the majority of modern instruments draw about 5 kW total power. This works out to be ~$500 for an instrument that is run 1,000 h per year.

25.3 CONSUMABLES

Because of the fundamental differences between the four AS techniques, it is important to understand that there are considerable differences in the cost of consumables. In addition, the cost of the same component used in different techniques can vary significantly between different vendors and suppliers. So, the data has been taken from a number of different sources and averaged.

25.3.1 FAA

The major consumable supplies used in flame AA are the hollow cathode lamps. Depending on usage, you should plan to replace three of them every year, at a cost of $300–500 for a good quality, single-element lamp. However, if a continuum source AA system is being used, there will not be a requirement to replace lamps on a regular basis. Other minor costs are nebulizer tubing and autosampler tubes. These are relatively inexpensive, but should be planned for. The total cost of lamps, nebulizer tubing, and a sufficient supply of autosampler tubes should not exceed $1,500–$2,000 per year, based on 1,000 h of instrument usage.

25.3.2 ETA

As long as the sample type is not too corrosive, a graphite furnace AA tube should last about 300 heating cycles (firings). Based on a normal heating program of 3 min per replicate, this represents 20 firings per hour. If the laboratory is running the instrument 1,000 h per year, it will carry out a total of 20,000 firings and use 70 graphite tubes in the process. There are many designs of graphite tubes, but for this exercise we will base the calculation on platform-based tubes that cost about $50 each when bought in bulk. If we add the cost of graphite contact cylinders, hollow cathode lamps and a sufficient supply of autosampler cups, the total cost of consumables for a graphite furnace will be approximately $5,000–$6,000 per year.

25.3.3 ICP-OES

The main consumable supplies in ICP-OES are in the plasma torch and in the sample-introduction area. The major consumable is the torch itself, which consists of two concentric quartz tubes and a sample injector either made of quartz or some ceramic material. In addition, a quartz bonnet normally protects the torch from the RF coil. There are many different demountable torch designs available, but they all cost about $600–700 for a complete system. Depending on sample workload and matrices being analyzed, it is normal to go through a torch every four to six months. In addition to the torch, other parts that need to be replaced or at least need to have spares include the nebulizer, spray chamber, and sample capillary and pump tubing. When all these items are added together, the annual cost of consumables for ICP-OES is on the order of $3,000–$3,200.

25.3.4 ICP-MS

In addition to the plasma torch and sample-introduction supplies, ICP-MS requires consumables that are situated inside the mass spectrometer. The first area is the interface region between the plasma and the mass spectrometer, which contains the sampler and skimmer cones. These are traditionally made of nickel, which is recommended for most matrices, or platinum for highly corrosive samples and organic matrices. A set of nickel cones costs $700–$1,000, whereas a set of platinum cones costs about $3,000–$4,000. Two sets of nickel cones and perhaps one set of platinum cones would be required per year. The other major consumable in ICP-MS is the detector, which has a lifetime of approximately 1 year, and costs about $1,200–$1,800. When all these are added together with the torch, the sample introduction components, and the vacuum pump consumables, investing in ICP-MS supplies represents an annual cost of $9,000–$11,000.

The approximate annual cost of gases, power and consumable supplies of the four AS techniques being operated for 1,000 h/year is shown in Table 25.1.

25.4 COST PER SAMPLE

We can take the data given in Table 25.1 a step further and use these numbers to calculate the operating costs per individual sample, assuming that a laboratory is determining ten analytes per sample. Let us now take a look at each technique to see how many samples can be analyzed, assuming the instrument runs 1,000 h per year.

25.4.1 FAA

A duplicate analysis for a single analyte in flame AA takes about 20s. This is equivalent to 180 analytes per hour or 180,000 analytes per year. For ten analytes, this represents 18,000 samples per year. Based on an annual operating cost of $6,550, this equates to $0.36 per sample.

25.4.2 ETA

A single analyte by ETA takes about 5–6 min for a duplicate analysis, which is equivalent to approximately 10 analytes per hour or 10,000 analytes per year. For 10 analytes/sample, this represents 1,000 samples per year. Based on an annual operating cost of $6,000, this equates to $6.00 per sample.

TABLE 25.1

Annual Operating Costs ($US) for the Four AS Techniques for a Laboratory Running an Instrument 1,000 h per Year (20 h per Week).

Technique	Gases ($)	Power ($)	Consumable supplies ($)	Total ($)
FAA	4,700	100	1,750	6,550
ETA	200	300	5,500	6,000
ICP-OES	2100 [1]	500	3,100	5,700
ICP-MS	2,750 [1,2]	500	10,000	13,250

Note: [1]using a liquid argon supply, [2]using a collision/reaction cell

TABLE 25.2

Operating Costs for a Sample Requiring Ten Analytes, Based on the Instrument Being Used for 1,000 h per Year

Technique	Operating Cost for Ten Analytes per Sample ($US)
FAA	0.36
ETA	6.50
ICP-OES [1]	0.29
ICP-MS [1,2]	0.66

Note: [1]using a liquid argon supply, [2]using a collision/reaction cell)

25.4.3 ICP-OES

A duplicate ICP-OES analysis for as many analytes as you require takes about 3 min. So for 10 analytes, this is equivalent to 20 samples per hour or 20,000 samples per year. Based on an annual operating cost of $5,700, this equates to $0.30 per sample.

25.4.4 ICP-MS

ICP-MS also takes about 3 min to carry out a duplicate analysis for 10 analytes, which is equivalent to 20,000 samples per year. Based on an annual operating cost of $13,250, this equates to $0.66 per sample.

Operating costs for all four AS techniques for the determination of ten analytes/sample are summarized in Table 25.2.

It must also be emphasized that this comparison does not take into account the detection limit requirements, but is based on instrument-operating costs alone. These figures have been generated for a typical workload using what would be considered the average cost of gases, power and consumables in the United States. Even though there will be geographical differences in the cost of these items in other parts of the world, the comparative costs should be very similar. Every laboratory's workload and analytical needs are unique, so this costing exercise should be treated with caution and only be used as a guideline for comparison purposes.

For a more accurate assessment of your lab's workload, the costing exercise should be carried using your sample throughput and analyte requirements. However, whatever the workload, it is a good exercise to show that there are running cost differences between the major AS techniques. If required, it can be taken a step further by also including the purchase price of the instrument, the cost of installing a clean room, the cost of sample preparation and the salary of the operator. This would be a very useful exercise as it would give a good approximation of the overall cost of analysis, and therefore it could be used as a guideline for calculating what a laboratory might charge for running samples on a commercial basis.

25.5 RUNNING COSTS OF ATOMIC FLUORESCENCE

Atomic Fluorescence has intentionally not been used in this comparison, because there are so many variables to consider. Even though it uses hollow cathode-type lamps similar to FAA and ETA, it will also depend on the type of heating used in the atomization source, particularly if the system is dedicated to hydride/cold-vapor generation. Some instruments use a flame to heat the quartz cell, while others use an electrically heated furnace. If combustion gases are used to generate the flame, running costs will vary based on whether the gas is acetylene or hydrogen (one system uses a hydrogen diffusion flame). If resistive electrical heating is used it will also depend on the current used to generate the ground-state atoms of the analyte elements. Additionally, if mercury is one of the analytes, atomization occurs at room temperature so there is no requirement for heating the cell. Finally, argon gas is typically used to sweep the gaseous hydride from the reaction vessel into the atomization cell, so this also has to be considered as a cost factor.

As a result, all these variables make it very complicated to calculate the cost of running an AFS system, especially when reducing chemicals such as stannous chloride or sodium borohydride have to be built into the calculation, which will clearly depend on the number of samples being run. This makes it very difficult to make a comparison with the other AS techniques, particularly as AFS can only measure a small group of elements at the same time. However, as a guideline, I think it's fair to say that the annual running costs of AFS will be more in line with FAA and ETA than either of the plasma spectrochemical techniques exemplified in Table 25.1

25.6   FINAL THOUGHTS

It can be seen from this evaluation that based on the annual operating costs, FAA, ETA and ICP-OES are all very similar, with ICP-MS being approximately twice as expensive to operate. However, when the number of samples/analytes is taken into consideration, the picture changes quite dramatically. It is also important to remember that there are many criteria to consider when selecting a trace element technique. Operating costs are just one of them, and they should not prevent you from choosing an instrument if your analytical requirements change, such as the need for lower detection limits. It's also important to emphasize that detection capability could be the most important selection criterion, particularly if you are trying to achieve the very low regulated limits of inhaled cannabinoid products However, if more than one of these techniques fulfills your analytical demands, then knowledge of the operating costs should help you make the right decision.

FURTHER READING

1. R. Thomas, *Practical Guide to ICP-MS: A Tutorial for Beginners*, 3rd ed., CRC Press, Boca Raton, FL, 2013, ISBN: 978-1-4665-5543-3.

26 How to Evaluate an ICP Mass Spectrometer
Some Important Analytical Considerations

Understanding the basic principles of ICP mass spectrometry is important but not absolutely essential to operate and use an instrument on a routine basis. However, understanding how these basic principles affect the performance of an instrument is a real benefit when evaluating the analytical capabilities of the technique. There are excellent commercial instruments on the market and are all capable of generating high-quality data. However, they all have their own strengths and weaknesses. For that reason, the better informed you are going into an instrument evaluation the better chance you have of selecting the right one for your organization. Having been involved in demonstrating ICP-MS equipment and running customer samples for over almost 25 years, I know the mistakes that people make when they get into the evaluation process. So, Chapter 26 is a set of evaluation guidelines which are a systematic approach to comparing different commercial instrumentation.

OK, you have convinced your boss that ICP-MS is the ideal technique to meet the analytical demands of your laboratory. Hopefully, the chapters on the fundamental principles have given you the basic knowledge and a good platform on which to go out and evaluate the marketplace. However, they do not really give you an insight into how to compare instrument designs, hardware components and software features, which are of critical importance when you have to make a decision regarding which instrument to purchase. There are a number of high-quality commercial systems available in the marketplace, which look very similar and have very similar specifications, but how do you know which is the best one that fits your needs? This chapter, supported by the other chapters in the book, presents a set of evaluation guidelines to help you decide the most important analytical figures of merit for your application. You might not need to run all the tests, but experience has told me over the years that each one will give you valuable information about the performance of the instrument, depending on you evaluation objectives.

26.1 EVALUATION OBJECTIVES

It is very important before you begin the selection process to decide what your analytical objectives are. This is particularly important if you are part of an evaluation committee. It is fine to have more than one objective, but it is essential that all the members of the group begin the evaluation process with the objectives clearly defined. For example, is detection-limit (DL) performance an important objective for your application, or is it more important to have an instrument that is easy to use? If the instrument is being used on a routine basis, maybe good reliability is also very critical. On the other hand, if the instrument is being used to generate revenue, perhaps sample throughput and cost of analysis are of greater importance. Every laboratory's scenario is unique, so it is important to prioritize before you begin the evaluation process. So, as well as looking at instrument features and components, the comparison should also be made with your analytical objectives in mind. Let us take a look at the most common ones that are used in the selection process. They typically include the following:

- Analytical performance
- Usability aspects

DOI: 10.1201/9781003187639-26

- Reliability issues
- Financial considerations

Let us examine these in greater detail.

26.1.1 Analytical Performance

Analytical performance can mean different things to different people. The major reason that the trace element community was attracted to ICP-MS over 30 years ago was its extremely low multielement DLs. Other multielement techniques, such as ICP-OES, offered very high throughput but just could not get down to ultratrace levels of ICP-MS. Even though ETA offered much better detection capability than ICP-OES, it did not offer the sample throughput capability that many applications demanded. In addition, ETA was predominantly a single-element technique and so was impractical for carrying out rapid multielement analysis. These limitations quickly led to the commercialization and acceptance of ICP-MS as a tool for rapid ultratrace element analysis. However, there are certain areas where ICP-MS is known to have weaknesses. For example, dissolved solids for most sample matrices must be kept below 0.2%, otherwise it can lead to serious drift problems and poor precision.

Polyatomic and isobaric interferences, even in simple acid matrices, can produce unexpected spectral overlaps, which will have a negative impact on your data. High-resolution instrumentation and collision/reaction cell/interface technology are helping to alleviate these spectral problems, but they also have their limitations. Depending on the types of samples being analyzed, matrix components can dramatically suppress analyte sensitivity and affect accuracy. These potential problems can all be reduced to a certain extent, but different instruments compensate for these problem areas in different ways. With a novice, it is often a basic lack of understanding of how a particular instrument works that makes the selection process more complicated than it really should be. So, any information that can help you prepare for the evaluation will put you in a much stronger position.

It should be emphasized that these evaluation guidelines are based on my personal experience and should be used in conjunction with other material in the open literature that has presented broad guidelines to compare figures of merit for commercial instrumentation.[1-3] In addition, you should talk with colleagues in your industry who might have carried out an evaluation or are using a particular instrument for the analysis. For example, if they have gone through a lengthy evaluation process, they can give you valuable pointers or even suggest an instrument that is better suited to your needs. Finally, before we begin, it is strongly suggested that you narrow the actual evaluation to two, or maybe three, commercial products. By carrying out some pre-evaluation research, you will have a better understanding of what ICP-MS technology or instrument to focus on. For example, if funds are limited and you are purchasing ICP-MS for the very first time to carry out high-throughput environmental testing, it is probably more cost-effective to focus on single-quadrupole technology. On the other hand, if you are investing in a second system to enhance the capabilities of your quadrupole instrument, it might be worth taking a look at magnetic-sector or triple-quadrupole collision/reaction-cell technology to enhance your lab capabilities. Or, if fast multielement transient signal analysis is one of your major reason for investing in ICP-MS, TOF technology should be given serious consideration (speciation studies would be included in this category). One final note I would like to add, although it is not strictly a technical issue. If you are prepared to forego an instrument demonstration or do not need any samples run, you will be in a much stronger position to negotiate price with the instrument vendor. You should keep that in mind before you decide to get involved in a lengthy selection process.

So, let us begin by looking at the most important aspects of instrument performance. Depending on the application, the major performance issues that need to be addressed include the following:

- Detection capability
- Precision/signal stability

- Accuracy
- Dynamic range
- Interference reduction
- Sample throughput
- Transient signal capability

26.1.1.1 Detection Capability

Detection capability is a term used to assess the overall detection performance of an ICP mass spectrometer. There are a number of different ways of looking at detection capability, including instrument detection limit (IDL), elemental sensitivity, background signal and background equivalent concentration (BEC). Of these four criteria, the IDL is generally thought to be the most accurate way of assessing instrument detection capability. It is often referred to as signal-to-background noise, and for a 99% confidence level is typically defined as 3× standard deviation (SD) of n replicates ($n = \sim 10$) of the sample blank and is calculated in the following manner:

$$IDL = \frac{3 \times Standard\,deviation\,of\,background\,signal}{Analyte\,intensity\,-\,background\,signal} \times analyte\,concentration$$

However, there are slight variations of both the definition and calculation of IDLs, so it is important to understand how different manufacturers quote their DLs if a comparison is to be made. They are usually run in single-element mode, using extremely long integration times (5–10 s) to achieve the highest-quality data. So, when comparing DLs of different instruments, it is important to know the measurement protocol used.

A more realistic way of calculating analyte DL performance in your sample matrices is to use method detection limit (MDL). The MDL is broadly defined as the minimum concentration of analyte that can be determined from zero with 99% confidence. MDLs are calculated in a similar manner to IDLs, except that the test solution is taken through the entire sample-preparation procedure before the analyte concentration is measured multiple times. This difference between MDL and IDL is exemplified in EPA Method 200.8, where a sample solution at two to five times the estimated IDL is taken through all the preparation steps and analyzed. The MDL is then calculated in the following manner:

$$MDL = (t) \times (S)$$

where t = Student's "t" value for a 95% confidence level and specifies a standard deviation estimate with $n - 1$ degrees of freedom ($t = 3.14$ for 7 replicates) and S = the SD of the replicate analyses.

Both IDL and MDL are very useful to understand the capability of ICP-MS. However, whatever method is used to compare DLs of different manufacturers' instrumentation, it is essential to carry out the test using realistic measurement times that reflect your analytical situation. For example, if you are determining a group of elements across the mass range in digested soil samples, it is important to know how much the sample matrix suppresses the analyte sensitivity, because the DL of each analyte will be impacted by the amount of suppression across the mass range. On the other hand, if you are carrying out high-throughput multielement analysis of drinking or wastewater samples, you probably need to be using relatively short integration times (1–2 s per analyte) to achieve the desired sample throughput. Or if you are dealing with transient signal from a chromatographic separation system, that only lasts 5–10 s, it is absolutely critical that you understand the impact that the time has on DLs compared to a continuous signal generated with a conventional nebulizer. (In fact, analysis time and DLs are very closely related to each other and will be discussed later on in this chapter.) In other words, when comparing IDLs, it is absolutely critical that the tests represent your real-world analytical situation.

Elemental sensitivity is also a useful assessment of instrument performance, but it should be viewed with caution. It is usually a measurement of background-corrected intensity at a defined mass and is typically specified as counts per second (cps) per concentration (ppb or ppm) of a mid-mass element such as $^{103}Rh^+$ or $^{115}In^+$. However, unlike DL, raw intensity usually does not tell you anything about the intensity of the background or the level of the background noise. It should be emphasized that instrument sensitivity can be enhanced by optimization of operating parameters such as RF power, nebulizer gas flows, torch-sampling position, interface pressure and sampler/skimmer-cone geometry, but usually comes at the expense of other performance criteria, including oxide levels, matrix tolerance or background intensity. So, be very cautious when you see an extremely high sensitivity specification, because there is a strong probability that the oxide or background specs might also be high. For this reason, it is unlikely there will be an improvement in DL unless the increase in sensitivity comes with no compromise in the background level. It is also important to understand the difference between background and background noise when comparing specifications (the background noise is a measure of the stability of the background and is defined as the square root of the background signal). Most modern quadrupole instruments today specify 150–200 million cps per ppm of rhodium ($^{103}Rh^+$) or indium ($^{115}In^+$) and < 1–2 cps of background (usually at 220 amu), whereas magnetic-sector instrument-sensitivity specifications are typically 10–20 times the higher with 10 times the lower background.

Another figure of merit that is being used more routinely nowadays is BEC. BEC is defined as the intensity of the background at the analyte mass, expressed as an apparent concentration and is typically calculated in the following manner:

$$BEC = \frac{Intensity\ of\ background\ signal}{Analyte\ intensity\ -\ background\ signal} \times analyte\ concentration$$

It is considered more of a realistic assessment of instrument performance in real-world sample matrices (especially if the analyte mass sits on a high background), because it gives an indication of the level of the background—defined as a concentration value. DLs alone can sometimes be misleading because they are influenced by the number of readings taken, integration time, cleanliness of the blank and at what mass the background is measured—and are rarely achievable in a real-world situation. Figure 26.1 emphasizes the difference between DL and BEC.

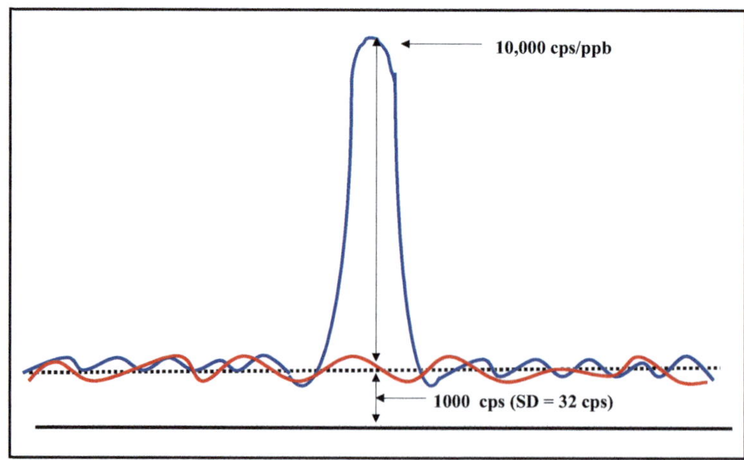

FIGURE 26.1 DL is calculated using the noise of the background, whereas BEC is calculated using the intensity of the background.

In this example, 1 ppb of an analyte produces a signal of 10,000 cps and a background of 1,000 cps. Based on the calculations defined earlier, the BEC is equal to 0.11 ppb because it is expressing the background intensity as a concentration value. On the other hand, the DL is ten times lower because it is using the SD of the background (i.e., the noise) in the calculation. For this reason, BECs are particularly useful when it comes to comparing the detection capabilities of techniques such as cool-plasma and collision/reaction-cell technology, because it gives you a very good indication of how efficient the background reduction process is.

It is also important to remember that peak measurement protocol will also have an impact on detection capability. As mentioned in Chapter 15, there are basically two approaches to measuring an isotopic signal in ICP-MS. There is the multichannel scanning approach, which uses a continuous smooth ramp of 1–20 channels per mass across the peak profile, and peak-hopping approach, where the mass analyzer power supply is driven to a discrete position on the peak, allowed to settle, and a measurement is taken for a fixed period of time. This is usually at the peak maximum but can be as many points as the operator selects. This process is simplistically shown in Figure 26.2.

The scanning approach is best for accumulating spectral and peak shape information when doing mass calibration and resolution scans. It is traditionally used as a classical method development tool to find out what elements are present in the sample and to assess spectral interferences on the masses of interest. However, when the best possible DLs are required, it is clear that the peak-hopping approach is best. It is important to understand that to get the full benefit of peak hopping, the best DLs are achieved when single-point peak hopping at the peak maximum is chosen. It is well accepted that measuring the signal at the peak maximum will always give the best signal-to-background noise for a given integration time, and there is no benefit in spreading your available integration time over more than one measurement point per mass.[4] Instruments that use more than one point per peak for quantitation are sacrificing measurement time on the sides of the peak, where the signal-to-noise ratio is worse. However, the ability of the mass analyzer to repeatedly scan to the same mass position every time during a multielement run is of paramount importance for peak hopping. If multiple points per peak are recommended, it is a strong indication that the spectrometer has poor mass calibration stability, because it cannot guarantee that it will always find the peak maximum with just one point. Mass-calibration specification, which is normally defined as a shift in peak position (in amu) over an eight-hour period, is a good indication of mass stability. However,

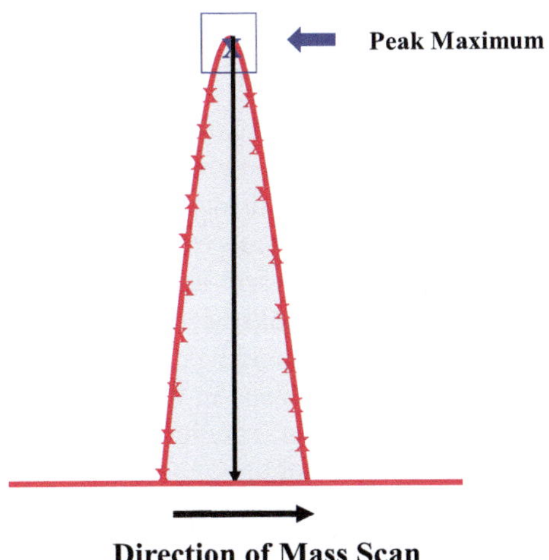

FIGURE 26.2 There are typically two approaches to peak quantitation—peak hopping (usually at peak maximum) and multichannel scanning (across the full width of the peak).

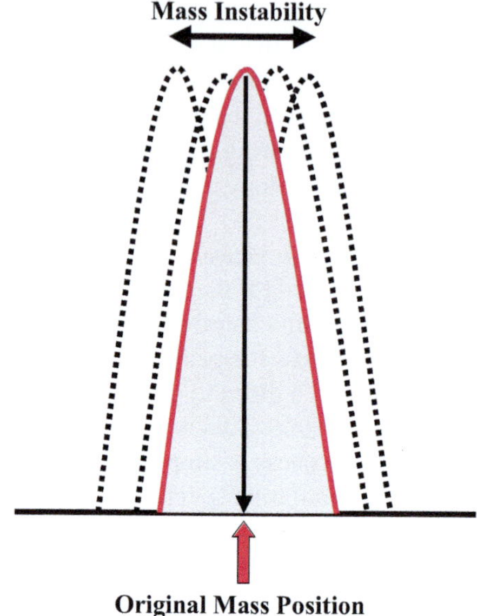

FIGURE 26.3   Good mass stability is critical for single-point, peak-hopping quantitation.

it is not always the best way to compare systems, because peak algorithms using multiple points are often used to calculate the peak position. A more accurate way is to assess the short-term and long-term mass stability by looking at relative peak positions over time. The short-term stability can be determined by aspirating a multielement solution containing four elements (across the mass range) and recording the spectral profiles using multichannel ramp scanning of 20 points per peak. Now repeat the multielement scan ten times and record the peak position of every individual scan. Calculate the average and relative standard deviation (RSD) of the scan positions. The long-term mass stability can then be determined by repeating the test 8 h later to see how far the peaks have moved. It is important, of course, that the mass-calibration procedure not be carried out during this time. Figure 26.3 shows what might happen to the peak position over time, if the analyzer's mass stability is poor.

26.1.1.2   Precision

Short- and long-term precision specifications are usually a good indication of how stable an instrument is (refer to Chapter 15). Short-term precision is typically specified as %RSD of ten replicates of 1–10 ppb of three elements across the mass range using 2–3 s integration times, whereas long-term precision is a similar test, but normally carried out every 5–10 min over a 4–8 h period. Typical short-term precision, assuming an instrument warm-up time of 30–40 min, should be approximately 1–3%, whereas long-term precision should be on the order of 3–5%—both determined without using internal standards. However, it should be emphasized that under these measurement protocols, it is unlikely you will see a big difference in the performance between different instruments in simple aqueous standards. A more accurate reflection of the stability of an instrument is to carry out the tests using a typical matrix that would be run in your laboratory at the concentrations you expect. It is also important that stability should be measured without the use of an internal standard. This will enable you to evaluate the instrument drift characteristics, without any type of signal-correction method being applied.

It is recognized that the major source of drift and imprecision in ICP-MS, particularly with real-world samples, is associated with either the sample-introduction area, design of the interface or the ion optics system. Some of the most common problems encountered with plant materials, organic matrices, agricultural samples and food-based products are as follows:

- Pulsations and fluctuations in the peristaltic pump (if one is used), leading to increased signal noise.
- Blockage of the nebulizer over time, resulting in signal drift—especially if the nebulizer does not have a tolerance for high dissolved solids.
- Poor drainage, producing pressure changes in the spray chamber and resulting in spikes in the signal.
- Buildup of solids in the sample injector, producing signal drift.
- Changes in the electrical characteristics of the plasma, generating a secondary discharge and increasing ion energies.
- Blockage of the sampler- and skimmer-cone orifice with sample material, causing instability, particularly challenging when aspirating carbon-based samples.
- Erosion of the sampler- and skimmer-cone orifice with high-concentration acids.
- Coating of the ion optics with matrix components, resulting in slight changes in the electrical characteristics the of ion-lens system.

These can all be somewhat problematic depending on the types of samples being analyzed. However, the most common and potentially serious problem with real-world matrices is the deposition of sample material on the interface cones and the ion optics over time. It does not impact short-term precision that much, because careful selection of internal standards matched to the analyte masses can compensate for slight instability problems. However, sample material, particularly matrix components found in environmental, botanical, biological and soil samples, can have a dramatic effect on long-term stability. The problem is exaggerated even more in a high-throughput laboratory, because poor stability will necessitate more regular recalibration and might even require some samples to be rerun if QC standards fall outside certain limits. There is no question that if an instrument has poor drift characteristics, it will take much longer to run an autosampler tray full of samples, and in the long term, this will result in much higher argon consumption.

For these reasons, it is critical that when short- and long-term precision are evaluated, you know all the potential sources of imprecision and drift. It is therefore important that you choose either a matrix that is representative of your pharmaceutical samples, whether a raw materials, active pharmaceutical ingredient or finished drug products and will genuinely test the instrument out.

Whatever matrices are chosen, it must be emphasized that for the stability test to be meaningful, no internal standards should be used, the sample should contain less than 0.2% total dissolved solids, and the representative elements should be at a reasonably high concentration (1–10 ppb) and be spread across the mass range. In addition, no recalibration should be carried out for the length of the test, which should reflect your real-world situation.[5] For example, if you plan to run your instrument in a high-throughput environment, you might want to carry out an 8-h or even an overnight (12–16-h) stability test. If you are not interested in such long runs, a 2–4-h stability test will probably suffice. However, just remember, plan the test beforehand and make sure you know how to evaluate the vast amount of data that will be generated. It will be hard work, but I guarantee it will be worth it in order to fully understand the short- and long-term drift characteristics of the instruments you are evaluating.

One word of caution. If a syringe or piston pump is being used to deliver the sample, it will almost certainly give you better precision than a peristaltic pump. So make sure you are comparing like with like. It is almost pointless comparing the stability of an instrument that uses a peristaltic pump with that of an instrument that uses a syringe pump. If you are unsure, make sure that the same pumping system is being used on both instruments.

26.1.1.3   Isotope-Ratio Precision

An important aspect of ICP-MS is its ability to carry out rapid isotope-precision analysis, particularly if multiple isotopes are being used for quantitation. For example, there is much information in the public domain about using carbon isotopes to determine where plants have been grown. By monitoring the C^{12}/C^{13} isotopic ratios by ICP-MS, it can be determined if the plant was grown outdoors, indoors using natural respiration or indoors using artificial CO_2 fertilization. With this example, two different isotopes of carbon are continuously measured over a fixed period of time. The ratio of the signal of one isotope to the other isotope is taken, and the precision of the ratios is then calculated. This ratio will be a very good indicator as to how the plant was grown. This is particularly useful when it comes to understanding how cannabis plants are cultivated—outdoors where the growing conditions are far more erratic and prone to environmental contaminants or indoors where the conditions are far more stable and controlled (https://atlasofscience.org/carbon-isotopes-the-chemists-tool-to-trace-marijuana-cultivation-environment/).

Analysts interested in isotope ratios are usually looking for the ultimate in precision. The optimum way to achieve this to get the best counting statistics would be to carry out the measurement simultaneously with a multicollector magnetic-sector instrument, or a TOF-ICP-MS system. However, a quadrupole mass spectrometer is a rapid sequential system, so the two isotopes are never measured at exactly the same moment. This means that the measurement protocol must be optimized to get the best precision. As discussed earlier, the best and most efficient use of measurement time is to carry out single-point peak hopping between the two isotopes. In addition, it is also beneficial to be able to vary the total measurement time of each isotope, depending on their relative abundance. The ability to optimize the dwell time and the number of sweeps of the mass analyzer ensures that the maximum amount of time is being spent on the top of each individual peak where the signal-to-noise ratio is at its highest.[6]

It is also critical to optimize the efficiency cycle of the measurement. With every sequential mass analyzer, there is an overhead time called a *settling time* to allow the power supply to settle before taking a measurement. This time is often called non-analytical time, because it does not contribute to the quality of the analytical signal. The only time that contributes to the analytical signal is the *dwell time*, or the time that is actually spent measuring the peak. The measurement efficiency cycle (MEC) is a ratio of the dwell time to the total analytical time (which includes settling time) and is expressed as follows:

$$MEC(\%) = \frac{\text{No. sweeps} \times \text{dwell time} \times 100}{\{\text{No. sweeps} \times (\text{dwell time} + \text{settling time})\}}$$

It is therefore obvious that to get the best precision over a fixed period of time, the settling time must be kept to an absolute minimum. The dwell time and the number of sweeps are operator selectable, but the settling time is usually fixed because it is a function of the mass analyzer electronics. For this reason, it is important to know what the settling time of the mass spectrometer is when carrying out peak hopping. Remember, a shorter settling time is more desirable because it will increase the measurement efficiency cycle and improve the quality of the analytical signal.[7]

In addition, if isotope ratios are being determined on vastly different concentrations of major and minor isotopes using the extended dynamic range of the system, it is important to know the settling time of the detector electronics. This settling time will affect the detector's ability to detect the analog and pulse signals (or in dynamic attenuation mode with a pulse-only EDR system) when switching between measurement of the major and minor isotopes, which could have a serious impact on the accuracy and precision of the isotope ratio. So, for that reason, no matter how the higher concentrations are handled, shorter settling times are more desirable, so that the switching or attenuation can be carried out as quickly as possible.

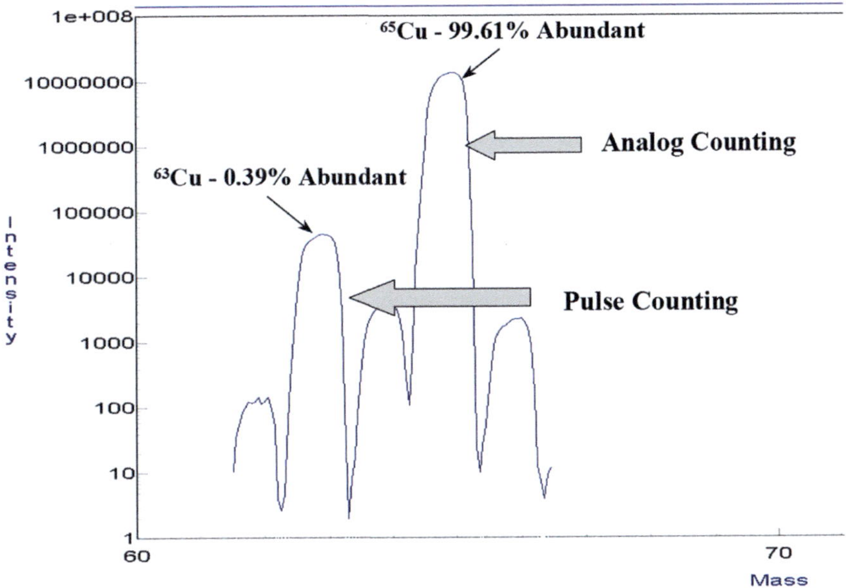

FIGURE 26.4 The detector electronics must be able to switch fast enough to detect isotope ratios that require both pulse and analog-counting modes.

(Courtesy of PerkinElmer Inc.)

This is shown in Figure 26.4, which shows a spectral scan of $^{63}Cu^+$ and $^{65}Cu^+$ using an automated pulse/analog EDR detection system. The natural abundance of these two isotopes is 69.17% and 26.83%, respectively. However, the ratio of these isotopes has been artificially altered to be 0.39% for $^{63}Cu^+$ and 99.61% for $^{65}Cu^+$. The intensity of ^{63}Cu is about 70,000 cps, which requires pulse counting, whereas the intensity of the $^{65}Cu^+$ is about 10 million cps, which necessitates analog counting. There is no question that the counting circuitry would miss many of the ions and generate erroneous concentration data if the switching between pulse and analog modes was not fast enough.

So, when evaluating isotopic ratio precision with a scanning device like a quadrupole, it is important that the measurement protocol and peak quantitation procedure are optimized. Isotope-precision specifications are a good indication regarding what the instrument is capable of, but once again, these will be defined in aqueous-type standards, using relatively short total measurement times (typically 5 min). For that reason, if the test is to be meaningful, it should be optimized to reflect your real-world analytical situation. That is why if very high precision (low RSDs) is a requirement of the analysis, then quadrupole ICP-MS is probably not the optimum technique to use.

26.1.1.4 Accuracy

Accuracy is a very difficult aspect of instrument performance to evaluate because it often reflects the skill of the person developing the method and analyzing the samples, instead of the capabilities of the instrument itself. If handled correctly, it is a very useful exercise to go through, particularly if you can get hold of reference material (ideally of similar matrices to your own) whose values are well defined. However, when attempting to compare the accuracy of different instruments, it is essential that you prepare every sample yourself, including the calibration standards, blanks, unknown samples, QC standards or certified reference material (CRM) if available. I suggest that you make up enough of each solution to give to each vendor for analysis. By doing this, you eliminate the uncertainty and errors associated with different people making up different solutions. It then becomes more of an assessment of the capability of the instrument, including its sample-introduction system, interface region, ion optics, mass analyzer, detector and measurement circuitry, to handle the unknown samples, minimize interferences and get the correct results.

A word of caution should be expressed at this point. Having worked in the field of ICP-MS for over 35 years, I know that the experience of the person developing the method, running the samples and performing the demonstration has a direct impact on the quality of the data generated in ICP-MS. There is no question in my mind that the analyst with the most application expertise has a much better chance of getting the right answer than someone who is either inexperienced or is not familiar with a particular type of sample. I think it is quite valid to compare the ability of the application specialist because this might be the person who is supporting you. However, if you want to assess the capabilities of the instrument alone, it is essential to take the skill of the operator out of the equation. This is not as straightforward as it sounds, but I have found that the best way to "level the playing field" is to send some of your sample matrices to each vendor before the actual demonstration. This allows the application person to spend time developing the method and become familiar with the samples. You can certainly hold back on your CRM or QC standards until you get to the demonstration, but at least it gives each vendor some uninterrupted time with your samples. This also allows you to spend most of the time at the demonstration evaluating the instrument, assessing hardware components, comparing features and getting a good look at the software. It is my opinion that most instruments on the market should get the right answer—at least for the majority of routine applications. So, even though the accuracy of different instruments should be compared, it is more important to understand how the result was generated, especially when it comes to the analysis of very difficult samples. This is especially true with a triple-quad ICP-MS system fitted with a collision/reaction cell. Method development with these kinds of instruments, whether it's carried out by a skilled operator, or using sophisticated decision-making software, is critical to achieving high-quality data.

26.1.1.5 Dynamic Range

When ICP-MS was first commercialized, it was primarily used to determine very low analyte concentrations. As a result, detection systems were only asked to measure concentration levels up to approximately five orders of magnitude. However, as the demand for greater flexibility grew, such systems were being called upon to extend their dynamic range to determine higher and higher concentrations. Today, the majority of commercial systems come standard with detectors that can measure signals up to ten orders of magnitude.

As mentioned in Chapter 14, there are subtle differences between how various detectors and detection systems achieve this, so it is important to understand how different instruments extend the dynamic range. The majority of quadrupole-based systems on the market extend the dynamic range by using a discrete-dynode detector operated either in pulse-only mode or a combination of pulse and analog mode. When evaluating this feature, it is important to know whether this is done in one or two scans because it will have an impact on the types of samples you can analyze. The different approaches have been described earlier, but it is worth briefly going through them again:

- **Two-scan approach:** Basically, two types of two-scan or prescan approaches have been used to extend the dynamic range. In the first one, a survey or prescan is used to determine what masses are at high concentrations and what masses are at trace levels. Then, the second scan actually measures the signals by switching rapidly between analog and pulse counting. In the second two-scan approach, the detector is first run in the analog mode to measure the high signals and then rescanned in pulse-counting mode to measure the trace levels.
- **One-scan approach:** This approach is used to measure both the high levels and trace concentrations simultaneously in one scan. This is typically achieved by measuring the ion flux as an analog signal at some midpoint on the detector. When more than a threshold

number of ions are detected, the ions are processed through the analog circuitry. When less than a threshold number of ions are detected, the ions cascade through the rest of the detector and are measured as a pulse signal in the conventional way.

- **Using pulse-only mode:** The most recent development in extending the dynamic range is to use the pulse-only signal. This is achieved by monitoring the ion flux at one of the first few dynodes of the detector (before extensive electron multiplication has taken place) and then attenuating the signal by applying a control voltage. Electron pulses passed by the attenuation section are then amplified to yield pulse heights that are typical in normal pulse-counting applications. Under normal circumstances, this approach requires only one scan, but if the samples are complete unknowns, dynamic attenuation might need to be performed, where an additional pre-measurement time is built into the settling time to determine the optimum detector attenuation for the selected dwell times used.

The methods that use a prescan or pre-measurement time work very well, but they do have certain limitations for some applications, compared to the one-scan approach. Some of these include the following:

- The additional scan/measurement time means it will use more of the sample. Ordinarily this will not pose a problem, but if sample volume is limited to a few hundred microliters, it might be an issue.
- If concentrations of analytes are vastly different, the measurement circuitry reaction time of a prescan system might struggle to switch quickly enough between high- and low-concentration elements. This is not such a major problem, unless the measurement circuitry has to switch rapidly between consecutive masses in a multielement run, or there are large differences in the concentrations of two isotopes of the same element when carrying out ratio studies. In both these situations, there is a possibility that the detection system will miss counting some of the ions and produce erroneous data.
- The other advantage of the one-scan approach is that more time can be spent measuring the peaks of interest in a transient peak generated by a flow injection or laser ablation system. With a detector that uses two scans or a prescan approach, much of the time will be used just to characterize the sample. It is exaggerated even more with a transient peak, especially if the analyst has no prior knowledge of elemental concentrations in the sample.

This final point is exemplified in Figure 26.5, which shows the measurement of a flow injection peak of NIST 1643C potable water CRM, using an automated simultaneous pulse/analog EDR system. It can be seen that the K and Ca are at ppm levels, which requires the use of the analog counting circuitry, whereas the Pb and Cd are at ppb levels, which requires pulse counting.[8] This would not be such a difficult analysis for a detector, except that the transient peak has only lasted 10 s. This means that to get the highest-quality data, you want to be spending all the available time quantifying the peak. In other words, you cannot afford the luxury of doing a pre-measurement, especially if you have no prior knowledge of the analyte concentrations.

For these reasons, it is important to understand how the detector handles high concentrations to evaluate them correctly. If you are truly interested in using ICP-MS to determine higher concentrations, you should check out the linearity of different masses across the mass range by measuring low-ppt (~10 ppt), low-ppb (~10 ppb) and high ppm (~100 ppm) levels. Do not be hesitant to analyze a standard reference material (SRM) sample such as one of the NIST 1643 series of drinking water reference standards, which has both high (ppm) and low (ppb) levels. Finally, if you know there are large concentration differences between the same analytes, make sure the detector is able to determine them with good accuracy and precision. On the other hand, if your instrument is only going to be used to carry out ultratrace analysis, it probably is not worth spending the time to evaluate the capability of the extended dynamic range feature.

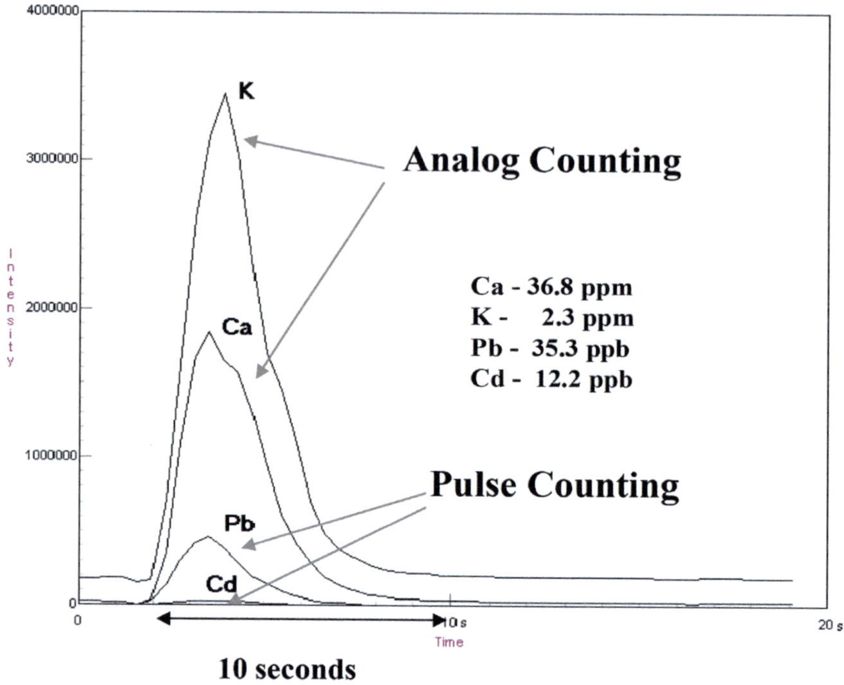

FIGURE 26.5 A one-scan approach to extending the dynamic range is more advantageous for handling a fast transient flow-injection peak, in this example, generated by NIST 1643C.

(From E. R. Denoyer, Q. H. Lu, *Atomic Spectroscopy,* **14**[6], 162–169, 1993)

However, it should be strongly emphasized that irrespective of which extended range technology is used, if low and high concentrations of the same analyte are expected in a suite of samples, it is unrealistic to think you can accurately quantitate down at the low end and at the top end of the linear range with the same calibration graph. If you want to achieve accurate and precise data at or near the limit of quanitation, you must run a set of appropriate calibration standards to cover your low-level samples. In addition, if you are expecting high and low concentrations in the same suite of samples, you have to be absolutely sure that a high concentration sample has been thoroughly washed out from the spray chamber/nebulizer system, before a low-level sample is introduced. For this reason, caution must be taken when setting up the method with an autosampler, because if the read delay/ integration times are not optimized for a suite of samples, erroneous results can be generated, which might necessitate a rerun under the manual supervision of the instrument operator.

26.1.1.6 Interference Reduction

As mentioned in Chapter 17, there are two major types of interferences that have to be compensated for: spectral and matrix (space charge and physical). Although most instruments approach the principles of interference reduction in a similar way, the practical aspect of compensating for them will be different, depending on the differences in hardware components and instrument design. Let us look at interference reduction in greater detail and compare the different approaches used.

26.1.1.6.1 Reducing Spectral Interferences

The majority of spectral interferences seen in ICP-MS are produced by the sample matrix, solvent, plasma gas or various combinations of them. If the interference is caused by the sample, the best approach might be to remove the matrix by some kind of ion-exchange column. However, this can be cumbersome and time consuming to do on a routine basis. If the interference is caused by solvent ions, simply desolvating the sample will have a positive effect on reducing the interference. For that reason, systems that come standard with chilled spray chambers to remove much of the solvent usually generate less sample-based oxide-, hydroxide- and hydride-induced spectral interferences. There are alternative ways to reduce these types of interferences, but cooling the spray chamber or using a membrane desolvation system can be a very effective way of reducing the intensity of the solvent-based ionic species.

Spectral interferences are an unfortunate reality in ICP-MS, and it is now generally accepted that instead of trying to reduce or minimize them, the best way is to resolve the problem away using high-resolution technology such as a double-focusing magnetic sector mass analyzer.[9] Even though they are not considered ideal for a routine, high-throughput laboratory, they offer the ultimate in resolving power and have found a niche in applications that require ultratrace detection and a high degree of flexibility for the analysis of complex sample matrices. If you use a quadrupole-based instrument and are looking to purchase a second system to enhance the flexibility of your laboratory, it might be worth taking a serious look at magnetic-sector technology. The full benefits of this type of mass analyzer for ICP-MS have been described in Chapter 11.

Let us now turn our attention to the different approaches used to reduce spectral interferences using quadrupole-based technology. Each approach should be evaluated on the basis of its suitability for the demands of your particular application.

26.1.1.6.2 Resolution Improvement

As described in Chapter 10, there are two very important performance specifications of a quadrupole—resolution and abundance sensitivity.[10] Although they both define the ability of a quadrupole to separate an analyte peak from a spectral interference, they are measured differently. Resolution reflects the shape of the top of the peak and is normally defined as the width of a peak at 10% of its height. Most instruments on the market have similar resolution specifications of 0.3–3.0 amu and typically use a nominal setting of 0.7–1.0 amu for all masses in a multielement run. For this reason, it is unlikely you will see any measurable difference when you make your comparison.

However, some systems allow you to change resolution settings on the fly on individual masses during a multielement analysis. Under normal analytical scenarios, this is rarely required, but at times it can be advantageous to improve the resolution for an analyte mass, particularly if it is close to a large interference and there is no other mass or isotope available for quantitation. This can be seen in Figure 26.6, which shows a spectral scan of 10 ppb $^{55}Mn^+$, which is monoisotopic, and 100 ppm of $^{56}Fe^+$. The left-hand spectra shows the scan using a resolution setting of 0.8 amu for both $^{55}Mn^+$ and $^{56}Fe^+$, whereas the right-hand spectra shows the same scan, but using a resolution setting of 0.3 amu for $^{55}Mn^+$ and 0.8 amu for $^{56}Fe^+$. Even though the $^{55}Mn^+$ peak intensity is about three times lower at 0.3 amu resolution, the background from the tail of the large $^{56}Fe^+$ is about seven times less, which translates into a fivefold improvement in the $^{55}Mn^+$ DL at a resolution of 0.3 amu, compared to 0.8 amu.

26.1.1.6.3 Higher Abundance-Sensitivity Specifications

The second important specification of a mass analyzer is abundance sensitivity, which is a measure of the width of a peak at its base. It is defined as the signal contribution of the tail of a peak at one mass lower and one mass higher than the analyte peak, and generally speaking, the lower the specification, the better the performance of the mass analyzer. The abundance sensitivity of a quadrupole is determined by a combination of factors, including shape, diameter and length of the rods; frequency of quadrupole power supply; and the slope of the applied RF/DC voltages. Even though there are differences between designs of quadrupoles in commercial ICP-MS systems, there appears to be very little difference in their practical performance.

When comparing abundance sensitivity, it is important to understand what the numbers mean. The trajectory of an ion through the analyzer means that the shape of the peak at one mass lower

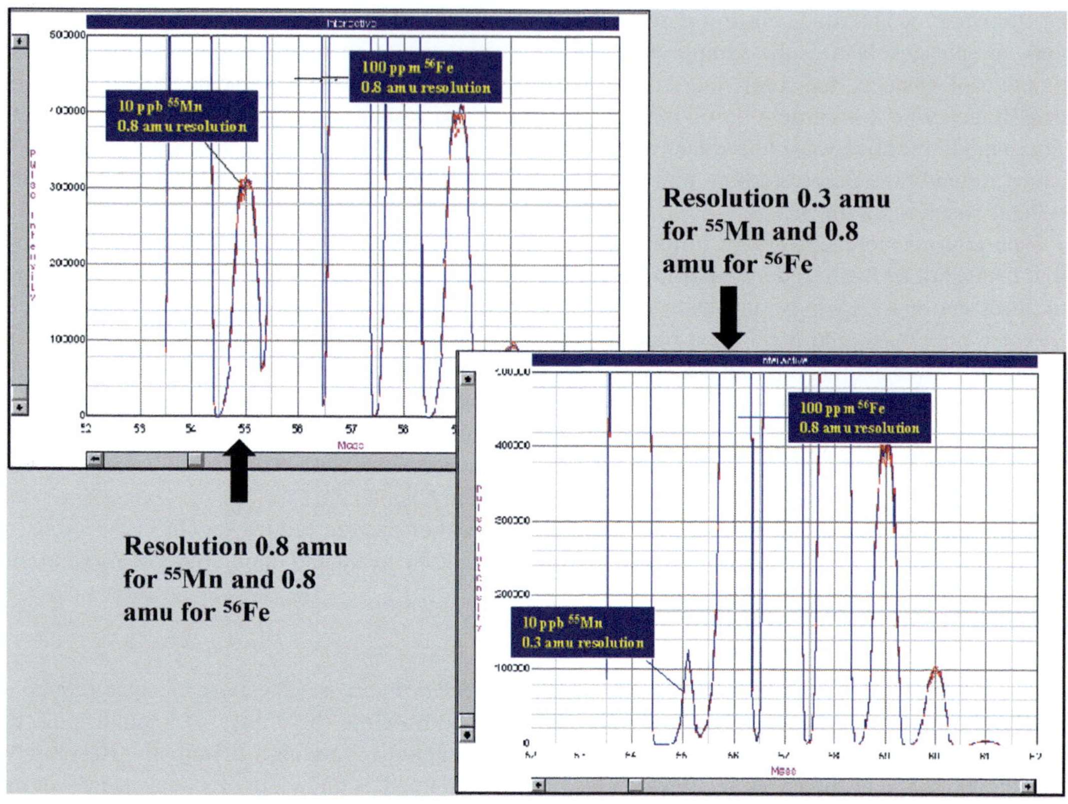

FIGURE 26.6   A resolution setting of 0.3 amu will improve the DL for $^{55}Mn^+$ in the presence of high concentrations of $^{56}Fe^+$.

(Courtesy of PerkinElmer Inc)

than the mass M, i.e., (M − 1), is slightly different from the other side of the peak at one mass higher, i.e., (M + 1). For this reason, the abundance sensitivity specification for all quadrupoles is always worse on the low-mass side (-) than the high-mass side (+), and is typically 1×10^{-6} at M − 1 and 1×10^{-7} at M + 1. In other words, an interfering peak of 1 million cps at M − 1 would produce a background of 1 cps at M, whereas it would take an interference of 10 million cps at M + 1 to produce a background of 1 cps at M. In theory, hyperbolic rods will demonstrate better abundance sensitivity than round ones, as will a quadrupole with longer rods and a power supply with higher frequency. However, you have to evaluate whether this produces any tangible benefits when it comes to the analysis of your real-world samples.

26.1.1.6.4   Use of Cool-Plasma Technology

All of the instruments on the market can be set up to operate under cool or cold plasma to achieve very low DLs for elements such as K, Ca and Fe. Cool-plasma conditions are achieved when the temperature of the plasma is cooled sufficiently low enough to reduce the formation of argon-induced polyatomic species.[11] This is typically achieved with a decrease in the RF power, an increase in the nebulizer gas flow, and sometimes a change in the sampling position of the plasma torch. Under these conditions, the formation of species such as $^{40}Ar^+$, $^{38}ArH^+$ and $^{40}Ar^{16}O^+$ is dramatically reduced, which allows the determination of low levels of $^{40}Ca^+$, $^{39}K^+$ and $^{56}Fe^+$, respectively.[12]

Under normal hot-plasma conditions (typically, RF power of 1,200–1,600 W and a nebulizer gas flow of 0.8–1.0 L/min), these isotopes would not be available for quantitation because of the argon-based interferences. Under cool-plasma conditions (typically, RF power of 600–800 W and a nebulizer gas flow of 1.2–1.6 L/min), the most sensitive isotopes can be used, offering low ppt detection in aqueous matrices. However, not all instruments offer the same level of cool-plasma performance, so if these elements are important to you, it is critical to understand what kind of detection capability is achievable. A simple way to test cool-plasma performance is to look at the BEC for iron at mass 56 with respect to cobalt at mass 59. This enables the background at mass 56 to be compared to a surrogate element such as Co, which has a similar ionization potential to Fe, without actually introducing Fe into the system and contributing to the ArO^+ background signal. When carrying out this test, it is important to use the cleanest deionized water to guarantee that there is no Fe in the blank. First, measure the background in counts/second at mass 56 aspirating deionized water. Then, record the analyte intensity of a 1-ppb Co solution at mass 59. The ArO^+ BEC can be calculated as follows:

$$\text{BEC (ArO}^-) = \frac{\text{Intensity of deionized water background at mass } 56 \times 1 \text{ ppb}}{\text{Intensity of 1 ppb of Co at mass } 59 - \text{background at mass } 56}$$

The ArO^+ BEC at mass 56 will be a good indication of the DL for $^{56}Fe^+$ under cool-plasma conditions. The BEC value will typically be about an order of magnitude greater than the DL.

Although most instruments offer cool-plasma capability, there are subtle differences in the way it is implemented. It is therefore very important to evaluate the ease of setup and how easy it is to switch from cool to normal plasma conditions and back in an automated multielement run. Also, remember that there will be an equilibrium time in switching from normal to cool-plasma conditions. Make sure you know what this is, because an equivalent read-delay will have to be built into the method, which could be an issue if speed of analysis is important to you. If in doubt, set up a test to determine the equilibrium time by carrying out a short stability run while switching back and forth between normal and cool-plasma conditions.

It is also critical to be aware that the electrical characteristics of a cool plasma are different from normal plasma. This means that unless there is a good grounding mechanism between the plasma and the RF coil, a secondary discharge can easily occur between the plasma and sampler cone. The result is an increased spread in kinetic energy of the ions entering the mass spectrometer, making them more difficult to control and steer through the ion optics into the mass analyzer. So, understand how this grounding mechanism is implemented and whether any hardware changes need to be made when going from cool to normal plasma conditions and vice versa (testing for a secondary discharge will be discussed later).

It should be noted that one of the disadvantages of the cool-plasma approach is that cool plasma contains much less energy than a normal, high-temperature plasma. As a result, elemental sensitivity for the majority of elements is severely affected by the matrix, which basically precludes its use for the analysis of samples with a real matrix, unless the necessary steps are taken. This is shown in Figure 26.7, which shows cool-plasma sensitivity for a selected group of elements in varying concentrations of nitric acid, and Figure 26.8, which shows the same group of elements under normal plasma conditions. It can be seen clearly that analyte sensitivity is dramatically reduced in a cool plasma as the acid concentration is increased, whereas, under normal plasma conditions, the sensitivity for most of the elements varies only slightly with increasing acid concentration.[13]

In addition, because a cool plasma contains much less energy than a normal plasma, chemical matrices and acids with a high boiling point are often difficult to decompose in the plasma, which has the potential to cause corrosion problems on the interface of the mass spectrometer. This is the inherent weakness of the cool-plasma approach—instrument performance is highly dependent on the sample being analyzed. As a result, unless simple aqueous-type samples are being analyzed,

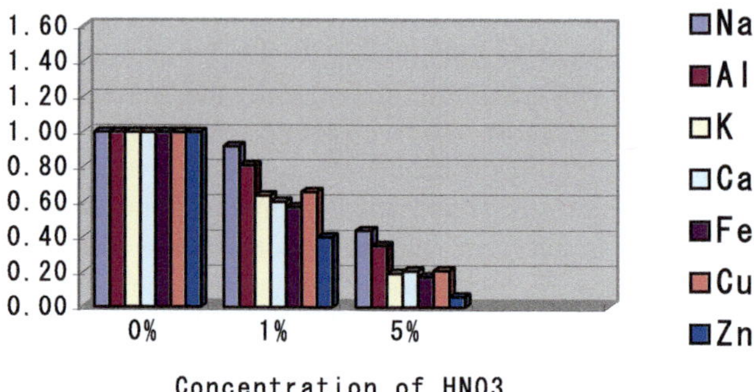

FIGURE 26.7   Sensitivity for a selected group of elements in varying concentrations of nitric acid, using cool plasma conditions (RF power—800 W, nebulizer gas—1.5 L/min).

(From J. M. Collard, K. Kawabata, Y. Kishi, R. Thomas, *Micro*, January 2002)

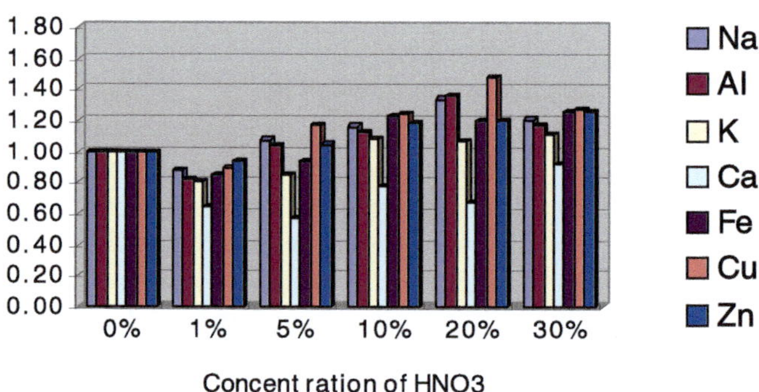

FIGURE 26.8   Sensitivity for a selected group of elements in varying concentrations of nitric acid, using normal plasma conditions (RF power—1600 W, nebulizer gas—1.0 L/min).

(From J. M. Collard, K. Kawabata, Y. Kishi, R. Thomas, *Micro*, January 2002)

cool-plasma operation often requires the use of standard additions or matrix matching to achieve satisfactory results. Additionally, to obtain the best performance for a full suite of elements, a multielement analysis often necessitates the use of two sets of operating conditions—one run for the cool-plasma elements and another for normal plasma elements—which can be both time and sample consuming.

In fact, these application limitations have led some vendors to reject the cool-plasma approach in favor of collision/reaction-cell technology. So, it could be that the cool-plasma capability of an instrument may not be that important if the equivalent elements are superior using the collision/reaction-cell option. However, you should proceed with caution in this area, because on the current evidence not all collision/reaction-cell instruments offer the same kind of performance. For some instruments, cool-plasma DLs are superior to the same group of elements determined in the collision cell mode. For that reason, an assessment of the suitability of using cool-plasma conditions or collision/reaction-cell technology for a particular application problem has to be made based on the vendor's recommendations.

For example, the recent development of a novel 34 MHz free-running designed RF generator using solid-state electronics, has enhanced the capability of ICP-MS to analyze some real-world samples, particularly when using cool-plasma conditions (refer to Chapter 14. on Interference

Reduction). This new design, which is based on an air-cooled plasma load coil, allows the matching network electronics to rapidly respond to changes in the plasma impedance produced by different sampling conditions and sample matrices, while still maintaining low plasma potential at the interface region. For these reasons, this technology appears to offer some benefits over traditional RF technology for some applications.

26.1.1.6.5 Using Collision/Reaction-Cell and Interface Technology

Collision and reaction cells and interfaces are an available option on all quadrupole-based instruments today and are used to reduce the formation of harmful polyatomic spectral interferences, such as $^{38}ArH^+$, $^{40}Ar^+$, $^{40}Ar^{12}C^+$, $^{40}Ar^{16}O^+$ and $^{40}Ar_2^+$, to improve detection capability for elements such as K, Ca, Cr, Fe and Se. However, when comparing systems, it is important to understand how the interference reduction is carried out, what types of collision/reaction gases are used and how the collision/reaction cell or interface deals with the many complex side reactions that take place—reactions that can potentially generate brand new interfering species and cause significant problems at other mass regions. The difference between collision cells and reaction cells and interfaces has been described in detail in Chapter 13. Two different approaches are used to reject these undesirable species. It can be done either by KED or by mass discrimination, depending on the type of multipole and the reaction gas used in the cell.

Unfortunately, the higher-order multipoles such as hexapoles or octopoles have less defined mass stability boundaries than lower-order multipoles, making them less than ideal to intercept these side reactions by mass discrimination. This means that some other mechanism has to be used to reject these unwanted species. The approach that has been traditionally used is to discriminate between them by kinetic energy. This is a well-accepted technique that is typically achieved by setting the collision cell potential slightly more negative than the mass filter potential. This means that the collision-product ions generated in the cell, which have a lower kinetic energy as a result of the collision process, are rejected; whereas the analyte ions, which have a higher kinetic energy, are transmitted. This method works very well but restricts their use to inert gases such as helium and less reactive gases such as hydrogen because of the limitations of higher-order multipoles in efficiently controlling the multitude of side reactions.

However, the use of highly reactive gases such as ammonia and methane can lead to more side reactions and potentially more interferences unless the by-products from these side reactions are rejected. The way around this problem is to utilize a lower-order multipole, such as a quadrupole, inside the reaction/collision cell and use it as a mass discrimination device. The advantages of using a quadrupole are that the stability boundaries are much better defined than a hexapole or an octapole, so it is relatively straightforward to operate the quadrupole inside the reaction cell as a mass or bandpass filter. Therefore, by careful optimization of the quadrupole electrical fields, unwanted reactions between the gas and the sample matrix or solvent, which could potentially lead to new interferences, are prevented. This means that every time an analyte and interfering ions enter the reaction cell, the bandpass of the quadrupole can be optimized for that specific problem and then changed on the fly for the next one.[14]

When assessing the capabilities of collision and reaction cells and interfaces, it is important to understand the level of interference rejection that is achievable, which will be reflected in the instrument's DL and BEC values for the particular analytes being determined. This has been described in greater detail in Chapter 13, but depending on the nature of interference being reduced, there will be differences between the collision/reaction-cell methods as well as with the collision/reaction-interface approach. It is therefore critical to evaluate the capabilities of commercial instrumentation on the basis of your sample matrices and particular analytes of interest.

On the evidence published to date, it seems that the use of highly reactive gases appears to offer a more efficient way of reducing some interfering ion background levels because the optimum reaction gas can be selected to create the most favorable ion–molecule reaction conditions for each analyte. In other words, the choice and flow of the reaction gas can be optimized for each and every application problem.

The benefit of using highly reactive gases to reduce interferences has been confirmed by the recent development of the "triple quadrupole" collision/reaction-cell instruments, where an additional quadrupole is placed prior to the collision/reaction cell-multipole and the analyzer quadrupole. This first quadrupole acts as a simple mass filter to allow only the analyte masses to enter the cell, while rejecting all other masses. With all non-analyte, plasma and sample matrix ions excluded from the cell, interference removal is then carried out using highly reactive gases in the collision/reaction cell. The analyte mass, free of the interference, is then passed into the analyzer quadrupole for separation and detection. The use of any kind of ion–molecule reaction chemistry is not as straightforward when it comes to developing methods, especially when new samples are encountered. In addition, if more than one reaction gas needs to be used, they might not be ideally suited for a high-sample-throughput environment because of the lengthy analysis times involved. However, I think it's fair to say that with the advent of intelligent, decision-making software routines, the choice of gases, cell conditions and the overall method-development process using a reaction cell has become relatively straight forward and user friendly. This is true not only with a single quadrupole, but also using reaction chemistry with a triple-quad system.

On the other hand, the use of inert or low-reactivity gases and KED appears to offer a much simpler approach to reducing polyatomic spectral interferences. Normally, only one gas is used for a particular application problem, which is much better suited to routine analysis. It is possible to use other low gases such as hydrogen when helium does not work, but for the majority of elements one gas is sufficient. For some applications, the collision gas is kept flowing all the time, even for elements that do not need a collision cell. However, its major analytical disadvantage is that its interference reduction capabilities are generally not as good as a system that uses highly reactive gases. Because more collisions are required with an inert gas to suppress the interfering ions, the analyte will also undergo more collisions, and as a result, fewer of the analyte ions will make it through the kinetic energy barrier at the exit of the cell. For that reason, DLs for the majority of the elements that benefit from a collision/reaction cell are generally poorer using inert gases and KED than with a system that uses highly reactive gases and selective bandpass (mass) tuning.

However, it should be emphasized that when you are comparing systems, it should be done with your particular analytical problem in mind. In other words, evaluate the interference suppression capabilities of the different collision and reaction cell interface approaches by measuring BEC and DL performance for the suite of elements and sample matrices you are interested in. In other words, make sure it works for your application problem. This is even more important with the newer "triple quadrupole" collision/reaction-cell approach, because it is complicated and at this present moment in time, there are very few applications in the public domain.

Every laboratory's analytical scenario is different, so it is almost impossible to determine which approach is better for a particular application problem. If you are not pushing DLs but are looking for a simplified approach to running samples on a routine basis, then maybe a collision cell using KED best suits your needs. However, if your samples are spectrally more complex and you are looking for more performance and flexibility because your DL requirements are more challenging, then either the dynamic-reaction cell using bandpass tuning or the "triple quadrupole" is probably the best way to go. Also, be aware that systems that use collisional mechanisms and KED will either have to use higher-purity gases or a gas purifier (getter) because of the potential for impurities in the gas generating unexpected reaction by-product ions, which could potentially interfere with other analyte ions (refer to Chapter 12).[15] This not only has the potential to affect the detection capability, but ultra-high-purity gases are typically two to three times more expensive than industrial- or laboratory-grade gases. But at the end of the day, if you are investing in brand-new technology, it also depends on what kind of funds are available to solve a particular application problem.

26.1.1.7 Reduction of Matrix-Induced Interferences

As discussed in Chapter 16, there are three major sources of matrix-induced problems in ICP-MS. The first and simplest to overcome is often called a *sample transport* or *viscosity effect* and is a physical suppression of the analyte signal brought about by the matrix components. It is caused by the sample matrices' impact on droplet formation in the nebulizer or droplet-size selection in the spray chamber. In some samples, it can also be caused by the variation in sample flow through the peristaltic pump. The second type of signal suppression is caused by the impact of the sample matrix on the ionization temperature of the plasma discharge. This typically occurs when different levels of matrix components or acids are aspirated into a cool/cold plasma. The ionization conditions in the low-temperature plasma are so fragile that higher concentrations of matrix components result in severe suppression of the analyte signal. The third major cause of matrix suppression is the result of poor transmission of ions through the ion optics owing to matrix-induced space charge effects.[16] This has the effect of defocusing the ion beam, which leads to poor sensitivity and DLs, especially when trace levels of low-mass elements are being determined in the presence of large concentrations of high-mass matrix elements. Unless an electrostatic compensation is made in the ion optic region, the high-mass element will dominate the ion beam, resulting in severe matrix suppression on the lighter ones. All these types of matrix interferences are compensated to varying degrees by the use of internal standardization, where the intensity of a spiked element that is not present in the sample is monitored in samples, standards and blank.

The single biggest difference in commercial instrumentation to focus the analyte ions into the mass analyzer is in the design of the ion-lens system. Although they all basically do the same job of transporting the maximum number of analyte ions through the system, there have been many different ways of implementing this fundamental process, including the use of an extraction lens, multicomponent lens systems, single-ion cylinder lens, right-angled reflectors or multipole ion guide systems. First, it is important to know how many lens voltages have to be optimized. If a system has many lens components, it is probably going to be more complex to carry out optimization on a routine basis. In addition, the cleaning and maintenance of a multicomponent lens system might be a little more time consuming. All of these are possible concerns, especially in a routine environment where maybe the skill level of the operator is not so high.

However, the design of the ion-focusing system or the number of lens components used is not as important as its ability to handle real-world matrices.[17] Most lens systems can operate in a simple aqueous sample because there are relatively few matrix ions to suppress the analyte ions. The test of the ion optics comes when samples with a real matrix are encountered. When a large number of matrix ions are present in the system, they can physically "knock" the analyte ions out of the ion beam. This shows itself as a suppression of the analyte ions, which means that fewer analyte ions are transmitted to the detector in the presence of a matrix. For this reason, it is important to measure the degree of matrix suppression of the instrument being evaluated across the full mass range. The best way to do this is to choose three or four of your typical analyte elements spread across the mass range (e.g., $^7Li^+$, $^{63}Cu^+$, $^{103}Rh^+$ and $^{138}Ba^+$). Run a calibration of a 20-ppb multielement standard in 1% HNO_3. Then make up a synthetic sample of 20 ppb of the same elements in one of your typical matrices. Measure this sample against the original calibration.

The percentage matrix suppression at each mass can then be calculated as follows:

$$\frac{20 \text{ ppb} - \text{Apparent concentration of 20 ppb analytes in your matrix}}{20 \text{ ppb}} \times 100$$

There is a strong possibility that your own samples will not really test the matrix suppression performance of the instrument, particularly if they are simple aqueous-type samples. If this is the case and you really would like to understand the matrix capabilities of your instrument, then make up a synthetic sample of your analytes in 500 ppm of a high-mass element such as thallium, lead or uranium. For this test to be meaningful, you should tell the manufacturers to set up the ion optic voltages that are best suited for multielement analysis across the full mass range. If the ion optics are designed correctly for minimum matrix interferences, it should not matter if it incorporates an extraction lens, uses a photon stop, has an off-axis mass analyzer or even whether it utilizes a single, multicomponent or right-angled ion-lens system.

It is also important to understand that an additional role of the ion optic system is to stop particulates and neutral species from making it through to the detector, which would increase the noise of the background signal. This will certainly impact the instrument's detection capability in the presence of complex matrices. Therefore, it is definitely worth carrying out a DL test in a difficult matrix such as rock digests, soil samples, biological specimens or metallurgical alloys, which tests the ability of the ion optics to transport the maximum number of analyte ions while rejecting the maximum number of matrix ions, neutral species and particulates.

Another aspect of an instrument's matrix capability is its ability to aspirate lots of different types of samples, using both conventional nebulization and sampling accessories that generate a dryer aerosol, such as chromatographic separation system or a desolvation device. When changing sample types similar to this on a regular basis, parameters such as RF power, nebulizer gas flow and sampling depth usually have to be changed. When this is done, there is an increased chance of altering the electrical characteristics of the plasma and producing a secondary discharge at the interface. All instruments should be able to handle this to some extent, but depending on how they compensate for the increase in plasma potential, parameters might need to be re-optimized because of the change in the spread of kinetic energy of the ions entering the mass spectrometer.[18] This may not be such a serious problem, but once again, it is important for you to be aware of this, especially if the instrument is running many different sample matrices on a routine basis.

Some of the repercussions of a secondary discharge, including increased doubly charged species, erosion of material from the skimmer cone, shorter lifetime of the sampler cone, a significantly different full-mass range response curve with laser ablation and the occurrence of two signal maximums when optimizing nebulizer gas flow, have been well reported in the literature.[19–21] On the other hand, systems that do not show signs of this phenomena have reported an absence of these deleterious effects.[22]

A simple way of testing for the possibility of a secondary discharge is to aspirate one of your typical matrices containing approximately 1 ppb of a small group of elements across the mass range (such as $^7Li^+$, $^{115}In^+$ and $^{208}Pb^+$) and continuously monitor the signals while changing the nebulizer gas flow. In the absence of a secondary discharge, all three elements, which have widely different masses and ion energies, should track each other and have the same optimum nebulizer gas flow. This can be seen in Figure 26.9, which shows the signals for $^7Li^+$, $^{115}In^+$ and $^{208}Pb^+$ changing as the nebulizer gas flow is first increased and then decreased.

If the signals do not track each other or there is an erratic behavior in the signals, it could indicate that the normal kinetic energy of the ions has been altered by the change in the nebulizer gas flow. There are many reasons for this kind of behavior, but it could point to a possible secondary discharge at the interface or that the RF coil-grounding mechanism is not working correctly.[23] It should be emphasized that Figure 26.9 is just a graphical representation of what the relative signals might look like and might not exactly be the same for all instruments.

26.1.1.8　Sample Throughput

In laboratories where high sample throughput is a requirement, the overall cost of analysis is a significant driving force determining what type of instrument is purchased. However, in a high-workload laboratory, there sometimes has to be a compromise between the number of samples analyzed and the DL performance required. For example, if the laboratory wants to analyze as many samples as possible, relatively short integration times have to be used for the suite of elements being determined. On the other hand, if DL performance is the driving force, longer integration times

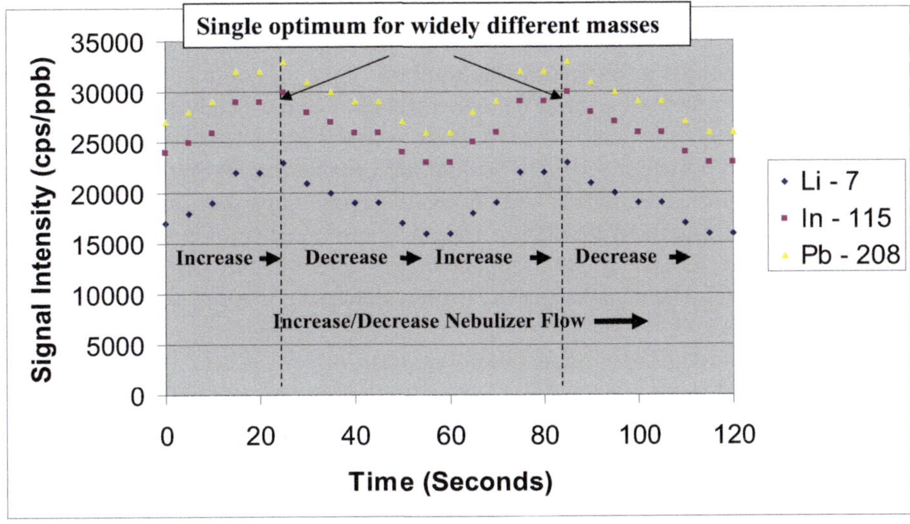

FIGURE 26.9 If the interface is grounded correctly, signals for 1-ppb $^7Li^+$, $^{115}In^+$ and $^{208}Pb^+$ should all track each other and have similar optimum values as the nebulizer gas flow is changed.

need to be used, which will significantly impact the total number of samples that can be analyzed in a given time. This was described in detail in Chapter 15, but it is worth revisiting to understand the full implications of achieving high sample throughput.

It is generally accepted that for a fixed integration time, peak hopping will always give the best DLs. As discussed earlier, measurement time is a combination of time spent on the peak taking measurements (dwell time) and the time taken to settle (settling time) before the measurement is taken. The ratio of the dwell time to the overall measurement time is often called the *measurement efficiency*. The settling time, as we now know, does not contribute to the analytical signal but definitely contributes to the analysis time. This means that every time the quadrupole sweeps to a mass and sits on the mass for the selected dwell time, there is an associated settling time. The greater the number of points that have been selected to quantitate the mass, the longer the total settling time and the worse the overall measurement efficiency.

For example, let us take a scenario where 20 elements need to be determined in duplicate. For argument's sake, let us use an integration time of 1 s per mass, comprising 20 sweeps at 50 ms per sweep. The total integration time that contributes to the analytical signal and the DL is therefore 20 s per replicate. However, every time the analyzer is swept to a mass, the associated scanning and settling time must be added to the dwell time. The greater the numbers of points that are taken to quantify the peak, the greater the magnitude of the settling time that must be added. For this scenario, let us assume that three points/peak are being used to quantify the peaks. Let us also assume for this case that the quadrupole and detector has a settling time of 5 ms. This means that a 15 ms settling time will be associated with every sweep of each individual mass. So, for 20 sweeps of 20 masses, this is equivalent to 6 s of non-analytical time every replicate, which translates into 12 s (plus 40 s of actual measurement time) for every duplicate analysis. This is equivalent to a 40/(12 + 40) × 100% or a 77% measurement efficiency cycle. It does not take long to realize that the fewer the number of points taken per peak and the shorter the settling time, the better the measurement cycle. Just by reducing the number of points to one per peak and cutting the detector settling time by half, the non-analytical time is reduced to 4 s, which is a 40/(2 + 40) × 100% or a 95% measurement efficiency per duplicate analysis. It is therefore very clear that the measurement protocol has a big impact on the speed of analysis and the number of the samples that can be analyzed in a given time. So, if sample throughput is important, you should understand how peak quantitation is carried out on each instrument.

The other aspect of sample throughput is the time taken for the sample to be aspirated through the sample-introduction system into the mass spectrometer, reach a steady-state signal and then be washed out when the analysis is complete. The wash-in and wash-out characteristics of the instrument will most definitely impact its sample throughput capabilities. Therefore, it is important for you to know what these times are for the system you are evaluating. You should also be aware that if the instrument uses a computer-controlled peristaltic pump to deliver the sample to the nebulizer and spray chamber, it can be speeded up to reduce the wash-in and wash-out times. So, this should also be taken into account when evaluating the memory characteristics of the sample-introduction system.

Therefore, if speed of analysis is important to your evaluation criteria, it is worth carrying out a sample throughput test. Choose a suite of elements that represents your analytical challenge. Assuming you are also interested in achieving good detection capability, let the manufacturer set the measurement protocol (integration time, dwell time, settling time, number of sweeps, points/peak, sample-introduction wash-in/wash-out times, etc.) to get their best DLs. If you are interested in measuring high and low concentrations, also make sure that the extended dynamic range feature is implemented. Then time how long it takes to achieve DL levels in duplicate from the time the sample probe goes into the sample to the time a result comes out on the screen or printer. If you have time, it might also be worth carrying out this test in an autosampler with a small number of your typical samples. It is important that the DL measurement protocol be used because factors such as integration times and wash-out times can be compromised to reduce the analysis time. All the measurement time issues discussed in this section and the memory characteristics of the sample-introduction system will be fully evaluated with this kind of test. (Note: If high sample throughput is important, there are automated productivity enhancement systems on the market that by efficient delivery and wash-out of the sample are realizing a two- to three-fold improvement in multielement analysis times—refer to Chapter 20 for details.)

26.1.1.9   Transient Signal Capability

The demands on an instrument to handle transient signals generated by sampling accessories, such as laser ablation, chromatography separation devices, flow injection or electrothermal vaporization systems, are very different from conventional multielement analysis using solution nebulization. Because the duration of a sampling accessory signal is short compared to a continuous signal generated by a pneumatic nebulizer, it is critical to optimize the measurement time to achieve the best multielement signal-to-noise ratio in the sampling time available. This was addressed in greater detail in Chapter 20, but basically the optimum design to capture the maximum amount of multielement data in a transient peak is to carry out the measurement in a simultaneous manner with a multicollector magnetic sector instrument, a TOF mass spectrometer or the Mattauch–Herzog simultaneous detection sector instrument.

However, a scanning system such as a quadrupole instrument can achieve good performance on a transient peak if the measurement time is maximized to get the best multielement signal-to-noise ratio. Therefore, instruments that utilize short settling times are more advantageous, because they achieve a higher measurement efficiency cycle. In addition, if the extended dynamic range is used to determine higher concentrations, the scanning and settling time of the detector will also have an impact on the quality of the signal. So, detectors that require two scans to characterize an unknown sample will use up valuable time in the quantitation process. For example, if a transient peak generated by a laser ablation device only lasts 10 s, a survey or prescan of 2 s will use up 20% of the available measurement time. This, of course, is a disadvantage when doing multielement analysis on a transient signal, especially if you have limited knowledge of the analyte concentration levels in your samples.

26.1.1.10 Single-Particle ICP-MS Transient Signals

Of all the applications involving transient peak measurements, single-particle ICP-MS is probably the most demanding, because the transient event only lasts a few milliseconds. Let's take a closer look at this rapidly emerging application area to better understand the measurement requirements. Single-particle ICP-MS (SP-ICP-MS) is a new technique that has recently been developed for detecting and sizing metallic nanoparticles (NP) at extremely low levels, in order to predict their environmental behavior. While this method is fairly new, it has shown a great deal of promise in several applications, including determining concentrations of nanoparticles in complex matrices, such as wastewater streams, potable waters and effluents (note: this type of analysis could be a future requirement for the cannabis industry as more studies are being published on how plants absorb nanoparticles through their roots system). The method involves introducing a liquid sample containing the nanoparticles at very dilute concentration into the ICP-MS. After the particles have been nebulized, ionized in the plasma and separated by the mass analyzer, the resulting ions are detected and collected as time-resolved pulses. The number of pulses generated are directly related to the population of nanoparticles in the sample, while intensity of the pulses are related to the size (and mass) of the nanoparticles.

In ICP-MS, dilute solutions of dissolved metals will produce relatively constant signals. If there are nanoparticles of various sizes suspended in that solution, they will appear as pulses, which deviates from a steady-state continuous signal generated by the background of the dissolved metals. By using relatively short dwell times of a few milliseconds, the packets of pulses can be quantified as long as the pulses can be adequately resolved in the time domain. The pulse height is then compared to the signal intensity of dissolved metal in a set of calibration standards. This is represented in Figure 26.10, which shows Sample A containing dissolved metal in solution being analyzed and measured in the conventional way and Sample B containing the same metal in the form of nanoparticles being analyzed and measured as a transient pulses of ions. The integration (dwell time), which is the same in both examples, is shown as the dotted box. By subtracting the background produced by the dissolved analyte in the sample from the nanoparticle pulse height, and comparing the net signal intensity against the calibration standard, the concentration and therefore the size (and mass) distribution of the particle can be calculated using well-understood single-particle ICP-MS theory.

However, for this approach to work effectively at low concentrations, the speed of data acquisition and the response time of the ICP-MS quadrupole and detector electronics must be fast enough to capture the time resolved nanoparticles pulses, which typically last only a few milliseconds or less. This is emphasized in Figure 26.11, which shows a real-world example of the time resolved analysis of 30 nm silver nanoparticles by SP-ICP-MS. It can be seen that the silver NP pulse has been resolved with approximately 15 data points in < 1 ms. For this application, the ICP-MS should be capable of using dwell times shorter than the particle transient time, thus avoiding false signals generated from clusters of particles. In practice, this means using a dwell time of a 10–100 µs so the pulse can be fully characterized.

For this kind of resolution, it is advantageous that the instrument-measurement electronics are capable of very fast data-acquisition rates, including dwell times that are as short as possible to capture the maximum number of data points within the transient event. It is also desirable that the quadrupole-settling time is extremely short, so there is no wasted time waiting for the quadrupole power supply to stabilize. Ideally, it would beneficial if there was no settling time, so the quadrupole could just park itself on the mass of interest and just take measurements. Unfortunately, this is not typically a standard feature of the measurement protocol on most ICP-MS systems, because for multielement analysis, there has to be a built-in settling time as the quadrupole scans from one mass to the next. However, for optimum measurement conditions when one element is under investigation, there is no question that the ability to run a method with no settling time, very short dwell times and fast data acquisition is extremely beneficial in order to increase the upper limit of the dynamic range when characterizing nanoparticles using single-particle ICP-MS measurements.[24]

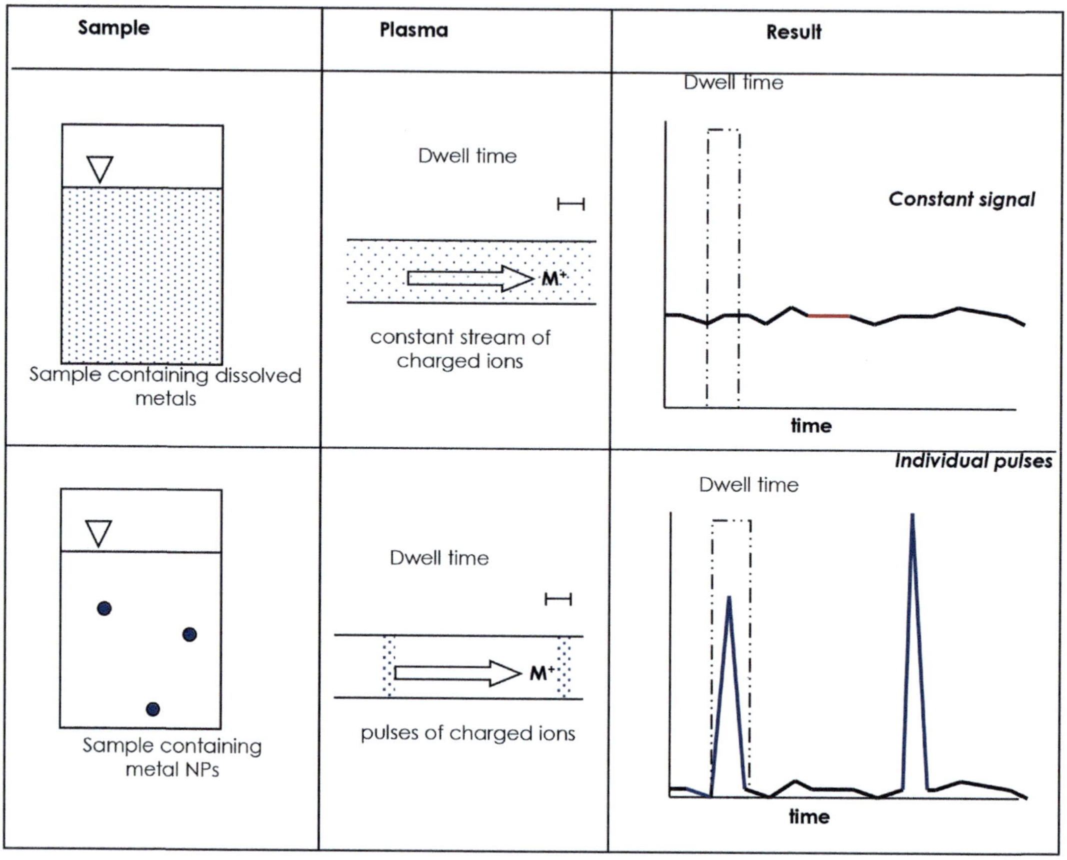

FIGURE 26.10 A comparison of a sample (A) containing dissolved metal in solution being analyzed and measured in the conventional way and another sample (B) containing the same metal in the form of nanoparticles being analyzed and measured as transient pulses of ions.

(Courtesy of the Colorado School of Mines)

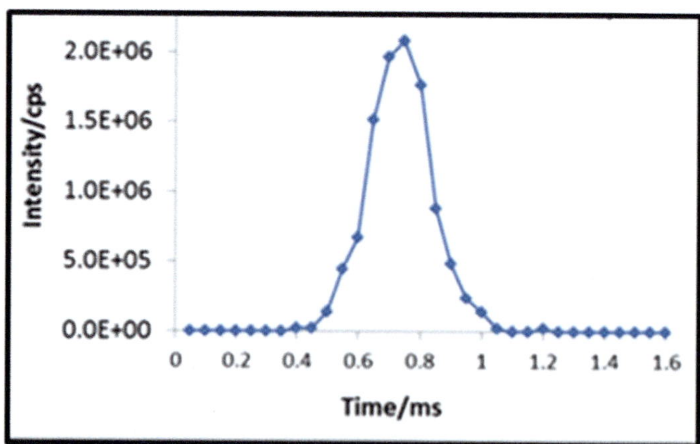

FIGURE 26.11 A time-resolved analysis of 30-nm gold particles by SP-ICP-MS, showing the pulse is fully characterized in less than 1 ms.

(Courtesy of PerkinElmer Inc)

26.1.2 Usability Aspects

In most applications, analytical performance is a very important consideration when deciding what instrument to purchase. However, the vast majority of instruments being used today are being operated by technician-level chemists. They usually have had some experience in the use of trace element techniques such as AA or ICP-OES, but in no way could be considered experts in ICP-MS. Therefore, the usability aspects might be competing with analytical performance as the most important selection criteria, particularly if the application does not demand the ultimate in detection capability. Even though usability is in the eye of the user, there are some general issues that need to be addressed. They include, but are not limited to, the following:

- Ease of use
- Routine maintenance
- Compatibility with sampling accessories
- Installation requirements
- Technical support
- Training

26.1.2.1 Ease of Use

First of all, you need to determine the skill level of the operator who is going to run the instrument. If the operator is a PhD-type chemist, then maybe it is not critical that the instrument be easy to use. However, if the instrument is going to be used in a high-workload environment and possibly operated around the clock, such as in the pharmaceutical industry, there is a strong possibility that the operators will not be highly skilled. Therefore, you should be looking at how easy the software is to use and how similar it is to other trace element techniques that are used in your laboratory. This will definitely have an impact on the time it takes to get a person fully trained on the instrument. Another issue to consider is whether the person who runs the instrument on a routine basis is the same person who will be developing the methods. Correct method development is critical because it impacts the quality of your data, and therefore usually requires more expertise than just running routine methods. This is most definitely the case with collision/reaction cells and interfaces, especially if the method has never been done before. It can take a significant amount of time and effort to select the best gases, gas flows and optimization of the cell parameters to maximize the reduction of interferences for certain analytes in a new sample matrix.

I am not going to get into different software features or operating systems, because it is a complicated criterion to evaluate and decisions tend to be made more on a personal preference or comfort level than on the actual functionality of ICP-MS software features. It is also a moving target, as instrument software is continually being modified and updated. However, there are differences in the way software feels. For example, if you have come from a mass spectrometry background, you are probably comfortable with fairly complex research-type software. Alternatively, if you have come from a trace element background and have used AA or ICP-OES, you are probably used to more routine software that is relatively easy to use. You will find that different vendors have come to ICP-MS from a variety of different analytical chemistry backgrounds, which is often reflected in the way they design their software. Depending on the way the instrument will be used, an appropriate amount of time should be spent looking at software features that are specific to your application needs. For example, if you are working in a high-throughput environmental, pharmaceutical or testing laboratory, you might be interested in turnkey methods such as SOPs that are used to run a particular application methodology, such as Method 200.8 for the determination of elements in water and wastes. Alternatively, with more and more instruments going into high-throughput production environments, look for customized methods for different applications. In addition, maybe you should also be looking very closely at all the features of the automated "Quality Control" (QC) software, or if you do not have the time to export your data to an external spreadsheet to create reports, you might be more interested in software with comprehensive reporting capabilities.

In highly regulated industries, the operating and reporting software needs to be compliant with a set of regulated standards and guidelines. For example, the pharmaceutical and food-manufacturing industry is dictated by federal regulations set down by the Food and Drug Authority (FDA) in Title 21 CFR Part 11, which gives detailed requirements that computerized systems need to fulfill in order to allow electronic signatures and records in lieu of handwritten signatures on paper records. In summary, the regulations apply to: validations for closed and open computerized systems; controlled access to the computerized system; content integrity; use of electronic signatures for authentication of electronic documents; audit trails for all records/signatures; and access to electronic records. Although not currently a regulatory requirement, it is only a matter of time before federal regulations become common practice. When that time eventually arrives, it is important that any ICP-MS used in a regulated environment has all the necessary software to be compliant.

26.1.2.2 Routine Maintenance

ICP mass spectrometers are complex pieces of equipment that, if not maintained correctly, have the potential to fail when you least expect them to. For that reason, a major aspect of instrument usability is how often routine maintenance has to be carried out, especially if complex sample matrices are being analyzed. You must not lose sight of the fact that your samples are being aspirated into the sample-introduction system and the resulting ions generated in the plasma are steered into the mass analyzer via the interface and ion optics. In other words, the sample, in one form or another, is in contact with many components inside the instrument. Even though modern instruments require much less routine maintenance than older-generation equipment, it is still essential to find out what components need to be changed and at what frequency to keep the instrument in good working order. Routine maintenance has been covered in great depth in Chapter 15, but you should be asking the vendor what needs to be changed or inspected on a regular basis, and what type of maintenance should be done on daily, weekly, monthly or yearly intervals. Some typical questions might include the following:

- If a peristaltic pump is being used to deliver the sample, how often should the tubing be changed?
- How often should the spray-chamber drain system be checked?
- Can components be changed if a nebulizer gets damaged or blocked?
- How long does the plasma torch last?
- Are ceramic torches available, which tend to have a much longer lifetime?
- Can the torch sample injector be changed without discarding the torch?
- How is a neutral plasma maintained, and if an external shield or sleeves are used for grounding purposes, how often do they last?
- Is the RF generator solid state or does it use a power amplifier (PA) tube? This is important because PA tubes are expensive consumable items that typically need replacing every one to two years.
- How often do you need to clean the interface cones, and what is involved in cleaning them and keeping the cone orifices free of deposits?
- How long do the cones last?
- Do you have a platinum cone trade-in service, and what is their trade-in value?
- Which type of pump is used on the interface, and if it is a rotary-type pump, how often should the oil be changed?
- What mechanism is used to keep the ion optics free of sample particulates or deposits?
- How often should the ion optics be cleaned?
- What is the cleaning procedure for the ion optics?

- Do the turbomolecular pumps require any maintenance?
- How long do the turbomolecular pumps typically last?
- Does the mass analyzer require any cleaning or maintenance?
- How long does the detector last and how easy is it to change?
- What spare parts do you recommend to keep on hand? This can often indicate the components that are likely to fail most frequently.
- What maintenance needs to be carried out by a qualified service engineer, and how long does it take?

This is not an exhaustive list, but it should give you a good idea about what is involved in keeping an instrument in good working order. I also encourage you to talk to real-world users of the equipment to make sure you get their perspective of these maintenance issues. This is particularly important if you are investing in a brand-new instrument that hasn't been on the market long. There won't be a large number of users in the field, so it's important you talk to them to get their real-world perspective of the routine maintenance issues. For more information about how to maximize performance and reduce routine maintenance, check out this reference by Brennan and co-workers. [25]

26.1.2.3 Compatibility with Alternative Sampling Accessories

Alternative sample-introduction techniques to enhance performance and productivity are becoming more necessary as ICP-MS is being utilized to analyze more complex sample types. Therefore, it is important to know if the sampling accessory is made by the ICP-MS instrument company or by a third-party vendor. Obviously, if it has been made by the same company, compatibility should not be an issue. However, if it is made by a third party, you will find that some sampling accessories work much better with some instruments than with others. It might be that the physical connection of coupling the accessory to the ICP-MS torch has been better thought out, or that the software "talks" to one system better than another. You should also understand how much routine maintenance a sampling accessory needs. The benefits of a rugged ICP-MS system, which requires very little maintenance is negated if the sampling accessory need to be cleaned every five to ten samples. You should refer to Chapter 20 on the ICP-MS Application Landscape for more details on their suitability for other application areas, but if they are required, software/hardware compatibility should be one of your evaluation objectives. You should also read Chapter 19 if you are thinking of interfacing an HPLC system to your ICP-MS for arsenic or mercury speciation studies.

26.1.2.4 Installation of Instrument

Installation of the instrument and where it is going to be located does not seem to be an obvious evaluation objective at first, but it could be important, particularly if space is limited. For example, is the instrument free-standing or bench-mounted? Maybe you have a bench available but no floor space, or vice versa? It could be that the instrument requires a temperature-controlled room to ensure good stability and mass calibration. If this is the case, have you budgeted for this kind of expense? If the instrument is being used for ultratrace detection levels, does it need to go into a class 1, 10 or 100 clean room? If it does, what is the size of the room and do the roughing pumps need to be placed in another room? In other words, it is important to fully understand the installation requirements for each instrument being evaluated and where it will be located. It is unlikely that a high-throughput testing laboratory is going to require a class 1 or class 10 environment. However, you cannot install an ICP-MS instrument in a warehouse and hope to get meaningful data at sub-ppb detection limits.

26.1.2.5 Technical Support

Technical and application support is a very important consideration, especially if you have had no previous experience with ICP-MS. You want to know that you are not going to be left on your own after you have made the purchase. Therefore, it is important to know not only the level of expertise of the specialist who is supporting you, but also whether they are local to you or located in the manufacturer's corporate headquarters. In other words, can the vendor give technical help whenever you need it? Another important aspect related to application support is the availability of application literature. Is there a wide selection of material available for you to read, either in the form of Web-based application reports or references in the open literature, to help you develop your methods? Also, find out if there are active user- or Internet-based discussion groups, because they will be an invaluable source of technical and application help. One such source of help in this area can be found on the PlasmaChem Listserver, a plasma spectrochemistry discussion forum out of Syracuse University.[26]

26.1.2.6   Training

Find out what kind of training course comes with the purchase of the instrument and how often it is run. Most instruments come with two- to three-day training courses for one person, but most vendors should be flexible regarding the number of people who can attend. Some manufacturers also offer application training, where they teach you how to optimize methods for major application areas. Historically, these customized application courses have included environmental, clinical, semiconductor and pharmaceutical analysis. However, with the rapid growth of the cannabis industry today, vendors are now beginning to offer ICP-MS training courses specifically designed for cannabis testing labs. So talk to other users in your field about the quality of the training they received when they purchased their instruments and also ask them what they thought of the vendor's operator manuals. You will often find that this is a good indication of how important customer training is to the vendor.

26.1.3   Reliability Issues

To a certain degree, instrument reliability is impacted by routine maintenance issues and the types of samples being analyzed, but it is generally considered more of a reflection of the design of an instrument. Most manufacturers will guarantee a minimum percentage uptime for their instrument, but this number (which is typically ~95%) is almost meaningless unless you really understand how it is calculated. Even when you know how it is calculated, it is still difficult to make the comparison, but at least you should understand the implications if the vendor fails to deliver. Good instrument reliability is taken for granted nowadays, but it has not always been the case. When ICP-MS was first commercialized, the early instruments were a little unpredictable, to say the least, and quite prone to frequent breakdowns. However, as the technique became more mature, the quality of instrument components improved, and therefore, the reliability improved. However, you should be aware that some components of the instrument are more problematic than others. This is particularly true when the design of an instrument is new or a model has had a major redesign. You will therefore find that in the life cycle of a newly designed instrument, the early years will be more susceptible to reliability problems than when the instrument is of an older design.

When we talk about instrument reliability, it is important to understand whether it is related to the samples being analyzed, the inexperience of the person operating the instrument, an unreliable component or an inherent weakness in the design of the instrument. For example, how does the instrument handle highly corrosive chemicals such as concentrated mineral acids? Some sample-introduction systems and interfaces will be more rugged than others and require less maintenance in this area. On the other hand, if the operator is not aware of the dissolved-solids limitation of the instrument, they might attempt to aspirate a sample, which will slowly block the interface cones, causing signal drift and, in the long term, possible instrument failure. Or, it could be something as unfortunate as a major component, such as the RF generator power amplifier tube, discrete-dynode detector or turbomolecular pump (which all have a finite lifetime) failing in the first year of use.

26.1.3.1 Service Support

Instrument reliability is very difficult to assess at the evaluation stage, so you have to look very carefully at the kind of service support offered by the manufacturer. For example, how close is a qualified support engineer to you or what is the maximum amount of time you will have to wait to get a support engineer at your laboratory, or at least to call you back to discuss the problem? Ask the vendor if they have the capability for remote diagnostics, where a service engineer can remotely run the instrument or check the status of a component by "talking to" your system computer via a modem. Even if this approach does not fix the problem, at least the service engineer can come to your laboratory with a very good indication of what the problem could be.

You should know up front what a service visit is going to cost you, irrespective of what component has failed. Also find out what routine maintenance jobs you can do and what requires an experienced service engineer. If it does require a service engineer, how long will it take them? because their time is not inexpensive. Most companies charge an hourly rate for a service engineer (which typically includes travel time as well), but if an overnight stay is required, fully understand what you are paying for (accommodations, meals, gas, etc.). Some companies might even charge for mileage between the service engineer's base and your laboratory. If you work in a commercial laboratory and cannot afford the instrument to be down for any length of time, find out what it is going to cost for 24/7 service coverage.

You can take a chance and just pay for each service visit, or you might want to budget for an annual preventative maintenance contract, where the service engineer checks out all the important instrumental components and systems frequently to make sure they are all working correctly. This might not be as critical if you work in an academic environment, where the instrument might be down for extended periods, but in my opinion, it is absolutely critical if you work in a commercial laboratory, which is using the instrument to generate revenue. Also find out what is included in the contract, because some also cover the cost of consumables or replacement parts, whereas others just cover the service visits. These annual preventative maintenance contracts typically make up about 10–15% of the cost of the instrument, but are well worth it if you do not have the expertise in-house, or if you just feel more comfortable having an insurance policy to cover instrument breakdowns.

Once again, talking to existing users will give you a very good perspective of the quality of the instrument and the service support offered by the manufacturer. There is no absolute guarantee that the instrument of choice is going to perform to your satisfaction 100% of the time, but if you work in a high-throughput, routine laboratory, make sure it will be down for the minimum amount of time. In other words, fully understand what it is going to cost you to maximize the uptime of all the instruments being evaluated.

26.2 FINANCIAL CONSIDERATIONS

The financial side of choosing an ICP mass spectrometer can often dominate the selection process. You may or may not have budgeted quite enough money to buy a top-of-the-line instrument or perhaps you had originally planned to buy another lower-cost trace element technique, or you could be using funds left over at the end of your financial year. All these scenarios dictate how much money you have available and what kind of instrument you can purchase. In my experience, you should proceed with caution in this kind of situation, because if only one manufacturer is willing to do a deal with you, there is no real need to carry out the evaluation process. Therefore, you should budget at least 12 months before you are going to make a purchase and add another 10–15% for inflation and any unforeseen price increases. In other words, if you want to get the right instrument for your application, never let price be the overriding factor in your decision. Always be wary of the vendor who will undercut everyone else to get your business. There could be a very good reason why they are doing this; for example, the instrument is being discontinued for a new model or it could be having some reliability problems that are affecting its sales.

This is not to say that price is unimportant, but what might appear to be the most expensive instrument to purchase might be the least expensive to run. Therefore, you must never forget the cost of ownership in the overall financial analysis of your purchase. So, by all means compare the price of the instrument, computer and any accessories you buy, but also factor in the cost of consumables, gases and electricity based on your usage. Maybe instrument consumables from one vendor are much less expensive than from another vendor. This is particularly the case with interface cones and plasma torches. Or maybe the purity of collision/reaction gases is more critical with one cell-based instrument than another. For example, there is a factor-of-four difference between the cost of high-purity (99.999%) hydrogen gas and ultra-high-purity (99.9999%) -grade. Supplies of high-purity helium gas are diminishing, so the price is increasing quite significantly, as well as becoming much more difficult to obtain from your gas supplier.

So, be diligent when you compare prices. Look at the overall picture, and not just the cost of the instrument. Also, be aware of differences in sample throughput. Perhaps you can analyze more samples with one instrument because its measurement protocol is faster or it does not need recalibrating as often (less drift). It therefore follows that if you can get through your daily allocation of samples much faster with one instrument than another, then your argon consumption will be reduced.

Another aspect that should be taken into consideration is the salary of the operator. Even though you might think that this is a constant, irrespective of the instrument, you must assess the expertise required to run it. For example, if you are thinking of purchasing more complex technology such as a magnetic-sector instrument or a "triple quad" collision/reaction-cell instrument for a research-type application, the operator needs to be of a much higher skill level than, say, someone who is being asked to run a routine application with a quadrupole-based instrument. As a result, the salary of that person will probably be higher.

Finally, if one instrument has to be installed in a temperature-controlled, air-conditioned environment for stability purposes, the cost of preparing or building this kind of specialized room must be taken into consideration when doing your financial analysis. In other words, when comparing systems, never automatically reject the most expensive instrument. You will find that over the ten years that you own the instrument, the cost of doing analysis and the overall cost of ownership are more important evaluation criteria.

26.3 THE EVALUATION PROCESS: A SUMMARY

As mentioned earlier in this chapter, it is not my intention to compare instrument designs and features, but to give you some general guidelines as to what are the most important evaluation criteria. Besides being a framework for your evaluation process, these guidelines should also be used in conjunction with the other chapters in this book and the cited referenced information.

However, if you want to find the best instrument for your application needs, be prepared to spend a few months evaluating the marketplace. Do not forget to prioritize your objectives and give each of them a weighting factor based on their degree of importance for the types of samples you analyze. Be careful to take the evaluation in the direction you want to go and not where the vendor wants to take it. In other words, it is important to compare apples with apples and not to be talked into comparing an apple with an orange that looks like an apple! However, be prepared that there might not be a clear-cut winner at the end of the evaluation. If this is the case, then decide what aspects of the evaluation are most important and ask the manufacturer to put them in writing. Some vendors might be hesitant to do this, especially if it is guaranteeing instrument performance with your samples.

Talk to as many users in your field as you possible can—not only ones given to you as references by the vendor, but ones chosen by yourself also. This will give you a very good indication of the real-world capabilities of the instrument, which can often be overlooked at a demonstration. You might find from talking to "typical" users that it becomes obvious which instrument to purchase. If that is the case and your organization allows it, ask the vendor what your options are if you do not have samples to run and you do not want a demonstration. I guarantee you will be in a much better position to negotiate a lower price.

Never forget that it is a very competitive marketplace, and your business is extremely important to each of the ICP-MS manufacturers. As mentioned at the beginning of this chapter, it is meant to offer guidance about what are the most important analytical considerations and to suggest a template by which you can compare different features, approaches and instrument designs, no matter what your testing requirements are.

Hopefully, this chapter has not only helped you understand the fundamentals of the technique a little better but has also given you some thoughts and ideas on how to find the best instrument for your needs. Refer to Chapter 28 for details on how to contact all the instrument vendors and consumables and accessories companies. And if you are still confused, I teach a half-day Short Course at the Pittsburgh Conference every year on "How to Evaluate ICP-MS: The Most Important Analytical Considerations". We talk about all these issues in more detail. But if I don't see you there, GOOD LUCK with your evaluation!

FURTHER READING

1. K. Nottingham, ICP-MS: It's Elemental, *Analytical Chemistry*, 35–38A, 2004, http://pubs.acs.org/subscribe/journals/ancham-a/76/i01/toc/toc_i01.html.
2. Royal Society of Chemistry, Report by the Analytical Methods Committee: Evaluation of Analytical Instrumentation—Part X Inductively Coupled Plasma Mass Spectrometers, *The Analyst*, **122**, 393–408, 1997.
3. Analytical Figures of Merit for ICP-MS, *Inductively Coupled Plasma Mass Spectrometry: An Introduction to ICP Spectrometries for Elemental Analysis*, A. Montasser (ed), Wiley-VCH, Weinheim, 1998, pp. 16–28, Chap. 1.4.
4. E. R. Denoyer, *Atomic Spectroscopy*, **13**(3), 93–98, 1992.
5. M. A. Thomsen, *Atomic Spectroscopy*, **13**(3), 93–98, 2000.
6. L. Halicz, Y. Erel, and A. Veron, *Atomic Spectroscopy*, **17**(5), 186–189, 1996.
7. R. Thomas, *Spectroscopy*, **17**(7), 44–48, 2002.
8. E. R. Denoyer and Q. H. Lu, *Atomic Spectroscopy*, **14**(6), 162–169, 1993.
9. R. Hutton, A. Walsh, D. Milton, and J. Cantle, *ChemSA*, **17**, 213–215, 1991.
10. P. H. Dawson, ed., *Quadrupole Mass Spectrometry and Its Applications*, Elsevier, Amsterdam, 1976; reissued by AIP Press, Woodbury, NY, 1995.
11. S. J. Jiang, R. S. Houk, and M. A. Stevens, *Analytical Chemistry*, **60**, 217, 1988.
12. K. Sakata and K. Kawabata, *Spectrochimica Acta*, **49B**, 1027, 1994.
13. J. M. Collard, K. Kawabata, Y. Kishi, and R. Thomas, *Micro*, January 2002.
14. S. D. Tanner and V. I. Baranov, *Atomic Spectroscopy*, **20**(2), 45–52, 1999.
15. B. Hattendorf and D. Günther, *Journal of Analytical Atomic Spectrometry*, **19**, 600–606, 2004.
16. S. D. Tanner, D. J. Douglas, and J. B. French, *Applied Spectroscopy*, **48**, 1373, 1994.
17. E. R. Denoyer, D. Jacques, E. Debrah, and S. D. Tanner, *Atomic Spectroscopy*, **16**(1), 1, 1995.
18. R. C. Hutton and A. N. Eaton, *Journal of Analytical Atomic Spectrometry*, **5**, 595, 1987.
19. A. L. Gray and A. Date, *Analyst*, **106**, 1255, 1981.
20. E. J. Wyse, D. W. Koppenal, M. R. Smith, and D. R. Fisher, *18th FACSS Meeting*, Anaheim, CA, October 1991, Paper No. 409.
21. W. G. Diegor and H. P. Longerich, *Atomic Spectroscopy*, **21**(3), 111, 2000.
22. D. J. Douglas and J. B. French, *Spectrochimica Acta*, **41B**(3), 197, 1986.
23. E. R. Denoyer, *Atomic Spectroscopy*, **12**, 215–224, 1991.
24. E. Heithmar and S. Pergantis, *Characterizing Concentrations and Size Distributions of Metal-Containing Nanoparticles in Waste Water*, EPA Report APM 272.
25. R. Brennan, G. Dulude, and R. Thomas, Approaches to Maximize Performance and Reduce the Frequency of Routine Maintenance in ICP-MS, *Spectroscopy Magazine*, **30**(10), 12–25, 2015.
26. *PlasmaChem Listserver: A Discussion Group for Plasma Spectrochemists Worldwide*, Syracuse University, New York, www.lsoft.com/scripts/wl.exe?SL1=PLASMACHEM-L&H=LISTSERV.SYR.EDU.

27 Glossary of Terms Used in Atomic Spectroscopy

In all my years of working in the field of plasma spectrochemistry, I had never come across any written material that included a basic dictionary of terms, primarily aimed at someone new to the technique. When I first became involved in the ICP-MS technique, most of the literature I read tended to give complicated descriptions of instrument components and explanations of fundamental principles that, more often than not, sailed over my head. It was not until I became more familiar with the technique that I began to get a better understanding of the complex jargon used in technical journals and presentations at scientific conferences. So, when I wrote my second ICP-MS textbook, I knew that a glossary of terms was an absolute necessity, and it has been in every subsequent book I've written since. In this book, the glossary has been expanded to also include all the atomic spectroscopy techniques described, including ICP-OES, AA, AF, XRF, XRD, LIBS, LALI-TOF-MS and MIP-AES terms. Even though the glossary is not exhaustive, it contains explanations and definitions of the most common AS words, expressions and terms used in this book. It should mainly be used as a quick reference guide. If you want more detailed information about the subject matter, you should use the index to find a more detailed explanation of the topic in the appropriate book chapter. Many of the same terms are used in different chapters, so to save duplication and where appropriate, those definitions have only been used once (note: the glossary does not include commercial names used by any of the instrument, accessories or consumables vendors). The order of the glossary is as follows:

- ICP-Mass Spectrometry
- ICP-Optical Emission Spectroscopy
- Atomic Absorption/Fluorescence
- Other Atomic Spectroscopy Techniques

INDUCTIVELY COUPLED PLASMA MASS SPECTROMETRY (ICP-MS) GLOSSARY

A

Abundance sensitivity

A way of assessing the ability of a mass-separation device, such as a quadrupole, to identify and measure a small analyte peak adjacent to a much larger interfering peak. An abundance-sensitivity specification is a combination of two measurements. The first is expressed as the ratio of the intensity of the peak at 1 amu (atomic mass unit) below the analyte peak to the intensity of the analyte peak, and the second is the ratio of the peak intensity 1 amu above the analyte mass to the intensity of the analyte peak. Because of the motion of the ion through the mass filter, the abundance-sensitivity specification of a mass-filtering device is always worse on the low-mass side compared to the high-mass side.

active film multipliers

Another name for **discrete-dynode multipliers**, which are used to detect, measure and convert ions into electrical pulses in ICP-MS. *Also refer to* **channel electron multiplier (CEM)** *and* **discrete-dynode detector (DDD)**.

DOI: 10.1201/9781003187639-27

addition calibration

A method of calibration in ICP-MS using standard additions. All samples are assumed to have a similar matrix, so spiking is only carried out on one representative sample and not the entire batch of samples, as per conventional standard additions used in graphite-furnace AA analysis.

AE

An abbreviation for **atomic emission**.

aerosol

The result of breaking up a liquid sample into small droplets by the nebulization process in the sample-introduction system. *Also refer to* **nebulizer** *and* **sample-introduction system**.

aerosol dilution

A way of introducing a flow of argon gas between the nebulizer and the torch, which has the effect of reducing the sample's solvent loading on the plasma, so it can tolerate much higher total dissolved solids.

AF4

Refer to Asymetrical Flow Field-Flow Fractionation.

alkylated metals

A metal complex containing an alkyl group. Typically detected by coupling liquid chromatography with ICP-MS. *Also refer to* **speciation analysis**.

alpha-counting spectrometry

A particle-counting technique that uses the measurement of the radioactive decay of alpha particles. *Also refer to* **particle-counting techniques**.

alternative sample-introduction accessories

Alternative ways of introducing samples into an ICP mass spectrometer other than conventional nebulization. Also known as **alternative sample-introduction devices**. Often used to describe desolvation techniques or laser ablation.

analog counting

A way of measuring high signals by changing the gain or voltage of the detector. *Also refer to* **pulse counting**.

argon

The gas used to generate the plasma in an ICP.

argon-based interferences

A polyatomic spectral interference generated by argon ions combining with ions from the matrix, solvent or any elements present in the sample.

array detectors

An ion detector based on solid-state, direct-charge arrays, similar to CID/CCD technology used in ICP optical emission. By projecting all the separated ions from a mass-separation device onto a two-dimensional array, these detectors can view the entire mass spectrum simultaneously. Used with the Mattauch–Herzog-sector technology.

ashing

A sample-preparation technique that involves heating the sample (typically in a muffle furnace) until the volatile material is driven off and an ash-like substance is left.

Asymmetrical flow field-flow fractionation (AF4)

Field-flow fractionation (FFF) is a single-phase chromatographic separation technique, where separation is achieved within a very thin channel, against which a perpendicular force field is applied. One of the most common forms of FFF is asymmetrical flow FFF (AF4), where the field is generated by a cross-flow applied perpendicular to the channel. Coupled with ICP-MS for the characterization of nano particles.

atom

A unit of matter. The smallest part of an element having all the characteristics of that element and consisting of a dense, central, positively charged nucleus surrounded by orbiting electrons. The entire structure has an approximate diameter of 10^{-8} cm and characteristically remains undivided in chemical reactions except for limited removal, transfer or exchange of certain electrons.

atom-counting techniques

A generic name given to techniques that use atom or ion counting to carry out elemental quantitation. Some common ones, besides ICP-MS, include secondary-ionization mass spectrometry (SIMS), thermal ionization mass spectrometry (TIMS), accelerator mass spectrometry (AMS) and fission-track analysis (FTA). *Also refer to* **ionizing radiation counting techniques**.

atomic absorption (AA)

An analytical technique for the measurement of trace elements that uses the principle of generating free atoms (of the element of interest) in a flame or electrothermal atomizer (ETA) and measuring the amount of light absorbed from a wavelength-specific light source, such as a hollow-cathode lamp (HCL) or electrode discharge lamp (EDL).

atomic emission (AE)

A trace element analytical technique that uses the principle of exciting atoms in a high-temperature source such as a plasma discharge and measuring the amount of light the atoms emit when electrons fall back down to a ground (stable) state.

atomic mass or weight

The average mass or weight of an atom of an element, usually expressed relative to the mass of carbon 12, which is assigned 12 atomic mass units.

atomic number

The number of protons in an atomic nucleus.

atomic structure

Describes the structural makeup of an atom. *Also refer to* **neutron, proton** *and* **electron**.

attenuation (of the detector)

Reduces the amplitude of the electrical signal generated by the detector, with little or no distortion. Usually carried out by applying a control voltage to extend the dynamic range of the detector. *Also refer to* **extended dynamic range**.

autocalibration

A way of carrying out calibration with an automated in-line sample-delivery system.

autodilution

A way of carrying out automatic in-line dilution of large numbers of samples with no manual intervention by the operator.

autosampler

A device to automatically introduce large numbers of samples into the ICP-MS system with no manual intervention by the operator.

axial view

An ICP-OES system in which the plasma torch is positioned horizontally (end-on) to the optical system as opposed to the conventional vertical (radial) configuration. It is generally accepted that viewing the end of the plasma improves emission intensity by a factor of approximately five- to ten-fold.

B

background equivalent concentration (BEC)

Defined as the apparent concentration of the background signal based on the sensitivity of the element at a specified mass. The lower the BEC value, the more easily a signal generated by an element can be discerned from the background. Many analysts believe BEC is a more accurate indicator of the performance of an ICP-MS system than detection limit, especially when making comparisons of background-reduction techniques, such as cool-plasma or collision/reaction cell and interface technology.

background noise

The square root of the intensity of the blank in counts per second (cps) anywhere of analytical interest on the mass range. Detection limit (DL) is a ratio of the analyte signal to the background noise at the analyte mass. Background noise as an instrumental specification is usually measured at mass 220 amu (where there are no spectral features), while aspirating deionized water. *Also refer to* **background signal, instrument background noise** *and* **detection limit**.

background signal

The signal intensity of the blank in counts per second (cps) anywhere of analytical interest on the mass range. Detection limit (DL) is a ratio of the analyte signal to the noise of the background at the analyte mass. Background as an instrumental specification is usually measured at mass 220 amu (where there are no spectral features), while aspirating deionized water. *Also refer to* **background noise, instrument background signal** *and* **detection limit**.

bandpass tuning/filtering

A mechanism used in a dynamic-reaction cell (DRC) to reject the by-products generated through secondary reactions utilizing the principle of mass discrimination. Achieved by optimizing the electrical fields of the reaction-cell multipole (typically a quadrupole) to allow transmission of the analyte ion, while rejecting the polyatomic interfering ion.

BEC

An abbreviation for **background equivalent concentration**.

by-product ions

Ionic species formed as a result of secondary reactions that take place in a reaction/collision cell. *Also refer to* **secondary (side) reactions**.

C

calibration

A plot, function or equation generated using calibration standards and a blank, which describes the relationship between the concentration of an element and the signal intensity produced at the analyte mass of interest. Once determined, this relationship can be used to determine the analyte concentration in an unknown sample.

calibration standard

A reference solution containing accurate and known concentrations of analytes for the purpose of generating a calibration curve or plot.

capacitive coupling

An undesired electrostatic (or capacitive) coupling between the voltage on the load coil and the plasma discharge, which produces a potential difference of a few hundred volts. This creates an electrical discharge or arcing between the plasma and sampler cone of the interface, commonly known as a "secondary discharge" or "pinch effect".

capillary electrophoresis (CE)

Refer to **capillary-zone electrophoresis (CZE)**.

capillary-zone electrophoresis (CZE or CE)

A chromatographic separation technique used to separate ionic species according to their charge and frictional forces. In traditional electrophoresis, electrically charged analytes move in a conductive liquid medium under the influence of an electric field. In capillary (zone) electrophoresis, species are separated based on their size-to-charge ratio inside a small capillary filled with an electrolyte. Its applicability to ICP-MS is mainly in the field of separation and detection of large biomolecules.

CCD

Charge-coupled-device detector.

CE

Refer to **capillary-zone electrophoresis**.

cell

In ICP-MS terminology, a cell usually refers to a collision or reaction cell.

ceramic torch

A plasma torch where either (or all) the sample injector, inner tube or outer tube is made of a ceramic material. Typically, has longer lifetime than a traditional quartz torch.

certified reference material (CRM)

Well-established reference matrix that comes with certified values and associated statistical data that have been analyzed by other complementary techniques. Its purpose is to check the validity of an analytical method, including sample preparation, instrument methodology and calibration routines to achieve sample results that are as accurate and precise as possible and can be defended when subjected to intense scrutiny.

channel electron multiplier (CEM)

A detector used in ICP-MS to convert ions into electrical pulses using the principle of multiplication of electrons via a potential gradient inside a sealed tube.

Channeltron®
Another name for a channel-electron-multiplier detector.

charge
transfer reaction
Sometimes referred to as "charge exchange". This is one of the ion–molecule reaction mechanisms that take place in a collision/reaction cell. Involves the transfer of a positive charge from the interfering ion to the reaction gas molecule, forming a neutral atom that is not seen by the mass analyzer. An example of this kind of reaction:

$$H_2 + {}^{40}Ar^+ = Ar + H_2{}^+.$$

charge-coupled-device detector (CCD)
A type of solid-state detector technology for converting photons into an electrical signal. Typically applied to ICP-OES.

charge-injection-device detector (CID)
A type of solid-state detector technology for converting photons into an electrical signal Typically applied to ICP-OES.

chemical modification
The process of chemically modifying the sample in electrothermal-vaporization (ETV) ICP-MS work to separate the analyte from the matrix. *Also refer to* **chemical modifier** *and* **electrothermal vaporization**.

chemical modifier
A chemical or substance that is added to the sample in an electrothermal vaporizer to change the volatility of the analyte or matrix. Typically added at the ashing stage of the heating program to separate the vaporization of the analyte away from the potential interferences of the matrix components. *Also refer to* **electrothermal vaporization**.

chromatographic separation device
Any device that separates analyte species according to their retention times or mobility through a stationary phase. When coupled with an ICP-MS system, it is used for the separation, detection and quantitation of speciated forms of trace elements. Examples include liquid, ion, gas, size exclusion and capillary electrophoresis chromatography. *Also refer to* **speciation analysis**.

chromatography terminology (as applied to trace element speciation)
The following are some of the most important terms used in the chapter on trace element speciation. For easy access, they are contained in one section and not distributed throughout the glossary.

buffer A mobile-phase solution that is resistant to extreme pH changes, even with additions of small amounts of acids or bases.

chromatogram The graphical output of the chromatographic separation. It is usually a plot of peak intensity of the separated species over time.

column The main component of the chromatographic separation. It is typically a tube containing the stationary-phase material that separates the species and an eluent that elutes the species off the column.

counterions The mobile phase contains a large number of ions that have a charge opposite to that of the surface-bound ions. These are known as counterions, which establish equilibrium with the stationary phase.

dead volume Usually refers to the volume of the mobile phase between the point of injection and the detector that is accessible to the sample species, minus the volume of mobile phase that is contained in any union or connecting tubing.

gradient elution Involves variation of the mobile-phase composition over time through a number of steps, such as changing the organic content, altering the pH, changing the concentration of the buffer or using a completely different buffer.

ion exchange A technique in which separation is based on the exchange of ions (anions or cations) between the mobile phase and the ionic sites on a stationary phase bound to a support material in the column.

ion pairing A type of separation that typically uses a reversed-phase column in conjunction with a special type of chemical in the mobile phase called an "ion-pairing reagent". *Also refer to* **reverse phase**.

isochratic elution An elution of the analytes or species using the same solvent throughout the analysis.

mobile phase A combination of the sample or species being separated or analyzed and the solvent that moves the sample through the column.

retention time The time taken for a particular analyte or species to be separated and pass through the column to the detector.

reverse phase A type of separation that is typically combined with ion pairing, and essentially means that the column's stationary phase is less polar and more organic than the mobile-phase solvents.

stationary phase A solid material, such as silica or a polymer, that is set in place and packed into the column for the chromatographic separation to take place.

clean room

The general description given to a dedicated room for the sample preparation and analysis of ultrapure materials. Usually associated with a number that describes the number of particulates per cubic foot of air (e.g., a class 100 clean room will contain 100 particles/ft^3 of air). It is commonly accepted that the semiconductor industry has the most stringent demands, which necessitates the use of class 10 and sometimes class 1 clean rooms.

cluster ions

Ions that are formed by two or more molecular ions combining together in a collision/reaction cell to form molecular clusters.

CMOS

Complementary-metal-oxide-semiconductor technology used in the direct-charge array detector, which is utilized in the Mattacuh–Herzog simultaneous-sector instrument.

cold-plasma technology

Cool- or cold-plasma technology uses low-temperature plasma to minimize the formation of certain argon-based polyatomic species. Under normal plasma conditions (approximately 1,000 W RF power and 1.0 L/min nebulizer gas flow), argon ions combine with matrix and solvent components to generate problematic spectral interferences, such as $^{38}ArH^+$, $^{40}Ar^+$ and $^{40}Ar^{16}O^+$, which impact the detection limits of a small number of elements including K, Ca and Fe. By using cool-plasma conditions (approximately 600 W RF power and 1.6 L/min nebulizer gas flow), the ionization conditions in the plasma are changed so that many of these interferences are dramatically reduced and detection limits are improved.

cold-vapor atomic absorption (CVAA)

An analytical approach to determine low levels of mercury by generating mercuric vapor in a quartz cell and measuring the number of mercury atoms produced, using the principle of atomic absorption. *Also refer to* **hydride-generation atomic absorption**.

collision cell

Specifically, a cell that predominantly uses the principle of collisional fragmentation to break apart polyatomic interfering ions generated in the plasma discharge. Collision cells typically utilize higher-order multipoles (such as hexapoles or octopoles) with inert or low-reactive gases (such as helium and hydrogen) to first stimulate ion–molecule collisions, and then kinetic energy discrimination to reject any undesirable by-product ionic species formed.

collision-induced dissociation (CID)

A basic principle, first used for the study of organic molecules using tandem mass spectrometry, that relies on using a nonreactive gas in a collision cell to stimulate ion–molecule collisions. The more collision-induced daughter species that are generated, the better the chance of identifying the structure of the parent molecule.

collision/reaction-cell (CRC) technology

A generic term applied to collision and reaction cells that use the principle of ion–molecule collisions and reactions to cleanse the ion beam of problematic polyatomic spectral interferences before they enter the mass analyzer. Both collision and reaction cells are positioned in the mass spectrometer vacuum chamber after the ion optics but prior to the mass analyzer. *Also refer to* **collision cell** *and* **reaction cell**.

collision/reaction-interface (CRI) technology

A collision/reaction-mechanism approach, which instead of using a pressurized cell, injects a gas directly into the interface between the sampler and skimmer cones. The injection of the collision/ reaction gas into this region of the ion beam produces high collision frequency between the argon gas and the injected gas molecules. This has the effect of removing argon-based polyatomic interferences before they are extracted into the ion optics.

collisional damping

A mechanism that describes the temporal broadening of ion packets in a quadrupole-based dynamic reaction cell to dampen out fluctuations in ion energy. By optimizing cell conditions such as gas pressure, RF stability boundary (q parameter), entrance/exit lens potentials and cell rod offsets, it has been shown that fluctuation in ion energies can be dampened sufficiently to carry out isotope-ratio precision measurements near their statistical limit.

collisional focusing

The mechanism of focusing ions toward the center of the ion beam in a collision/reaction cell. By using a neutral collision gas of lower molecular weight than the analyte, the analyte ions will lose kinetic energy and migrate toward the axis as a result of the collisions with the gaseous molecules. Therefore, the number of ions exiting the cell and reaching the detector will increase. *Also refer to* **collision cell** *and* **reaction cell**.

collisional fragmentation

The mechanism of breaking apart (fragmenting) a polyatomic interfering ion in a collision/reaction cell using collisions with a gaseous molecule. The predominant mechanism used in a collision cell, as opposed to a reaction cell. *Also refer to* **collision cell** *and* **reaction cell**.

collisional mechanisms

The mechanisms by which the interfering ion is reduced or minimized to allow the determination of the analyte ion. The most common collisional mechanisms seen in collision/reaction cells include collisional focusing, dissociation and fragmentation, whereas the major reaction mechanisms include exothermic/endothermic associations, charge transfer, molecular associations and proton transfer.

collisional retardation

A mechanism in a collision/reaction cell where the gas atoms/molecules undergo multiple collisions with the polyatomic interfering ion, retarding or lowering its kinetic energy. Because the interfering ion has a larger cross-sectional area than the analyte ion, it undergoes more collisions, and as a result, can be separated or discriminated from the analyte ion based on their kinetic energy differences.

concentric nebulizer

A nebulizer that uses two narrow concentric capillary tubes (one inside the other) to aspirate a liquid into the ICP-MS spray chamber. Argon gas is usually passed through the outer tube, which creates a Venturi effect, and as a result, the liquid is sucked up through the inner capillary tube.

cones

Refer to **interface cones**.

cool-plasma technology

Refer to **cold-plasma technology**.

cooled spray chamber

A spray chamber that is cooled in order to reduce the amount of solvent entering the plasma discharge. Used for a variety of reasons, including reducing oxide species, minimizing solvent-based spectral interferences, and allowing the trouble-free aspiration of organic solvents.

correction equation

A mathematical approach used to compensate for isobaric and polyatomic spectral overlaps. It works on the principle of measuring the intensity of the interfering species at another mass, which is ideally free of any interference. A correction is then applied by knowing the ratio of the intensity of the interfering species at the analyte mass to its intensity at the alternate mass.

counts per second (cps)

Units of signal intensity used in ICP-MS. Number of detector electronic pulses counted per second.

cps

An abbreviation for **counts per second**.

CRC

An abbreviation for **collision/reaction-cell technology**.

CRI

An abbreviation for **collision/reaction-interface technology**.

CRM

An abbreviation for certified reference materials.

cross-calibration

A calibration method that is used to correlate both pulse (low levels) and analog (high levels) signals in a dual-mode detector. This is possible because the analog and pulse outputs can be defined in identical terms (of incoming pulse counts per second) based on knowing the voltage at the first analog stage, the output current and a conversion factor defined by the detection circuitry electronics. By carrying out a cross-calibration across the mass range, a dual-mode detector is capable of achieving approximately eight to nine orders of dynamic range in one simultaneous scan.

cross-flow nebulizer

A nebulizer that is designed for samples that contain a heavier matrix or small amounts of undissolved solids. In this design, the argon gas flow is directed at right angles to the tip of a capillary tube through which the sample is drawn up with a peristaltic pump.

CVAA

An abbreviation for **cold-vapor atomic absorption**.

cyclonic spray chamber

A spray chamber that operates using the principle of centrifugal force. Droplets are discriminated according to their size by means of a vortex produced by the tangential flow of the sample aerosol and argon gas inside the spray chamber. Smaller droplets are carried with the gas stream into the ICP-MS, while the larger droplets impinge on the walls and fall out through the drain.

cylinder lens

A type of lens component used in the ion optics.

CZE

An abbreviation for **capillary-zone electrophoresis**.

D

data-quality objectives

A term used to describe the quality goals of the analytical result. Typically achieved by optimizing the measurement protocol to achieve the desired accuracy/precision/sample throughput required for the analysis.

DCD

Refer to direct-charge detector.

dead time correction

Sometimes ions hit the detector too fast for the measurement circuitry to handle in an efficient manner. This is caused by ions arriving at the detector during the output pulse of the preceding ion and not being detected by the counting system. This "dead time", as it is known, is a fundamental limitation of the multiplier detector and is typically 30–50 ns, depending on the detection system. A compensation or "dead time correction" has to be made in the measurement circuitry in order to count the maximum number of ions hitting the detector.

Debye length

The distance over which ions exert an electrostatic influence over one another as they move from the interface region into the ion optics. In the ion-sampling process, this distance is small compared to the orifice diameter of the sampler or skimmer cone. As a result, there is little electrical interaction between the ion beam and the cones, and relatively little interaction between the individual ions within the ion beam. In this way, the compositional integrity of the ion beam is maintained throughout the interface region.

desolvating microconcentric nebulizer

A microconcentric nebulizer that uses some type of desolvation system to remove the sample solvent. *Also refer to* **desolvation device** *and* **membrane desolvation**.

desolvating spray chamber

A general name given to a spray chamber that removes or reduces the amount of solvent from a sample using the principle of desolvation. Some of the approaches that are typically used include conventional water cooling, heating with cooling condensers, Peltier (thermoelectric) cooling, or membrane-based desolvation techniques.

desolvation device

A general name given to a device that removes or reduces the amount of solvent from a sample using the principle of desolvation. Some of the approaches that are typically used include conventional water cooling, heating/condensing units, Peltier (thermoelectric) cooling or membrane-based desolvation techniques.

detection capability

A generic term used to assess the overall detection performance of an ICP mass spectrometer. There are a number of different ways of evaluating detection capability, including instrument detection limit (IDL), method detection limit (MDL), element sensitivity and background equivalent concentration (BEC).

detection limit

Most often refers to the instrument detection limit (IDL) and is typically defined as a ratio of the analyte signal to the noise of the background at a particular mass. For a 99% confidence level, it is usually calculated as 3× standard deviation (SD) of 10 replicates (measurements) of the sample blank expressed as concentration units.

detector

A generic name used for a device that converts ions into electrical pulses in ICP-MS.

detector dead time

Refer to **dead time correction**.

devitrification

Crystalline breakdown of glass or quartz by a combination of chemical attack and elevated temperatures, typically associated with the plasma torch.

digital counting

Refers to the process of counting the number of pulses generated by the conversion of ions into an electrical signal by the detector measurement circuitry.

DIHEN

An abbreviation for **direct-injection high-efficiency nebulizer**.

DIN

An abbreviation for **direct-injection nebulizer**.

direct-charge detector (DCD)

A detector technology used to convert photons into an electric current. A type of CMOS array ion detector used in the Mattacuh–Herzog simultaneous-sector instrument.

direct-injection high-efficiency nebulizer (DIHEN)

A more recent refinement of the direct-injection nebulizer (DIN), which appears to have overcome many of the limitations of the original design.

direct-injection nebulizer (DIN)

A nebulizer that injects a liquid sample under high pressure directly into the base of the plasma torch. The benefit of this approach is that no spray chamber is required, which means that an extremely small volume of sample can be introduced directly into the ICP-MS with virtually no carryover or memory effects from the previous sample.

discrete-dynode detector (DDD)

The most common type of detector used in ICP-MS. As ions emerge from the quadrupole rods onto the detector, they strike the first dynode, liberating secondary electrons. The electron-optic design of the dynode produces acceleration of these secondary electrons to the next dynode, where they generate more electrons. This process is repeated at each dynode, generating a pulse of electrons that are finally captured by the multiplier anode. *Also refer to* **active film multipliers**.

double-focusing magnetic-sector mass spectrometer (analyzer)

A mass spectrometer that uses a very powerful magnet combined with an electrostatic analyzer (ESA) to produce a system with very high resolving power. This approach, known as "double focusing", samples the ions from the plasma. The ions are accelerated in the plasma to a few kilovolts into the ion-optic region before they enter the mass analyzer. The magnetic field, which is dispersive with respect to ion energy and mass, then focuses all the ions with diverging angles of motion from the entrance slit. The ESA, which is only dispersive with respect to ion energy, then focuses all the ions onto the exit slit, where the detector is positioned. If the energy dispersions of the magnet and ESA are equal in magnitude but opposite in direction, they will focus both ion angles (first focusing) and ion energies (second focusing) when combined together. *Also refer to* **electrostatic analyzer**.

double-pass spray chamber

A spray chamber that comprises an inner (central) tube inside the main body of the spray chamber. The smaller droplets are selected by directing the aerosol from the nebulizer into the central tube. The aerosol emerges from the tube, where the larger droplets fall out (because of gravity) through a drain tube at the rear of the spray chamber. The smaller droplets then travel back between the outer wall and the central tube into the sample injector of the plasma torch. The most common type of double-pass spray chamber is the Scott design.

doubly charged ion

A species that is formed when an ion is generated with a double positive charge as opposed to a normal single charge and produces an isotopic peak at half its mass. For example, the major isotope of barium at mass 138 amu also exhibits a doubly charged ion at mass 69 amu, which can potentially interfere with gallium at mass 69. Some elements, such as the rare earths, readily form doubly charged species, whereas others do not. Formation of doubly charged ions is also impacted by the ionization conditions (RF power, nebulizer gas flow, etc.) in the plasma discharge.

DRC

An abbreviation for **dynamic reaction cell**.

droplet

Refers to individual particles (either small or large) that make up an aerosol generated by the nebulizer.

dry plasma

When a sample is introduced into the plasma that does not contain any liquid or solvent, such as laser ablation, ETV or desolvation sample-introduction systems.

duty cycle (%)

Also known as the "measurement duty cycle". It refers to the actual peak measurement time and is expressed as a percentage of the overall integration time. It is calculated by dividing the total peak quantitation time (dwell time × number of sweeps × replicates × elements) by the total integration time ([dwell time + settling/scanning time] × number of sweeps × replicates × elements).

dwell time

The time spent sitting (dwelling) on top of the analytical peak (mass) and taking measurements.

dynamic reaction cell (DRC)

A type of collision/reaction cell. Unlike a simple collision cell, a quadrupole is used instead of a hexapole or octapole. A highly reactive gas such as ammonia or methane is bled into the cell, which is a catalyst for ion–molecule chemistry to take place. By a number of different reaction mechanisms, the gaseous molecules react with the interfering ions to convert them either into an innocuous species different from the analyte mass or a harmless neutral species. The analyte mass then emerges from the dynamic reaction cell, free of its interference, and is steered into the analyzer quadrupole for conventional mass separation. Through careful optimization of the quadrupole electrical fields, unwanted reactions between the gas and the sample matrix or solvent, which could potentially lead to new interferences, are prevented. Therefore, every time an analyte and interfering ions enter the dynamic reaction cell, the bandpass of the quadrupole can be optimized for that specific problem and then changed on the fly for the next one.

dynamically scanned ion lens

A commercial ion-optic approach to focus the maximum number of ions into the mass analyzer. In this design, the voltage is dynamically ramped on the fly in concert with the mass scan of the analyzer. The benefit is that the optimum lens voltage is placed on every mass in a multielement run to allow the maximum number of analyte ions through, while keeping the matrix ions down to an absolute minimum. This is typically used in conjunction with a grounded stop acting as a physical barrier to reduce particulates, neutral species and photons from reaching the mass analyzer and detector.

E

EDR

An abbreviation for the term **extended dynamic range**, used in detector technology.

Electrodynamic forces

Flow of the ion beam through the interface region, where the positively charged ions of varying mass-to-charge ratios exert no electrical influence on each other.

electron

A negatively charged fundamental particle orbiting the nucleus of an atom. It has a mass equal to 1/1,836 of a proton's mass. Removal of an electron by excitation in the plasma discharge generates a positively charged ion.

electrostatic analyzer (ESA)

An ion-focusing device (utilizing a series of electrostatic lens components) that varies the electric field to allow the passage of ions of certain energy. In ICP-MS, it is typically used in combination with a conventional electromagnet to focus ions based on their angular motion and their kinetic energy to produce very high resolving power. *Also refer to* **double-focusing magnetic-sector mass spectrometer (analyzer)**.

electrothermal atomization (ETA)

An atomic absorption (AA) analytical technique that uses a heated metal filament or graphite tube (in place of the normal flame) to generate ground-state analyte atoms. The sample is first injected into the filament or tube, which is heated up slowly to remove the matrix components. Further heating then generates ground-state atoms of the analyte, which absorb light of a particular wavelength from an element-specific, hollow-cathode lamp source. The amount of light absorbed is measured by a monochromator (optical system) and detected by a photomultiplier or solid-state detector, which converts the photons into an electrical pulse. This absorbance signal is used to determine the concentration of that element in the sample. Typically used for ppb-level determinations.

electrothermal vaporization (ETV)

A sample-pretreatment technique used in ICP-MS. Based on the principle of electrothermal atomization (ETA) used in atomic absorption (AA), ETV is not used to generate ground-state atoms, but instead uses a carbon furnace (tube) or metal filament to thermally separate the analytes from the matrix components and then sweep them into the ICP mass spectrometer for analysis. This is achieved by injecting a small amount of the sample into a graphite tube or onto a metal filament. After the sample is introduced, drying, charring and vaporization are achieved by slowly heating the graphite tube or metal filament. The sample material is vaporized into a flowing stream of carrier gas, which passes through the furnace or over the filament during the heating cycle. The analyte vapor recondenses in the carrier gas and is then swept into the plasma for ionization.

elemental fractionation

A term used in laser ablation. It is typically defined as the variation in intensity of a particular element over time compared to the total amount of dry aerosol generated by the sample. It is generally sample and element specific, but there is evidence to suggest that the shorter-wavelength excimer lasers exhibit better elemental fractionation characteristics than the longer-wavelength Nd:YAG design because they produce smaller particles that are easier to volatilize.

endothermic reaction

In thermodynamics, this describes a chemical reaction that absorbs energy in the form of heat. In ICP-MS, it generally refers to an ion–molecule reaction in a collision/reaction cell that is not allowed to proceed because the ionization potential of the analyte ion is significantly less than that of the reaction gas molecule. *Also refer to* **exothermic reaction**.

engineered nanomaterials (ENM)

These are man-made materials made of particles with < 100-nm diameter, that can be made to exhibit: greater physical strength, enhanced magnetic properties, conduction of heat or electricity, greater chemical reactivity or size-dependent optical properties. An example of an engineered nanomaterial is silver nanoparticles, which are added to detergents as a bactericide.

ENM

Refer to engineered nanomaterials.

ESA

An abbreviation for **electrostatic analyzer**.

ETA

An abbreviation for **electrothermal atomization**.

ETV

An abbreviation for **electrothermal vaporization**.

excimer laser

A gas-filled laser in which a very short electrical pulse excites a mixture containing a halogen such as fluorine and a rare gas such as argon or krypton. It produces a brief, intense pulse of UV light. The output of an excimer laser is used for writing patterns on semiconductor chips because the short wavelength can write very fine lines. In ICP-MS, the most common excimer laser used is ArF at 193 nm and is typically used to ablate material with a very small size, such as inclusions on the surface of a geological sample.

exothermic reaction

In thermodynamics, this describes a chemical reaction that releases energy in the form of heat. In ICP-MS, it generally refers to an ion–molecule reaction in a collision/reaction cell that is spontaneous because the ionization potential of the interfering ion is much greater than the reaction gas molecule. *Also refer to* **endothermic reaction**.

extended dynamic range (EDR)

An approach used in ICP-MS to extend the linear dynamic range of the detector from five orders of magnitude up to eight or nine orders of magnitude. *Also refer to* **discrete-dynode detector** *and* **Farady-cup detector**.

external standardization

The normal mode of calibration used in ICP-MS by comparing the analyte intensity of unknown samples to the intensity of known calibration or reference standards.

extraction lens

An ion lens used to electrostatically extract the ions out of the interface region.

F

FAA

An abbreviation for **flame atomic absorption**.

Faraday collector

Another name for a Faraday-cup detector.

Faraday-cup detector

A simple metal-electrode detector used to measure high ion counts. When the ion beam hits the metal electrode, it will be charged, whereas the ions are neutralized. The electrode is then discharged to measure a small current equivalent to the number of discharged ions. By measuring the ion current on the metal part of the circuit, the number of ions in the circuit can be determined. Unfortunately, with this approach, there is no control over the applied voltage (gain). So, it can only be used for high ion counts, and therefore is not suitable for ultratrace determinations.

field-flow fractionation (FFF)

Field-flow fractionation (FFF) is a single-phase chromatographic separation technique, where separation is achieved within a very thin channel, against which a perpendicular force field is applied. One of the most common forms of FFF is asymmetrical flow FFF (AF4), where the field is generated by a cross-flow applied perpendicular to the channel. Coupled with ICP-MS for the characterization of nanoparticles.

FFF

Refer to field-flow fractionation.

FGDW

Refer to flue gas desulfurization wastewaters.

FIA

An abbreviation for **flow-injection analysis**.

flame atomic absorption (FAA)

An atomic absorption analytical technique that uses a flame (usually air–acetylene or nitrous oxide–acetylene) to generate ground-state atoms. The sample solution is aspirated into the flame via a nebulizer and a spray chamber. The ground-state atoms of the sample absorb light of a particular wavelength from an element-specific, hollow-cathode-lamp source. The amount of light absorbed is measured by a monochromator (optical system) and detected by a photomultiplier or solid-state detector, which converts the photons into an electrical pulse. This absorbance signal is used to determine the concentration of the element in the sample. Typically used for ppm-level determinations.

flatapole

A quadrupole with rods that have flat corners. Used in a particular commercial design of collision/reaction cells.

flight tube

A generic name given to the housing that contains a series of optical components that focus ions onto the detector of a time-of-flight (TOF) mass analyzer. There are basically two different kinds of flight tubes that are used in commercial TOF mass analyzers. One is the orthogonal design, where the flight tube is positioned at right angles to the sampled ion beam, and the other, the axial design, where the flight tube is in the same axis as the ion beam. In both designs, all ions are sampled through the interface region, but instead of being focused into the mass filter in the conventional sequential way, packets (groups) of ions are electrostatically injected into the flight tube at exactly the same time.

flow-injection analysis (FIA)

A powerful front-end sampling accessory for ICP-MS that can be used for preparation, pretreatment and delivery of the sample. It involves the introduction of a discrete sample aliquot into a flowing carrier stream. Using a series of automated pumps and valves, procedures can be carried out online to physically or chemically change the sample or analyte before introduction into the mass spectrometer for detection.

flue gas desulfurization wastewaters (FGDW)

This is one of the most widely used technologies for removing pollutants, such as sulfur dioxide, from flue gas emissions produced by coal-fired power plants. Sometimes called the limestone-forced-oxidation scrubbing system, but more commonly known as flue gas desulfurization (FGD), this process employs gas scrubbers to spray limestone slurry over the flue gas to convert gaseous sulfur dioxide to calcium sulfate.

fractogram

The separated particles that exit the outlet port of a field-flow-fractionation device into the detection system (e.g., UV/Vis or ICP-MS) are displayed as a temporal signal called a fractogram (similar to a chromatogram in chromatographic separation).

fringe rods

A set of four short rods operated in the RF-only mode, positioned at the entrance of a quadrupole mass analyzer. Their function is to minimize the effect of the fringing fields at the entrance of a quadrupole mass analyzer and thus improve the efficiency of transmission of ions into the mass analyzer. They are usually straight, but it has been suggested that curved fringe rods might reduce background levels.

fusion mixture

A compound or mixture added to solid samples as an aid to get them into solution. Fusion mixtures are usually alkaline salts (e.g., lithium metaborate, sodium carbonate) that are mixed with the sample (in powdered form) and heated in a muffle furnace to create a chemical/thermal reaction between the sample and the salt. The fused mixture is then dissolved in a weak mineral acid to get the analytes into solution.

G

gamma-counting spectrometry

A particle-counting technique that uses the measurement of the radioactive decay of gamma particles. *Also refer to* **particle-counting techniques.**

gas dynamics

In ICP-MS, it refers to the flow and velocity of the plasma gas through the interface region. It dictates that the composition of the ion beam immediately behind the sampler cone be the same as the composition in front of the cone because the expansion of the gas at this stage is not controlled by electrodynamics. This happens because the distance over which ions exert influence on one another (the Debye length) is small compared to the orifice diameter of the sampler or skimmer cone. Consequently, there is little electrical interaction between the ion beam and the cone and relatively little interaction between the individual ions in the beam. In this way, gas dynamics ensures that the compositional integrity of the ion beam is maintained throughout the interface region.

getter (gas purifier)

A device that "cleans up" inorganic and organic contaminants in pure gases. The getter usually refers to a metal that oxidizes quickly, and when heated to a high temperature (usually by means of RF induction), evaporates and absorbs/reacts with any residual impurities in the gas.

GFAA

An abbreviation for **graphite-furnace atomic absorption.**

graphite-furnace atomic absorption (GFAA)

An electrothermal-atomization (ETA) analytical technique that specifically uses a graphite tube (in place of the normal flame) to generate ground-state analyte atoms. The sample is first injected into the tube, which is heated up slowly to remove the matrix components. Further heating then generates ground-state atoms of the analyte, which absorb light of a particular wavelength from an element-specific, hollow-cathode-lamp source. The amount of light absorbed is measured by a monochromator (optical system) and detected by a photomultiplier or solid-state detector, which converts the photons into an electrical pulse. This absorbance signal is used to determine the concentration of that element in the sample. Typically used for ppb-level determinations. *Also refer to* **electrothermal atomization (ETA).**

grounding mechanism

A way of eliminating the secondary discharge (pinch effect) produced by capacitive (RF) coupling of the load coil to the plasma. This undesired coupling between the RF voltage on the load coil and the plasma discharge produces a potential difference of a few hundred volts, which creates an electrical discharge (arcing) between the plasma and sampler cone of the interface. This mechanism varies with different instrument designs, but basically involves grounding the load coil to make sure the interface region is maintained at zero potential.

H

half-life

The time required for half the atoms of a given amount of a radioactive substance to disintegrate. This principle is used in particle-counting measuring techniques.

heating zones

The zones that describe the different temperature regions within a plasma discharge, where the sample passes through. The most common zones include the preheating zone (PHZ), where the sample is desolvated; the initial-radiation zone (IRZ), where the sample is broken down into its molecular form; and the normal analytical zone (NAZ), where the sample is first atomized and then ionized.

HEN

An abbreviation for **high-efficiency nebulizer.**

hexapole

A multipole containing six rods, used in collision/reaction-cell technology.

HGAA

An abbreviation for **hydrige-generation atomic absorption.**

high-efficiency nebulizer (HEN)

A generic name given to a nebulizer that is very efficient, with very little wastage. Usually used to describe direct-injection or microconcentric-designed systems, which deliver all or a very high percentage of the sample aerosol into the plasma discharge.

high-resolution mass analyzer

A generic name given to a mass spectrometer with very high resolving power. Commercial designs are usually based on the double-focusing magnetic-sector design.

high-sensitivity interface (HSI)

High-sensitivity interfaces (HSI) are offered as an option with most commercial ICP-MS systems. They all work slightly differently but share similar components. By using a slightly different cone geometry, higher vacuum at the interface, one or more extraction lenses or modified ion-optic design, they offer up to ten times the sensitivity of a traditional interface. However, their limitations are that background levels are often elevated, particularly when analyzing samples with a heavy matrix. Therefore, they are more suited for the analysis of clean solutions.

high-solids nebulizers

Nebulizers that are used to aspirate higher concentrations of dissolved solids into the ICP-MS. The most common types used are the Babbington, V-groove and cone-spray designs. Not widely used for ICP-MS because of the dissolved-solids limitations of the technique, but are sometimes used with flow-injection sample-introduction techniques.

hollow ion mirror

A more recent development in ion-focusing optics. The ion mirror, which has a hollow center, creates a parabolic electrostatic field to reflect and refocus the ion beam at right angles to the ion source. This allows photons, neutrals and solid particles to pass through it, while allowing ions to be reflected at right angles into the mass analyzer. The major benefit of this design is the highly efficient way the ions are refocused, offering extremely high sensitivity and low background across the mass range.

homogenized sample beam
The laser beam in an excimer laser, which produces a much flatter beam profile and more precise control of the ablation process.

hydride-generation atomic absorption (HGAA)
A very sensitive analytical technique for determining trace levels of volatile elements such as As, Bi, Sb, Se and Te. Generation of the elemental hydride is carried out in a closed vessel by the addition of a reducing agent, such as sodium borohydride, to the acidic sample. The resulting gaseous hydride is swept into a special heated quartz cell (in place of the traditional flame burner head), where atomization occurs. Atomic absorption quantitation is then carried out in the conventional way, by comparing the absorbance of unknown samples against known calibration or reference standards.

hydrogen-atom transfer
An ion–molecule reaction mechanism in a collision/reaction cell where a hydrogen atom is transferred to the interfering ion, which is converted to an ion at one mass higher.

hyperbolic fields
The four rods that make up a quadrupole are usually cylindrical or elliptical in shape. The electrical fields produced by these rods are typically hyperbolic in shape.

hyper skimmer cone
The name for an additional cone used in one commercial ICP-MS interface design, used in order to tighten the ion beam entering the ion optics.

I

ICP
An abbreviation for **inductively coupled plasma**.

ICP-OES
An abbreviation for **inductively coupled plasma optical emission spectrometry**.

IDL
Refer to instrument detection limit.

impact bead (nebulizer)
A type of spray chamber more commonly used in atomic absorption spectrometers. The aerosol from the nebulizer is directed onto a spherical bead, where the impact breaks the sample into large and small droplets. The large droplets fall out due to gravitational force and the smaller droplets are directed by the nebulizer gas flow into the atomization/excitation/ionization source.

inductively coupled plasma (ICP)
The high-temperature source used to generate ions in ICP-MS. It is formed when a tangential (spiral) flow of argon gas is directed between the outer and middle tube of a quartz torch. A load coil (usually copper) surrounds the top end of the torch and is connected to an RF generator. When RF

power (typically, 750–1,500 W) is applied to the load coil, an alternating current oscillates within the coil at a rate corresponding to the frequency of the generator. The RF oscillation of the current in the coil creates an intense electromagnetic field in the area at the top of the torch. With argon gas flowing through the torch, a high-voltage spark is applied to the gas, causing some electrons to be stripped from their argon atoms. These electrons, which are caught up and accelerated in the magnetic field, then collide with other argon atoms, stripping off still more electrons. This collision-induced ionization of the argon continues in a chain reaction, breaking down the gas into argon atoms, argon ions and electrons, forming what is known as an "inductively coupled plasma (ICP) discharge" at the open end of the plasma torch.

inductively coupled plasma optical emission spectrometry (ICP-OES)

A multielement technique that uses an inductively coupled plasma to excite ground-state atoms to the point where they emit wavelength-specific photons of light, characteristic of a particular element. The number of photons produced at an element-specific wavelength is measured using high-resolving optical components to separate the analyte wavelengths and a photon-sensitive detection system to measure the intensity of the emission signal produced. This emission signal is directly related to the concentration of that element in the sample. Commercial instrumentation comes in two configurations: a traditional radial view, where the plasma is vertical and is viewed from the side (side-on viewing), and an axial view, where the plasma is positioned horizontally and is viewed from the end (end-on viewing).

infrared (IR) lasers

Laser ablation systems that operate in the IR region of the electromagnetic spectrum, such as the Nd:YAG laser, which has its primary wavelength at 1,064 nm.

instrument background noise

Square root of the spectral background of the instrument (in cps), usually measured at mass 220 amu, where there are no spectral features. *Also refer to* **background signal, background noise** *and* **detection limit**.

instrument background signal

Spectral background of the instrument (in cps), usually measured at mass 220 amu, where there are no spectral features. *Also refer to* **background signal, background noise** *and* **detection limit**.

instrument detection limit (IDL)

It's a way of assessing an instrument's detection capability. It is often referred to as signal-to-background noise, and for a 99% confidence level is typically defined as 3× standard deviation (SD) of 10 replicates of the sample blank.

integration time

The total time spent measuring an analyte mass (peak). Comprising the time spent dwelling (sitting) on the peak multiplied by the number of points used for peak quantitation multiplied by the number of scans used in the measurement protocol. *Also refer to* **duty cycle, measurement duty cycle, peak measurement protocol, settling time** *and* **dwell time**.

interface

The plasma discharge is coupled to the mass spectrometer via the interface. The interface region comprises a water-cooled metal housing containing the sampler cone and the skimmer cone, which directs the ion beam from the central channel of the plasma into the ion-optic region.

interface cones

Refer to the sampler and skimmer cones housed in the interface region. *Also refer to* **interface** *and* **interface region**.

interface pressure

The pressure between the sampler cone and skimmer cone. This region is maintained at a pressure of approximately 1–2 torr by a mechanical roughing pump.

interface region

A region comprising a water-cooled metal housing, containing the sampler cone and the skimmer cone, which directs and focuses the ion beam from the central channel of the plasma into the ion-optic region.

interferences

A generic term given to a non-analyte component that enhances or suppresses the signal intensity of the analyte mass. The most common interferences in ICP-MS are spectral, matrix or sample transport in nature.

internal standardization (IS)

A quantitation technique used to correct for changes in analyte sensitivity caused by variations in the concentration and type of matrix components found in the sample. An internal standard is a non-analyte isotope that is added to the blank solution, standards and samples before analysis. It is typical to add three or four internal standard elements to the samples to cover all the analyte elements of interest across the mass range. The software adjusts the analyte concentration in the unknown samples by comparing the intensity values of the internal standard elements in the unknown sample to those in the calibration standards. Because ICP-MS is prone to many matrix- and sample-transport-based interferences, internal standardization is considered necessary to analyze most sample types.

ion

An electrically charged atom or group of atoms formed by the loss or gain of one or more electrons. A cation (positively charged ion) is created by the loss of an electron, and an anion (negatively charged ion) is created by the gain of an electron. The valency of an ion is equal to the number of electrons lost or gained and is indicated by a plus sign for cations and a minus sign for anions. ICP-MS typically involves the detection and measurement of positively charged ions generated in a plasma discharge.

ion chromatography (IC)

A chromatographic separation technique used for determination of anionic species such as nitrates, chlorides and sulfates. When coupled with ICP-MS, it becomes a very sensitive hyphenated technique for the determination of a wide variety of elemental ionic species.

ion energy

In ICP-MS, it refers to the kinetic energy of the ion, in electron volts (eV). It is a function of both the mass and velocity of the ion ($KE = \frac{1}{2}MV^2$). It is generally accepted that the spread of kinetic energies of all the ions in the ion beam entering the mass spectrometer must be on the order of a few electron volts to be efficiently focused by the ion optics and resolved by the mass analyzer.

ion energy spread

The variation in kinetic energy of all the ions in the ion beam emerging from the ionization source (plasma discharge). It is generally accepted that this variation (spread) of kinetic energies must be on the order of a few electron volts to be efficiently focused by the ion optics and resolved by the mass analyzer.

ion flow

The flow of ions from the interface region through the ion optics into the mass analyzer.

ion-focusing guide

An alternative name for the **ion optics**.

ion-focusing system

An alternative name for the **ion optics**.

ion formation

The transfer of energy from the plasma discharge to the sample aerosol to form an ion. By traveling through the different heating zones in the plasma, where the sample is first dried, vaporized and atomized, then finally converted to an ion.

ion kinetic energy

Refer to **ion energy**.

ion lens

Often referred to as a single-lens component in the ion-optic system. *Also refer to* **ion optics**.

ion-lens voltages

The voltages put on one or more lens components in the ion-optic system to electrostatically steer the ion beam into the mass analyzer. *Also refer to* **ion optics**.

ion mirror

A more recent development in ion-focusing optics. With this design, a parabolic electrostatic field is created with a hollow ion mirror to reflect and refocus the ion beam at right angles to the ion source. The ion mirror is an electrostatically charged ring, which is hollow in the center. This allows photons, neutrals and solid particles to pass through it, while allowing ions to be reflected at right angles into the mass analyzer. The major benefit of this design is the highly efficient way the ions are refocused, offering extremely high sensitivity across the mass range with very little compromise in oxide performance. In addition, there is very little contamination of the ion optics because a vacuum pump sits behind the ion mirror to immediately remove these particles before they have a chance to penetrate further into the mass spectrometer.

ion optics

Comprises one or more electrostatically charged lens components that are positioned immediately after the skimmer cone. They are made up of a series of metallic plates, barrels or cylinders, which have a voltage placed on them. The function of the ion-optic system is to take ions after they emerge from the interface region and steer them into the mass analyzer. Another function of the ion optics is to reject the nonionic species such as particulates, neutral species and photons, and prevent them from reaching the detector. Depending on the design, this is achieved by using some kind of physical barrier, positioning the mass analyzer off axis relative to the ion beam, or electrostatically bending the ions by 90° into the mass analyzer.

ion packet

A "slice of ions" that is sampled from the ion beam in a time-of-flight (TOF) mass analyzer. In the TOF design, all ions are sampled through the interface cones, but instead of being focused into the mass filter in the conventional way, packets (groups) of ions are electrostatically injected into the flight tube at exactly the same time. Whether the orthogonal (right-angle) or axial (straight-on) approach is used, an accelerating potential is applied to the continuous ion beam. The ion beam is then "chopped" by using a pulsed voltage supply to provide repetitive voltage "slices" at a frequency of a few kilohertz. The "sliced" packets of ions are then allowed to "drift" into the flight tube, where the individual ions are temporally resolved according to their differing velocities.

ion repulsion

The degree to which positively charged ions repel each other as they enter the ion optics. The generation of a positively charged ion beam is the first stage in the charge-separation process. Unfortunately, the net positive charge of the ion beam means that there is now a natural tendency for the ions to repel one another. If nothing is done to compensate for this repulsion, ions of higher mass-to-charge ratio will dominate the center of the ion beam and force the lighter ions to the outside. The degree of loss will depend on the kinetic energy of the ions—ions with high kinetic energy (high-mass elements) will be transmitted in preference to ions with medium (midmass elements) or low kinetic energy (low-mass elements).

ionization source

In ICP-MS, the ionization source is the plasma discharge, which reaches temperatures of up to 10,000 K to ionize the liquid sample.

ionizing radiation counting techniques

Particle-counting techniques such as alpha, gamma and scintillation counters that are used to measure the isotopic composition of radioactive materials. However, the limitation of particle-counting techniques is that the half-life of the analyte isotope has a significant impact on the method detection limit. This implies that they are better suited for the determination of short-lived radioisotopes, because meaningful data can be obtained in a realistic amount of time. They have also been successfully applied to the quantitation of long-lived radionuclides, but unfortunately require a combination of extremely long counting times and large amounts of sample to achieve low levels of quantitation.

ion–molecule chemistry

A chemical reaction between the analyte or interfering ion and molecules of the reaction gas in a collision/reaction cell. A reactive gas, such as hydrogen, ammonia, oxygen, methane or gas mixtures, is bled into the cell, which is a catalyst for ion–molecule chemistry to take place. By a number of different reaction mechanisms, the gaseous molecules react with the interfering ions to convert them into either an innocuous species different from the analyte mass or a harmless neutral species. The analyte mass then emerges from the cell free of its interference and is steered into the analyzer quadrupole for conventional mass separation. In some cases, the chemistry can take place between the gaseous molecule and the analyte to form a new analyte ion free of the interfering species.

isobar (or isobaric)

Used in the context of atomic principles, it refers to two or more atoms with the same atomic mass (same number of neutrons) but different atomic number (different number of protons). *Also refer to* **isobaric interferences.**

isobaric interferences

The word "isobaric" is used in the context of atomic principles and refers to two or more atoms with the same atomic mass but different atomic number. In ICP-MS, they are a classification of spectrally induced interferences produced mainly by different isotopes of other elements in the sample, creating spectral interferences at the same mass as the analyte.

isotope

A different form of an element having the same number of protons in the nucleus (i.e., same atomic number) but a different number of neutrons (i.e., different atomic mass). There are 275 isotopes of the 81 stable elements in the periodic table, in addition to over 800 radioactive isotopes. Isotopes of a single element possess very similar properties.

isotope dilution

An absolute means of quantitation in ICP-MS based on altering the natural abundance of two isotopes of an element by adding a known amount of one of the isotopes. The principle works by spiking the sample solution with a known weight of an enriched stable isotope. By knowing the natural abundance of the two isotopes being measured, the abundance of the spiked enriched isotope, the weight of the spike and the weight of the sample, it is possible to determine the original trace element concentration. It is considered one of the most accurate and precise quantitation techniques for elemental analysis by ICP-MS.

isotope ratio

The ability of ICP-MS to determine individual isotopes makes it suitable for an isotopic measurement technique called "isotope ratio analysis". The ratio of two or more isotopes in a sample can be used to generate very useful information, such as an indication of the age of a geological formation, a better understanding of animal metabolism and the identification of sources of environmental contamination. Similar to isotope dilution, isotope-ratio analysis uses the principle of measuring the exact ratio of two isotopes of an element in the sample. With this approach, the isotope of interest is typically compared to a reference isotope of the same element but can also be referenced to an isotope of another element.

isotope-ratio precision

The reproducibility or precision of measurement of isotope ratios is very critical for some applications. For the highest-quality isotope-ratio precision measurements, it is generally acknowledged that either magnetic-sector or time-of-flight (TOF) instrumentation offers the best approach over quadrupole ICP-MS.

isotopic abundance

The percentage abundance of an isotope compared to the element's total abundance in nature. *Also refer to* **natural abundance** *and* **relative abundance of natural isotopes**.

K

KE

An abbreviation for **kinetic energy**.

KED

An abbreviation for **kinetic energy discrimination**.

kinetic energy (KE)

The energy possessed by a moving body due to its motion. It is equal to one-half the mass of the body times the square of its speed (velocity): $KE = \frac{1}{2}MV^2$. For kinetic energy as applied to moving ions, *refer to* **ion energy**.

kinetic energy discrimination (KED)

In collision/reaction-cell technology, it is one way to separate the newly formed by-product ions from the analyte ions. It is typically achieved by setting the collision-cell potential (voltage) slightly more negative than the mass-filter potential. This means that the collision by-product ions generated in the cell, which have a lower kinetic energy as a result of the collision process, are rejected, whereas the analyte ions, which have a higher kinetic energy, are transmitted to the mass analyzer.

L

laser ablation

A sample-preparation technique that uses a high-powered laser beam to vaporize the surface of a solid sample and sweep it directly into the ICP-MS system for analysis. It is mainly used for samples that are extremely difficult to get into solution or for samples that require the analysis of small spots or inclusions on the surface.

laser absorption

The "coupling" efficiency of the sample with the laser beam in laser ablation work. The more light the sample absorbs, the more efficient the ablation process becomes. It is generally accepted that the shorter-wavelength excimer lasers have better absorption characteristics than the longer-wavelength IR laser systems for UV-transparent/opaque materials such as calcites, fluorites and silicates, and, as a result, generate smaller particle size and higher flow of ablated material.

laser fluence

A term used to describe the power density of a laser beam in laser ablation studies. It is defined as the laser pulse energy per focal spot area, measured in J/cm^2. It is related to laser irradiance, which is the ratio of the fluence to the width of the laser pulse.

laser irradience

A term used to describe the power density of a laser beam in laser ablation studies. Laser irradiance is the ratio of the laser pulse energy per focal spot area (i.e., fluence) to the width of the laser pulse. *Also refer to* **laser fluence**.

laser sampling

Refer to **laser ablation**.

laser vaporization

Refer to **laser ablation**.

laser wavelength

The primary wavelength of the optical components used in the design of a laser ablation system.

linear-plane array detectors

solid-state-detector technology used to measure the mass spectrum in a simultaneous manner (recently commercialized in the Mattauch–Herzog magnetic-sector instrument)

load coil

Another name for the RF coil used to generate a plasma discharge. *Also refer to* **RF generator**.

low-mass cut off

This is a variation on bandpass filtering in a collision/reaction cell, which uses slightly different control of the filtering process. By operating the cell in the RF-only mode, the quadrupole's stability boundaries can be tuned to cut off low masses where the majority of the interferences occur.

low-temperature plasma

An alternative name for cool or cold plasma.

M

magnetic field

A region around a magnet, an electric current or a moving charged particle that is characterized by the existence of a detectable magnetic force at every point in the region and by the existence of magnetic poles. In ICP-MS, it usually refers to the magnetic field around the RF coil of the plasma discharge or the magnetic field produced by a quadrupole or an electromagnet.

magnetic-sector mass analyzer

A design of mass spectrometer used in ICP-MS to generate very high resolving power as a way of reducing spectral interferences. Commercial designs typically utilize a very powerful magnet combined with an electrostatic analyzer (ESA). In this approach, known as the double-focusing design, the ions from the plasma are sampled. In the plasma, the ions are accelerated to a few kilovolts into the ion-optic region before they enter the mass analyzer. The magnetic field, which is dispersive with respect to ion energy and mass, then focuses all the ions with diverging angles of motion from the entrance slit. The ESA, which is only dispersive with respect to ion energy, then focuses all the ions onto the exit slit, where the detector is positioned. If the energy dispersion of the magnet and ESA are equal in magnitude but opposite in direction, they will focus both ion angles (first focusing) and ion energies (second focusing) when combined together. *Also refer to* **electrostatic analyzer**.

mass analyzer

The part of the mass spectrometer where the separation of ions (based on their mass-to-charge ratio) takes place. In ICP-MS, the most common type of mass analyzers are quadrupole, magnetic-sector and time-of-flight (TOF) systems.

mass calibration

The ability of the mass spectrometer to repeatedly scan to the same mass position every time during a multielement analysis. Instrument manufacturers typically quote a mass-calibration stability specification for their design of mass analyzer based on the drift or movement of the peak (in atomic mass units) position over a fixed period of time (usually 8 h).

mass-calibration stability

Refer to **mass calibration**.

mass discrimination

Sometimes called "mass bias". In ICP-MS, it occurs when a higher-concentration isotope is suppressing the signal of the lower-concentration isotope, producing a biased result. The effect is not so obvious if the concentrations of the isotopes in the sample are similar but can be quite significant if the concentrations of the two isotopes are vastly different. If that is the case, it is recommended to run a standard of known isotopic composition to compensate for the effects of the suppression.

mass filter

Another name for a mass analyzer.

mass-filtering discrimination

A way of discriminating between analyte ions and the unwanted by-product interference ions generated in a collision/reaction cell.

mass resolution

A measure of a mass analyzer's ability to separate an analyte peak from a spectral interference. The resolution of a quadrupole is nominally 1 amu and is traditionally defined as the width of a peak at 10% of its height.

mass scanning

The process of electronically scanning the mass-separation device to the peak of interest and taking analytical measurements. Basically, two approaches are used: single-point peak hopping, in which a measurement is typically taken at the peak maximum, and the multipoint-scanning approach, in which a number of measurements are taken across the full width of the peak. *Also refer to* **ramp scanning, integration time, dwell time, settling time, peak measurement protocol** *and* **peak hopping**.

mass separation

The process of separating the analyte ions from the non-analyte, matrix, solvent and interfering ions with the mass analyzer.

mass-separation device

Another name for a mass analyzer.

mass shift mode

A mode used with a triple-quadrupole collision/reaction-cell instrument. In this configuration, Q1 and Q2 are set to different masses. Similar to the "On-mass mode", Q1 is set to the precursor-ion mass (analyte and on-mass polyatomic interfering ions), controlling the ions that enter octapole collision/reaction cells. However, in the mass-shift mode, Q2 is then set to mass of a target reaction-product ion containing the original analyte. Mass-shift mode is typically used when the analyte ion is reactive, while the interfering ions are unreactive with a particular collision/reaction-cell gas.

mass spectrometer

The mass spectrometer section of an ICP-MS system is generally considered to be everything in the vacuum chamber from the interface region to the detector, including the interface cones, ion optics, mass analyzer, detector and vacuum pumps.

matching network (RF)

The matching network of the RF generator compensates for changes in impedance (a material's resistance to the flow of an electric current) produced by the sample's matrix components or differences in solvent volatility. In crystal-controlled generators, this is usually done with mechanically driven servo-type capacitors. With free-running generators, the matching network is based on electronic tuning of small changes in the RF brought about by the sample, solvent or matrix components.

mathematical correction equations

Used to compensate or correct for spectral interference in ICP-MS. Similar to interelement corrections (IECs) used in ICP-OES, they work on the principle of measuring the intensity of the interfering isotope or interfering species at another mass, which is ideally free of any interferences. A correction is then applied, depending on the ratio of the intensity of the interfering species at the analyte mass to its intensity at the alternate mass.

Mathieu stability plot

A graphical representation of the stability of an ion as it passes through the rods of a multipole mass-separation device. It is a function of the ratio of the RF to the DC current placed on each pair of rods. A plot of these ratios of multiple ions traveling through the multipole shows which ions are stable and make it through the rods to the detector and which ions are unstable and get ejected from the multipole. The most well-defined stability boundaries are obtained with a quadrupole and become more diffuse with higher-order multipoles such as hexapoles and octopoles.

matrix interferences

There are basically three types of matrix-induced interferences. The first, and simplest to overcome, is often called a "sample transport or viscosity effect" and is a physical suppression of the analyte signal brought on by the level of dissolved solids or acid concentration in the sample. The second type of matrix suppression is caused when the sample matrix affects the ionization conditions of the plasma discharge, which results in varying amounts of signal suppression depending on the concentration of the matrix components. The third type of matrix interference is often called "space charge matrix suppression". This occurs mainly when low-mass analytes are being determined in the presence of larger concentrations of high-mass matrix components. It has the effect of defocusing the ion beam, and unless any compensation is made, the high-mass matrix element will dominate the ion beam, pushing the lighter elements out of the way, leading to low sensitivity and poor detection limits. The classical way to compensate for matrix interferences is to use internal standardization.

matrix separation

Usually refers to some kind of chromatographic column technology to remove the matrix components from the sample before it is introduced into the ICP-MS system.

Mattauch–Herzog magnetic-sector design

One of the earliest designs of double-focusing magnetic-sector mass spectrometers. In this design, which was named after the German scientists who invented it, two or more ions of different mass-to-charge ratios are deflected in opposite directions in the electrostatic and magnetic fields. The divergent monoenergetic ion beams are then brought together along the same focal plane. Recently commercialized using a simultaneous-based direct-charge array detector.

MDL

Refer to **method detection limits**.

measurement duty cycle

Also known as the duty cycle, it refers to a percentage of actual quantitation time compared to total integration time. It is calculated by dividing the total quantitation time (dwell time × number of sweeps × replicates × elements) by total integration time ([dwell time + settling/scanning time] × number of sweeps × replicates × elements).

membrane desolvation

Can be used with any sample-introduction technique to remove solvent vapors. However, it is typically used with an ultrasonic or microconcentric nebulizer to remove the solvent from a liquid sample. In this design, the sample aerosol enters the membrane desolvator, where the solvent vapor passes through the walls of a tubular micro-porous PTFE or Nafion membrane. A flow of argon gas removes the volatile vapor from the exterior of the membrane, while the analyte aerosol remains inside the tube and is carried into the plasma for ionization.

method detection limit (MDL)

The MDL is broadly defined as the minimum concentration of analyte that can be determined from zero with 99% confidence. MDLs are calculated in a similar manner to IDLs, except that the test solution is taken through the entire sample-preparation procedure before the analyte concentration is measured multiple times.

microconcentric nebulizer

Is based on the concentric nebulizer design but operates at much lower flow rates. Conventional nebulizers have a sample uptake rate of about 1 mL/min with an argon gas pressure of 1 L/min, whereas microconcentric nebulizers typically run at less than 0.1 mL/min and typically operate at much higher gas pressure to accommodate the lower sample flow rates.

microflow nebulizer

A generic name for nebulizers that operate at much lower flow rates than conventional concentric or cross-flow designs. *Also refer to* **microconcentric nebulizer**.

micro-porous membrane

A tubular membrane made of an organic micro-porous material such as Teflon or Nafion, used in membrane desolvation. The sample aerosol enters the desolvation system, where the solvent vapor passes through the walls of the tubular membrane. A flow of argon gas then removes the volatile vapor from the exterior of the membrane, while the analyte aerosol remains inside the tube and is carried into the plasma for ionization.

microsampling

A generic name given to any front-end sampling device in atomic spectrometry that can be used for the preparation, pretreatment and delivery of the sample to the spectrometric analyzer. The most common type of microsampling device used in ICP-MS is the flow-injection technique, which involves the introduction of a discrete sample aliquot into a flowing carrier stream. Using a series of automated pumps and valves, procedures can be carried out online to physically or chemically change the sample or analyte, before introduction into the mass spectrometer for detection.

microwave digestion

A method of digesting difficult-to-dissolve solid samples using microwave technology. Typically, a dissolution reagent such as a concentrated mineral acid is added to the sample in a closed acid-resistant vessel contained in a specially designed microwave oven. By optimizing the current, temperature and pressure settings, difficult samples can be dissolved in a relatively short time compared to traditional hot-plate sample-digestion techniques.

microwave dissolution

An alternative name for **microwave digestion**.

molecular association reaction

An ion–molecule reaction mechanism in a collision/reaction cell, where an interfering ion associates with a neutral species (atom or molecule) to form a molecular ion.

molecular cluster ions

Species that are formed by two or more molecular ions combining together in a reaction cell to form molecular clusters.

molecular spectral interferences

Another name for polyatomic spectral interferences, which are typically generated in the plasma by the combination of two or more atomic ions. They are caused by a variety of factors but are usually associated with the argon plasma/nebulizer gas used, matrix components in the solvent/sample, other elements in the sample or entrained oxygen/nitrogen from the surrounding air.

Monodisperse particulates

Nanoparticles, which are predominantly one size.

M/S mode

A mode used in a "triple quadrupole" collision/reaction cell where the first quadrupole acts as a simple ion guide, allowing all ions through to the collision/reaction cell, similar to a traditional single-quad ICP-MS system that uses a collision cell.

M/S M/S mode

A mode used in a "triple quadrupole" collision/reaction cell, where the first quadrupole is operated with a 1-amu fixed bandpass window, allowing only the target ions to enter the collision/reaction cell. This process can be implemented in two different ways: either the "on-mass" or "mass shift" modes.

multicomponent ion lens

An ion-lens system consisting of several lens components, all of which have a specific role to play in the transmission of the analyte ions into the mass filter. To achieve the desired analyte specificity, the voltage can be optimized on every ion lens and is usually combined with an off-axis mass analyzer to reject unwanted photons and neutral species.

multichannel analyzer

The data-acquisition system that stores and counts the ions as they strike the detector. As the ions emerge from the end of the quadrupole rods, they are converted into electrical pulses by the detector and stored by the multichannel analyzer. This multichannel data-acquisition system typically has 20 channels per mass, and as the electrical pulses are counted in each channel, a profile of the mass is built up over the 20 channels, corresponding to the spectral peaks of the analyte masses being determined.

multichannel data acquisition

The process of storing and counting ions in ICP-MS. *Also refer to* **multichannel analyzer**.

multipole

The generic name given to a mass filter that isolates an ion of interest by applying DC or RF currents to pairs of rods. The most common type of mass-analyzer multipole used in ICP-MS is the quadrupole (four rods). However, other higher orders of multipoles used in collision/reaction-cell technology include hexapoles (six rods) and octopoles (eight rods).

m/z

Another way of expressing mass-to-charge ratio.

N

nanomaterials

Nanomaterials can occur in nature, such as clay minerals and humic acids; they can be incidentally produced by human activity such as diesel emissions, or welding fumes; or they can be specifically engineered to exhibit unique optical, electrical, physical or chemical characteristics. Also refer to engineered nanomaterials (ENMs).

nanometrology

The measurement and characterization of nanoparticles.

nanoparticles (NP)

Particles that are released from engineered nanomaterials when they enter the environment.

natural abundance

The natural amount of an isotope occurring in nature. *Also refer to* **isotopic abundance**.

natural isotopes

Different isotopic forms of an element that occur naturally on or beneath the earth's crust.

Nd:YAG laser

Nd:YAG is an acronym for neodymium-doped yttrium aluminum garnet, a compound that is used as the lasing medium for certain solid-state lasers. In this design, the YAG host is typically doped with around 1% neodymium by weight. Nd:YAG lasers are optically pumped using a flashlamp or laser diodes and emit light with a wavelength of 1,064 nm in the infrared region. However, for many applications, the infrared light is frequency-doubled, -tripled, -quadrupled or -quintupled by using additional optical components to generate output wavelengths in the visible and UV regions. Typical wavelengths used for laser ablation/ICP-MS work include 532 nm (doubled), 266 nm (quadrupled), and 213 nm (quintupled). Pulsed Nd:YAG lasers are usually operated in the so-called "Q-switching" mode, where an optical switch is inserted in the laser cavity, waiting for a maximum population inversion in the neodymium ions before it opens. Then the light wave can run through the cavity, depopulating the excited laser medium at maximum population inversion. In this Q-switched mode, output powers of 20 MW and pulse durations of less than 10 ns are achieved.

nebulizer

The component of the sample-introduction system that takes the liquid sample and pneumatically breaks it down into an aerosol using the pressure created by a flow of argon gas. The concentric and cross-flow designs are the most common in ICP-MS.

neutral species

Species generated in the plasma torch that have no positive or negative charge associated with them. If they are not eliminated, they can find their way into the detector and produce elevated background levels.

neutron

A fundamental particle that is neutral in charge, found in the nucleus of an atom. It has a mass equal to that of a proton. The number of neutrons in the atomic nucleus defines the isotopic composition of that element.

Nier–Johnson magnetic-sector design

Nier–Johnson double-focusing magnetic-sector instrumentation is the technology that all modern magnetic-sector instrumentation is based on. Named after the scientists who developed it, Nier–Johnson geometry comes in two different designs, the "standard" and "reverse" Nier–Johnson geometry. Both these designs, which use the same basic principles, consist of two analyzers: a traditional electromagnet analyzer and an electrostatic analyzer (ESA). In the standard (sometimes called "forward") design, the ESA is positioned before the magnet, and in the reverse design, it is positioned after the magnet.

ninety- (90) degree ion lens

A design of ion optics that bends the ion beam 90°.

NP

Refer to **nanoparticles**.

O

octopole

A multipole mass-filtering device containing eight rods. In ICP-MS, octopoles are typically used in collision/reaction-cell technology.

off-axis ion lens

An ion-lens system that is not on the same axis as the mass analyzer. Designed to stop particulates, neutral species and photons from hitting the detector.

on-mass mode

A mode used with the "triple quadrupole" collision/reaction cell. In this configuration, Q1 and Q2 are both set to the target mass. Q1 allows only the precursor ion mass to enter the cell (analyte and on-mass polyatomic interfering ions). The octopole collision/reaction cell then separates the analyte ion from the interferences using the reaction chemistry of a reactive gas, while Q2 measures the analyte ion at the target mass after the on-mass interferences have been removed by reactions in the cell.

oxide ions

Polyatomic ions that are formed between oxygen and other elemental components in the plasma gas, sample matrix or solvent. They are generally not desirable because they can cause spectral overlaps on the analyte ions. Oxide formation is typically worse in the cooler zones of the plasma, and as a result, can be reduced by optimizing the RF power, nebulizer gas flow and sampling position.

P

parabolic field

Shape of the magnetic fields produced by a quadrupole.

particle-counting techniques

Include alpha, gamma and scintillation counters that are used to measure the isotopic composition of radioactive materials. However, the limitation of particle-counting techniques is that the half-life of the analyte isotope has a significant impact on the method's detection limit. This means that to get meaningful data in a realistic amount of time, they are better suited for the determination of short-lived radioisotopes. They have been successfully applied to the quantitation of long-lived radionuclides, but unfortunately require a combination of extremely long counting times and large amounts of sample to achieve low levels of quantitation.

peak hopping

A quantitation approach in which the quadrupole power supply is driven to a discrete position on the analyte mass (normally the maximum point), allowed to settle (settling time), and a measurement taken for a fixed amount of time (dwell time). The integration time for that peak is the dwell time multiplied by the number of scans (scan time). Multielement peak quantitation involves peak hopping to every mass in the multielement run. *Also refer to* **measurement duty cycle** *and* **peak measurement protocol**.

peak integration

The process of integrating an analytical peak (mass). *Also refer to* **integration time, peak measurement protocol** *and* **measurement duty cycle**.

peak measurement protocol

The protocol of scanning the quadrupole and measuring a peak in ICP-MS. In multielement analysis, the quadrupole is scanned to the first mass. The electronics are allowed to settle (settling time), left to dwell for a fixed period of time at one or multiple points on the peak (dwell time), and signal-intensity measurements are taken (based on the dwell time). The quadrupole is then scanned to the next mass and the measurement protocol repeated. The complete multielement measurement cycle (sweep) is repeated as many times as is needed to make up the total integration per peak and the number of required replicate measurements per sample analysis.

peak quantitation

The process of quantifying the peak in ICP-MS using calibration standards. *Also refer to* **peak hopping, peak integration** *and* **peak measurement protocol**.

Peltier cooler

A thermoelectric cooler using the principle of generating a cold environment by creating a temperature gradient between two different materials. It uses electrical energy via a solid-state heat pump to transfer heat from a material on one side of the device to a different material on the other side, thus producing a temperature gradient across the device (similar to a household air-conditioning system).

peristaltic pump

A small pump in the sample-introduction system that contains a set of mini rollers (typically, 12) all rotating at the same speed. The constant motion and pressure of the rollers on the pump tubing feeds the sample through to the nebulizer. Peristaltic pumps are usually used with cross-flow nebulizers.

photon stop

A grounded metal disk in the ion-lens system that is used as a physical barrier to stop particulate matter, neutral species and photons from getting to the detector.

physical interferences

An alternative term used to describe sample transport- or viscosity-based suppression interferences.

pinch effect

An effect caused by an undesired electrostatic (capacitive) coupling between the voltage on the load coil and the plasma discharge, which produces a potential difference of a few hundred volts. This capacitive coupling is commonly referred to as the pinch effect and shows itself as a secondary discharge (arcing) in the region where the plasma is in contact with the sampler cone.

piston pump

A pump using a small piston or syringe to introduce the sample to the nebulizer (used instead of a peristaltic pump). They typically produce a more stable signal.

plasma discharge

Another name for an inductively coupled plasma (ICP).

plasma source

Refers to the RF hardware components that create the plasma discharge, including the RF generator, matching network, plasma torch and argon-gas pneumatics.

plasma torch

Another name for the quartz torch that is used to generate the plasma discharge. The plasma torch consists of three concentric tubes: an outer tube, middle tube and sample injector. The torch can either be one piece, where all three tubes are connected, or it can have a demountable design, in which the tubes and the sample injector are separate. The gas (usually argon) that is used to form the plasma (plasma gas) is passed between the outer and middle tubes at a flow rate of 12–17 L/min. A second gas flow (auxiliary gas) passes between the middle tube and the sample injector at 1 L/min, and is used to change the position of the base of the plasma relative to the tube and the injector. A third gas flow (nebulizer gas), also at 1 L/min brings the sample, in the form of a fine-droplet aerosol, from the sample-introduction system and physically punches a channel through the center of the plasma. The sample injector is often made from other materials besides quartz, such as alumina, platinum and sapphire, if highly corrosive materials need to be analyzed.

polyatomic spectral interferences

Another name for molecular-based spectral interferences, which are typically generated in the plasma by the combination of two or more atomic ions. They are caused by a variety of factors, but are usually associated with the argon plasma/nebulizer gas used, matrix components in the solvent/sample, other elements in the sample or entrained oxygen/nitrogen from the surrounding air.

polydisperse particulates

Nanoparticles that are many different sizes and dimensions.

precursor ion

Usually refers to a polyatomic or isobaric interfering ion that is formed in the plasma as opposed to a product (or by-product) ion that is formed in the collision/reaction cell.

product ion

Usually refers to a product (or by-product) ion that is formed in the collision/reaction cell as opposed to a precursor interfering ion (polyatomic or isobaric) that is formed in the plasma.

proton

A stable, positively charged fundamental particle that shares the atomic nucleus with a neutron. It has a mass 1,836 times that of the electron.

proton transfer

A reaction mechanism in a collision/reaction cell in which the interfering polyatomic species gives up a proton, which is then transferred to the reaction gas molecule to form a neutral atom.

pulse counting

Refers to the conventional mode of counting ions with the detector measurement circuitry. Depending on the type of detection system that is used, an ion emerges from the quadrupole and strikes the ion-sensitive surface (discrete dynode, Channeltron, etc.) of the detector to generate electrons. These electrons move down the detector and generate more secondary electrons. This process is repeated at each stage of the detector, producing a pulse of electrons that is finally captured by the detector's collecting and counting circuitry.

Q

QID

Refer to quadrupole ion deflector.

quadrupole

The most common type of mass-separation device used in commercial ICP-MS systems. It consists of four cylindrical or hyperbolic metallic rods of the same length (15–20 cm) and diameter (approximately 1 cm). The rods are typically made of stainless steel or molybdenum and sometimes coated with a ceramic coating for corrosion resistance. A quadrupole operates by placing both a DC field and a time-dependent AC of RF 2–3 MHz on opposite pairs of the four rods. By selecting the optimum AC/DC ratio on each pair of rods, ions of a selected mass are then allowed to pass through the rods to the detector, while the others are unstable and ejected from the quadrupole.

quadrupole ion deflector (QID)

A ion-optics design that bends the ion beam at right angles.

quadrupole power supply

Another name for the electronic components that control the RF and DC voltages to change the mass-filtering characteristics.

quadrupole scan rate

Scan rates of commercial quadrupole mass analyzers are on the order of 2,500 amu/s. The quadrupole scan rate and the slope at which the RF and DC voltages of the quadrupole power supply are scanned will determine the desired resolution setting. A steeper slope translates to higher resolution, whereas a shallower slope means poorer resolution.

quadrupole stability regions

The region of the Mathieu stability plot where the trajectory of an ion is stable and makes it through to the end of the quadrupole rods. All commercial ICP-MS systems that utilize quadrupole technology as the mass-separation device operate in the first stability region, where resolving power is typically on the order of 500–600. If the quadrupole is operated in the second or third stability regions, resolving powers of 4,000 and 9,000, respectively, have been achieved. However, improving resolution using this approach has resulted in a significant loss of signal and higher background levels.

quantitative methods

The different kinds of quantitative analyses available in ICP-MS, which include traditional quantitative analysis (using external calibration, standard additions or addition calibration), semiquantitative routines (semiquant), isotope-dilution (ID) methods, isotope-ratio (IR) measurements and classical internal standardization (IS).

quartz torch

The standard plasma torch used in ICP-MS. *Also refer to* **plasma torch**.

R

radio frequency (RF) generator

The power supply used to create the plasma discharge. Hardware includes the RF generator, matching network, plasma torch and argon-gas pneumatics.

radioactive isotope

Sometimes known as "radioisotope", a radioactive isotope is a natural or artificially created isotope of an element having an unstable nucleus that decays, emitting alpha, beta or gamma rays until stability is reached. The stable end product is typically a nonradioactive isotope of another element.

ramp scanning

One of the two approaches for quantifying a peak in ICP-MS (peak hopping being the other). In the multichannel-ramp-scanning approach, a continuous smooth ramp of 1-n channels (where n is typically 20) per mass is made across the peak profile. Mainly used for accumulating spectral and peak shape information when doing mass scans. It is normally used for doing mass calibration and resolution checks and as a classical qualitative method-development tool to find out what elements are present in the sample and to assess their spectral implications on the masses of interest. Full-peak ramp scanning is not normally used for doing rapid quantitative analysis, because valuable analytical time is wasted taking data on the wings and valleys of the peak where the signal-to-noise ratio is poorest. For this kind of work, peak hopping is normally chosen.

reaction cell

A collision/reaction cell that specifically uses ion–molecule reactions to eliminate the spectral interference. Often used to describe a dynamic reaction cell (DRC).

reaction mechanism

The mechanism by which the interfering ion is reduced or minimized to allow the determination of the analyte ion. The most common collisional mechanisms seen in collision/reaction cells include collisional focusing, dissociation and fragmentation, whereas the major reaction mechanisms include exothermic/endothermic associations, charge transfer, molecular associations and proton transfer.

reactive gases

In ICP-MS, the term refers to gases such as hydrogen, ammonia, oxygen, methane or those used to stimulate ion–molecule reactions in a collision/reaction cell (CRC) or collision/reaction interface (CRI).

relative abundance of natural isotopes

The isotopic composition expressed as a percentage of the total abundance of that element found in nature.

resolution

A measure of the ability of a mass analyzer to separate an analyte peak from a spectral interference. The resolution of a quadrupole is nominally 1 amu and is traditionally defined as the width of a peak at 10% of its height.

resolving power

Although resolving power and resolution are both a measure of a mass analyzer's ability to separate an analyte peak from a spectral interference, the term "resolving power" is normally associated with magnetic-sector technology and is represented by the equation $R = m/\Delta m$, where m is the nominal mass at which the peak occurs and Δm is the mass difference between two resolved peaks. The resolving power of commercial double-focusing magnetic-sector mass analyzers is on the order of 1,000–10,000, depending on the resolution setting chosen.

response tables

The intensity values for known concentrations of every elemental isotope stored in the instrument's calibration software. When semiquanitative analysis is carried out, the signal intensity of an unknown sample is compared against the stored response tables. By correcting for common spectral interferences and applying heuristic, knowledge-driven routines in combination with numerical calculations, a positive or negative confirmation can be made for each element present in the sample.

reverse Nier–Johnson double-focusing magnetic-sector instrumentation

The technology that all modern magnetic-sector instrumentation is based on. Named after the scientists who developed it, Nier–Johnson geometry comes in two different designs, the "standard" and "reverse" Nier–Johnson geometry. Both these designs, which use the same basic principles, consist of two analyzers: a traditional electromagnet analyzer and an electrostatic analyzer (ESA). In the standard (sometimes called "forward") design, the ESA is positioned before the magnet, and in the reverse design, it is positioned after the magnet.

RF generator

An alternative name for radio frequency generator.

right-angled ion-lens design

A recent development in ion-focusing optics, which utilizes a parabolic ion mirror to bend and refocus the ion beam at right angles to the ion source. The ion mirror incorporates a hollow structure that allows photons, neutrals and solid particles to pass through it, while allowing ions to be deflected at right angles into the mass analyzer.

roughing pump

Traditional mechanical roughing or oil-based pumps are used in ICP-MS to pump the interface region down to approximately 1–2 torr and also to back up the turbomolecular pump used in the ion optics region of the mass spectrometer.

ruby laser

Ruby laser systems operate at 694 nm in the visible region of the electromagnetic spectrum.

S

S/B

An abbreviation for **signal-to-background ratio**.

sample aerosol

Refer to **aerosol**.

sample digestion

The process of digesting a sample by traditional hot plate, fusion techniques or microwave technology to get the matrix and analytes into solution.

sample dissolution

The process of dissolving a sample by traditional hot plate, fusion techniques or microwave technology to get the matrix and analytes into solution.

sample injector

The central tube of the plasma torch that carries the sample aerosol mixed with the nebulizer gas. It can be a fixed part of the quartz torch, or it can be separate (demountable) and be made from other materials, such as alumina, platinum and sapphire, for the analysis of highly corrosive materials.

sample-introduction system

The part of the instrument that takes the liquid sample and puts it into the plasma torch as a fine-droplet aerosol. It comprises a nebulizer to generate the aerosol and a spray chamber to reject the larger droplets and allow only the smaller droplets into the plasma discharge.

sample preparation

The entire process of preparing the sample for aspiration into the ICP mass spectrometer.

sample throughput

The rate at which samples can be analyzed.

sample transport interferences

A term used to describe a physical suppression of the analyte signal caused by matrix components in the sample. It is more exaggerated with samples having high levels of dissolved solids, because they are transported less efficiently through the sample-introduction system than aqueous-type samples. *Also refer to* **physical interferences**.

sampler cone

A part of the mass spectrometer interface region, where the ion beam from the plasma discharge first enters. The sampler cone, which is the first cone of the interface, is typically made of nickel or platinum and contains a small orifice of approximately 0.8–1.2 mm diameter, depending on the design. The sampler cone is much more pointed than the skimmer cone.

sampling accessories

Customized sample-introduction techniques optimized for a particular application problem or sample type. The most common types used today include the following: laser ablation/sampling (LA/S), flow-injection analysis (FIA), electrothermal vaporization (ETV), desolvation systems, direct-injection nebulizers (DIN) and chromatography separation techniques.

scan time

The mass analyzer scan time is the time it takes to scan from one isotope to the next.

Scott spray chamber

A sealed spray chamber with an inner tube inside a larger tube. The sample aerosol from the nebulizer is first directed into the inner tube. The aerosol then travels the length of the inner tube, where the larger droplets fall out by gravity into a drain tube and the smaller droplets return between the inner and outer tube, where they eventually exit into the sample injector of the plasma torch.

secondary discharge

Another term used for the **pinch effect**.

secondary (side) reactions

Reactions that occur in a collision/reaction cell that are not a part of the main interference-reduction mechanism. If not anticipated and compensated for, secondary reactions can lead to erroneous results.

semiquant

An abbreviated name used to describe **semiquantitative analysis**.

semiquantitative analysis

A method for assessing the approximate concentration of up to 70 elements in an unknown sample. It is based on comparing the intensity of a small group of elements against known response tables stored in the instrument's calibration software. By correcting for common spectral interferences and applying heuristic, knowledge-driven routines in combination with numerical calculations, a positive or negative confirmation can be made for each element present in the sample.

settling time

The time taken for the mass-analyzer electronics to settle before a peak-intensity measurement is taken for the operator-selected dwell time. The dwell time can usually be selected on an individual mass basis, but the settling time is normally fixed because it is a function of the mass analyzer and detector electronics.

shadow stop

A grounded metal disk that stops particulate matter, neutral species and photons from getting to the detector. It is considered a part of the ion optics and is sometimes called a photon stop.

side reactions

Reactions that occur in a collision/reaction cell that are not a part of the main interference-reduction mechanism. If not anticipated and compensated for, secondary reactions can lead to erroneous results.

signal-to-background ratio (S/B)

The ratio of the signal intensity of an analyte to its background level at a particular mass. When considering the noise of the background signal (standard deviation of the signal), it is typically used as an assessment of the detection limit for that element. *Also refer to* **detection limit**, **background signal** *and* **background noise**.

single-particle ICP-MS

A technique used to characterize nanoparticles. The method involves introducing nanoparticle (NP) -containing samples, at very dilute concentration, into the ICP-MS and collecting time-resolved data.

single-point peak hopping

A quantitation in which the quadrupole power supply is driven to a discrete position on the analyte mass (normally the maximum point), allowed to settle (settling time), and a measurement is taken for a fixed amount of time (dwell time). The integration time for that peak is the dwell time multiplied by the number of scans (scan time). Multielement peak quantitation involves peak hopping to every mass in the multielement run. *Also refer to* **measurement duty cycle** *and* **peak measurement protocol**.

skimmer cone

A part of the mass spectrometer interface region where the ion beam from the plasma discharge first enters. The skimmer cone, which is the second cone of the interface, is typically made of nickel or platinum and contains a small orifice of approximately 0.5–0.8-mm diameter, depending on the design. The skimmer cone is much less pointed than the sampler cone.

solvent-based interferences

Spectral interferences derived from an elemental ion in the solvent (e.g., water, acid, etc.) combining with another ion from either the sample matrix or plasma gas (argon) to produce a polyatomic ion that interferes with the analyte mass.

SOP

Standard operating procedure.

space charge effect

A type of matrix-induced interference that produces a suppression of the analyte signal. This occurs mainly when low-mass analytes are being determined in the presence of larger concentrations of high-mass matrix components. It has the effect of defocusing the ion beam, and without compensation, the high-mass matrix element will dominate the ion beam, pushing the lighter elements out of the way, leading to low sensitivity and poor detection limits. The classical way to compensate for a space charge matrix interference is to use an internal standard of similar mass to the analyte.

speciation analysis

In ICP-MS, it is the study and quantification of different species or forms of an element using a chromatographic separation device coupled to an ICP mass spectrometer. In this configuration, the instrument becomes a very sensitive detector for trace element speciation studies when coupled with high-performance liquid chromatography (HPLC), ion chromatography (IC), gas chromatography (GC) or capillary electrophoresis (CE). In these hybrid techniques, element species are separated on the basis of their chromatograph retention/mobility times and then eluted/passed into the ICP mass spectrometer for detection. The intensity of the eluted peaks is then displayed for each isotopic mass of interest in the time domain.

spectral interferences

A generic name given to interferences that produce a spectral overlap at or near the analyte mass of interest. In ICP-MS, there are two main types of spectral interference that have to be taken into account. Polyatomic spectral interferences (or molecular-based spectral interferences) are typically generated in the plasma by the combination of two or more atomic ions. They are caused by a variety of factors but are usually associated with the argon plasma/nebulizer gas used, matrix components in the solvent/sample, other elements in the sample or entrained oxygen/nitrogen from the surrounding air. The other type is an isobaric spectral interference, which is caused by different isotopes of other elements in the sample creating spectral interferences at the same mass as the analyte.

SP-ICP-MS

Refer to single-particle ICP-MS.

spray chamber

The component of the sample-introduction system that takes the aerosol generated by the nebulizer and rejects the larger droplets for the more desirable smaller droplets.

SRM

An abbreviation for **standard reference materials**.

stability

The ability of a measuring device to consistently replicate a measurement. In ICP-MS, it usually refers to the capability of the instrument to reproduce the signal intensity of the calibration standards over a fixed period of time without the use of internal standardization. Short-term stability is generally defined as the precision (as % RSD [relative standard deviation]) of ten replicates of a single or multielement solution, whereas long-term stability is defined as the precision (as % RSD) of a fixed number of measurements over a 4–8-h time period of a single or multielement solution. However, stability in mass spectrometry can *also refer to* mass-calibration stability, which is the ability of the mass spectrometer to repeatedly scan to the same mass position every time during a multielement analysis.

stability boundaries/regions

The RF/DC boundaries of the Mathieu stability plot where an ion is stable as it passes through a quadrupole mass-filtering device. *Also refer to* **Mathieu stability plot**.

standard additions

A method of calibration that provides an effective way to minimize sample-specific matrix effects by spiking samples with known concentrations of analytes. In standard-addition calibration, the intensity of a blank solution is first measured. Next, the sample solution is "spiked" with known concentrations of each element to be determined. The instrument measures the response for the spiked samples and creates a calibration curve for each element for which a spike has been added. The calibration curve is a plot of the blank-subtracted intensity of each spiked element against its concentration value. After creating the calibration curve, the unspiked sample solutions are then analyzed and compared to the calibration curve. Depending on the slope of the calibration curve and where it intercepts the x-axis, the instrument software determines the unspiked concentration of the analytes in the unknown samples.

standard reference materials (SRM)

Well-established reference matrices that come with certified values and associated statistical data that have been analyzed by other complementary techniques. Their purpose is to check the validity of an analytical method, including sample preparation, instrument methodology and calibration routines, to achieve sample results that are as accurate and precise as possible and can be defended under intense scrutiny.

standardization methods

Refers to the different types of calibration routines available in ICP-MS, including quantitative analysis (external calibration and standard additions), semiquantitative analysis, isotope dilution, isotope ratio and internal-standardization methods.

syringe pump

Refer to **piston pump**.

T

thermoelectric cooling device

Better known as a Peltier cooler, it generates a cold environment by creating a temperature gradient between two different materials. It uses electrical energy via a solid-state heat pump to transfer heat from a material on one side of the device to a different material on the other side, thus producing a temperature gradient across the device (similar to a household air conditioner).

thermoelectric flow meter

A device to measure the liquid flow through a nebulizer to check for any blockages or breakages

time-of-flight mass spectrometry (TOF-MS)

A mass spectrometry technique based on the principle that the kinetic energy (KE) of an ion is directly proportional to its mass (m) and velocity (V), which can be represented by the equation $KE = \frac{1}{2}MV^2$. Therefore, if a population of ions with different masses is given the same KE by an accelerating voltage (U), the velocities of the ions will all be different, depending on their masses. This principle is then used to separate ions of different mass-to-charge (m/z) in the time (t) domain, over a fixed flight path distance (D), represented by the equation $m/z = 2Ut^2/D^2$. The simultaneous nature of sampling ions in TOF offers distinct advantages over traditional scanning (sequential) quadrupole technology for ICP-MS applications, where large amounts of data need to be captured in a short amount of time, such as the multielement analysis of transient peaks (laser ablation, flow injection, etc.).

time-of-flight (TOF) mass spectrometry (axial design)

There are basically two different sampling approaches that are used in commercial TOF mass analyzers: the axial and orthogonal designs. In the axial design, the flight tube is in the same axis as the ion beam, whereas in the orthogonal design, the flight tube is positioned at right angles to the sampled ion beam. The axial approach applies an accelerating potential in the same axis as the incoming ion beam as it enters the extraction region. Because the ions are in the same plane as the detector, the beam has to be modulated using an electrode grid to repel the "gated" packet of ions into the flight tube. This kind of modulation generates an ion packet that is long and thin in cross section (in the horizontal plane), which is then resolved in the time domain according to the different ionic masses. *Also refer to* **time-of-flight mass spectrometry**.

time-of-flight (TOF) mass spectrometry (orthogonal design)

There are basically two different sampling approaches that are used in commercial TOF mass analyzers: the axial and orthogonal designs. In the axial design, the flight tube is in the same axis as the ion beam, whereas in the orthogonal design, the flight tube is positioned at right angles to the sampled ion beam. With the orthogonal approach, an accelerating potential is applied at right angles to the continuous ion beam from the plasma source. The ion beam is then "chopped" by using a pulsed voltage supply coupled to the orthogonal accelerator to provide repetitive voltage "slices" at a frequency of a few kilohertz. The "sliced" packets of ions, which are typically tall and thin in cross section (in the vertical plane), are then allowed to "drift" into the flight tube, where the ions are temporally resolved according to their differing velocities. *Also refer to* **time-of-flight mass spectrometry.**

TOF-MS

An abbreviation for **time-of-flight mass spectrometry.**

torch design

Refers to the different kinds of commercially available torch designs.

trace-metal speciation studies

Refer to
speciation analysis.

transient signal (peak)

A signal that lasts for a finite amount of time, compared to a continuous signal that lasts for as long as the sample is being aspirated. Transient peaks are typically generated by alternative sampling devices such as laser ablation, flow injection or chromatographic separation systems where discrete amounts of sample are introduced into the ICP mass spectrometer.

triple-cone interface

A commercial design of an ICP-MS interface that uses three cones. Also refer to hyper skimmer cone.

"triple quadrupole" collision/reaction cell

A commercial collision/reaction-cell design that has an additional quadrupole prior to the collision/reaction-cell multipole and the analyzer quadrupole. This first quadrupole acts as a simple mass filter to allow only the analyte masses to enter the cell, while rejecting all other masses. With all non-analyte, plasma and sample matrix ions excluded from the cell, sensitivity and interference removal efficiency is significantly improved compared to traditional collision/reaction-cell technology coupled with a single-quadrupole mass analyzer.

turbomolecular pump (turbo pump)

A type of vacuum pump used to maintain a high vacuum in the ion-optics and mass-analyzer regions of the ICP mass spectrometer. These pumps work on the principle that gas molecules can be given momentum in a desired direction by repeated collision with a moving solid surface. In a turbo pump, a rapidly spinning turbine rotor strikes gas (argon) molecules from the inlet of the pump towards the exhaust, creating and maintaining a vacuum. In the case of ICP-MS, two pumps are normally used, a large pump for the ion-optic region, which creates a vacuum of approximately 10^{-3} torr, and another small pump for the mass-analyzer region, which generates a vacuum of 10^{-6} torr. However, some designs use a twin-throated turbo pump, in which one powerful pump is used with two outlets, one for the ion optics and one for the mass-analyzer region.

twin-throated turbomolecular pump

In some designs of ICP mass spectrometers, a single twin-throated turbo pump is used instead of two separate pumps. In this design, one powerful pump is used with two outlets, one for the ion optics and one for the mass-analyzer region.

U

ultrasonic nebulizer (USN)

A type of desolvating nebulizer that generates an extremely fine-droplet aerosol for introduction into the ICP mass spectrometer. The principle of aerosol generation using this approach is based on a sample being pumped onto a quartz plate of a piezoelectric transducer. Electrical energy of 1–2 MHz is coupled to the transducer, which causes it to vibrate at high frequency. These vibrations disperse the sample into a fine-droplet aerosol, which is carried in a stream of argon. With a conventional ultrasonic nebulizer, the aerosol is passed through a heating tube and a cooling chamber, where most of the sample solvent is removed as a condensate before it enters the plasma. However, commercial ultrasonic nebulizers are also available with membrane-desolvation systems.

universal cell

The name applied to a commercial collision/reaction cell that offers the capability of either a collision cell using inert gases and KED or a dynamic reaction cell using highly reactive gases.

USN

An abbreviation for **ultrasonic nebulizer**.

UV laser

A generic name given to a laser ablation system that works in the ultraviolet region of the electromagnetic spectrum. The three most common wavelengths used in commercial equipment are all UV lasers. They include the 266-nm (frequency-quadrupled) Nd:YAG laser, the 213-nm (frequency-quintupled) Nd:YAG laser, and the 193-nm ArF excimer laser system. *Also refer to* **excimer laser** *and* **Nd:YAG laser**.

V

vacuum chamber

The region of the mass spectrometer that is under negative pressure created by a combination of roughing and turbomolecular pumps. As the ion beam moves from the plasma, which is at atmospheric pressure (760 torr), it enters the interface region between the sampler and skimmer cone (1–2 torr) before it is focused through the ion-optic vacuum chamber region (10^{-3} torr) and eventually goes through the mass-analyzer vacuum chamber (at 10^{-6} torr). *Also refer to* **turbomolecular pump**.

vacuum gauge

Used to measure the pressure in the different vacuum chambers of the mass spectrometer.

vacuum pump

A number of vacuum pumps are used to create the vacuum in an ICP mass spectrometer. Two roughing pumps are used, one for the interface region and another to back up the first turbomolecular pump of the ion-optic region. Also, two turbomolecular pumps (or in some designs, one twin-throated pump) are used, one for the ion optics and another for the main mass-analyzer region. *Also refer to* **roughing pump, turbomolecular pump** *and* **twin-throated turbomolecular pump**.

vapor phase decomposition (VPD)

A technique used to dissolve and collect trace metal impurities on the surface of a silicon wafer, by rolling a few hundred microliters of hydrofluoric acid (HF) over the surface.

visible laser

A laser that operates in the visible region of the electromagnetic spectrum. An example is the ruby laser, which operates at 694 nm.

W

wavelength

A name commonly used to identify an emission line in nanometers (nm) used in ICP optical emission spectroscopy. Also used to describe a type of laser ablation system (e.g., a ruby laser operating at a 694-nm wavelength) used to couple to an ICP-MS to analyze solid samples directly.

wet plasma

When a liquid sample is introduced or aspirated into the plasma, the plasma is called a "wet plasma".

X

X-rays

A part of the electromagnetic spectrum that has a wavelength in the range of 0.01 to 10 nanometers (nm).

X-ray fluorescence

An analytical technique that uses the emission of characteristic "secondary" (or fluorescent) X-rays from a material that has been excited by bombarding with high-energy X-rays or gamma rays. The phenomenon is widely used for elemental analysis and chemical analysis, particularly in the study of solid materials such as metals, glass, ceramics, soils and rocks.

"x" position

Refers to the alignment of the plasma torch in the lateral (sideways) position. Typically carried out to maximize sensitivity or to optimize sampling conditions for cool-plasma use.

Y

"y" position

Refers to the alignment of the plasma torch in the longitudinal (vertical) position. Typically carried out to maximize sensitivity or to optimize sampling conditions for cool-plasma use.

Z

"z" position

Refers to the alignment of the plasma torch in relation to the distance from the interface cone (in and out position). Typically carried out to maximize sensitivity or to optimize sampling conditions for cool-plasma use.

INDUCTIVELY COUPLED PLASMA OPTICAL EMISSION SPECTROMETRY (ICP-OES) GLOSSARY OF TERMS

A

aerosol

A group of small-diameter liquid droplets that are formed when the nebulizer breaks up a continuously pumped solution.

argon

One of the noble gases in the periodic table. This gas is used to generate a plasma in ICP-based instruments.

array detector

A group of photoactive elements, arranged in a linear or two-dimensional configuration, which is used to measure photons from a plasma. Arrays can consist of a variety of detectors, including photodiodes, CCD image sensors and CID image sensors.

atomic emission (AE)

This is sometimes used to refer to the technique ICP-AES and refers to the emission from atoms in a sample. This term has been replaced with the term "optical emission", which is more accurate (refer to Optical Emission definition).

axial view

This describes the measurement of light down the central channel of the plasma, which includes emission from the entire length of the plasma. Currently, technology allows axial measurements to be made, regardless of whether the torch is mounted vertically or horizontally in the instrument.

B

background equivalent concentration (BEC)

For a given analyte, this is the concentration of that analyte, which produces a net emission signal equal to the background signal.

background noise

The uncertainty affiliated with measuring a blank solution. This value is often calculated by taking the square root of the background signal.

background signal

The intensity of the signal emitted when a blank solution is measured.

C

ceramic torch

A torch for an ICP-OES that is made from a ceramic material that makes the torch more robust and typically increases its lifetime. The outer, intermediate and injector tubes can all be made out of ceramic.

charge-coupled device (CCD)

A type of solid-state detector that can convert photons into photoelectric signals. This type of detector is used in many commercially available ICP-OES instruments.

charge-injection device (CID)

A type of solid-state detector that can convert photons into photoelectric signals. This type of detector is used in many commercially available ICP-OES instruments.

chemical interferences

These occur when the standards behave differently from the samples as they enter the plasma. These types of interferences typically result from changes in temperature within the plasma. They can be overcome by utilizing ionization suppressants or using a radially configured plasma.

complementary metal-oxide semiconductor (CMOS)

This is a type of technology that is used to construct circuits that are used in many applications, including imaging sensors for detectors in ICP-OES instruments.

concentric nebulizer

A nebulizer design that uses a stream of high-velocity argon gas to create an area of low pressure at its tip, causing a constant flow of solution to break apart into small droplets as it leaves the nebulizer. This nebulizer can be fed via a pumped solution, or it can operate via free aspiration.

cooled spray chamber

A spray chamber that is housed inside a cooled chamber, which lowers the temperature inside the spray chamber and reduces the amount of aerosol that gets transported from the spray chamber to the plasma.

cyclonic spray chamber

A spray-chamber design that is commonly used in ICP-OES instruments. As the name implies, this spray chamber accepts aerosol from the nebulizer and forms it into a virtual cyclone before transferring the smallest droplets into the plasma. This spray chamber is manufactured with or without a center tube, which acts as an additional impact surface for the aerosol.

D

desolvating nebulizer

A nebulizer that desolvates (removes a significant amount of the solvent from) the aerosol prior to its passage into the plasma. This system utilizes a heated or cooled spray chamber and a solvent membrane to remove either aqueous or organic solvent from the nebulized aerosol.

detection limit (DL)

Sometimes referred to as the LOD, the detection limit is the lowest concentration of an analyte that can be distinguished from a solution that does not contain that analyte (i.e., a blank solution).

detector

A device used to convert photons from the instrument's plasma to a digital signal that can be read out by a microprocessor.

devitrification

A chemical and temperature-related process that breaks down the structure of a glass or quartz component (typically a torch).

direct-injection high-efficiency nebulizer (DIHEN)

This is a modification of the original direct-injection nebulizer, which incorporates some design improvements.

direct-injection nebulizer (DIN)

A nebulizer design that injects sample solution directly into the base of the instrument's plasma, eliminating the need for a spray chamber. This nebulizer maximizes the transport efficiency of the aerosol and minimizes solution waste and memory effects between samples.

double pass spray chamber

A spray-chamber design.

droplet

A small drop of solution. During the process of introducing samples into the instrument's plasma, samples must be converted from continuous solutions to a group of small-diameter droplets.

H

high-efficiency nebulizer (HEN)

A term used to describe nebulizers that have been designed to convert a significant percentage of the pumped solution into a useable aerosol. Very little solution is sent to waste

high-solids nebulizers

These nebulizers are designed with the intent to create a useable aerosol from solutions with high levels of dissolved solids. These include V-groove and Babington nebulizer designs.

I

inductively coupled plasma (ICP)

A body of gas ions that are formed and maintained via the interaction between radio frequency and magnetic fields. This type of plasma is used as an atomization/ionization source in inductively coupled plasma optical emission spectrometry.

inductively coupled plasma optical emission spectrometry (ICP-OES)

An elemental analysis technique that uses an inductively coupled plasma to desolvate, vaporize, atomize and excite liquid samples. The resulting emitted light is collected, measured and used to quantify the elements present in the original sample.

injector

Also known as the sample injector, this is the center tube in the plasma torch and is used to transport the sample aerosol and inject it into the center of the plasma.

instrument detection limit (IDL)

This is often used synonymously with the limit of detection (LOD) and refers to the signal-to-background noise capability of the instrument, for a given analyte. This is often calculated in accordance with the IUPAC definition as three times the standard deviation of 10 replicate measurements of a blank solution.

integration time

The amount of time the instrument allows light to pass to the detector to collect emission from the plasma.

interferences

In plasma-based techniques, this is a broad term that refers to something that hinders the instrument's ability to precisely and accurately quantify elemental impurities in a sample. Interferences can be chemical, physical or spectral in nature and sometimes involve more than one correction technique to ensure quality results.

internal standardization

A technique used to overcome some physical and chemical interferences in ICP-based techniques. This technique involves the addition of an analyte to all blanks, standards and samples, which will behave in a similar manner to the analytes being measured. Sample results are reported as a ratio between the analyte and the internal standard, which corrects for small changes in intensity that are due to physical/chemical changes in the solutions. These small changes, if left uncorrected, would result in errors in the reported results.

L

load coil

A term that refers to the radiofrequency (RF) coil in a plasma-based instrument. This coil maintains the radiofrequency signal being generated by the instrument's RF generator, which helps sustain the plasma.

M

method detection limit (MDL)

A conservative estimate of the instrument's detection limits. The MDL can be calculated in a number of different ways; however, it is the minimum concentration of an analyte that can be distinguished from the absence of the analyte and quantified with a high degree of confidence (95%–99%).

microconcentric nebulizer

Based on the concentric design, this nebulizer operates at significantly lower solution flow rates than the standard concentric nebulizer.

microwave-induced plasma (MIP)

The most basic form of electrodeless plasma discharge. In this device, microwave energy (typically, 100–200 W) is supplied to the plasma gas from an excitation cavity around a glass/quartz tube. The plasma discharge in the form of a ring is generated inside the tube. Unfortunately, even though the discharge achieves a very high-power density, the high excitation temperatures only exist along a central filament. The bulk of the MIP never goes above 2,000–3,000 K, which means it is prone to very severe matrix effects. In addition, it is easily extinguished when aspirating liquid samples, so it has a found a niche as a detection system for gas chromatography.

MIP

An abbreviation for **microwave-induced plasma**.

MICAP-OES

An abbreviation for Microwave Inductively Coupled Atmospheric Plasma Optical Emission Spectrometer

microwave inductively coupled atmospheric plasma optical emission spectrometer

A commercially-available microwave-induced ICP-OES system that utilizes an atmospheric plasma using nitrogen gas and Cerawave™ RF technology for excitation of the sample and a novel Echelle grating to separate the analyte wavelenghths of interest.

N

nebulizer

One of the components in the sample-introduction system, the nebulizer takes a continuous solution and breaks it up into a body of small-diameter droplets (aerosol).

P

peltier cooler

A thermoelectric cooler, which is used to cool the spray chamber and maintain a specific cooled temperature during a sample run.

peristaltic pump
A type of pump that uses rollers to positively displace sample solutions, pushing them through flexible tubing. These pumps are used to move solution through the sample uptake and drain tubing that is used on many commercially available ICP-OES instruments.

physical interferences
Occur when the nebulization and/or transport efficiency of the standards differs from that of the samples. Physical interferences are usually overcome by utilizing internal standardization and preparing the calibration standards in a matrix that matches that of the samples.

Q

quartz torch
A torch for an ICP-OES that is made from quartz. The outer, intermediate and injector tubes can all be made out of quartz.

R

radial view
This describes the measurement of light perpendicular to the direction of the plasma.

radiofrequency (RF) generator
The RF generator typically consists of a DC power supply, an RF generator board and a number of other electrical components that work together to generate a radiofrequency field, which is necessary for maintaining the plasma in an ICP-based instrument.

S

sample-introduction system
This refers to the components that transport a sample solution into the instrument's plasma, while converting it from a continuous solution to a fine aerosol. The components that make up the sample-introduction system include the nebulizer, spray chamber and the torch, as well as the pump and affiliated pump tubing.

scott spray chamber
This is a type of double-pass spray chamber.

signal-to-background ratio (S/B or SBR)
For a given analyte, this is the intensity ratio of the signal and the background at a particular concentration. When the S/B is calculated in a blank solution, this is often used to represent the detection limit for that analyte.

spray chamber
One of the components in the sample-introduction system, the spray chamber takes the aerosol from the nebulizer and filters out the larger-diameter droplets such that only small-diameter droplets are transferred to the plasma. The larger droplets are sent directly to waste.

T

torch
One of the components in the sample-introduction system, the torch houses the plasma and receives the sample aerosol that was generated and transported by the nebulizer and spray chamber.

torch injector

Also known as the injector or the sample injector, this is the center tube in the plasma torch and is used to transport the sample aerosol and inject it into the center of the plasma.

U

ultrasonic nebulizer (USN)

A nebulizer design that uses high-frequency vibrations to convert a solution into an aerosol with small-diameter droplets. Unlike conventional nebulizers, an ultrasonic nebulizer generates a higher percentage of smaller droplets, which improves the transport of solution to the plasma and increases the analyte signal. Most commercially available ultrasonic nebulizers use a heated/cooled chamber and/or a membrane-desolvation system to remove excess solvent from the aerosol prior to being transported to the plasma.

V

viewing height

This refers to where the emitted light is being collected from the plasma when making measurements in radial-view mode. This viewing position is usually in reference to the height above the top turn of the load (RF) coil.

W

wavelength

In ICP-OES, wavelength-specific photons are emitted from excited atoms and ions and measured by the instrument's detector.

ATOMIC ABSORPTION AND ATOMIC FLUORESCENCE

A

AAS

An abbreviation for atomic absorption spectrometry, which can be either flame AAS or graphite-furnace AAS.

absorbance

Defined as the log of the ratio of the lamp intensity after the atomizer, to the lamp intensity before the atomizer. This term follows a linear relationship with concentration according to Beer's law.

acetylene

Flammable gas used to create the flame used as the atomization source.

aerosol

Fine particles generated by the nebulizer transported via the spray chamber into the flame burner head.

AFS

An abbreviation for atomic fluorescence spectrometry.

B

background absorption

An artefact other than the analyte prevents the light from the hollow-cathode lamp reaching the detector.

background correction
This is a way of compensating for background absorption and is carried out by either continuum source or Zeeman background correction.

Beer's law
Also called Beer–Lambert law, defines the relationship between the absorption of radiant energy by an absorbing medium. It's expressed as an equation: Absorbance = (concentration) * (an absorption coefficient) * (path length).

C

cold vapor
A technique used for the determination of mercury, using a reducing agent to generate elemental-mercury vapor and passing it into a heated tube for atomization.

continuum background correction
This correction technique utilizes a secondary deuterium lamp installed in the instrument, which emits continuously over a broad wavelength range, rather than at a discreet wavelength. The instrument then monitors intensities from both the deuterium lamp and the element lamp and then subtracts the one from the other.

E

electrothermal atomization
Another term used for graphite-furnace AAS.

ETA
Abbreviation for electrothermal atomization.

F

flame
An atomization source used in AAS, usually produced by air/acetylene or nitrous oxide/acetylene.

FAAS
An abbreviation of flame atomic absorption.

furnace program
Heating cycle used in a GFAAS.

G

GFAAS
Abbreviation for graphite-furnace AAS.

graphite-furnace atomic absorption
Atomization source in AAS that uses a small graphite tube to atomize the sample by resistive heating.

graphite tube
The device/tube used for atomization in GFAAS.

H

hollow-cathode lamp

Source of the photons used in AAS, generated by passing an electric current between two electrodes and heating/exciting the cathode material, which is made from the analyte element being determined.

hydride generation

A technique used to generate volatile hydrides to improve detection limits for a small group of elements, including Pb, As, Cd, Hg Sb, Se, Sn, Te.

I

interferences

In atomic absorption, interferences are divided into two main categories: spectral and non-spectral.

M

matrix modifiers

A chemical that changes the volatility of either the analyte or the matrix in order to stabilize the analyte so that no losses of the element occur before the atomization or to increase volatility of the matrix to drive it off before atomization.

N

nebulizer

The nebulizer mixes the liquid sample with air to form an aerosol mist that is sprayed into the spray chamber.

nitrous oxide

Oxidant used with acetylene to create a hotter atomization flame than air/acetylene.

non-spectral interferences

Relates to how the atom population is achieved and includes transport (or physical), chemical and ionization interferences.

P

peak area

One of the measurement protocols used in quantitation in GAAS that measures the area under the peak.

peak height

One of the measurement protocols used in quantitation in GAAS, the other being peak area.

pyrolysis

Slow decomposition of the sample in a graphite furnace, typically referred to by the pyrolysis temperature.

S

spectral interferences

Interferences that contribute to the element signal producing an incorrect result and are primarily background absorption and emission interferences from other elements in the sample matrix.

spray chamber
Device used to eliminate larger particles from the aerosol generated by the nebulizer.

T

transmittance
Defined as the ratio of the lamp intensity after the atomizer to the lamp intensity before the atomizer.

V

volatilization
This is the breaking of molecular bonds as the sample is being preheated in a graphite furnace prior to atomization.

W

wavelength
Energy of photons generated by a hollow-cathode lamp, which are specific and unique to the analyte element being measured.

zeeman background correction
This mode of correction carries out the background measurement at the exact analyte wavelength when a magnetic field is applied. Since the background is measured at the analyte wavelength and not averaged as in deuterium background systems, structural molecular background and spectral interferences are easily corrected.

OTHER ATOMIC SPECTROSCOPY TECHNIQUES

E

EDXRF
An abbreviation for energy-dispersive XRF.

energy-dispersive (EDXRF)
In EDXRF, the intensity of the photon energy of the individual X-rays is detected and measured simultaneously using multi-channel data electronics.

F

fiber optics
A way of transferring photons generated by an excitation source into a spectrometer where the individual analyte wavelengths are separated and identified

H

hand-held XRF
Small portable XRF systems for on-site or field studies.

L

LALI-TOF-MS
An abbreviation for Laser Ablation, Laser Ionization TOF Mass Spectrometry.

laser ablation, laser ionization tof mass spectrometry (LALI-TOF-MS)

This technique utilizes one laser to remove material from a solid sample and a second laser to subsequently ionize neutrals. The wavelength and power of the first laser can be selected to accommodate both ablations for inorganic elemental analysis or desorption, for organic analysis. This approach removes much of the matrix effects that plague other techniques and allows quantitative analysis across heterogenous matrices without matrix-matched reference standards.

laser induced breakdown spectroscopy (LIBS)

LIBS is an atomic-emission spectroscopic technique that utilizes a small plasma generated by a focused pulsed laser beam. The energy from the plasma excites the vaporized sample, which results in a characteristic emission spectrum of the elements present in the sample, which is then optically dispersed and detected by an emission spectrometer.

LIBS

An abbreviation for Laser Induced Breakdown Spectroscopy.

M

microwave induced plasma atomic emission spectrometry (MIP-AES)

In MIP-AES, a magnetron is coupled into a nitrogen plasma. By using the magnetic field rather than the electric field (used in ICP-OES), it concentrates both the axial magnetic and radial electrical fields around the torch, and as a result an extremely robust plasma is formed, which can be used to excite the sample and to study elemental emission profiles using a conventional optical spectrometer.

MIP-AES or MP-AES or MIP-OES

An abbreviation for microwave induced plasma atomic/optical emission spectrometry.

P

primary X-rays

The primary X-rays generated by the X-ray source (typically a tube) that irradiates the sample.

S

secondary X-rays

X-rays that are emitted and measured as a result of the excitation of the sample atoms.

W

wavelength-dispersive (WDXRF)

In WDXRF, the polychromatic beam emerging from a sample surface is dispersed into its monochromatic components or wavelengths with an analyzing crystal. A specific wavelength is then calculated from knowledge of the dispersion characteristics of the crystal.

WDXRF

An abbreviation for wavelength-dispersive XRF.

X

X-ray diffraction (XRD)

XRD is a comparative technique to XRF that looks at the X-ray scattering from crystalline or polymorphic materials. Each material produces a unique X-ray "fingerprint" of X-ray intensity versus scattering angle that is characteristic of its crystalline atomic structure.

X-ray fluorescence spectrometry (XRFS)

X-ray fluorescence (XRF) is a nondestructive analytical technique used to determine the elemental composition of materials, by measuring the fluorescent (or secondary) X-ray emitted from a sample when it is excited by a primary X-ray source.

XRF

An abbreviation for X-ray fluorescence.

28 Useful Contact Information

The final chapter of the book is dedicated to providing you with useful contact information related to atomic spectroscopy. It includes contact details for major instrumentation manufacturers, instrument consumables, sample-introduction components and alternative sources of sampling accessories, together with suppliers of laboratory chemicals, calibration standards, certified reference materials, high-purity gases, deionized water systems and clean-room equipment. I have also included information about the major scientific conferences, professional societies, publishing houses, Internet discussion groups and the most popular ICP-related journals. It is sorted alphabetically by category. However, some vendors sell many different products, so they are listed under the category represented by their major product line.

Certified reference materials/calibration standards

National Research Council (NRC)
of Canada
1500 Montreal Road
Ottawa, Ontario, K1A 0R9, Canada
Phone: 613–993–9391
Fax: 613–952–8239
www.nrc-cnrc.gc.ca

NIST
100 Bureau Drive, Stop 200
Gaithersburg, MD 20899
Phone: 301–975–6776
Fax: 301–975–2183
www.nist.gov

VHG Labs
276 Abby Road
Manchester, NH 03103
Phone: 603–622–7660
Fax: 603–622–5180
www.vhglabs.com

High Purity Standards
P.O. Box 41727
Charleston, SC 29423
Phone: 843–767–7900
Fax: 843–767–7906
www.highpuritystandards.com

Inorganic Ventures, Inc.
Inorganic Ventures
300 Technology Drive
Christiansburg, VA 24073
Phone: 540–585–3030
Fax: 540–585–3012
www.inorganicventures.com

SPEX CertiPrep
203 Norcross Avenue
Metuchen, NJ 08840
Phone: 800–522–7739
Fax: 732–603–9647
www.spexcertiprep.com

SCP Science
21800 Clark Graham
Baie D'urfe, H9X 4B6, Canada
Phone: 800–361–6820
Fax: 514–457–4499
www.scpscience.com

Conostan Standards
A division of SCP Science
348 Route 11,
Champlain, NY 12919
Phone: 800–361–6820
Fax: 800–253–5549
www.conostan.com

LGC Standards
276 Abby Road
Manchester
NH 03103
Tel: 603 206–0799
Toll-free: 1–850-LGC-USA1
www.lgcgroup.com

United States Pharmacopeial
Convention (USP)
12601 Twinbrook Parkway
Rockville, MD 20852–1790
Phone: 1-800-227-8772
www.usp.org/reference-standards

DOI: 10.1201/9781003187639-28

High-purity/reagent chemicals

Sigma-Aldrich Chemicals
Customer Support
PO Box 14508
St. Louis, MO 63178
Phone: 800–325–3010
Fax: 800–325–5052
www.sigmaaldrich.com

Fisher Scientific, Inc.
300 Industry Drive,
Pittsburgh, PA 15275
Phone: 800–766–7000
Fax: 800–926–1166
www.fishersci.com

Eichrom Technologies, LLC
1955 University Lane
Lisle, IL 60532
Phone: 630–963–0320
Fax: 630–963–1928
www.eichrom.com

Mallinckrodt Baker
222 Red School Lane
Phillipsburg, NJ 08865
Phone: 908–859–9315
Fax: 908–859–9385
www.mbigloballabcatalog.com

Chromatographic separation equipment for trace element speciation

Agilent Technologies
2850 Centerville Road
Wilmington, DE 19808
Phone: 302–633–8264
Fax: 302–633–8916
www.chem.agilent.com

PerkinElmer Inc.
710 Bridgeport Avenue
Shelton, CT 06484
Phone: 800–762–4000
Fax: 203–944–4914
www.perkinelmer.com

Thermo Fisher Scientific
81 Wyman Street
Waltham
MA 02454
Phone: 781–622–1000
Fax: 781–622–1207
www.thermoscientific.com

Dionex Corporation
A Thermo Company
1228 Titan Way
P.O. Box 3603
Sunnyvale, CA, 94088
Phone: 408–737–0700
Fax: 408–730–9403
www.dionex.com

Shimadzu Scientific Instruments
(North America)
7102, Riverwood Drive
Columbia, MD 21046
Phone: 410–381–1227
Fax: 410–381–122
www.ssi.shimadzu.com/

Waters Corporation
34 Maple Street
Milford, MA 01757
Phone: 508–478–2000
Fax: 508–872–1990
www.waters.com

Clean rooms and equipment

Microzone Corp.
86 Harry Douglas Drive
Ottawa, Ontario,
Canada, K2S 2C7
Phone: 613–831–8318
Fax: 613–831–8321
www.microzone.com

Clestra Hauserman
259 Veterans Lane, Suite 201
Doylestown, PA 18901
Phone: 267 880 3700
Fax: 267 880 3705
www.clestra-cleanroom.com

Consumables—ICP-MS detectors

SGE, Inc.
2007 Kramer Lane
Austin, TX 78758
Phone: 800–945–6154
Fax: 512–836–9159
www.sge.com

Spectron, Inc.
1601 Eastman Avenue, Suite 205
Ventura, CA 93003
Phone: 805.642.0400
Fax: 805.642.0300
www.spectronus.com

Consumables (sample-introduction/interface components)

Burgener Research, Inc.
1680–2 Lakeshore Rd. W.
Mississauga, Ontario, L5J 1J5, Canada
Phone: 905–823–3535
Fax: 905–823–2717
www.burgenerresearch.com

CPI International
5580 Skylane Blvd.
Santa Rosa, CA 95403
Phone: 800–878–7654
Fax: 707–545–7901
www.cpiinternational.com

Elemental Scientific, Inc. (ESI)
2440 Cumming Street
Omaha, NE 68131
Phone: 402–991–7800
Fax: 402–997–7799
www.elementalscientific.com

Glass Expansion Pty.
4 Barlows Landing Road, Unit #2
Pocasset, MA 02559
Phone: 505–563–1800
Fax: 505–563–1802
www.geicp.com

Meinhard Glass Products
An Elemental Scientific Company
700 Corporate Circle, Suite A
Golden, CO 80401
Phone: 303–277–9776
Fax: 303–216–2649
www.meinhard.com

Precision Glassblowing
14775 E. Hindsdale Avenue
Centennial, CO 80112
Phone: 303–693–7329
Fax: 303–699–6815
www.precisionglassblowing.com

SCP Science
21800 Clark Graham
Baie D'urfe, H9X 4B6, Canada
Phone: 800–361–6820
Fax: 514–457–4499
www.scpscience.com

Spectron, Inc.
1601 Eastman Avenue, Suite 205
Ventura, CA 93003
Phone: 805.642.0400
Fax: 805.642.0300
www.spectronus.com

Savillex LLC c/o Spectron, Inc.
1601 Eastman Avenue, Suite 205
Ventura, CA 93003
Phone: 805.642.0400
Fax: 805.642.0300
www.spectronus.com

Texas Scientific Products, LLC
204 E 4th St Justin, TX 76247
Phone: 888-268-6037 Fax:
972-850-7659 www.txscientific.
com/

Expositions and conferences

Eastern Analytical
P.O. Box 633
Montchanin, DE 19710
Phone: 610–485–4633
Fax: 610–485–9467
www.eas.org

FACSS (SCIX)
1201 Don Diego Avenue
Santa Fe, NM 87505
Phone: 505–820–1648
Fax: 505–989–1073
www.facss.org

Expositions and conferences

Pittsburgh Conference and Exposition
300 Penn Center Blvd., Suite 332
Pittsburgh, PA 15235
Phone: 412–825–3220
Fax: 412–825–3224
www.pittcon.org

Plasma Winter Conference
c/o Dr. Ramon Barnes
ICP Information Newsletter
P.O. Box 666
Hadley MA 01035–0666
Phone: 239–674–9430
Fax: 239–674–9431
http://icpinformation.org/

High-purity gases

Air Liquide Specialty Gases LLC
Corporate Headquarters
6141 Easton Road
Plumsteadville, PA 18949
Phone: 215.766.8860
Fax: 215.766.2476
www.airliquide.com

Air Products and Chemicals
7201 Hamilton Blvd.
Allentown, PA 18195
Phone: 610–481–4911
Fax: 610–481–5900
www.airproducts.com

Praxair Inc.
Worldwide Headquarters
39 Old Ridgebury Road
Danbury, CT 06810
Phone: -800–772–9247
Fax: 716–879–2040
www.praxair.com

Scott Specialty Gases
A Division of Air Liquide
6141 Easton Road, P.O. Box 310
Plumsteadville, PA 18949
Phone: 215–766–8861
Fax: 215–766–2476
www.scottgas.com

ICP-MS instrumentation (Magnetic Sector Technology, inc. ICP-MS. GD-MS, TIMS, IRMS, NGMS)

Thermo Fisher Scientific
81 Wyman Street
Waltham
MA 02454
Phone: 781–622–1000
Fax: 781–622–1207
www.thermoscientific.com

Nu Ametek Instruments Ltd
Unit 74
Clywedog Road South
Wrexham, LL13 9XS
United Kingdom
Phone: +44 (0)1978 661304
www.nu-ins.com

ICP-MS instrumentation (quadrupole technology)

Agilent Technologies
ICP-MS Systems
2850 Centerville Road
Wilmington, DE 19808
Phone: 302–633–8264
Fax: 302–633–8916
www.chem.agilent.com

Analytic Jena
100 Cummings Center
Suite 234-N
Beverly, MA, 01915
Phone:781 376 9899
Fax:781 376 9897
http://us.analytik-jena.com

ICP-MS instrumentation (quadrupole technology)

PerkinElmer Inc.
710 Bridgeport Avenue
Shelton, CT 06484
Phone: 800–762–4000
Fax: 203–944–4914
www.perkinelmer.com

**Shimadzu Scientific Instruments
(North America)
7102, Riverwood Drive
Columbia, MD 21046
Phone: 410–381–1227
Fax: 410–381–122
www.ssi.shimadzu.com/**

Thermo Fisher Scientific
81 Wyman Street
Waltham
MA 02454
Phone: 781–622–1000
Fax: 781–622–1207
www.thermoscientific.com

Advion Interchim Scientific
61 Brown Rd, Suite 100
Ithaca,
NY 14850
Phone: 607.266.9162
Fax: 607.257.5761
www.advion.com/

ICP-MS instrumentation (time-of-flight technology)

GBC Scientific
151A North State Street
PO Box 339
Hampshire IL 60140
Phone: 847–683–9870
Fax: 847–683–9871
www.gbcsci.com

TOFWERK, USA
2760 29th St, Suite 1F
Boulder, Colorado, USA 80301
Phone: 303 524 2205
www.tofwerk.com/contact-us/

Nu Ametek Instruments Ltd
Unit 74
Clywedog Road South
Wrexham, LL13 9XS
United Kingdom
Phone: +44 (0)1978 661304
www.nu-ins.com

Exum Instruments
747 Sheridan Boulevard Unit 9A,
Denver, CO 80214, USA
**info@builtbyexum.com
www.exuminstruments.com/**

ICP-OES/MIP-AES/MICAP-OES/AA instrumentation

Agilent Technologies
(ICP-OES, MIP-AES and AA)
2850 Centerville Road
Wilmington, DE 19808
Phone: 302–633–8264
Fax: 302–633–8916
www.chem.agilent.com

Analytic Jena
100 Cummings Center
Suite 234-N
Beverly, MA, 01915
Phone:781 376 9899
Fax:781 376 9897
http://us.analytik-jena.com

**Jobin Yvon/Horiba Scientific
(Not AA—ICP-OES only)
3880 Park Avenue
Edison, NJ 08820–3097
Phone: 732–494–8660
Fax: 732–549–5125
www.horiba.com/**

PerkinElmer Inc.
710 Bridgeport Avenue
Shelton, CT 06484
Phone: 800–762–4000
Fax: 203–944–4914
www.perkinelmer.com

ICP-OES/MIP-AES/MICAP-OES/AA instrumentation

Shimadzu Scientific Instruments
7102, Riverwood Drive
Columbia, MD 21046
Phone: 410–381–1227
Fax: 410–381–122
www.ssi.shimadzu.com/

Spectro Ametek
(Not AA—ICP-OES only)
Boschstrasse. 10
47533 Kleve
Germany
Phone: +49 / 2821 / 8 92–0
Fax: +49 / 2821 / 8 92–22 00
www.spectro.com/contact/contact-us

Teledyne Leeman Labs
110 Lowell Road
Hudson, NH 03051
Phone: 603–886–8400
Fax: 603–886–4322
www.teledyneleemanlabs.com/

Thermo Fisher Scientific
81 Wyman Street
Waltham
MA 02454 USA
Phone: 781–622–1000
Fax: 781–622–1207
www.thermoscientific.com

Radom Corp. (MICAP-OES Only)
1560 Sawgrass Corporate Parkway 4th Floor,
Suite 494 Sunrise, FL 333
Phone: 877-977-2366
https://radomcorp.com

Internet discussion group

PLASMACHEM List Server
312 Heroy Geology Laboratory
University of Syracuse
Syracuse, NY 13244
Phone: 315–443–1261 (Michael Cheatham)
Fax: 315–443–3363
To subscribe:
Email:
mmcheath@mailbox.syr.edu

European Virtual Institute for
Speciation Analysis (EVISA)
A forum dedicated to trace element
speciation analysis
www.speciation.net/

Journals/magazines

Analytical Chemistry
1155 16th Street, NW
Washington, DC 20036
Phone: 202–872–4570
Fax: 202–872–4574
www.pubs.acs.org/ac

Applied Spectroscopy
201b Broadway Street
Frederick, MD 21701
Phone: 301–694–8122
Fax: 301–694–6860
www.s-a-s.org

International Labmate Ltd
Oak Court Business Center
Sandridge Park
Porters Wood, St. Albans
Hertfordshire, AL3 6PH, UK
+44 (0)1727 840 310
www.labmate-online.com

Journal of Analytical Atomic
Spectrometry (JAAS)
Thomas Graham House
Science Park
Milton Rd.
Cambridge, CB4 4WF, UK
Phone: 44–1223–420066
Fax: 44–1223–420247
www.rsc.org

Journals/magazines

Pharmaceutical Technology
485 Route One South
Building F, Suite 210
Iselin, NJ 08830
Phone: 728.596.0276
Fax: 728.647.1235
www.pharmtech.com/

Spectroscopy Magazine
485 Route One South
Building F, First Floor
Iselin, NJ 08830
Phone: 732–596–0276
Fax: 732–225–0211
www.spectroscopyonline.com

Laser ablation/LALI/LIBS equipment

Applied Spectra Inc.
46665 Fremont Blvd.
Fremont, CA 94538
Phone: 510–657–7679
http://appliedspectra.com/

Elemental Scientific, Inc.
2440 Cumming Street
Omaha, NE 68131
Phone: 402–991–7800
Fax: 402–997–7799
www.elementalscientific.com

Teledyne CETAC Technologies
14306 Industrial Road
Omaha, NE 68144
Phone: 402–733–2829
Fax: 402–733–5292
www.teledynecetac.com/

TSI Inc.
500 Cardigan Road
Shoreview, MN 55126
Phone: 651–483–0900
Fax: 651–490–3824
www.tsi.com

Exum Instruments
747 Sheridan Boulevard Unit 9A,
Denver, CO 80214, USA
info@builtbyexum.com
www.exuminstruments.com/

Ocean Insight
500 Quadrangle Blvd.
Orlando, FL 32817 USA
Phone: 407–673–0041
www.oceaninsight.com/

Microwave digestion equipment

CEM Corporation
3100 Smith Farm Road
Matthews, NC 62810
Phone: 800–726–3331
Fax: 704–821–5185
www.cem.com

Milestone, Inc.
160 B Shelton Road
Monroe, CT 06468
Phone: 203–925–4240
Fax: 203–925–4241
www.milestonesci.com

SCP Science
21800 Clark Graham
Baie D'urfe, H9X 4B6, Canada
Phone: 800–361–6820
Fax: 514–457–4499
www.scpscience.com

Anton Paar USA Inc.
10215 Timber Ridge Drive
Ashland, VA 23005
Phone: 804–5501051
Fax: 804–5501051
www.info.us@anton-paar.com

PerkinElmer Inc.
710 Bridgeport Avenue
Shelton, CT 06484
Phone: 800–762–4000
Fax: 203–944–4914
www.perkinelmer.com

Questron Technologies Corp.
6660 Kennedy Rd #14A,
Mississauga,
Ontario, L5T 2M9
Canada
Phone: 905–362–1225
Fax: 905–362–1229
https://questron.ca/
microwave-digestion-system/

Performance/productivity-enhancement technology

Teledyne CETAC Technologies
14306 Industrial Road
Omaha, NE 68144
Phone: 402–733–2829
Fax: 402–733–5292
www.cetac.com

Glass Expansion Pty.
4 Barlows Landing Road, Unit #2
Pocasset, MA 02559
Phone: 505–563–1800
Fax: 505–563–1802
www.geicp.com

Elemental Scientific, Inc. (ESI)
1500 N. 24th St.
Omaha, NE 68110
Phone: 402–991–7800
Fax: 402–997–7799
www.elementalscientific.com

Professional societies/regulatory agencies/pharmacopeias

American Chemical Society (ACS
1155 16th Street NW
Washington, DC 20036
Phone: 800–227–5558
Fax: 202–872–4615
www.pubs.acs.org

American Society for Testing Materials (ASTM)
100 Barr Harbor Drive
West Conshohocken, PA 19428
Phone: 610–832–9605
Fax: 610–834–3642
www.astm.org

Society for Applied Spectroscopy (SAS)
5320 Spectrum Drive, Suite C
Frederick, MD 21703
Phone: 301–694–8122
Fax: 301–694–6860
www.s-a-s.org

U.S. Food and Drug Administration (FDA)
10903 New Hampshire Avenue
Silver Spring, MD 20993
Phone:1-888-463-6332)
www.fda.gov/

American Society for Mass Spectrometry (ASMS)
2019 Galisteo St, Bldg I-1
Santa Fe, NM 87505
Phone: 505–989–4517
Fax: 505–989–1073
www.asms.org

International Society for Pharmaceutical Engineering (ISPE)
7200 Wisconsin Ave., Suite 305
Bethesda, MD 20814
Phone: 301–364–9201
Fax:240–204–6024
www.ispe.org/

United States Pharmacopeial Convention (USP)
12601 Twinbrook Parkway
Rockville, MD 20852–1790
Phone: 1-800-227-8772
www.usp.org/

European Pharmacopeia (EP)www.edqm.eu/en/european-pharmacopoeia.html
Japanese Pharmacopeia (JP)http://jpdb.nihs.go.jp/jp14e/
International Convention for Harmonization (ICH)www.ich.org/home.html
European Medicines Agency (EMA):www.ema.europa.eu/ema

Publishers

CRC Press/Taylor and Francis
6000 Broken Sound Parkway NW, Suite 300
Boca Raton, FL 33487
Phone: 561–994–0555
Fax: 561–989–9732
www.crcpress.com

John Wiley and Sons
Corporate Headquarters
111 River Street
Hoboken, NJ 07030–5774
Phone: 201.748.6000
Fax: 201.748.6088
www.wiley.com

Elsevier Science Publishing
Journals Customer Service
3251 Riverport Lane
Maryland Heights, MO 63043
Phone: 800–222–9570
www.elsevier.com

American Laboratory
395 Oyster Point Blvd., #321
South San Francisco, CA 94080
Phone: 650–234–5600
www.americanlaboratory.com

Sample-preparation equipment (cryogenic grinding, fusion systems, pelletizing equipment)

SPEX SamplePrep
203 Norcross Avenue
Metuchen, NJ 08840
Phone: 800–522–7739
Fax: 732–603–9647
www.spexsampleprep.com

SCP Science
21800 Clark Graham
Baie D'urfe, H9X 4B6, Canada
Phone: 800–361–6820
Fax: 514–457–4499
www.scpscience.com

Vacuum pumps and components

OerlikonLeybold Vacuum (USA), Inc.
5700 Mellon Road
Export, PA 15632
Phone: 800–764–5369
Fax: 724–733–1217
www.leyboldvacuum.com

Agilent Technologies
Vacuum Products Division
121 Hartwell Avenue
Lexington, MA 02421
Phone: 781 861 7200
Fax: 781 860 5405
www.chem.agilent.com/en-US/
ProductsServices/Instruments-
Systems/Vacuum-Technologies/
Pages/default.aspx

XRF/XRD equipment

Shimadzu Scientific Instruments (North America)
7102, Riverwood Drive
Columbia, MD 21046
Phone: 410–381–1227
Fax: 410–381–122
www.ssi.shimadzu.com/

SPECTRO Analytical Instruments Inc.
91 McKee Drive
Mahwah, NJ 07430
Phone: 201.642.3000
Fax: 201.642.3091
www.spectro.com

XRF/XRD equipment

Bruker Daltonics
Chemical Analysis Group
3500 West Warren Avenue
Fremont, CA 94538, USA
Phone +1 (510) 683–4300
Fax +1 (510) 687–1217
www.bdal.com/chemicalanalysis

Thermo Fisher Scientific
81 Wyman Street
Waltham
MA 02454 USA
Phone: 781–622–1000
Fax: 781–622–1207
www.thermoscientific.com

Applied Rigaku Technologies
9825 Spectrum Drive, Bldg. 4, Suite 475
Austin, TX 78717
Phone: 512–225–1796
Fax: 512–225–1797
www.rigakuedxrf.com

Oxford Instruments
X-Ray Technology
360 El Pueblo Road
Scotts Valley
CA 95066, USA
Phone: 831 439 9729
Fax: 831 439 9729
https://xray.oxinst.com

Index

Note: Page locators in **bold** indicate a table and page locators in *italics* indicate a figure.s